Elastic Wave Propagation in Structures and Materials

Elastic Wave Propagation in Structures and Materials initiates with a brief introduction to wave propagation, different wave equations, integral transforms including fundamentals of Fourier Transform, and its numerical implementation. The concept of spectral analysis and procedure to compute the wave parameters, wave propagation in 1-D isotropic waveguides, and wave dispersion in 2-D waveguides are explained. Wave propagation in different media such as laminated composites, functionally graded structures, granular soils, and nonlocal elasticity models is addressed. Separate chapters on signal processing and wave propagation in viscoelastic waveguides are included in this book. The entire book is written in modular form for easy understanding of the basic concepts.

Features:

- Brings out the idea of wave dispersion and its utility in the dynamic responses
- Introduces concepts such as Negative Group Speeds, Einstein's Causality, and escape frequencies using solid mathematical framework
- Discusses the propagation of waves in materials such as laminated composites and functionally graded materials
- Proposes spectral finite element as analysis tool for wave propagation
- Supports each concept/chapter with homework problems and MATLAB® codes

This book is aimed at senior undergraduates and advanced graduates in all fields of engineering, especially Mechanical and Aerospace Engineering.

Elastic Wave Propagation in Structures and Materials

Srinivasan Gopalakrishnan

CRC Press
Taylor & Francis Group
Boca Raton London New York

CRC Press is an imprint of the
Taylor & Francis Group, an **informa** business

First edition published 2023
by CRC Press
6000 Broken Sound Parkway NW, Suite 300, Boca Raton, FL 33487-2742

and by CRC Press
4 Park Square, Milton Park, Abingdon, Oxon, OX14 4RN

CRC Press is an imprint of Taylor & Francis Group, LLC

ISBN: 9780367637576 (hbk)
ISBN: 9780367637637 (pbk)
ISBN: 9781003120568 (ebk)

DOI: 10.1201/9781003120568

Publisher's note: This book has been prepared from camera-ready copy provided by the authors.

This book is dedicated to my Late mother
Saraswathi, my father Srinivasan,
my wife Anu and
my Children
Karthik and Keerthana

Contents

Preface

In 2016, Taylor & Francis/CRC Press published my book titled *Wave Propagation in Materials and Structures* [54]. This 990-page book contained both basic and advanced topics of wave propagation under one umbrella. The thought process in writing this large book which we refer to as "Unified book" in this text is that the reader will have both basic and advanced topics of the subject in one place.

The Publisher(Engineering) of CRC Press, Dr. Gangandeep Singh, informed me that although the unified book is well received by the community, the shear size of this book made handling very difficult, especially among those interested in the basic concepts such as senior undergraduates or entry-level graduate students. Hence, he requested me to pull out all the chapters dealing with basic concepts in the unified book and write a separate edition of the book having only the basic concepts by adding a few more topics not covered in the unified book. Hence, the present book can be considered as the second edition of my unified book. Reading this book first will enable the reader to smoothly transit to advanced topics covered in the unified book.

The second edition of the book has 12 chapters. Here, Chapters 1 to 4, 7 to 9, and 12, which were in the unified book, are completely rewritten to bring more emphasis on the fundamentals of wave propagation. Chapter 4 is a very big chapter on wave propagation in isotropic waveguides and is central to all the other chapters that follow. This chapter is thoroughly rewritten with several new examples. In addition, in this edition of the book, only analysis based on Fourier-based transform is covered to make the concepts of wave propagation analysis more understandable to beginners in the subject. As in the case of unified book, the central theme on which the wave propagation analysis is based in this book is the spectral analysis. In order to extend the reach of spectral analysis, three new contemporary topics in wave propagation are introduced in this edition of the book. These three chapters are the following: (1) Chapter 5 on Wave propagation in viscoelastic waveguides, (2) Chapter 10 on wave propagation in granular materials, and (3) Chapter 11 on wave propagation in non-local waveguides. All of these new chapters are treated within the framework of spectral analysis. Since wave propagation deals with different signals, a new chapter (Chapter 6) on Signal Processing is introduced in this edition of the book, thus making the book comprehensive in its coverage of different and important topics necessary for basic understanding of wave propagation in materials and structures. In addition, at the end of each chapter, several *practice exercises* are given to enable readers to understand the subject better. Last, but not least, most of the chapters are provided with several MATLAB® scripts to assist the readers in understanding the material presented in the text. These scripts can not only help in solving all the examples presented in this text, but also help the readers in solving all the exercise problems given at the end of each chapter by modifying these codes. The source code listing on the MATLAB scripts is not provided in this text. However, these scripts are housed on the publisher's website

Unlike many other subjects, the subject of wave propagation requires not only the solution to the problem under consideration, but also visualization of the obtained solution to understand the physics of the problem better. Most of the wave propagation problems are computationally intensive and coding in computer to obtain solutions is required. In this book, we have used extensively MATLAB programming to obtain solutions to the problems presented in this text. Hence, basic understanding of MATLAB programming is necessary to understand the MATLAB scripts provided in this book. The material provided in this book is ideal to start a full course in wave propagation, which can be taken by senior undergraduates and entry level graduate students. The text contains extensive references, which will be very useful to those performing research in this exciting area.

I started writing this book during the peak of the COVID-19 pandemic. Endless lockdowns and working from home enabled me to dedicate more time than usual to finish this book. This extra time enabled me to write several MATLAB codes to be provided along with this text. The content of this book is a result of my teaching this course and performing research in this exciting area of wave propagation for over 24 years at Indian Institute of Science. There are several people I would like to thank for enabling me to write this book. First, I would like to thank my Ph.D. research supervisor Professor James Francis Doyle, who introduced me to this exciting field during my graduate studies at Purdue University, West Lafayette, USA. Many of my former graduate students, with whom I learnt this subject together, contributed significantly to my understanding of this subject and I thank each one of them. In particular, I would like to thank my graduate student Dr. Venkat Mutnuri for helping me in writing the MATLAB scripts for doubly bounded media given in Chapter 7. Finally, I would like to deeply thank my late mother Saraswathi and my father Srinivasan, who are responsible for what I am today, and my family, especially my wife Anu and my children Karthik and Keerthana for showing deep understanding and patience and putting up with my hectic and erratic schedule in realizing this book.

S. Gopalakrishnan
December 2021

Author Biography

Professor Srinivasan Gopalakrishnan received his Master's Degree in Engineering Mechanics from the Indian Institute of Technology, Madras, Chennai and Ph.D. from School of Aeronautics and Astronautics from Purdue University, USA, in the year December 1992. After his Ph.D., he was a Postdoctoral Fellow in the Department of Mechanical Engineering at the Georgia Institute of Technology. In the year November 1997, he joined the Department of Aerospace Engineering at the Indian Institute of Science, Bangalore, where currently he is a senior Professor. His main areas of interest are Wave Propagation in complex media, Computational Mechanics, Smart Structures, and Structural Health Monitoring. He has published 220 refereed international journal publications and has authored 9 textbooks and monographs and has an h-index of 51 on Google scholar. He is on the editorial board of 6 international journals and is the Associate Editor for Smart Materials and Structures and Structural Health Monitoring international journals. Professor Gopalakrishnan is decorated with many awards and honors, which include the International Structural Health Monitoring Person of the year awards 2016 instituted by SAGE Publications, Fellow of Indian National Academy of Engineering, Fellow of Indian Academy of Sciences, Associate Fellows AIAA, Distinguished Alumnus Award, Indian Institute of Technology, Madras, Chennai, Satish Dhawan Young Scientist Award by Government of Karnataka, Biren Roy Trust award of Aeronautical Society of India, Alumni Award of excellence in research at IISc and the Royal Academy of Engineering, UK Distinguished visiting Fellowship. He was elected Fellow of the Institute of Mechanical Engineers, UK, in the year 2020.

CHAPTER 1

Introduction to Wave Propagation

Wave propagation is an exciting area that has far reaching applications in all branches of science and engineering. In fact every discipline can be seen through the prism of wave propagation. We see things and hear only because of waves. Electromagnetic waves have applications ranging from FM radios, TVs, mobile phones, and X-ray machines. Sound travels as waves through air. When an artiste plays violin, guitar, or mandolin, standing waves are generated in the vibrating musical column, which produces a pressure change or a sound wave, which makes the sound audible. Here are some more examples of wave propagation in different disciplines. For example, waves can propagate on solid surfaces such as the earth surfaces (for example, in earthquakes) or in bulk solids as in the case of metallic body. Throwing a stone on a pool of water causes circular ripples normally called *circularly crested water waves.* A different kind of water waves occurs during a Tsunami wherein the high-energy wave produced has so much of energy that it can capsize large ships and uproot buildings and other crucial infrastructure. Hence, we see that waves propagating in a medium can have both beneficial and devastating effects.

The above-mentioned situations are those where we very well know the existence of wave propagation phenomenon. There are many other situations where wave propagation phenomenon exists which are not obvious. The example of traffic flow on the road supports a variety of wave-like disturbances, and one can experience this if one travels in slow-moving traffic. Our heart beats are nothing but regulated spiral waves that swirl around the surface. The movement of human bodies is controlled by electro-chemical waves stimulated by the nervous system. The last but not the least, according to the latest research in quantum physics, everything around us can be described in terms of waves. In this chapter, we will show the existence of wave phenomenon on a number of systems through their partial differential equations that govern their physics.

DOI: 10.1201/9781003120568-1

Physicists normally study only two types of waves, namely *mechanical* or *electromagnetic.* The main differences between these two wave types are the following: In a mechanical wave, displacement (or velocity or acceleration) or stress fields oscillate about an equilibrium point called the *mechanical equilibrium* point, while in the case of an electromagnetic wave, both electric and magnetic fields oscillate about an equilibrium point. In both cases, only the energy is transported but not the particles themselves.

The two terms that are normally used in the context of wave propagation are *plane waves* and *standing waves.* Although waves can propagate in three-dimensional space, if they are confined to only one plane, then such waves are called plane waves. Mathematically, the parameter defining the wave field in a given plane will have a constant value in a plane that is perpendicular to this given plane. Standing waves are the ones whose position in the time axis remains the same. It is caused by the interaction of two different waves traveling in opposite directions.

1.1 ESSENTIAL COMPONENTS OF A WAVE

At this point, we should look at the definition of a *wave.* The first thing that will strike to everyone is to look at the definition given in standard dictionaries. Oxford dictionary defines wave as *a periodic disturbance of the particles of a substance which may be propagated without net movement of the particles, such as in the passage of undulating motion, heat, or sound.* This may not be a proper definition since waves such as standing waves in vibrating musical instruments do not propagate in the medium. A better definition can be found from Wikipedia on waves [62]. According to this website, a *wave is defined as is a disturbance (change from equilibrium) of one or more fields such that the field values oscillate repeatedly about a stable equilibrium (resting) value.* If the relative amplitude of oscillation at different points in the field remains constant, the wave is said to be a *standing wave.* If the relative amplitude at different points in the field changes, the wave is said to be a *traveling wave.* Waves can only exist in fields when there is a force that tends to restore the field to equilibrium.

Another feature that makes the definition of a wave complex is its interaction with the media in which it is propagating. For example, waves on the surface of the pond due to, say, a drop of stone propagate in the water without changing the medium. On the other hand, if the wave interacts in a mechanical solid say with cracks, it creates additional waves. In the case of anisotropic materials such as composites, there will be stiffness and inertial coupling, due to which there will again be the generation of new waves. Hence, it can be stated that there is no single definition of waves and one has to view them as a generic set of phenomenon with many similarities. Covering all the aspects of wave propagation is indeed a formidable task. There are exclusive texts for understanding wave propagation in different media. In this book, we will confine ourselves to elementary concepts of mechanical waves in structural materials systems. The previous book by the author [54] was a complete treatise of wave propagation in structures and materials, and the current book is the reduced version of this book [54], wherein not only the content of the elementary chapters are retained with enhanced content but also three new chapters are added to give readers the broader view on the subject of wave propagation in structures and materials.

Unlike electromagnetic waves, which can propagate in vacuum and air, mechanical waves need a medium for them to propagate. This means that the speed of the mechanical wave will be different for different media. That is, steel and aluminum will propagate mechanical waves at different speeds. Let us elaborate this a little further. Any one-dimensional wave, mechanical or otherwise, and moving with a velocity c_0 in the positive x-direction can be represented as

$$u_i(x,t) = \mathbf{A}f(x - c_0 t) \tag{1.1}$$

In the above equation, $u_i(x,t)$ normally represents deformation in the case of mechanical waves, which is a function of both space x and time t, and $\mathbf{A}$ is the amplitude of the wave. The function $f(x - c_0 t)$ can be any function that represents the character of the wave. An alternative way to represent waves can be in terms of *wavenumber* and *frequency*. This form of the wave can be written as

$$u_i(x,t) = \mathbf{A}f(kx - \omega t) \tag{1.2}$$

In the above equation k is the wavenumber, which has a unit of $1/length$, and ω is the circular frequency, which has a unit $cycles/second$ or $radians/second$.

When the wave is represented as in Eqn. (1.1) or (1.2), then such wave is called the *traveling wave* since the relative amplitudes of the wave at different points in the domain change. If we consider a finite one-dimensional medium of length, say L, then the incident wave traveling in the positive x-direction is given by Eqn. (1.1). This wave travels the entire length L and gets reflected back as a new wave in the negative x-direction given by

$$u_r(x,t) = \mathbf{B}g(x + c_0 t) \qquad \text{or} \qquad u_r(x,t) = \mathbf{B}g(kx + \omega t) \tag{1.3}$$

where $u_r(x,t)$ represents the deformation of the reflected wave and the function $g(x+c_0 t)$ or $g(kx+\omega t)$ represents the reflected wave. Since we are considering only linear behavior, the principle of superposition holds, which means that the total wave representation is given by the sum of the incident and the reflected waves. That is,

$$\begin{aligned} u(x,t) = u_i(x,t) + u_r(x,t) &= \mathbf{A}f(x - c_0 t) + \mathbf{B}g(x + c_0 t) \\ &\quad \text{or} \\ &= \mathbf{A}f(kx - \omega t) + \mathbf{B}g(kx + \omega t) \end{aligned} \tag{1.4}$$

Eqn. (1.4) is normally referred to as *D'Alembert Solution*, which is the solution of second-order 1-D hyperbolic wave equation. We will deal with this equation in more detail in Chapter 4. Such solutions are not available for fourth or even higher-order partial differential equations. If the functions representing the incident and reflected waves are given by *sine* or *cosine* functions, then such functions are oscillatory, and they are normally referred to as *time harmonic waves.* In this case, Eqn. (1.4) becomes

$$u(x,t) = \mathbf{A}\cos(kx - \omega t) + \mathbf{B}\cos(kx + \omega t) \tag{1.5}$$

The reader will recognize in the course of reading this book, the parameter k, which represents wavenumber, is a very important parameter that governs the physics of wave propagation of any wave propagating in a medium. The incident and reflected wave amplitudes $\mathbf{A}$ and $\mathbf{B}$ are determined based on the boundary conditions of the medium in which the waves are propagating.

Let us now consider only the first part of Eqn. (1.5), which is essentially the oscillatory cosine harmonic wave propagating with a wavenumber k. For some spatial locations, the wave will be represented by a simple cosine function $u(x,t) = \mathbf{A}\cos(\bar{\omega}t)$, where $\bar{\omega} = 2\pi f$ and f is the frequency in *Hertz*. This is shown in Fig. 1.1.

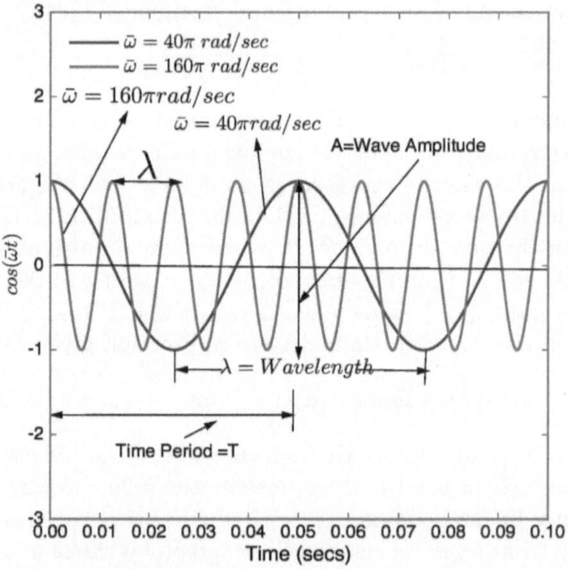

FIGURE 1.1: Anatomy of an oscillatory wave

Waves are characterized by a sequence of peaks and troughs as shown in Fig. 1.1, propagating in x-direction. We will use this figure to describe the anatomy of a wave and define some definitions connected with the wave motion:

1. **Amplitude**: It is the maximum displacement that the particles undergo about their mean position as the wave propagates through the medium. It is denoted by **A** in Fig. 1.1.

2. **Period**: The time period is defined as the time taken to complete one cycle of motion. It is shown as T in Fig. 1.1. Two different waves are shown in Fig. 1.1, and we see that the period is different for the two different waves.

3. **Frequency**: The frequency of a wave is the number of cycles or the vibrations the wave undergoes per unit time in a periodic motion. The frequency is denoted by f and the unit of frequency is *Hertz* (Hz). It is related to period T as $f = 1/T$. That is, when the frequency is large, the period is small. This can be clearly seen for the two waves plotted in Fig. 1.1.

4. **Angular Frequency**: The angular frequency is defined as the angular displacement the wave undergoes as it propagates per unit time. It is normally denoted as ω, and it is related to period T as $\omega = 2\pi/T$.

5. **Wavelength**: Wavelength can be defined as the distance between two successive crests or troughs of a wave. It is normally denoted as λ. Wavelength is inversely proportional to frequency f. This means the longer the wavelength, the lower the frequency. These can be clearly seen in Fig 1.1, wherein the first wave (shown in blue) having a longer wavelength is of smaller frequency compared to the second wave shown in red.

6. **Wavenumber**: The number of waves present in a unit distance of the medium in which the wave is propagating is called the Wavenumber. It is normally denoted as k, and it is inversely proportional to wavelength. It is related to wavelength λ as $k = 2\pi/\lambda$.

7. **Phase Velocity**: The distance covered by the wave per unit time in the direction of the wave propagation is called the *phase velocity.* It is normally denoted as c_0 and is related to frequency and wavenumber as $c_0 = \omega/k$.

1.2 INTERFERENCE OF WAVES

When two or more waves interact with each other, they give rise to new waves. For example, an incident wave, when it interacts with a boundary, generates a new wave called the reflected wave. When these two waves interact with each other, they create a new wave which has the combined property of both the incident and reflected waves. The combined wave is given by Eqn. (1.4). We could combine waves this way because the system in which the wave is propagating is assumed linear, and hence, the *principle of superposition* can be applied to combine the effects of the two waves. The interference of waves in second- and fourth-order systems are quite different. As shown later in this book, in a fourth-order system, four different waves will be generated, of which two will be the incident wave components, and the rest two will be the reflected wave components. We will see that some of these waves do not propagate and such waves are called *evanescent waves.* We will next discuss two important cases of interference of waves.

1.2.1 Interference of Two Similar Waves Propagating in the Same Direction

Let us consider two similar waves y_1 and y_2 traveling in the same direction x. These waves are cosine functions and they are given by

$$y_1 = A_1 \cos(kx - \omega t), \qquad y_2 = A_2 \cos(kx - \omega t + \delta) \tag{1.6}$$

where the second wave y_2 is the same as the first wave y_1, but they are offset by a phase difference of δ, and both the waves are propagating in the positive x-direction. The objective here is to determine the form of the new wave that will be generated due to the interaction of these two waves. The interaction of these two waves can be written as

$$\begin{aligned} y = y_1 + y_2 &= A_1 cos(kx - \omega t) + A_2 cos(kx - \omega t + \delta) \\ &= A_1 cos(kx - \omega t) + A_2[cos(kx - \omega t)cos\delta - sin(kx - \omega t)sin\delta] \\ &= cos(kx - \omega t)[A_1 + A_2 cos\delta] - sin(kx - \omega t)[A_2 sin\delta] \end{aligned} \tag{1.7}$$

Let

$$A_1 + A_2 cos\delta = Acos\epsilon, \qquad \text{and} \qquad A_2 sin\delta = Asin\epsilon \tag{1.8}$$

where A is the amplitude of the combined wave and ϵ is the phase difference of the combined wave with the original wave y_1. Using Eqn. (1.8) in Eqn. (1.7), we can write the expression of the combined wave as

$$\begin{aligned} y &= A[cos(kx - \omega t)cos\epsilon - sin(kx - \omega t)sin\epsilon \\ &= Acos(kx - \omega t + \epsilon) \end{aligned} \tag{1.9}$$

where

$$A = \sqrt{A_1^2 + A_2^2 + 2A_1A_2cos\delta}, \qquad tan\epsilon = \frac{A_2 sin\delta}{A_1 + A_2 cos\delta} \tag{1.10}$$

From Eqn. (1.9), we see that the combined wave y is similar to the first wave y_1, but shifted by a phase of ϵ and propagating with a combined amplitude of A, whose expression is given by Eqn. (1.10). The resultant amplitude A is maximum when $cos\delta = +1$ or $\delta = 2n\pi, n = 1, 2, 3.....n$ and minimum when $cos\delta = -1$ or $\delta = (2n+1)\pi, n = 1, 2, 3.....n$. In the first case, the amplitude $A = A_1 + A_2$, while in the second case, the amplitude $A = A_1 - A_2$. The former is called *constructive interference* and the latter is called *destructive interference.* At these two interferences, the phase angle $\epsilon = 0$. These two interference phenomena are shown in Fig. 1.2. This figure is created using the following parameters:

$$(kx - \omega t) = 6\pi t, \quad A_1 = 0.05\,m, \quad A_2 = 0.03\,m, \quad \delta = 2\pi$$

and

$$(kx - \omega t) = 6\pi t, \quad A_1 = 0.05\,m, \quad A_2 = 0.03\,m, \quad \delta = 3\pi$$

The first corresponds to constructive interference, while the second case refers to destructive interference.

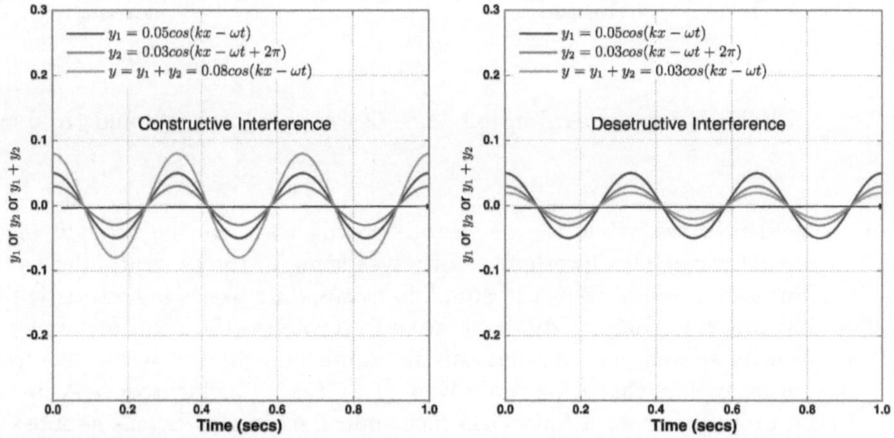

(a) (b)

FIGURE 1.2: Interference of waves. (a) Constructive interference and (b) Destructive interference

From Fig. 1.2, we can conclude that when the wave interference is constructive or destructive, the phase angle of the combined wave ϵ is equal to zero, which means that the combined wave is in phase with the wave y_1. Next, we will plot the interference for different values of $\delta = \pi/2$ and $\delta = 3\pi/2$ with all other parameters of y_1 and y_2 being the same. In both these cases, $\cos\delta = 0$ and the amplitude of the combined wave $y = \sqrt{A_1^2 + A_2^2}$. The general wave interference for these models is plotted in Fig. 1.3.

1.2.2 Standing Waves

We will look at the interference of two waves of equal magnitude but moving in opposite directions. The forward-moving wave can be thought of as an

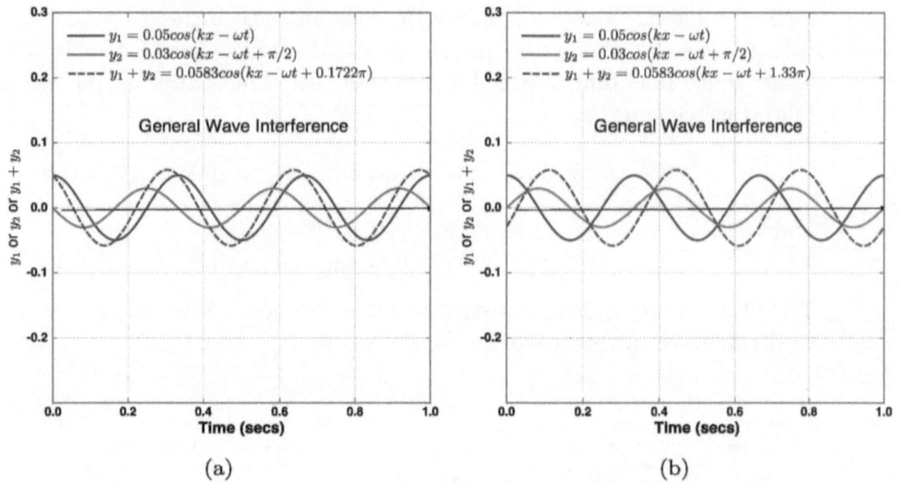

FIGURE 1.3: General interference of waves (a) $\delta = \pi/2$ and (b) $\delta = 3\pi/2$

incident wave, while the backward moving wave can be thought of as a reflected wave. This interference of waves leads to the generation of new waves, normally referred to as the *standing waves.* The basic characteristic of standing waves is that for different wave frequencies, the resulting waves will be fixed in spatial position and only its amplitude changes with wave frequency. Let us explain this using again Eqn. (1.5). Let an incident wave $\mathbf{A}\cos(kx - \omega t)$ be propagating in a finite 1-D mechanical structure. Let us suppose this incident wave generates a reflected wave of the same amplitude as that of the incident wave. That is, the reflected wave is given by $\mathbf{A}\cos(kx + \omega t)$. The total deformation field is given by

$$\begin{aligned} y = & \quad \mathbf{A}\left[\cos(kx - \omega t) + \cos(kx + \omega t)\right] \\ = & \quad \mathbf{A}\left[\cos kx \cos \omega t + \sin kx \sin \omega t + \cos kx \cos \omega t - \sin kx \sin \omega t\right] \\ = & \quad 2\mathbf{A} \cos kx \cos \omega t \end{aligned} \tag{1.11}$$

Eqn. (1.11) represents a standing wave that is generated due to the combined action of the incident and reflected wave. We can plot this function in terms of variable x for a given ω and time t, or we can plot this function as a function of t for a given wavenumber k and spatial location x. This wave has an amplitude of $(2\mathbf{A} \cos kx)$, and its position for different wave frequencies (shown as wave **A**, **B**, **C** and **D** curves in Fig. 1.4) is constant spatially.

Standing waves are very common in mechanical or metallic structures. For example, in metallic rod structures, when an incident wave interacts with a free boundary, it generates a reflected wave, whose amplitude is same as that of the incident wave. On the other hand, if this wave interacts with a fixed boundary, then the reflected wave is negative of the incident wave. This

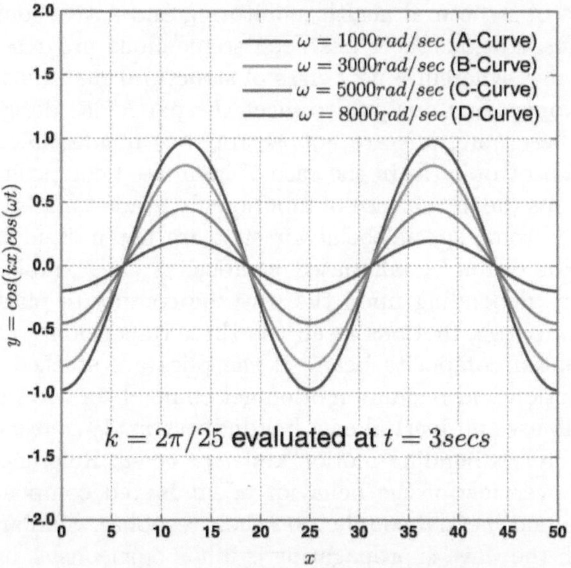

FIGURE 1.4: Example of a standing wave

interaction will also generate a standing wave. However, its wave profile will be quite different when compared to the earlier case.

Mechanical waves are categorized into two categories, namely the *Longitudinal waves* and *Transverse waves.* This classification is purely on the basis that the medium on which the wave propagation is based is one-dimensional. If the dimension of the medium is either 2-D or 3-D, the wave propagation in such media is quite complex, and such clear-cut definitions as longitudinal or transverse waves may not be possible. The wave propagation in such 2-D or higher dimensional media is categorized differently based on the nature and direction of the wave propagation, and these are explained in detail in subsequent chapters. There are many other introductory aspects of mechanical wave propagation, which are quite well known and are not covered in this book. There are many classic texts and notes in physics, for example [42] and [107], which cover the introductory part extensively.

1.3 NEED FOR WAVE PROPAGATION ANALYSIS IN STRUCTURES AND MATERIALS

As mentioned earlier, wave propagation is an exciting field having applications cutting across many disciplines. In the area of structures and materials, wave propagation-based tools have found increasing applications, especially in the area of structural health monitoring and active control of vibrations and noise. In addition, there has been tremendous progress in the area of material science, wherein a new class of structural materials is designed based on wave propagation analysis to meet the particular design standards. In most cases, these materials are not isotropic as in metallic structures. These are either anisotropic (as in the case of laminated composite structures) or inhomogeneous (as in the case of functionally graded materials). These structures introduce some special behavior that are not present in conventional structures. For example, laminated composites can exhibit stiffness and inertial coupling depending upon the way the composite plies are stacked. Due to these couplings, motions in all the three dimensions get coupled. That is, in a laminated composite beam, if the plies are stacked unsymmetrically, the axial motion and bending motion get coupled. In wave propagation terminology, stiffness and inertial coupling induces *mode conversion*. That is, an axial impact causes bending motion and *vice versa*. References [69] and [117] give a good overview of the behavior of laminated composites and the effect of stiffness and inertial coupling on their response. Similarly, when the medium in which the wave is propagating is inhomogeneous as in the case of functionally graded material structures, the traditional definition of plane wave gets destroyed and the waves in such a medium propagate with attenuation. In these material systems, a systematic wave propagation analysis will help us in understanding the behavioral physics of these systems better.

Wave propagation analysis involves performing spectral analysis, which helps us to understand the behavior of the waves. It also provides the frequency bands in certain materials where the wave responses are retarded or even blocked. Many acoustic meta materials are designed based on the wave propagation analysis [128]. Dynamic analysis in structural engineering falls under two different classes: one involving low-frequency loading and the other involving high-frequency loading. Note that the frequency content of loading can be determined by taking the Fourier Transform of the input time history. Low-frequency problems are categorized under *Structural dynamics* problems, while those involving high-frequency loading fall under the category of *Wave propagation* problems. In structural dynamic problems, the frequency content of the dynamic load is of the order of few hundred Hertz, and the designer will be mostly interested in the long-term (or steady-state) effects of the dynamic load on the structures. Hence, the first few normal modes and the natural frequencies are sufficient to assess the performance of the structure. The phase information of the response is not critical here. Most dynamic problems in structure will fall under this category. On the other hand, for wave propagation problems, the frequency content of the input loading is very high (of the order of kilo Hertz or higher), and hence the short-term effects (transient response) become very critical. Many higher-order modes will participate in amplifying the dynamic response. Impact and blast types of loading fall under this category. The multi-modal phenomenon of wave propagation makes one parameter very important and that is the phase information. A good discussion on the spectral analysis of motion for wave propagation is given in [47].

For many scientists/engineers, the clear difference between structural dynamics and wave propagation is not quite evident. Traditionally, a structural designer will not be interested in the behavior of structures beyond certain frequencies, which are essentially at the lower side of the frequency scale. For such situations, available general-purpose analysis tools such as finite element method will satisfy the designer's requirement. However, in the recent times, structures are required to be designed to sustain very complex and harsh loading conditions. These loading histories are essentially a multi-modal phenomenon and fall under the category of wave propagation. That is, for such loading environments, in addition to the dynamic response, phase information contained in the response will also become important. Evaluation of structural integrity of such structures subjected to harsh loadings is a complex process. The currently available analysis tools are highly inadequate to handle the modeling of these structures. In this book, we present techniques such as *spectral finite element method*, which we believe will address some of the shortcomings of the existing analysis tools.

Another area where understanding wave propagation phenomenon is becoming increasingly important is in the design of sand bunkers for resisting blast loads. Sand bunkers are normally made of dry sands, which are essentially low-modulus materials that have a high ability to absorb high-velocity impact. The amount of energy absorption in a dry sand bunker depends on the type of dry sand that fills this bunker. The material property of the granular materials such as dry sands is dependent on various parameters such as void ratio of the soil, confined hydrostatic pressure, density, etc. The energy absorption capacity of a granular soil bunker mainly depends on its shear modulus value, which depends on the host of parameters mentioned above. Hence, understanding wave propagation behavior in these materials is very important if the geo-technical engineer needs to design a good sand bunker for resisting blast loads. Understanding the wave propagation behavior in granular soil is also a requirement in earthquake engineering.

Increasing emphasis of miniature devices has made scientists look for newer and novel materials which can be handled at the atomistic scales. In this regard, nanoscale materials and structures with nano thicknesses have attracted considerable interest from the scientific community in the fields of microelectronics and nanotechnology. More and more nanostructures, such as ultra-thin films, nanowires, and nanotubes, have been fabricated and served as the basic building blocks for nano-electro-mechanical-systems (NEMS). For long-term stability and reliability of various devices at nanoscale, researchers should possess a deep understanding and knowledge of mechanical properties of nanomaterials and structures, especially the time-dependent or dynamic properties. Among many available techniques, the high-frequency acoustic wave technique is regarded as one of the very efficient nondestructive methods to characterize elastic media with nanostructures. Hernaandez *et al.* [59] used high-frequency laser-excited guided acoustic waves to estimate the in-plane mechanical properties of silicon nitride membranes. Mechanical properties and residual stresses in the membranes were evaluated from measured acoustic dispersion curves. The mean values of Young's modulus and density of three nanocrystalline diamond films and a free-standing diamond plate were determined by analyzing the dispersion of laser-generated surface waves by Philip *et al.* [108].

Nanostructures such as Carbon nanotubes (CNTs) can propagate waves of the order of tera Hertz (THz). THz waves in nanoscale materials and nano-photonic or nano-phononic devices have opened a new topic on the wave characteristics of nano-materials [35, 116, 123]. As dimensions of the material become smaller, however, their resistance to deformation is increasingly determined by internal or external discontinuities (such as surfaces, grain boundary, strain gradient, and dislocation). Although many sophisticated approaches for predicting the mechanical properties of nano-materials have been reported, few addressed the challenges posed by interior nanostructures, such as the surfaces, interfaces, structural discontinuities, and deformation gradient of the nano-materials under extreme loading conditions. The use of atomistic simulation may be a potential solution in the long run. However, it is well known that the capability of this approach is limited by its need of prohibitive computing time and an astronomical amount of data generated in the calculations. Wave propagation analysis using continuum models, especially using non-local elasticity models can be used to address the above problems.

Wave propagation studies mainly include the estimation of wavenumber and wave speeds such as phase and group speeds. The concept of wavenumber and group velocity will be useful in understanding the dynamics of structures at both normal and atomistic scales, since parameters are related to the energy transportation during the propagation of the waves. One of the many objectives of this book is to study the wave propagation in many material systems including nanostructures, composites, graded materials, and granular materials that can be handled within the framework of spectral analysis.

The above discussion makes it clear that the analysis of wave propagation in different materials and structures is indeed a complex one, which requires an understanding of material science, mechanics, and solution methods. There are many classic textbooks such [5], [34], [48], [55], and [121] that address mostly the basic wave propagation concepts on only some specified material systems. The previous book by the author [54] is a comprehensive book that contains both the elementary and advanced topics of the wave propagation. This book can be thought of as an abridged version of the previous unified book [54], wherein all the chapters concerning elementary topics are retained, enhanced with three new chapters, and some of the sections are rewritten to bring in more clarity and understanding of the topic of wave propagation in structures and materials.

1.4 SOME WAVE PROPAGATION PROBLEMS IN SCIENCE AND ENGINEERING

As mentioned earlier, wave propagation application finds application in diverse disciplines. Each of these applications is governed by a partial differential equation in terms of a field variable(s) which can be functions of three spatial variables, namely x, y, z, and time t. These PDEs are derived using Laws of Physics such as Newton's second law (Conservation of Momentum theory) and Laws of Conservation of Mass (Continuity Equation) coupled with other conditions such as constitutive laws (Hook's law) and compatibility conditions. In this section, we provide the governing partial differential equations for a few applications without going into details of how these equations were derived. The reader can refer to many classic books on physics to know more about how these equations were derived. The objective of giving these equations here is for the reason that if the reader wants, he/she can apply the techniques described in this book for these problems.

Next, how will one classify a problem (or governing PDE) that it indeed falls under the category of wave propagation. Except for very few cases, it is very difficult to do this by merely looking at the governing partial differential equation. Most wave propagation problems associated with structures and materials deal with short duration input (of the order of micro to millisecond duration), which excites many higher-order vibrational modes. Hence, most wave propagation is a multi-modal phenomenon involving very-high frequencies. This is one way of identifying problems involving the wave propagation phenomenon. Alternatively, the governing equation of a system can be transformed into frequency domain using integral transforms such as Fourier, Wavelet, or Laplace transforms (or using variable separable type solutions). This can be done only when the governing PDE is linear and the domain is regular. This will result in a single PDE transforming into a set of ODEs (especially in the case of one-dimensional problem). The resulting ODEs can be either constant coefficient ODE or variable coefficient ODEs. If the ODEs are of constant coefficient type, then the solutions of the governing equations will be in terms of complex exponentials, which essentially consist of wave functions such as sine or cosine functions. However, when the governing equations are variable coefficient type, then the solutions to the problem will be in terms of special functions such as Bessel functions, which again have a wavy profile with distance.

Most PDEs require numerical methods such as finite element method, finite difference method, etc. for its solution. These methods are necessarily solved in time domain, and they lead to very large problem sizes, especially for problems involving short-duration input as we encounter in wave propagation problems. These methods are not covered in this book. A very detailed aspect of these methods is covered in the earlier book by the author [54]. The present book exclusively uses frequency domain spectral analysis to study wave propagation in different material systems. This method can be applied to most linear systems. Next, we will present the governing partial differential equations of some material systems which have wave propagation type solutions. Only governing equations are presented without going into details of how the equations were derived

1. **Transverse Vibrations of a Taut String:**

$$\frac{\partial^2 w}{\partial x^2} - \frac{1}{c^2}\frac{\partial^2 w}{\partial t^2} = \frac{p}{T} \tag{1.12}$$

   where w is the transverse displacement, c is the speed of the wave given by $\sqrt{T/\rho}$, T is the tension in the string, ρ is the mass per unit length, and p is the loading per unit length. The above equation is a hyperbolic equation that resembles the governing equation for longitudinal wave motion in rods, which we cover extensively in Chapter 4. If the right-hand side is zero, then Eqn. (1.12) can be solved using variable separable method or methods based on integral transforms. One can derive the traveling wave solution to this problem using variable separable solution method for this PDE using the two boundary conditions at the end of the sting and two initial condition of $w = w_0$ at $t = 0$ and $\dot{w} = v_0$ at $t = 0$.

2. **Long Waves on Canals:**
   Here, the canal is assumed to be shallow, where the depth of the water is less than the wavelength of the water wave as shown in Fig. 1.5. In this

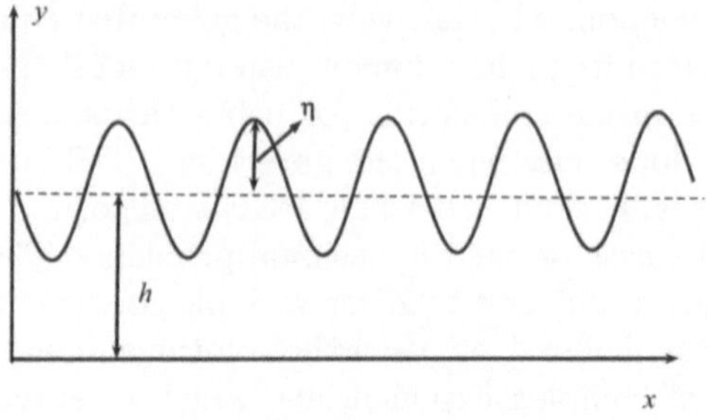

FIGURE 1.5: A shallow canal

   figure, we assume that the depth of water h in the canal is much less than the wavelength of surface water wave. Hence, the vertical acceleration in the y-direction can be neglected. In Fig. 1.5, η is the surface displacement as the wave propagates in the canal. Using the linearized Navier-Stokes equation (please see [20] for detailed derivation of governing equation), we can write the governing PDE for water wave propagation in shallow canal as

$$\frac{\partial^2 \eta}{\partial t^2} = c^2 \frac{\partial^2 \eta}{\partial x^2} \tag{1.13}$$

Clearly, the above form looks very similar to Eqn. (1.12), except that $c = \sqrt{gh}$, where g is the acceleration due to gravity. One can derive the traveling wave solution for this case as is done for the case of the string.

3. **Surface Water Waves in Deep Canals:**
In this case, considering Fig. 1.5, we assume that the depth of the water h in the canal is very deep when compared to the wavelength of the surface wave. Hence, in this case, we cannot ignore the vertical acceleration in the y-direction. The details of the governing equation derivation are given in [20]. The governing PDE is derived using the assumption that the fluid in the deep canal is irrotational as well as equi-voluminal and if u and v are the velocities in the two coordinate direction, which are related to the stream function ϕ as $u = -\partial\phi/dx$ and $v = -\partial\phi/dy$, the governing equation for this case is the elliptic partial differential governing equation given by

$$\frac{\partial^2 \phi}{\partial x^2} + \frac{\partial^2 \phi}{\partial y^2} = 0 \tag{1.14}$$

In the above equation, we see that the equation does not have any time dependency. It comes from the following boundary conditions.

$$\frac{\partial \phi}{\partial y} = 0, \qquad \text{on}\, y = -h \frac{\partial^2 \phi}{\partial t^2} + g\frac{\partial \phi}{\partial y} = 0 \qquad \text{on}\, y = 0$$

Reference [20] has shown that the above equation with boundary condition does have a traveling wave solution.

4. **Traffic Flow on a Freeway:**
This theory was first formulated by Lighthill and Whitham [83]. It is a simple theory defined by first-order non-linear PDE, which is capable of describing many real-life features of highway traffic. The governing equation that governs the flow of traffic is given by

$$\frac{\partial \rho}{\partial t} + \left(\frac{dq}{d\rho}\right)\frac{\partial \rho}{\partial x} = 0 \tag{1.15}$$

where $\rho(x, t)$ is the density of cars on the road (or number of cars per unit length) and $q(x, t)$ is the number of cars crossing a point x per unit time. In the above equation, q is normally expressed as a function of ρ, the car density on the road, then we can either use variable separable or transform technique to solve the resulting equation. Different variation of q as a function of ρ will yield different traffic condition.

5. **Transmission Lines:**
The transmission lines is a model of a co-axial cable having a conductor with a flowing current I and voltage across the cable being V. This

conductor is shown in Fig. 1.6(a), while Fig. 1.6(b) shows the corresponding electric parameters per unit length over a small length of dx, which are the inductance L, capacitance C, and the leakage conductance G between the two conductors. The governing equation for the determi-

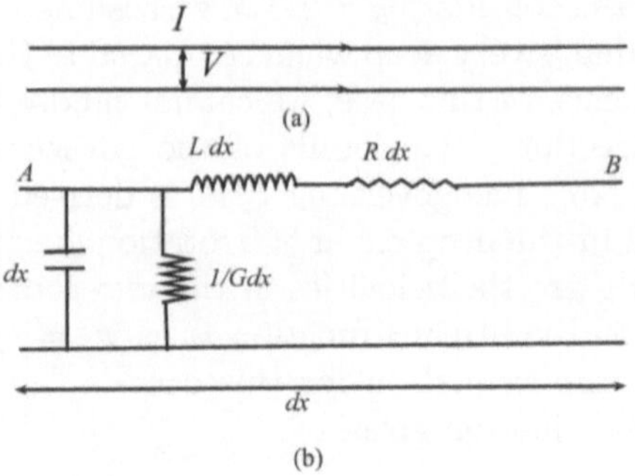

FIGURE 1.6: A schematic of a transmission line with its parameters

nation of the unknown voltage V in terms the electrical parameters is given by

$$LC\frac{\partial^2 V}{\partial t^2} + (LG + RC)\frac{\partial V}{\partial t} - \frac{\partial^2 V}{\partial x^2} + RGV = 0 \tag{1.16}$$

The above equation is also called the *telegraph equation*, which governs the behavior of transmission lines. The Eqn. (1.16) is very similar to the string PDE (Eqn. (1.12)) except that it has first derivative in time, which normally provides the dissipative component to the response. As before, the above equation also admits a traveling wave solution, and the equation can be solved using two spatial boundary conditions and two initial conditions as is done for solving string PDE (Eqn. (1.12)).

We have provided a number of different wave equations in different disciplines, and for all of these equations, although the physics of the problem they represent is different, the method of solution are all same. In other words, the spectral methodology presented in this book can also be applied to the above PDEs.

1.5 ORGANIZATION AND SCOPE OF THE BOOK

Wave propagation in a material system is a dynamic phenomenon involving propagation, reflection, transmission, and generation of new stress waves in the media in which it is propagating. The wave propagation in a media is characterized by two important parameters, namely the wavenumber and group speed. These two-parameters help us to identify the nature of waves, that is dispersive or non-dispersive waves. Dispersive waves have an important characteristic that their speeds change with frequency, and as a result, they change their wave profiles as they propagate. On the other hand, non-dispersive waves force their wavenumbers to have a linear relationship with frequency, and as a result, the wave profile does not change as they propagate. In addition, phase information is very crucial for wave propagation. That is, the propagation of waves in the time domain is associated with phase changes in the frequency domain. The wavenumbers and group speeds depend on two important factors, namely the material properties of the media in which they are propagating and the governing differential equation that govern the motion of the material system. These aspects will be made clear in the subsequent chapters. Hence, the main objective of this book is to capture the physics of wave propagation in different material system, study their interaction with different boundaries and understand the generation of any new waves that would result as an outcome of these interactions. Using general and unified principles, the physics of wave propagation in different material system will be brought out in this book in such a manner that the same principles will be sufficient to address the wave propagation problem discussed in the previous section. As mentioned earlier, this book is the abridged version of the earlier book by the author [54]. In this book, all the advanced topic of the unified book is left out, and the basic material is reinforced with the addition of three new chapters. There are in all 12 chapters in this version of the book, which tries to bring out clearly the physics associated with wave propagation in different structures and material systems. Most chapters will have practice problems at the end of the chapter and also some MATLAB® scripts are developed specifically to solve some of the problems presented in this book. These MATLAB scripts will be housed in the publisher's website. These MATLAB codes will aid in a better understanding of the subject.

It is assumed that the reader of this book has a good background in the solution of differential and partial differential equations, linear algebra, and also good understanding of theory of elasticity and constitutive modeling. Unlike many other wave propagation texts, this text presents the complete wave propagation analysis entirely in the frequency domain. Hence, it is appropriate to begin the text (Chapter 2) with a detailed introduction to the fundamentals of Fourier Transforms and its numerical implementation.

The heart of wave propagation is spectral analysis, which enables the computation of wavenumbers and group speeds. These computations are simple for some elementary waveguides and become increasingly difficult for complex waveguides with the increase of number of possible wave motions (or degrees of freedom) such as, say the composite or graded structures. Hence, an exclusive chapter (Chapter 3) is presented to introduce the reader the concept of spectral analysis and the detailed procedure to compute the wave parameters is outlined in this chapter.

Chapter 4 introduces the reader the concepts of wave propagation in isotropic 1-D waveguides, wherein the wave propagation in rods, beams, and frames are discussed in detail. Also, wave propagation in variable coefficient differential systems such as tapered waveguides and also rotating waveguides are presented in this chapter. The studies presented includes both elementary and higher-order waveguides.

One of the important material systems for wave propagation is the viscoelastic material system since these material waveguides are normally used to attenuate the waves in structural system. Wave propagation in linear viscoelastic material waveguides is presented in Chapter 5. These materials not only deform due to loading but also experience viscous type behavior; that is, when stressed, they experience forces that are proportional to the velocities. Modeling of viscoelastic waveguides are quite different as well as complex compared to isotropic waveguides since the constitutive models are dependent on time and their derivatives. As a result, the material properties become frequency dependent, and they are complex in the frequency domain. The imaginary part of material property causes the waves to attenuate as it propagates. This chapter discusses different viscoelastic material models and presents wave propagation in these models in greater detail. This is again a new chapter, which is not available in the unified book [54].

Processing and handling wave signals (especially the dispersive signals) is indeed a very complex issue, especially when the wave propagation analysis is performed using numerical Fourier Transforms or Fast Four Transforms (FFT). This is due to the periodicity of the signals and the finite time window associated with FFT. Hence, a separate chapter on Signal Processing is presented in Chapter 6. This is gain a new chapter, which is not available in the unified book [54].

In Chapter 7, wave propagation in two-dimensional isotropic waveguides is presented. This chapter will outline the procedure of computing the wavenumbers, group speeds, and responses of P-wave, SH-Wave, and SV-waves. In addition to this, wave propagation in doubly bounded media or *Lamb wave* and surface wave propagation or *Rayleigh wave propagation* is discussed in detail.

In Chapter 8, wave propagation in laminated composite waveguides is discussed. In first few pages, laminated composites and its constitutive models at the lamina and at laminate levels are discussed. Here, as in Chapters 4 and 7, the discussion will be centered on wave behavior in 1-D and 2-D composite waveguides. In composite waveguides, due to material anisotropy, there is no clear distinction between P, SV, and SH waves, and they are usually coupled. Hence, the traditional Helmholtz decomposition method of solving 2-D wave equation, which is normally used for isotropic solids, cannot be used here. Here, we will show the use of Partial Wave Technique in obtaining the wave modes in a 2-D composite waveguides. This chapter will end with the methodology to obtain Lamb wave modes in 2-D composite waveguide.

In Chapter 9, we introduce yet another material system called the *functionally graded material system* (FGM), which are inhomogeneous structures. That is, the material property in these structures vary spatially. That is, the material property grading can be done either depth-wise or length-wise. As before, we will present the wave propagation in both 1-D and 2-D FGM waveguides in this chapter.

In Chapter 10, the wave propagation in granular soils is presented. Soil as such is very complex inhomogeneous material and the material properties depend on host of parameters. First, the material properties for different dry sands are discussed as a function of soil parameters. Using the characterized material properties, the wave propagation concepts developed in Chapter 4 are applied for the longitudinal and flexural 1-D wave propagation in different granular soils. Next, the chapter presents a method of obtaining blast pressure profiles as a function of quantity of explosives used in the blast and the distance of the structure from the blast site. Following this, the response characteristic due to the estimated blast loads on different soil systems is investigated. Finally, the chapter also briefly outlines the design concepts for sand bunker for maximum attenuation of the transmitted wave in the attached structure. This is again a new chapter, which is not available in the unified book [54].

Chapter 11 deals with wave propagation in non-local one-dimensional waveguides. This chapter again is a new chapter, which is not available in the unified book [54]. The first part begins with a brief introduction of non-local elasticity, and different non-local theories, such as Strain Gradient Theories and Eringen's Stress Gradient Theory are explained in detail. The chapter presents not only the wave propagation characteristics of some of these waveguide models but also discuss the stability of these models from the point of view of wave propagation. Some of the new physics associated with wave propagation at nanoscales is brought out clearly in this chapter.

In all the above chapters, wave propagation analysis in structures is performed in an ad hoc manner, wherein keeping track of myriad of reflections and transmissions in multiply connected waveguides becomes very difficult. Hence, there is a need to automate the analysis process. In this regard, Chapter 12 introduces to the reader, a matrix-based frequency domain method called the *Spectral Finite Element formulation.* Here, the method of formulation of different spectral finite elements (both 1-D and 2-D elements) based on Fourier Transform is outlined in this chapter.

The entire book is written in modular form for easy understanding. The topics are organized in increasing order of complexity for better appreciation of the subject. Several MATLAB scripts are provided, which will enable the reader to not only verify the examples provided in this book but also develop new scripts for solving many new wave propagation problems not covered in this book. Each chapter is provided with exercise problems, which will enable the reader to further enhance the understanding of the subject.

Notations used in the book

In this book, all bold-face letter indicates either a matrix or vector; $\dot{(.)}$ indicates first derivative in time; $\ddot{(.)}$ indicates second derivative in time; and $\hat{(.)}$ indicates that the quantity is frequency dependent.

CHAPTER 2

Introduction to Fourier Transforms

Wave propagation analysis involves two important steps, namely the deriving the governing equation of motion and performing spectral analysis. The governing differential equation of motion for structures made from different material systems can be derived using the concepts outlined in Chapters 2 and 3 of the unified book [54]. The details of spectral analysis will be discussed in the next chapter. In spectral analysis, the governing differential equation need to be transformed to frequency domain for determination of wave parameters such as wavenumbers, phase speeds, group speeds, cut-off frequencies and band gaps. In addition, the wave propagation analysis also deals with many different and complex experimentally obtained time signals, which may require fine tuning and manipulation, such as removing white noise, filtering unwanted signals, etc. Some of these operations can be effectively handled if one understands the concept of handling time signals and frequency spectrums using time series analysis. Also, if one needs to understand the physics of wave propagation, one has to perform spectral analysis in the frequency domain. The governing equations can be transformed into frequency domain using any of the available *Integral Transforms.* There are a number of integral transforms available to transform a time domain variable into frequency domain. Among those, four important integral transforms, namely, the *Fourier Transforms*, *Short Term Fourier Transforms*, *Wavelet Transforms*, and *Laplace Transforms*, are available for signal analysis. However, in this chapter as well as in this book, we will discuss only Fourier Transform as all our analysis in this book will be based on Fourier Transform. The use of other integral transforms for wave propagation analysis is available in the author's unified book [54].

DOI: 10.1201/9781003120568-2

2.1 FOURIER TRANSFORMS

Wave propagation analysis always deals with time signals. A time signal can be represented in the Fourier (frequency) domain in three possible ways, namely the *Continuous Fourier Transform* (CFT), the *Fourier Series* (FS) and the *Discrete Fourier Transform* (DFT). In this section, only brief descriptions of the above transforms are given. The reader is encouraged to refer to [34] for more details.

2.1.1 Continuous Fourier Transforms (CFT)

Consider any time signal $F(t)$. The inverse and the forward CFTs, normally referred to as the transform pair, are given by

$$F(t) = \frac{1}{2\pi}\int_{-\infty}^{\infty} \hat{F}(\omega)e^{i\omega t}d\omega, \quad \hat{F}(\omega) = \int_{-\infty}^{\infty} F(t)e^{-i\omega t}dt\,, \tag{2.1}$$

where $\hat{F}(\omega)$ is the CFT of the time signal, ω is the angular frequency and i $(i = \sqrt{-1})$ is the complex number. $\hat{F}(\omega)$ is always complex and a plot of the amplitude of this function against frequency will give the frequency content of the time signal. As an example, let us consider a rectangular time signal of pulse width d. Mathematically, this function can be represented as

$$\begin{aligned} F(t) &= F_0 \qquad -d/2 \le t \le d/2 \\ &= 0 \qquad \text{otherwise}\,. \end{aligned} \tag{2.2}$$

This time signal is symmetrical about the origin. If this expression is substituted in Eqn. (2.1), we get

$$\hat{F}(\omega) = F_0 d\left\{\frac{\sin(\omega d/2)}{\omega d/2}\right\}. \tag{2.3}$$

The CFT for this function is a real number and symmetric about $\omega = 0$. The term inside the curly brace is called the *sinc* function. Also, the value of the CFT at $\omega = 0$ is equal to the area under the time signal.

Now let the pulse be allowed to propagate in the time domain by an amount t_0 seconds. Mathematically such a signal can be written as

$$\begin{aligned} F(t) &= F_0 \qquad t_0 \le t \le t_0 + d \\ &= 0 \qquad \text{otherwise.} \end{aligned} \tag{2.4}$$

Substituting the above function in Eqn. (2.1) and integrating, we get

$$\hat{F}(\omega) = F_0 d\left\{\frac{\sin(\omega d/2)}{\omega d/2}\right\} e^{-i\omega(t_0+d/2)}\,. \tag{2.5}$$

We see that now the CFT is complex, that is, it has both real and imaginary parts. These are also plotted in Fig. 2.1. From Eqns. (2.3 & 2.5), we see that the magnitude of both these transforms are the same, however, the second transform is phase shifted. That is, we see that the propagation of the signal in the time domain is associated with the change of phase in the frequency domain. Waves propagating in the medium are always associated with phase changes as the signal propagates.

Next, we will look at the *spread* of the signal both in the time and the frequency domain. This can be obtained if one determines the frequency values at which CFT becomes zero. This occurs when

$$\sin\left(\frac{\omega_n d}{2}\right) = 0, \quad \text{or } \frac{\omega_n d}{2} = 2n\pi\,, \quad \text{or } \omega_n = \frac{4n\pi}{d}\,,$$
$$\omega_2 - \omega_1 = \Delta\omega = \frac{4\pi}{d}\,. \tag{2.6}$$

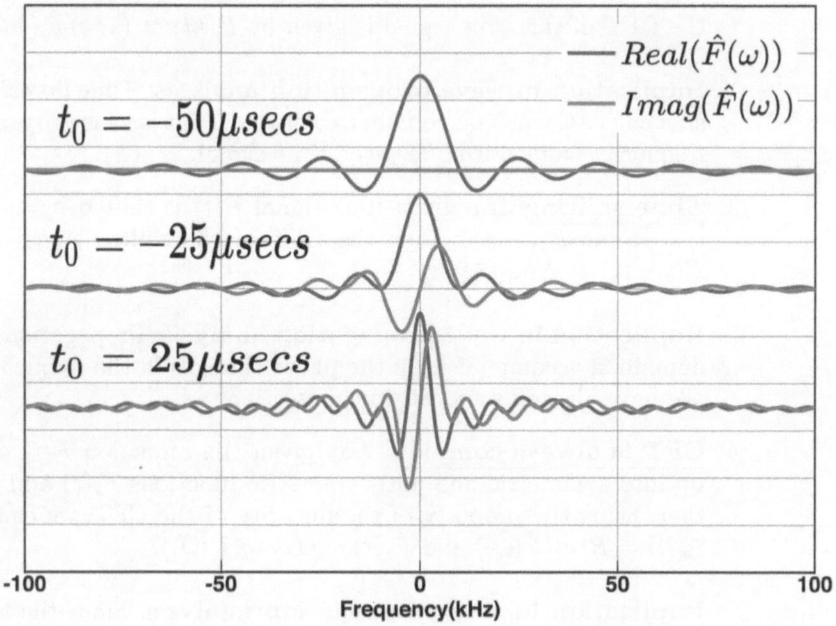

FIGURE 2.1: Continuous Fourier Transforms for various propagating time t_0 for a given pulse width

That is, if the spread of the signal in the time domain is d, then the spread in the frequency domain is $\Delta\omega = 4\pi/d$. Here, $\Delta\omega$ represents the frequency bandwidth. Hence, a *Dirac delta function, which has infinitesimal width in the time domain, will have infinite bandwidth in the frequency domain.*

Next, we will look at properties of CFT that has significant implications in wave propagation analysis. Following are some of the properties of the CFT.

1. **Linearity**: Let us consider two time functions $F_1(t)$ and $F_2(t)$. The CFT of these functions are given by $\hat{F}_1(\omega)$ and $\hat{F}_2(\omega)$, then the Fourier Transform of combined function is $F_1(t)+F_2(t) \Leftrightarrow \hat{F}_1(\omega)+\hat{F}_2(\omega)$. Here, the symbol $\Leftrightarrow$ is used to denote the CFT of a time signal.

   Implication in wave propagation analysis: Here, $F_1(t)$ and $F_2(t)$ can be thought of as the incident and reflected waves, respectively. The linearity property states that the combined transform of incident and reflected waves is equal to the individual transform of these obtained separately. This essentially comes from the Principle of Superposition, which was discussed in Chapter 1. As discussed earlier, interaction two or more waves will lead to standing waves.

2. **Scaling**: If a time signal is multiplied by a factor k to become $F(kt)$, the CFT of this time signal is given by $F(kt) \Leftrightarrow (1/k)\,\hat{F}\,(\omega/k)$

   Implication in wave propagation analysis: Time domain compression causes frequency domain expansion. That is, this property fixes the frequency bandwidth of a given time signal.

3. **Time shifting**: If a given time signal $F(t)$ is shifted by an amount t_s to become $F(t-t_s)$, then the CFT of the shifted signal is given by $F(t-t_s) \Leftrightarrow \hat{F}(\omega)e^{-i\omega t}$.

   Implication in wave propagation analysis: Propagation in the time domain is accompanied by the phase changes in the frequency domain. We have already seen this in this section.

4. **CFT is always complex**: Any given time function $F(t)$ can be split up into symmetric and anti-symmetric functions $F_s(t)$ and $F_a(t)$. Further, using the property of the linearity of the CFT, we can show that $F_s(t) = Real(\hat{F}(\omega))$ and $F_a(t) = iImag(\hat{F}(\omega))$.

   Implication in wave propagation analysis: Since the time signals encountered in wave mechanics is neither symmetric (even) or anti-symmetric (odd) in nature, the CFT is necessarily complex in nature. Hence, the wave propagation problems are always associated with phase changes.

5. **Symmetric property of the CFT**: Since the CFT of a time signal $F(t)$ is complex, it can be split up into real and imaginary parts as $\hat{F}(\omega) = \hat{F}_R(\omega) + i\hat{F}_I(\omega)$. Substituting this in the second equation of Eqn. (2.1) and expanding the complex exponential in terms of sine and

the cosine function, we can write real and imaginary part of the transform as

$$\hat{F}_R(\omega) = \int_{-\infty}^{\infty} F(t)\cos(\omega t)dt, \quad \hat{F}_I(\omega) = \int_{-\infty}^{\infty} F(t)\sin(\omega t)dt$$

The first integral is an even function and the second is the odd function, that is $\hat{F}_R(\omega) = \hat{F}_R(-\omega)$,and $\hat{F}_I(\omega) = -\hat{F}_I(-\omega)$. Now, if we consider the CFT about a point $\omega = \omega_0$, the transform on the right of ω_0 can be written as $\hat{F}(\omega_0) = \hat{F}_R(\omega_0) + i\hat{F}_I(\omega_0)$. Similarly, the transforms to left of ω_0 can be written as $\hat{F}(-\omega_0) = \hat{F}_R(-\omega_0) + i\hat{F}_I(-\omega_0) = \hat{F}_R(\omega_0) - i\hat{F}_I(\omega_0) = \hat{F}^*(\omega_0)$, which is the complex conjugate of the transform on the right side of ω_0. The frequency point about which this happens is called the *Nyquist Frequency.*

Implication in wave propagation analysis: The Nyquist frequency is an important parameter in the wave propagation analysis, especially in the context of using Fast Fourier Transform (to be introduced later in this chapter), since it is only up to this frequency that all the analysis will be performed.

6. **Convolution**: This is a property relating to the product of two time signals $F_1(t)$ and $F_2(t)$. The CFT of the product of these two functions can be written as

$$\hat{F}_{12}(\omega) = \int_{-\infty}^{\infty} F_1(t)F_2(t)e^{-i\omega t}dt$$

Now, substituting Eqn. (2.1) for both these functions in the above equation, we can write

$$\hat{F}_{12}(\omega) = \int_{-\infty}^{\infty} \hat{F}_1(\bar{\omega}) \int_{-\infty}^{\infty} F_2(t)e^{-i(\omega-\bar{\omega})t}dtd\bar{\omega} = \int_{-\infty}^{\infty} \hat{F}_1(\bar{\omega})\hat{F}_2(\omega - \bar{\omega})d\bar{\omega}$$

or

$$F_1(t)F_2(t) \Leftrightarrow \int_{-\infty}^{\infty} \hat{F}_1(\bar{\omega})\hat{F}_2(\omega - \bar{\omega})d\bar{\omega}$$

The above form of CFT is called the Convolution. Conversely, we can also write

$$\hat{F}_1(\omega)\hat{F}_2(\omega) \Leftrightarrow \int_{-\infty}^{\infty} F_1(\tau)F_2(t-\tau)d\tau$$

Implication in wave propagation analysis: The first property of using the product of two time domain signals has its use in understanding some signal processing aspects. For example, a truncated signal in the time domain is equal to the product of the original signal and the truncated signal. The second (or the converse) property is also of great utility in the wave propagation analysis. That is, all the responses (output) obtained in mechanical or other waveguides subjected to applied loadings can be represented as the frequency domain product of the input times the system transfer function. Thus, the time responses are obtained by convolving the transfer functions with the load spectrum.

Next, we will verify one of the properties of CFT (property number 3), namely the time domain compression is frequency domain expansion. Here again we will consider the same rectangular pulse of pulse width d and we will propagate this signal to a time of 25 μsecs. Fig. 1.2 shows the amplitude of $\hat{F}(\omega)$ for different pulse width d.

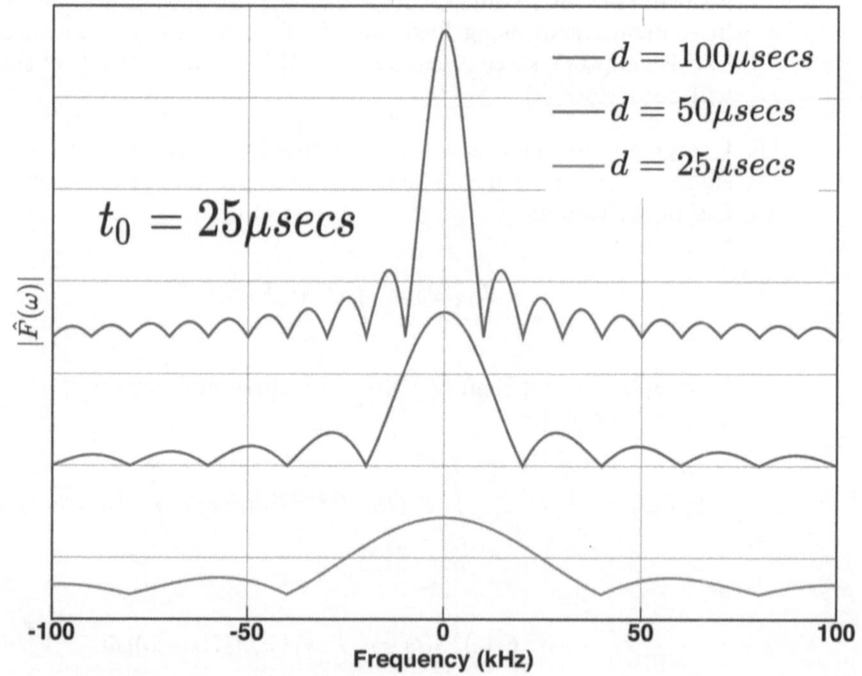

FIGURE 2.2: Continuous Fourier Transforms for different pulse width d for a given $t_0 = 25\mu$sec

From Fig. 2.2, we can see as the pulse width d reduces, the frequency band increases with increase in number of side lobes. In addition, the transform amplitude also increases with decrease in pulse width d.

2.1.2 Fourier Series

Both forward and inverse CFT requires mathematical description of time signal as well as their integration. In most cases, the time signals are point data acquired during experimentation. Hence, what we require is the numerical representation for the transform pair (Eqn. (2.1)), which is called *Discrete Fourier Transforms* (DFT). DFT is introduced in detail in the next subsection. Fourier Series (FS) is another method of representing time signals, which is in between CFT and DFT, wherein the inverse transform is represented by a series, while the forward transform is still in the integral form as in CFT. That is, one still needs the mathematical description of the time signal for getting the transforms.

The FS of a given time signal can be represented as

$$F(t) = \frac{a_0}{2} + \sum_{n=1}^{\infty} \left[a_n \cos\left(2\pi n \frac{t}{T}\right) + b_n \sin\left(2\pi n \frac{t}{T}\right) \right] \tag{2.7}$$

where $(n = 0, 1, 2, \ldots)$

$$a_n = \frac{2}{T} \int_0^T F(t) \cos\left(\frac{2\pi n t}{T}\right) dt, \quad b_n = \frac{2}{T} \int_0^T F(t) \sin\left(\frac{2\pi n t}{T}\right) dt, \tag{2.8}$$

Eqn. (2.7) corresponds to inverse transform, while Eqn. (2.8) corresponds to the forward transforms. Here, T is the period of the time signal. That is, the discrete representation of a continuous time signal $F(t)$ introduces periodicity of the time signal. The FS given in Eqn. (2.7) can also be written in terms of complex exponentials, which can give one-to-one comparison with CFT. That is, Eqns. (2.7 & 2.8) can be rewritten as

$$\begin{aligned} F(t) &= \tfrac{1}{2} \sum_{-\infty}^{\infty} (a_n - b_n) e^{i\omega_n t} = \sum_{-\infty}^{\infty} \hat{F}_n e^{i\omega_n t}, n = 0, \pm 1, \pm 2, \ldots\ldots \\ \hat{F}_n &= \tfrac{1}{2}(a_n - b_n) = \tfrac{1}{T} \int_0^T F(t) e^{-i\omega_n t} dt, \; \omega_n = \tfrac{2\pi n}{T}, \end{aligned} \tag{2.9}$$

Because of enforced periodicity, the signal repeats itself after every T seconds. Hence, we can define the fundamental frequency either in radians/per second (ω_0) or Hertz ($f_0 = \omega_0 / 2\pi = 1/T$). We can now express the time signal in terms fundamental frequency as

$$F(t) = \sum_{-\infty}^{\infty} \hat{F}_n e^{i2\pi n f_0 t} = \sum_{-\infty}^{\infty} \hat{F}_n e^{in\omega_0 t} \tag{2.10}$$

From Eqn. (2.10), it is clear that, unlike in CFT, the transform given by FS is discrete in frequency. To understand the behavior of FS as apposed to CFT, the same rectangular time signal used earlier is again considered here. The

FS coefficients (or transform) are obtained by substituting the time signal variation in Eqn. (2.9). This is given by

$$\hat{F}_n = \frac{F_0}{T}\left[\frac{\sin(\frac{\pi n d}{T})}{(n\pi d/T)}\right] e^{-i(t_0+d/2)2\pi n/T} \tag{2.11}$$

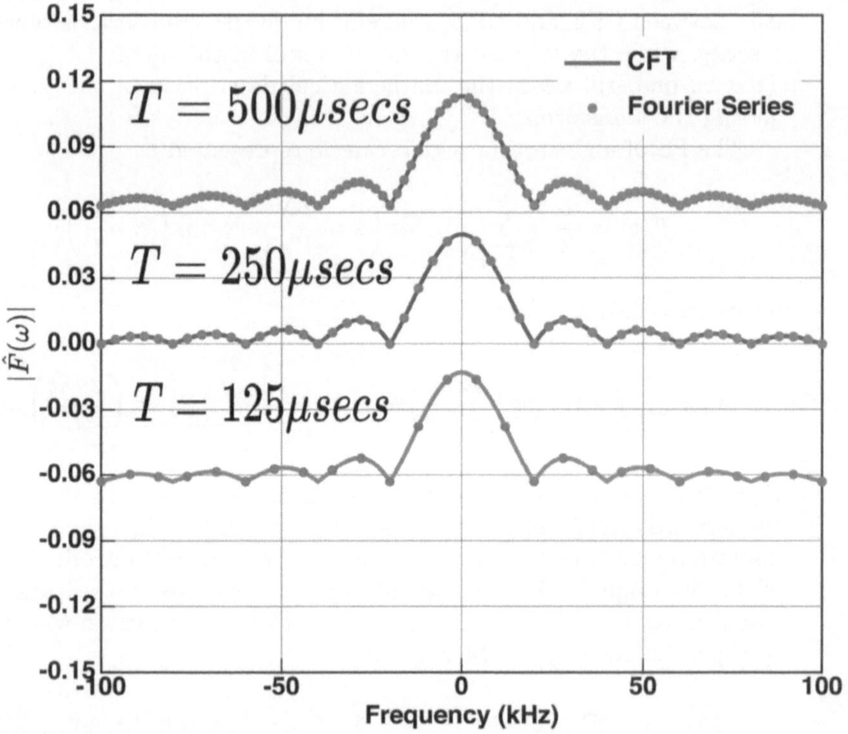

FIGURE 2.3: Comparison of Fourier Series with Continuous Fourier Transforms: $\hat{F}(\omega)$ amplitude for $d = 50$ μsecs and $t_0 = 25$ μsecs

The plot of transform amplitude obtained from CFT and FS is shown in Fig. 2.4. The figure shows that the value of transform obtained by FS at discrete frequencies falls exactly on the transform obtained by CFT. The figure also shows the transform value for different time periods T. We see from the figure that larger the time period, more close are the frequency spacing. The frequency spacing is given by $1/T$, which is 2 kHz, 4 kHz and 8 kHz for the periods of $T = 500$ μsecs, $T = 250$ μsecs, and $T = 125$ μsecs, respectively. Hence, if the period tends to infinity, the transform obtained by FS will be exactly equal to the transform obtained by CFT.

2.1.3 Discrete Fourier Transform

The Discrete Fourier Transform (DFT) is yet another way of mathematically representing the CFT in terms of summations. Here, both the forward and inverse CFT given in Eqn. (2.1) are represented by summations. This will completely do away with all complex integration involved in the computation of CFT. In addition, it is not necessary to represent the time signals mathematically and the great advantage of this is that one can use directly the time data obtained from experimentation. Numerical implementation of the DFT is done using the famous Fast Fourier Transforms (FFT) algorithm.

We begin here with Eqn. (2.9), which is the FS representation of the time signal. The main objective here is to replace the integral involved in computation of the Fourier coefficients by summation. For this, the plot of time signal shown in Fig. 2.4 is considered.

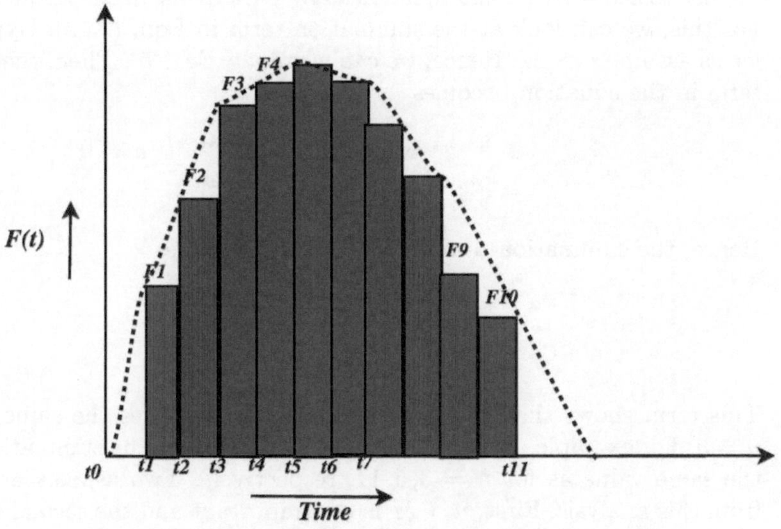

FIGURE 2.4: Time signal discretization for DFT

The time signal is divided into M piecewise constant rectangles, whose height is given by F_m and the width of these rectangles is equal to $\Delta T = T/M$. We have derived earlier that the continuous transform of a rectangle is a *sinc* function. By rectangular idealization of the signal, the DFT of the signal will be the summation of M *sinc* functions of pulse width ΔT and hence second integral in Eqn. (2.9) can now be written as

$$\hat{F}_n = \Delta T \left[\frac{\sin(\omega_n \Delta T/2)}{(\omega_n \Delta T/2)} \right] \sum_{m=0}^{M} F_m e^{-i\omega_n t_m} \tag{2.12}$$

Let us now look at the *sinc* function in Eqn. (2.12). Its value depends on the width of the rectangle ΔT. That is, as the width of the rectangle becomes smaller, the term inside the bracket of Eqn. (2.12) tends to reach the unit value. This will happen for all values of $n < M$. It can be easily shown that for values of $n \geq M$, the values of the transform is approximately equal to zero. Hence, the DFT transform pairs can now be written as

$$\begin{aligned} F_m = F(t_m) = \tfrac{1}{T} \sum_{n=0}^{N-1} \hat{F}_m e^{i\omega_n t_m} = \tfrac{1}{T} \sum_{n=0}^{N-1} \hat{F}_m e^{i2\pi nm/N} \\ \hat{F}_n = \hat{F}(\omega_n) = \Delta T \sum_{n=0}^{N-1} F_m e^{-i\omega_n t_m} = \Delta T \sum_{n=0}^{N-1} F_m e^{-i2\pi nm/N} \end{aligned} \tag{2.13}$$

Here, both m and n range from *0* to $N-1$.

The periodicity of time signal is necessary for DFT as we began its derivation from FS representation of the time signal. Now, we can probe a little further to see whether the signal has any periodicity in the frequency domain. For this, we can look at the summation term in Eqn. (2.12). Hypothetically, let us assume $n > M$. Hence, we can write $n = M + \bar{n}$. Then, the exponential term in the equation becomes

$$\begin{aligned} e^{-i\omega_n t_m} &= e^{-in\omega_0 t_m} = e^{-iM\omega_0 t_m} e^{-i\bar{n}\omega_0 t_m} \\ &= e^{-i2\pi m} e^{-i\bar{n}\omega_0 t_m} = e^{-i\bar{n}\omega_0 t_m} \end{aligned}$$

Hence, the summation term in Eqn. (2.12) becomes

$$\Delta T \sum_{m=0}^{M-1} F_m e^{-i\bar{n}\omega_0 t_m}$$

This term shows that the above summation evaluates the same value when $n = \bar{n}$. For example, if $M = 6$, then for $n = 9, 11, 17$, the summation evaluates the same value as for $n = 3, 5, 11$, respectively. Two aspects are very clear from this analysis. First, $n > M$ is not important and the second is that there is a forced periodicity in both time and frequency domain while representing a signal in DFT. This periodicity occurs about a frequency where the transform goes to zero. This frequency can be obtained if one looks at the *sinc* function given in Eqn. (2.12). That is, the argument of the *sinc* function is given by

$$\frac{\omega_n \Delta T}{2} = \pi n \Delta T = \frac{\pi n}{M}$$

where, we have used the relation $\Delta T = T/M$ in the above relation.

Here, we see that the *sinc* function goes to zero when $n = M$. It is at this value of n, the periodicity is enforced and the frequency corresponding to this value is called the *Nyquist* frequency. As mentioned earlier, this happens due to the time signal being real only and the transform beyond the Nyquist frequency is the complex conjugate of the transform before this frequency.

Thus, N real points are transformed to $N/2$ complex points. Knowing the sampling rate ΔT, we can compute the Nyquist frequency from the expression

$$f_{Nquist}=\frac{1}{2\Delta T} \tag{2.14}$$

Numerical implementation of the DFT is the FFT. It was developed by Cooley and Tukey [30]. Their algorithm is the foundation for Digital Signal Processing area, which we will be using extensively in this book. There are number of issues in the numerical implementation of the DFT. These are not discussed here. However, the reader is encouraged to refer [30] and [34] to get more information on these aspects.

In order to see the difference in different transform representation, the same rectangular pulse is again used here. The pulse width d is assumed to be 50 μsecs and the pulse is propagated to a time of $t_0 = 50\mu$ sec. There are two-parameters on which the accuracy of the transforms obtained from DFT depends, namely the sampling rate ΔT and the time window parameter N. Figs. 2.5(a) and (b) shows the transform obtained for various time windows N for a given ΔT of 1 μsecs and for various sampling rates ΔT for a given time window $N = 512$. First, for a given sampling rate ΔT, the time window is varied through the parameter N. The transform for this case is shown in Fig. 2.5(a). Here, the Nyquist frequency given by $1/2\Delta T$ occurs at 500 kHz, which is beyond the frequency scale plotted in the figure. However, for larger N, frequency spacing becomes smaller and hence we get denser frequency distribution. Next, from Fig. 2.5(b), we can clearly see the periodicity about the Nyquist frequency ($1/2\Delta T$). For a given time window $N = 512$, the figure shows that frequency spacing increases with the decrease in the sampling rate. Also, Nyquist frequency shifts to the higher value. Higher the sampling rate ΔT, we see the CFT and FFT responses do not match very well, and this again is due to enforced periodicity. Hence, proper care should be taken while choosing the two-parameters ΔT and N, while characterizing signal in the Fourier domain.

2.2 COMPARATIVE MERITS AND DEMERITS OF FFT

The Fourier Transform is extensively used in wave propagation studies due its simplicity in transforming a time signal from time domain to frequency domain and its *vice versa.* Due to induced periodicity both in time and frequency domain, FFT is always associated with time windows. Hence, if the measured signal does not die out within the chosen time window, the remaining part of the signal, will start appearing at the start of the time history, distorting the signal completely. This problem is quite severe in finite structures where multiple reflections from the boundaries do not die down within the chosen time window even in the presence of damping. Such problem is called the *signal wraparound*, and this is discussed in detail in Chapter 6. In addition, application of FFT to initial value problems is not straight forward. The

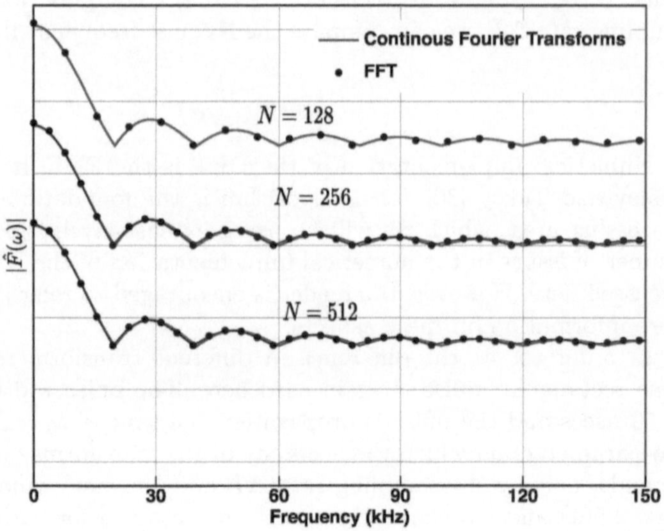

(a)

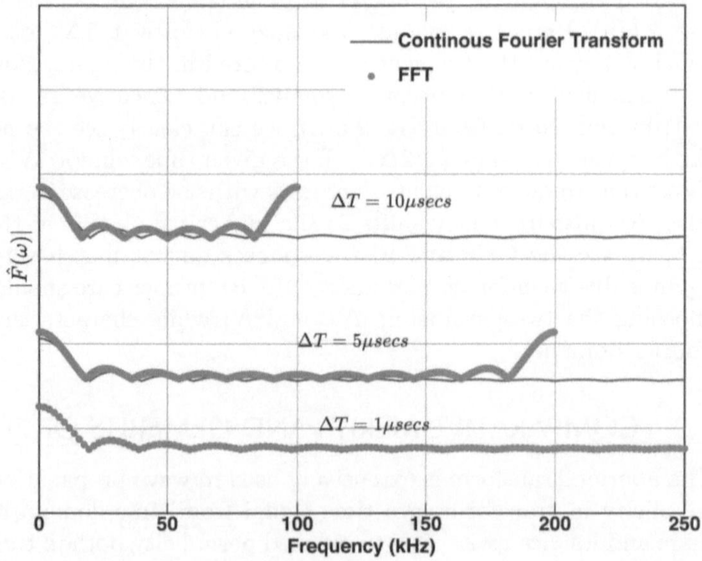

(b)

FIGURE 2.5: (a) Comparison of FFT and Continuous Transform for a sampling rate $\Delta T = 1\mu$ s for different number of FFT point N. (b) Comparison of FFT and Continuous Transform for various sampling rates for a given number of point $N = 512$

detailed signal analysis using FFT is given in [34]. Some of these problems can be avoided if one uses other transforms such as Laplace transforms or Wavelet transforms. Usage of these transforms for wave propagation problems is given in the author's unified book [54].

Note on MATLAB® scripts provided in this chapter

This chapter is provided with MATLAB script *all signal.m*, which generates (a) Gaussian signal, (b) tone burst signal and (c) linear chirp signal. These signals will be extensively used in the reminder of the text for studying wave propagation in different media.

Summary

This chapter gives a comprehensive review of Fourier Transforms used in this book to study wave propagation in different media. However, Fourier Transforms are susceptible to host of signal processing issues, which will be discussed in Chapter 6. The chapter begins with a overview of different ways of representing time signals, namely the CFT, Fourier series and DFT and their correlation with each other. This chapter also discuss the various properties of CFT and their implications in wave propagation studies. The importance of Nyquist frequency in the wave propagation analysis is also brought forward in this chapter. Toward the end of the chapter, advantages and disadvantages of Fourier Transforms in solving wave propagation problems is summarized.

Exercises

2.1 The figure below shows a one sided-exponential signal, mathematically defined as $F(t) = e^{-at}$, where $a > 0$ is a constant that determines the pulse width. Choose the appropriate value of a such that the pulse width is 50 μsecs, 100 μsecs and 200 μsecs, respectively. For these three values of a,

1. Determine and plot CFT amplitude
2. Determine the Fourier series representation for $T = 250\,\mu$secs, $T = 500\,\mu$secs and $T = 1000\,\mu$secs and plot the same. What do you observe?
3. Determine and plot FFT amplitude of this signal for sampling rate $\Delta T = 1.0\,\mu$sec, $\Delta T = 5.0\,\mu$secs and $\Delta T = 10.0\,\mu$secs for $N = 512$ FFT points. What is your observation?
4. Determine and plot FFT for $\Delta T = 1.0\,\mu$secs for $N = 512$, $N = 1024$ and $N = 2048$ FFT points

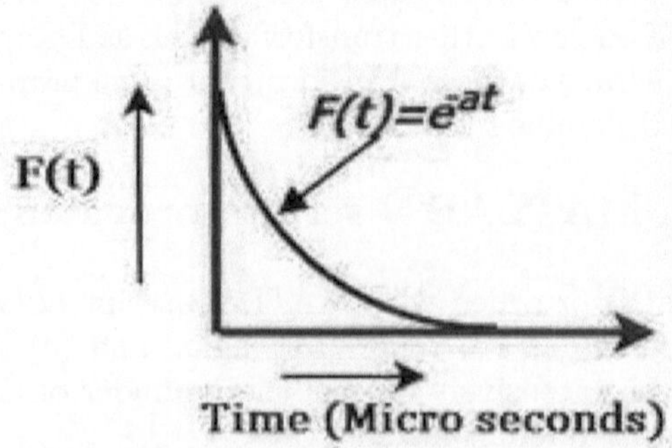

Figure for Problem 2.1

2.2 Let us consider a two sided exponential function shown below. The function is mathematically expressed as $F(t) = e^{-a|t|}$, where $a > 0$. For the

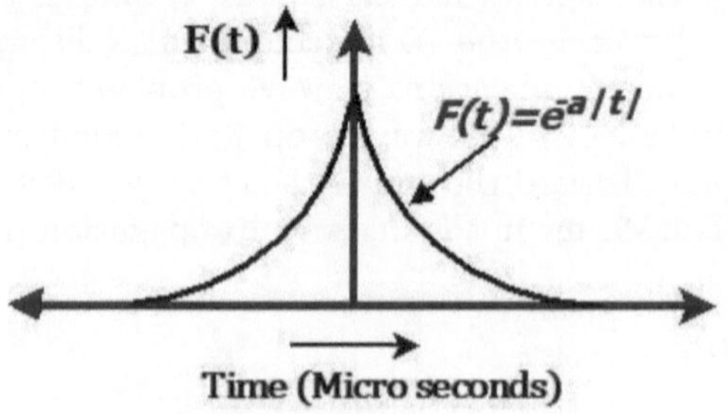

Figure for Problem 2.2

three chosen values of a in Problem 2.1, obtain the CFT, Fourier Series and FFT for this signal for the signal parameters given in Problem 2.1 and plot the frequency domain for all the three values of a and compare your results. In addition, also plot the phase of the signal as a function of frequency. Do you see any phase in this signal?

2.3 If the same signal used in Problem 2 is allowed to propagate by $t_0 = 100\,\mu$secs as shown in the figure below: For this signal perform all the operations as outlined in Problem 2.1. In addition, plot the phase as a function of frequency. What do observe now?

2.4 Obtain the CFT of the function $F(t) = \sin qt$, where $q = 50\,$kHz.

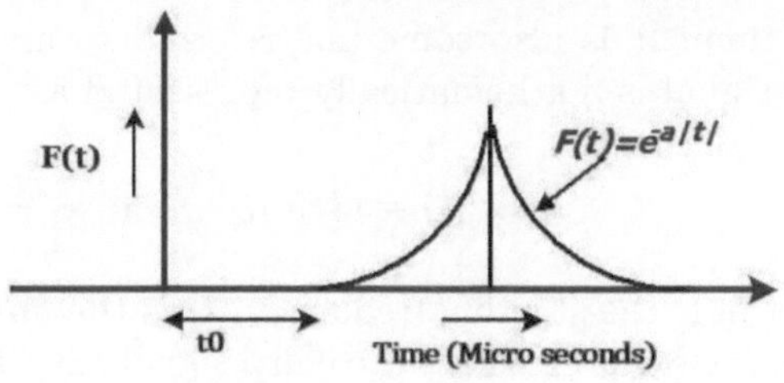

Figure for Problem 2.3

1. For this function plot FFT, CFT amplitude and FFT amplitude? You will see that the results does not superimpose. This is due to signal leakage, which is dealt with in detail in Chapter 6.
2. Next, we will window the function with a rectangular window (window aspects are discussed again in Chapter 6) and modify the signal as follows

$$\begin{aligned} F(t) &= \sin qt \qquad t_0 < t < t_0 + a \\ &= 0 \qquad \text{otherwise} \end{aligned}$$

where $a = 50\,\mu$secs. For this function obtain CFT and FFT amplitudes and see if they compare. Here, use $N = 2048$ and $\Delta T = 1.0\,\mu$secs. This example will illustrate how signal window has convolution effect in the frequency domain.

2.5 In elastic wave propagation studies, *modulated sine signal* or *Tone burst signal* is extensively used. The FFT of such signal dominates only at a single frequency called the *central frequency* denoted by q. It is created by windowing a sine signal with a window named *Hanning window*, also discussed in Chapter 6. MATLAB function *gauspuls* generates such a modulated sine signal having a central frequency $q\,kHz$ and window bandwidth ratio b_w. Using this MATLAB function *gauspuls*, generate this tone burst signal for the following cases and plot them

1. $q = 50$ kHz, $b_w = 0.6$
2. $q = 100$ kHz, $b_w = 0.6$, and
3. $q = 200$ kHz, $b_w = 0.6$

For all these cases, obtain FFT for $N = 2048$ with $\Delta T = 1.0\,\mu$ sec, plot them together and state your observation.

2.6 A chirp signal is the one in which the frequency increases or decreases with time. It is also sometime referred to as *sweep signal.* The linear chirp signal is mathematically represented as

$$F(t) = ct + f_0, \qquad c = \frac{f_1 - f_0}{T}$$

where f_0 is the starting frequency, f_1 is the final frequency and T is the time in μsecs over which the chirp signal acts. In this type of signal, the phase of the signal $\frac{d\phi(t)}{dt} = 2\pi F(t)$.

1. First, show that the phase varies as quadratic function of time.
2. Next, using the *chirp* function in MATLAB, vary T, f_0 and f_1 and obtain the FFT of this signal and plot its amplitude for the different cases and comment on your results. Here use $N = 2048$ and $\Delta T = 1.0\,\mu$secs. Comment on the usefulness of this signal for wave propagation studies in structures.

2.7 The duality property of CFT is a very useful property that was not studied in this chapter. It is used in obtaining CFT of certain functions if it resembles similar to the CFT of another function. This property reveals the effect when we interchange the roles of t and ω. It states that every property of CFT has a dual function, given as follows:

$$\hat{F}(t) = 2\pi * CFT\,(F(\omega))$$

Using the CFT of the problem obtained in Problem 2.2, obtain the CFT of the function $F(t) = \frac{2}{t^2+1}$

CHAPTER 3

Introduction to Wave Propagation in Structures

In Chapter 1, we defined wave as a medium to transport energy for one location to another and the propagation of mechanical waves requires a medium. It was also mentioned that the speed of the wave depends upon the material properties of the medium. The nature of waves or the physics of wave propagation is different for different medium, and they are a function of both space and time. In this chapter we will elaborate on these aspects in more detail especially in the context of materials and structures.

A material system, when subjected to dynamic loads, will experience stresses of varying degree of severity depending upon the load magnitude and its duration. If the temporal variation of load is of long duration, the intensity of the load felt by the structure will usually be of lower severity and such problems falls under the category of *structural dynamics*. For these problems, the two-parameters which are of importance in its response determination are the natural frequency of the system and its normal modes, which are normally referred to as mode shapes. The total response of structure is obtained by the superposition of the normal modes. Large duration of the load (of the order of few hundreds of milliseconds and more) makes it low on the frequency content, and hence the load will excite only the first few modes. Hence, the structure could be idealized with fewer unknowns (which we call it as degrees of freedom, a terminology which is normally used in structural dynamics). However, when the duration of the load is very small typically of the order of $\mu secs$ and lower, stress waves are set up, which starts propagating in the medium with a certain velocity. Hence, the response is always transient in nature and in the process, many normal modes will get excited. Hence, the model sizes will be many orders larger than what is required for structural dynamics problem.

DOI: 10.1201/9781003120568-3

Such problems come under the category of *wave propagation.* The key factor in the wave propagation is the propagating velocity, level of attenuation of the responses and its wavelengths. In addition, phase information is one of the important parameters in wave propagation analysis.

From the above discussion, we see that the wave propagation is a multi-modal phenomenon. Now, the question is what method need to be adopted to solve wave propagation problems? Wave propagation problems are always governed by PDE and the solution can be performed either in time or frequency domain [34]. There are several advantages in solving the wave propagation problem in frequency domain, which were highlighted in the previous two chapters. All the governing equations, boundary conditions and the variables are transformed to the frequency domain using Fourier Transforms discussed in Chapter 2, although one can use other transforms such as Wavelet or Laplace transforms. The most common Integral transform used for transformation of variables to frequency domain is the FFT. All of these transforms has the discrete representation and hence these are amenable for numerical implementation, which makes their use very attractive for wave propagation analysis. By transforming the problem to frequency domain, the complexity of the governing partial differential equation is reduced by removing the time variable out of the formulation, thus making the solution of the resulting ODE (in the 1-D case) much simpler than the original PDE. In wave propagation problems, two-parameters are very important, namely the wavenumber and the speeds of the propagation and this chapter will provide a general methodology to compute these quantities for a material system. There are many types of waves that can be generated in structure. Wavenumber expression reveals the type of waves that are generated. Hence, in wave propagation problems, two relations are very important, namely the *Spectrum Relations*, which is plot of the wavenumber as a function of frequency and the *Dispersion Relations*, which is a plot of wave velocity with the frequency. These relations reveal the characteristics of different waves that are generated in a given material system.

3.1 CONCEPT OF WAVENUMBER, GROUP SPEEDS AND PHASE SPEEDS

A wave propagating in a medium can be represented as

$$u(x,t) = e^{i(kx-\omega t)} \tag{3.1}$$

Eqn. (3.1) is an alternate way of representing wave, which is given in Chapter 1 (Eqn. (1.1)). In the above equation k is the wavenumber, which specifies the behavior of the wave. The exponent in Eqn. (3.1) $i(kx - \omega t)$ is called the *phase* of the wave. If the wave moves with constant phase, then we have

$$\frac{d(i(kx-\omega t)}{dt} = 0, \qquad \text{or} \qquad \frac{dx}{dt} = C_p = \frac{\omega}{k} \tag{3.2}$$

In the above equation, C_p represents the speed of the wave that moves with constant phase and hence it is called *Phase speed.* From the point of view of sending information, these waves are not useful. They are the same throughout the time and the space. Some quantities must therefore be modulated, such as frequency or amplitude, in order to convey information. The resulting wave may be a perturbation that acts over a short distance, which is called a *wave packet.* This wave packet can be considered to be a superposition of a number of harmonic waves.

The concept of *wave packet* can be explained as follows. It can be explained using the concept of wave interference explained in Chapter 1 (see Section 1.2.1). In order that the wave conveys energy (or information), something more than a simple harmonic wave is needed. However, the superposition of many such waves of varying frequencies can result in an *envelope* wave (or a wave packet) and a carrier wave within the envelope. The envelope can transmit data. As a simple example, let us consider the superposition of two harmonic waves with frequencies that are very close ($\omega_1 \sim \omega_2$) to each other and having the same amplitude A_0. These waves can be written as

$$u(x,t) = A_0 e^{i(k_1 x - \omega_1 t)} + A_0 e^{i(k_2 x - \omega_2 t)} \tag{3.3}$$

At this point, we introduce following variables

$$k = \frac{k_1 + k_2}{2}, \qquad \Delta k = \frac{k_1 - k_2}{2}, \qquad \omega = \frac{\omega_1 + \omega_2}{2},$$
$$\Delta\omega = \frac{\omega_1 - \omega_2}{2} \tag{3.4}$$

Simplifying Eqn. (3.4), we can write the two wavenumbers and frequencies as

$$k_1 = k + \Delta k, \qquad k_2 = k - \Delta k, \qquad \omega_1 = \omega + \Delta\omega, \qquad \omega_2 = \omega - \Delta\omega \tag{3.5}$$

Substituting Eqn. (3.5) in Eqn. (3.3), we get

$$u(x,t) = A_0 e^{-i(k+\Delta k)x} e^{i(\omega+\Delta\omega)t} + A_0 e^{-i(k-\Delta k)x} e^{i(\omega-\Delta\omega)t} \tag{3.6}$$

Expanding and simplifying the above equation, we get

$$u(x,t) = A_0 e^{-i(kx-\omega t)} \left[e^{-i(\Delta k x - \Delta\omega t)} + e^{i(\Delta k x - \Delta\omega t)} \right] \tag{3.7}$$

The term inside the brackets is a cosine term and is given by $2\cos(\Delta k x - \Delta\omega t)$ and Eqn. (3.7) can be simplified as

$$u(x,t) = 2A_0 e^{-i(kx-\omega t)} \times \cos(\Delta k x - \Delta\omega t) \tag{3.8}$$

Hence the total response is the product of two waves, one traveling at a speed of ω/k (which we call it as the *Phase speed* C_p), and the second wave traveling at a speed of $\Delta\omega/\Delta k = (\omega_1 - \omega_2)/(k_1 - k_2) = d\omega/dk$. This speed, we denote

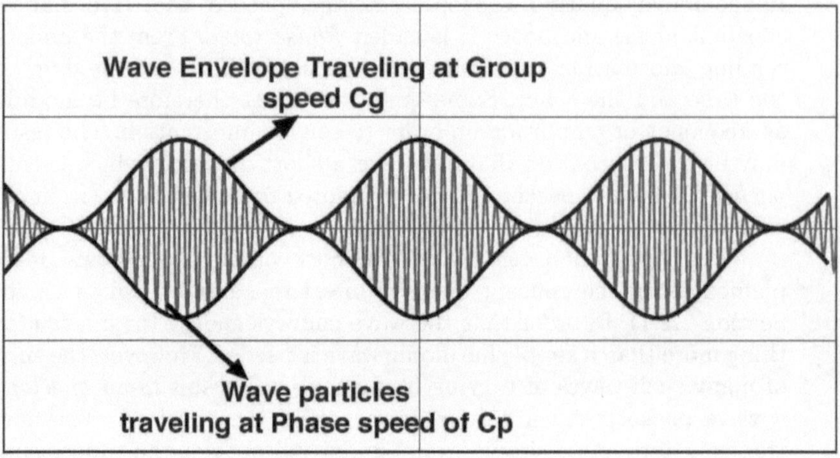

FIGURE 3.1: The concept of *Phase velocity* and *Group velocity* of a propagating 1-D waves

it as C_g, and it is called *Group speed.* The main difference between the phase speed and group speed of a 1-D propagating wave is represented in Fig. 3.1.

The envelope shown in Fig 3.1 travels at the group speed C_g. The carrier wave shown inside the envelope travels at the phase speed of C_p. The complete wave packet moves at the group speed C_g. It is this envelope that carries response information. Normally group and phase speeds of a wave are not necessarily the same. They are related by

$$C_g = C_p \left[1 - \frac{\omega}{C_p}\frac{dC_p}{d\omega}\right] \tag{3.9}$$

If the wavenumbers k_1 and k_2 are very small and are nearly equal in addition to frequencies ω_1 and ω_2 being nearly equal, then Eqn. (3.8) resembles that of the equation for the *beat phenomenon* and this is also seen in Fig. 3.1.

From Eqn. (3.9), it is clear that $C_p = C_g$, when the phase speed C_p is not a function of frequency ω. This is possible only when the wavenumber will be a linear function of frequency ω. Such waves, wherein the phase speeds and group speeds are equal, are called *non-dispersive waves.* These waves retain their shapes as they propagate. Alternatively, if the phase speeds are functions of frequency ω, then phase speeds and group speeds are not equal and the group speed C_g will also be function of frequency ω. This means that the high-frequency components will travel faster compared to low frequency components and as a result, the wave profile completely changes during propagation. In this case, the wavenumber will have a nonlinear relationship with the frequency ω. Such waves are called *dispersive waves.* The examples of dispersive and non-dispersive waves are shown in Fig. 3.2

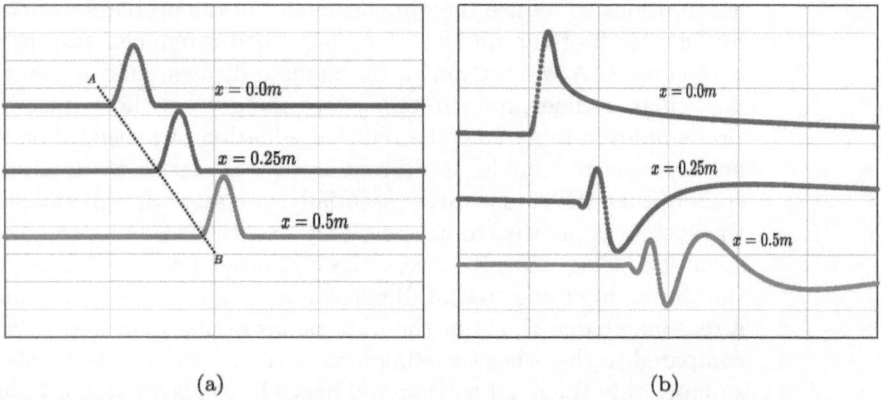

FIGURE 3.2: (a) An example of non-dispersive wave. (b) An example of dispersive waves

In Fig. 3.2(a) we see a triangular wave as it propagates along the positive x-direction, retains its shape indicating that the wave is traveling in a non-dispersive medium. The line AB is the line connecting the tip of triangle for different propagating distances. The slope of the line AB will give the speed of the group wave (or phase speed). Most second-order systems defined by hyperbolic partial differential equations, such as mechanical wave equation, equations governing vibration of string, etc., behave in a non-dispersive manner. Fig. 3.2(b) is an example of a propagation of a dispersive wave, wherein the same triangular input, completely looses its shape after propagating even a small distance of 0.25 m. Most fourth order systems such as equation governing the dynamics of beams, behave in this manner. In this book, in the later chapters, we will see waves in different material systems which have different level of dispersiveness, that is, some are weakly dispersive while the waveguides defined by fourth order systems such as the one shown in Fig. 3.2(b), are strongly dispersive. The level of dispersiveness depends on the order of governing differential equation governing the behavior of the material system.

3.2 WAVE PROPAGATION TERMINOLOGIES

In this section, we will define some of the commonly used terminologies in wave propagation analysis. These are summarized below:

1. **Waveguide:**
   Any material system is governed by equation(s) of motion, which will be function of three coordinate directions and time. Such 3-D partial

differential equations are very difficult to solve. Hence, many simplified assumptions are normally made on the behavior of the material system in order to reduce the dimensionality of problem. Some of these assumptions to reduce the dimensionality of the problem were discussed in Chapter 2 of the unified book [54]. Such a reduced system is called a *waveguide.* A waveguide, as the name indicates, guides the wave in a particular manner and direction depending upon the nature of reduction employed in getting the reduced equation of motion. For example, an isotropic rod model is derived from 3-D elasticity equation by assuming that the structure is essentially in a state of uniaxial stress and the system is allowed to support only one motion along the axis of the structure. This will give rise to second-order partial differential equation function of one spatial dimension and time. Such an assumption is a reasonable one if one of the dimensions of the structure is very large compared to the other two dimensions. Hence, the 1-D isotropic rod can support only the axial motion and hence it is called axial or *longitudinal waveguide.* Similarly, if the same 1-D structure of the previous case is allowed to move in the lateral or transverse direction, then, the structure can undergo rotation in addition to the lateral motion. Such a waveguide is called the *flexural waveguide* and its reduced governing equation is fourth order in space and the waves in such waveguides are highly dispersive in nature. Likewise, there are many different waveguides. For example, in the case of shafts normally employed in many machines, the only possible motion is the twist and hence they are called *torsional waveguide.* In case of laminated composite beam, due to stiffness coupling, both axial and flexural motions are possible. In general, if there are **n** highly coupled governing partial differential equations, then such a waveguide can support **n** different motions, which means there will be **n** different wave modes.

2. **Propagating and Evanescent waves:**
Any 1-D wave having a field variable $y(x,t)$ in the frequency domain is represented as $\hat{y}(x,\omega) = Ae^{ikx}$, where $\hat{y}(x,\omega)$ is the frequency domain amplitude of $y(x,t)$, ω is the frequency, and k is the wavenumber, which can be purely real, purely imaginary or complex. If the wavenumber is real, then such waves will exhibit oscillating behavior and hence they are called *propagating waves.* On the other hand, if the wavenumber is purely imaginary, then the term e^{ikx}, will tend to zero for large values of x. In such situations, no wave exists and such wave mode is called *evanescent wave.* Many waveguides have some of their wave modes exhibit evanescent behavior and we will study some of them in this book. If the wavenumber is complex, that is they have both the real and imaginary part, then such waves is called *inhomogeneous waves.* The behavior of these waves are such that, as they propagate, the amplitude of these waves attenuate. The real part of the wavenumber allows the

propagation of the waves, while the imaginary part causes the waves to reduce its amplitude. As we will see in the later part of this book, waves in composites and inhomogeneous waveguides are always inhomogeneous in nature.

3. **Cut-off Frequency**:
   In some waveguides, the wavenumber will be purely imaginary to start with and becomes real or complex after certain frequency. In such cases, the wave will be evanescent to start with and will start propagating only after certain frequency. This frequency at which the change from evanescent mode to propagating mode happens is called the *cut-off frequency.* The cut-off frequency has great significance in understanding the physics of wave propagation of waveguide systems, as we will see in the later chapters of this book.

3.3 SPECTRAL ANALYSIS OF MOTION

The fundamental objective of the wave propagation analysis in any waveguide is to understand its physics. This requires first the deriving of the partial differential equation governing the behavior of the waveguide, which is then used to obtain the wave parameters such as its wavenumber, its speeds and other features such as existence of cut-off frequencies, band gaps, etc. To obtain the above wave parameters, we need to perform spectral analysis on the derived governing equation(s) of motion. Spectral analysis yields two distinct relations, namely the *spectrum relations*, which a relation between wavenumber and frequency and the *dispersion relations*, which is a relation between the group velocity with the frequency. In the next few paragraphs, we will outline the procedure of obtaining the wave parameter for a generalized 1-D second-order and fourth order systems and discuss at length the physics of wave propagation in these two generalized systems. Similar procedure can be extended for higher-dimensional waveguides.

3.3.1 Second-Order System

Consider a generalized second-order partial differential equation given by

$$a\frac{\partial^2 u}{\partial x^2} + b\frac{\partial u}{\partial x} = c\frac{\partial^2 u}{\partial t^2} \tag{3.10}$$

where, a, b, c are known constants and u(x,t) is the field variable, x is the spatial variable and t is the temporal variable. We first approximate or transform the above partial differential equation to the frequency domain using DFT, which is given by

$$u(x,t) = \sum_{n=0}^{N} \hat{u_n}(x,\omega_n)e^{i\omega_n t} \tag{3.11}$$

where ω_n is the circular frequency and N is the total number of frequency points used in the approximation. Here $\hat{u}$ is the frequency dependent Fourier Transform of the field variable. Substituting Eqn. (3.11) into Eqn. (3.10), we get

$$a\frac{d^2\hat{u_n}}{dx^2} + b\frac{d\hat{u_n}}{dx} + c\omega_n^2\hat{u_n} = 0 \tag{3.12}$$

From the above equation, we see that a partial differential equation is reduced to a set of ordinary differential equation with time variation removed and instead frequency is introduced as a parameter. The summation is omitted in the above equation for brevity. Eqn. (3.12) is a constant coefficient differential equation which has solution of the type $\hat{u}_n(x,\omega) = A_n e^{ikx}$, where A_n is some unknown constant and k is the wavenumber of the medium. Substituting the above solution in Eqn. (3.12), we get the following characteristic solution for determination of the wavenumber k.

$$(k^2 - \frac{bi}{a}k + \frac{c\omega_n^2}{a})A_n = 0 \tag{3.13}$$

The above equation is quadratic in k and has two roots corresponding to the two modes of wave propagation. These two modes correspond to incident and reflected waves. If the wavenumbers are real, then the wave modes are called *propagating mode*. On the other hand, if the wavenumbers are imaginary, then the wave modes damp out the responses and hence they are called *evanescent modes*. If the wavenumber is complex, then the waves, as they propagate, they attenuate. In other words, such waves are called the *Inhomogeneous waves.* These concepts were explained in the previous section.The roots of the equation are given by

$$k_{1,2} = \frac{bi}{2a} \pm \sqrt{\frac{-b^2}{4a^2} + \frac{c\omega_n^2}{a}} \tag{3.14}$$

Eqn. (3.14) is the generalized expression for the determination of the wavenumbers. Different wave behavior is possible depending upon the values of $a, b,$ and c. The behavior depends on the numerical value of the radical $\sqrt{c\omega_n^2/a - b^2/4a^2}$. Let us consider a simple case of $b = 0$. The two wavenumbers are given by

$$k_1 = \omega_n\frac{c}{a}, \qquad k_2 = -\omega_n\frac{c}{a} \tag{3.15}$$

From the above expression, we find that wavenumbers are real and hence they are propagating modes. The wavenumbers are linear function of frequency ω. Next, we will obtain the wave speeds namely the phase speeds C_p and group speeds C_g . These are given by

$$C_p = \frac{\omega_n}{Real(k)}, \qquad C_g = \frac{d\omega_n}{dk} \tag{3.16}$$

Note the difference between Eqn. (3.16) for C_p and Eqn. (3.2). Since there is a possibility of k being complex, we need to find the speeds of only the propagating modes and hence for computing the speeds, we will consider only the real part of k. For the wavenumbers given in Eqn. (3.15), these parameters are given by

$$C_p = C_g = \frac{a}{c} \tag{3.17}$$

We find that both group and phase speeds are constants and equal. Hence, when the wavenumbers vary linearly with the frequency ω and the phase speeds and group speeds are constants and equal, then the wave, as it propagates, retains its shape. Hence, the waves will be non-dispersive in nature. Longitudinal waves in elementary rod waveguide are of this type. If the wavenumber varies in a non-linear manner with respect to frequency, then the phase and group speeds will not be constants and these will be a function of frequency ω. That is, each frequency component of the wave travels with different speeds and as a result the wave changes its shape as it propagates thus making the wave dispersive. This was explained in the previous section.

Next, let us now again consider Eqn. (3.14) with all constants non-zero. The wavenumber no longer varies linearly with the frequency. Hence, one can expect dispersive behavior of the waves and the level of dispersion will depend up the numerical value of the radical. We will investigate this aspect in little more detail. There can be following three situations.

$$1.\quad \frac{b^2}{4a^2} > \frac{c\omega_n^2}{a}, \qquad 2.\quad \frac{b^2}{4a^2} < \frac{c\omega_n^2}{a}, \quad \text{and} \quad 3.\quad \frac{b^2}{4a^2} = \frac{c\omega_n^2}{a}$$

Let us now consider **Case 1**. When $(b^2)/(4a^2) > (c\omega_n^2)/(a)$, then the radical will be an imaginary number and hence the wavenumber will be purely imaginary, implying that the wave modes will not propagate and they would damp out very fast. For the **Case 2**, where $(b^2)/(4a^2) < (c\omega_n^2)/(a)$, the value of the radical will be positive and real and hence the wavenumber will have both real and imaginary parts and it takes the form $k = p + iq$. Therefore, the waves having this feature will attenuate as it propagates, that is such wave will behave as inhomogeneous wave. The phase and group speeds for this case are given by

$$C_p = \frac{\omega_n}{k} = \frac{\omega_n}{\sqrt{\frac{c\omega_n^2}{a} - \frac{b^2}{4a^2}}} \tag{3.18}$$

$$C_g = \frac{d\omega_n}{dk} = \frac{a\sqrt{\frac{c\omega_n^2}{a} - \frac{b^2}{4a^2}}}{c\omega_n} \tag{3.19}$$

It is quite obvious that these are not same and hence the waves could be dispersive in nature. One can obtain non-dispersive solution by substituting $b = 0$ in Eqn. (3.19). Now let us see **Case 3** where the value of radical will be zero and hence the wavenumber is purely imaginary indicating that the wave mode is a damping mode. The interesting aspect is to find the frequency of transition of the propagating mode becoming evanescent or damping mode. This frequency called the *transition frequency*, and it can be determined by equating the radical to zero. The transition frequency ω_t is given by

$$\omega_t = \frac{b}{2\sqrt{ac}} \tag{3.20}$$

After the wavenumbers are determined, the solution to the governing wave equation (Eqn. (3.12)) in the frequency domain can be written as

$$\hat{u_n}(x, \omega_n) = \mathbf{A}_n e^{-ik_n x} + \mathbf{B}_n e^{ik_n x}, \qquad k_n = \omega_n \sqrt{\frac{c}{a}} \tag{3.21}$$

In the above equation, n represents the frequency sampling point, $\mathbf{A}_n$ represent the incident wave coefficient at the nth frequency sampling point, while $\mathbf{B}_n$ represents the reflected wave coefficient. It is clearly seen how the value of the constants in the governing differential equation play an important part in dictating the type of wave propagation in the medium.

3.3.2 Fourth-Order System

Now let us consider a fourth-order system and study the wave behavior in such systems. Wave propagation in flexural waveguides such as beams are the examples of the fourth order systems. Consider the following governing partial differential equation of motion.

$$A\frac{\partial^4 w}{\partial x^4} + Bw + C\frac{\partial^2 w}{\partial t^2} \tag{3.22}$$

The above PDE is the fourth-order in spatial dimension and second-order in time. Here w is the field variable, and A, B, C are some known constants. The above equation is similar to the equation of motion of a beam on elastic foundation. Let us now assume the spectral form of solution to the field variable, which is given by

$$w(x,t) = \sum_{n=0}^{N} \hat{w_n}(x, \omega_n) e^{i\omega_n t} \tag{3.23}$$

Using Eqn. (3.23) in Eqn. (3.22), the partial differential equation is transformed into ordinary differential equation and is given by

$$A\frac{d^4 \hat{w_n}}{dx^4} - (C\omega_n^2 - B)\hat{w_n} = 0 \tag{3.24}$$

Again, this equation is a differential equation with constant coefficients and it will have solutions of the form $\hat{w}_n = P_n e^{ikx}$. Using this solution in Eqn. (3.24), we get the characteristic equation for the solution of the wavenumber, which is given by

$$k^4 - \beta^4 = 0, \qquad \beta^4 = \left(\frac{C}{A}\omega_n^2 - \frac{B}{A}\right) \tag{3.25}$$

The above is a fourth-order equation corresponding to four wave modes, two of which is for incident wave and the other two is for reflected wave. Also, the type of wave is dependent upon the numerical value of $C\omega_n^2/A - B/A$. Let us now assume that $C\omega_n^2/A > B/A$. For this case, solution of Eqn. (3.25) will give the following wavenumbers:

$$k_1 = \beta, \qquad k_2 = -\beta, \qquad k_3 = i\beta, \qquad k_4 = -i\beta \tag{3.26}$$

In the above equation, k_1 and k_2 are the propagating modes, while k_3 and k_4 are the damping or evanescent modes. From the above equations, we find that the wavenumbers are nonlinear functions of frequency and hence the waves are expected to be highly dispersive in nature. Also, using the above expression, we can find the phase and group speeds for the propagating modes using Eqns. (3.16 & 3.17), respectively.

Next, consider the case when $C\omega_n^2/A < B/A$. For this case, the characteristic equation and hence the wavenumbers are given by

$$k^4 + \beta^4 = 0 \tag{3.27}$$

$$k_1 = \left[\frac{1}{\sqrt{2}} + i\frac{1}{\sqrt{2}}\right]\beta, \qquad k_2 = -\left[\frac{1}{\sqrt{2}} + i\frac{1}{\sqrt{2}}\right]\beta \tag{3.28}$$

$$k_3 = \left[-\frac{1}{\sqrt{2}} + i\frac{1}{\sqrt{2}}\right]\beta, \qquad k_4 = -\left[-\frac{1}{\sqrt{2}} + i\frac{1}{\sqrt{2}}\right]\beta \tag{3.29}$$

From the equation above, we see that the change of sign of $C\omega_n^2/A - B/A$ has brought about completely changed wave behavior. We find that all the wavenumbers have both real and imaginary parts and hence all the modes are inhomogeneous wave modes. That is, the waves, while propagating will also attenuate. Here, what has happened is that the initial evanescent mode after a certain frequency, becomes a propagating mode, giving completely a different wave behavior. The frequency at which this transition takes place is called the *Cut-off frequency*, which was explained in the previous section. The expression for cut-off frequency can be obtained if we equate $C\omega_n^2/A - B/A$ to zero, which is given by $\omega_{cut-off} = \sqrt{B/C}$. We can see that when $B = 0$, the cut-off frequency vanishes and the wave behavior is similar to the first case, that is, it will have two propagating and two damping modes. In all the cases, the waves will be highly dispersive in nature.

The solution of the fourth order governing equation in the frequency domain (Eqn. (3.24)) can be written (with $B = 0$) as

$$\hat{w_n}(x, \omega_n) = \mathbf{A}_n e^{-i\beta x} + \mathbf{B}_n e^{-\beta x} + \mathbf{C}_n e^{i\beta x} + \mathbf{D} e^{\beta x} \quad (3.30)$$

As in the previous case, $\mathbf{A}_n$ and $\mathbf{B}_n$ are the incident wave coefficients and $\mathbf{C}_n$ and $\mathbf{D}_n$ are the reflected wave coefficients. These can be determined using the boundary conditions of the problem.

From the above study, we see that the spectral analysis gives us the deep insight into wave mechanics of the material system defined by its differential equation. The direct output of the spectral analysis is the *spectrum relations*, which is plot frequency and the wavenumbers and the *dispersion relations*, which is the plot of speeds with the frequency.

What is presented here is the spectral analysis of 1-D waveguides. In the case of 2-D waveguides, the governing differential equations will be function of two spatial coordinates and a temporal coordinate. Transforming the governing equation to the frequency domain will only eliminate time from the governing equation and the resulting equation will still be function of two spatial variables. In order to determine the spectrum and dispersion relations, the governing equation requires one more transform in one of the coordinate directions, thus introducing an additional wavenumber in the direction where the transform is taken. Once the second transform is taken, resulting equation is an ordinary differential equation, and the procedure to determine the dominant wavenumber is same as what is outlined here. This procedure is adopted to determine the wavenumbers for a few 2-D waveguides explained in Chapters 7 and 8.

3.4 GENERAL FORM OF WAVE EQUATION AND THEIR CHARACTERISTICS

In this book, we will be looking at the propagation of elastic waves in different material systems. In all we will be considering six different material systems, namely isotropic, viscoelastic, anisotropic, inhomogeneous, granular and small scale materials systems. The characteristics of the waves in each of these systems will be different and the details of these differences will be highlighted in their respective chapters,. The waves in isotropic systems are extensively presented in Chapters 4 and 7, while the propagation of waves in granular material (dealt in Chapter 10) can be derived from the equations of isotropic material except that their material models are mostly empirical derived from extensive experimentation. Hence, in this chapter, we will only outline briefly the characteristics of the wave in three different material systems, namely the anisotropic material system, inhomogeneous material system and the nanostructures material (or small scale) systems. First, the general form of the wave equation in any material systems is first presented followed by a general discussion on the characteristics of waves in the above three material systems.

3.4.1 General Form of Wave Equations

The general form of linear wave equation assuming the local theory of elasticity is given by

$$\frac{\partial^2 u}{\partial t^2} = \sum_{\alpha,\beta} u_{\alpha\beta} \tag{3.31}$$

where $u(x, y, z, t)$ is the field variable governing the material system with $u_{\alpha\beta}$ being the derivatives of the field variable. The wave equation in the structural mechanics context is the conservation of momentum (dynamic equilibrium) equation

$$\nabla \cdot \mathbf{T} = \rho\ddot{\mathbf{u}} \tag{3.32}$$

where $\mathbf{T}$ is the stress measure at any point in the body, which is in general a nonlinear function of the displacement vector $\mathbf{u} = \{u_x, u_y, u_z\}$. A nonlinear relation between $\mathbf{T}$ and $\mathbf{u}$ results nonlinear wave equation. Most of the analysis presented in this book assumes a linear relationship between $\mathbf{T}$ and strain (that is, displacement gradient), either in time domain (linear elastic material) or in frequency domain (viscoelastic material). However, the linear coefficient (of constitutive relation) can be the direction (anisotropy) and/or position (inhomogeneity) dependent. In both the cases, response of homogeneous isotropic materials can be retrieved easily from these general material models. However, in the case of nanostructures, the scale effects become prominent and hence their constitutive models (the constitutive models will contain either the stress or strain gradients and hence the wave responses will be a function of the scale parameter. We will briefly discuss the wave characteristics in these waveguides next.

3.4.2 Characteristics of Waves in Anisotropic Media

One important characteristic that separates the waves in an anisotropic media from its isotropic media counterpart is the direction of energy flow (group velocity) [105]. For isotropic material, incident and reflected waves are purely longitudinal (normally called *P*-waves) or shear dominated (normally called *Shear Waves*). That is, for the wave vector $\mathbf{k} = (k_x, k_y)$, the wave propagation directions would be (k_x, k_y) and $(-k_y, k_x)$ for P and S waves in two dimensions. However, the situation is much more complex in the anisotropic case, where the wave directions are material property dependent and they can no longer be thought as purely P or S waves. They are called *Quasi-P*-wave and *Quasi-S*-wave, which has two components, namely *QSV-wave* polarized vertically and *QSH-wave* polarized horizontally. In this case, all the wave modes (in three Cartesian coordinate directions) are coupled, and in order to identify them, one needs to solve a sixth-order characteristic polynomial equation. Thus, the simplification of analysis as seen in isotropic case, due to uncoupled P and S motions, is not possible in the anisotropic case. The wave velocity amplitude and direction in anisotropic material can be obtained from the governing equation and plane wave assumption. The governing equation for general homogeneous anisotropic media is given by

$$\frac{\partial \sigma_{ik}}{\partial x_k} = \rho \ddot{u}_i \,, \quad \sigma_{ik} = C_{ik\ell m}\epsilon_{\ell m} \tag{3.33}$$

where the constitutive matrix $C_{ik\ell m}$ is symmetric with respect to ℓ and m. For plane wave solution,

$$u_i = A\alpha_i e^{[ik(n_m x_m - ct)]} \tag{3.34}$$

where n_m is the direction cosines of the normal to the wave front and α_i is direction cosines of particle displacement. Substitution of the assumed form in the governing equation results in an eigenvalue problem given by

$$(\Gamma_{im} - \rho c^2 \delta_{im})\alpha_m = 0 \,, \quad \Gamma_{im} = C_{ik\ell m} n_k n_\ell \tag{3.35}$$

where Γ_{im} is called the Christoffel symbol. Solving Eqn. (3.35), the wave phase velocity and wave directions are obtained. From the previous discussion, it can be said that for a general anisotropic media, vector product, $\alpha \times \mathbf{n}$ and $\alpha \cdot \mathbf{n}$ are never zero.

3.4.3 Characteristics of Waves in Inhomogeneous Media

Compared to the wave propagation in anisotropic media, the waves in inhomogeneous media have totally different characteristics. Typically, the wave motion in the direction of inhomogeneity is characterized by reduction in the amplitude of the wave while it is propagating (see [21], [144]). As the name indicates, the waves in such waveguides will be inhomogeneous, that is the wavenumbers will be complex in nature. The homogeneous plane wave has the form

$$f = \Phi(\omega)\exp[i(k_x x + k_y y + k_z z - \omega t)] = \Phi(\omega)\exp[i(\mathbf{k}\cdot\mathbf{r} - \omega t)] \,, \tag{3.36}$$

where $\Phi(\omega)$ is the wave amplitude (real) and $\mathbf{k}$ is the wave vector (real). This form describes wave propagation phenomena in homogeneous, linear and non-dissipative solids. Inhomogeneous wave is described by the above form with complex Φ and $\mathbf{k}$ (see [24]). These waves were initially observed in media with inbuilt dissipative properties (like the viscoelastic materials). However, inhomogeneous materials are also conducive for this kind of wave, when the inhomogeneity is in the direction of wave propagation. The literature on this subject is vast. These waves successfully describe wave modes, and their dispersive character, whereby propagation speed, attenuation constant and pertinent angles are in general dependent on the frequency (see [65], [85] and [31]). It is known that time harmonic waves in dissipative media have a complex valued wavenumber. But it is less known that, there can be waves with a complex valued wave vector, with real and imaginary parts $\mathbf{k}_1$ and $\mathbf{k}_2$ not necessarily parallel. Thus, more precisely, inhomogeneous waves ([109]) are the waves where the wave vector $\mathbf{k}$ is complex valued ($\mathbf{k} = \mathbf{k}_1 + i\mathbf{k}_2$), and $\mathbf{k}_1$ and $\mathbf{k}_2$ are not parallel. Thus, the general form becomes

$$\begin{aligned} f &= \Phi(\omega)e^{[i(k_{1x}x+k_{1y}y+k_{1z}z-\omega t)-(k_{2x}x+k_{2y}y+k_{2z}z)]} \\ &= \Phi(\omega)e^{[i(\mathbf{k}_1\cdot\mathbf{r}-\omega t)-\mathbf{k}_2\cdot\mathbf{r}]}, \end{aligned} \tag{3.37}$$

which describes a plane wave of varying amplitude [23]. It can be shown that $\mathbf{k}_1 \cdot \mathbf{k}_2 = 0$, which indicates that this wave propagates in the direction given by the vector $\mathbf{k}_1$, and its amplitude decreases in the perpendicular direction.

3.4.4 Characteristics of Waves in Non-local Waveguides

Non-local waveguides are essentially small scale structures and hence its behavior cannot be predicted accurately using continuum theories normally used in isotropic or anisotropic structures. Typically, small scale structures need to be modeled using any of the models such as *ab-initio* models, molecular dynamics models or density functional models. These models are explained in the context of nanostructures modeling for wave propagation analysis in [53]. However, all of these models are computationally expensive to implement and hence as an alternative to the above atomistic models, non-local elasticity theory can be used to model these small scale structures, wherein the scale parameter is introduced into to the continuum model using either gradient elasticity theories or other non-local theories such as peridynamics theory [129]. Some of these concepts are briefly introduced in Chapter 11 in the context of 1-D wave propagation. While the governing equations for gradient theories have higher-order stress or strain gradients as the part of governing equation, the equation governing the peridynamic model is an integral partial differential equation, wherein the spatial part of the equation are expressed in terms of an integral. The scale parameter enter the governing equations in these two models differently. While in the gradient theories, the scale parameter enters the governing equations through the constitutive models, while in the peridynamic models, they enter through a parameter called the horizon length. In the gradient theory models, as mentioned earlier, the constitutive models contain higher-order stress or strain gradients depending upon the type of gradient theory used. Hence, the governing wave equations are normally few orders higher than the equations obtained from local theories and obtaining solutions for wavenumbers and wave speeds are more involved. In this book, we will cover only the wave propagation on systems governed by gradient theories hence, and the general form of the wave equation is only reviewed for gradient theories.

The wave equation governing the behavior of non-local waveguides dictated by the gradient elasticity theory is of the form

$$\frac{\partial^2 u}{\partial t^2} = \sum_{\alpha,\beta} u_{\alpha\beta} \pm g^2 \sum_{\gamma,\delta} u_{\gamma\delta} \tag{3.38}$$

The second term in Eqn. (3.38) represents the contribution coming from the scale effects and g is the term associated with the small scale effects. When g is equal to zero, then we recover the wave equation (Eqn. (3.31)) obtained by local theory of elasticity. Due to the presence of non-local parameter, the wavenumbers in most cases are complex indicating that the wave propagating in this media is always inhomogeneous. In addition, due to higher-order governing equations, the waves are always dispersive in nature. One of the key features that the scale parameter introduces in the spectrum and dispersion relations is the fact that for some non-local waveguides, the scale parameter forces the group velocity to go to zero at a frequency where the wavenumber escapes to infinity. This frequency at which the wavenumber escapes to infinity is called the *Escape frequency.* In other words, scale parameter introduces band gaps even though the system as such is not periodic. Chapter 11 addresses these concepts in more detail.

3.5 DIFFERENT METHODS OF COMPUTING WAVENUMBERS AND WAVE AMPLITUDES

Understanding wave propagation in complex media requires accurate computation of wavenumbers and group speeds. The main difficulty is in the wavenumber computation, where the order of the characteristic polynomial for computation of wavenumbers is increased when the material waveguide supports multiple degrees of freedom. For determining the wave propagation response, computation of wave amplitudes are also required. The conventional method of wavenumber and wave amplitude computation (see [34]) is not adequate to tackle this kind of situation. The development towards this end started with elementary composite beam [89], which needs a sixth order polynomial to be solved for wavenumber computation. The wavenumbers were computed numerically, where the Newton-Raphson (NR) method was used to find a single real root. Rest of the roots were expressed in terms of this real wavenumber. The wave amplitude vectors were evaluated analytically. The situation became a little complicated in the first order shear deformable composite beam case [90], where again the characteristic polynomial for the solution of k was again sixth order with the expressions being more complicated. As was done before, the spectrum relation was solved numerically (by tracking the real root first using NR algorithm) and wave amplitude vectors were evaluated analytically. If the order of characteristic polynomial for computation of k increases further, these ad hoc method of computing the wavenumber and the wave amplitude will be highly inadequate. Hence, what we need are the following:

1. A robust, generally applicable and accurate wavenumber solving algorithm, which will be applicable to all the above problems and for any model of any order of polynomial for the computation of k . Further, the solver must be efficient, since the job is to be performed at each frequency step.

2. A robust, accurate and efficient numerical scheme for computing the wave amplitude vectors is also needed for a given value of the wavenumber.

In this section, we outline two methods, which are quite well known among those knowledgeable in the area of linear algebra, for the computation of the wavenumbers and wave amplitudes. The problem statement is that the characteristic polynomial for computation of wavenumber k can be represented in terms of *wave matrix* **W** and vector **v** as

$$\mathbf{W}(k)\mathbf{v} = (\sum_{i=0}^{p} k^i \mathbf{A}_i)\mathbf{v} = 0\,, \mathbf{A}_i \in \mathbb{C}^{N_v \times N_v}\,, \mathbf{v} \in \mathbb{C}^{N_v \times 1}\,, \tag{3.39}$$

where N_v is the number of independent variables and p is the order of the polynomial of the characteristic equation. Each $\mathbf{A}_i$ depends upon the material properties and wavenumbers (Note in 2-D waveguide case, we will have two wavenumbers corresponding to two coordinate directions.) We will now outline two methods for computing the wavenumbers when the characteristic polynomial is expressed in the form shown in Eqn. (3.39).

3.5.1 Method - 1 : The Companion Matrix & the SVD Technique

In the first method, it is noted that the desired eigenvalues are the latent roots, which satisfy the condition $det(\mathbf{W}(k)) = 0$ (see [81]). Further, if k_i is any such root, then there is at least one non-trivial solution for **v**, which is known as the latent eigenvector. To find the latent root, the determinant is expanded in a polynomial of k, $p(k)$ and solved by the companion matrix method. In this method, the companion matrix $L(p)$ corresponding to $p(k)$ is formed, which is defined as

$$\mathbf{L(p)} = \begin{bmatrix} 0 & 1 & 0 & \cdots & 0 \\ 0 & 0 & 1 & & . \\ \vdots & \vdots & \vdots & \ddots & \vdots \\ 0 & 0 & & 0 & 1 \\ -\alpha_m & -\alpha_{m-1} & \cdots & -\alpha_2 & -\alpha_1 \end{bmatrix}, \tag{3.40}$$

where $p(\lambda)$ is given by

$$p(\lambda) = \lambda^m + \alpha_1 \lambda^{m-1} + \cdots + \alpha_m\,. \tag{3.41}$$

One of the many important properties of the companion matrix is that the characteristic polynomial of $\mathbf{L}(\mathbf{p})$ is $p(k)$ itself ([82]. Thus, eigenvalues of $\mathbf{L}(\mathbf{p})$ are the roots of $p(k)$, which are obtained readily using standard subroutines of LAPACK (xGEEV group).

Once the eigenvalues are obtained, they can be used to obtain the eigenvectors. To do so, it is to be noted that the eigenvectors are the elements of the null space of $\mathbf{W}(k)$ and the eigenvalues make this null space non-trivial by rendering $\mathbf{W}(k)$ singular. Hence, computation of the eigenvectors is equivalent to the computation of the null space of a matrix. To this end, the singular value decomposition (SVD) method is the most effective. Any matrix $\mathbf{A} \in \mathbb{C}^{m\times n}$ can be decomposed in terms of unitary matrices $\mathbf{U}$ and $\mathbf{V}$ and the diagonal matrix $\mathbf{S}$ as $\mathbf{A} = \mathbf{USV}^H$, where H in the superscript denotes the Hermitian conjugate, (see [45]). The $\mathbf{S}$ is the matrix of singular values. For singular matrices, one or more of the singular values will be zero and the required property of the unitary matrix $\mathbf{V}$ is that the columns of $\mathbf{V}$ that correspond to zero singular values (zero diagonal elements of $\mathbf{S}$) are the elements of the null space of $\mathbf{A}$. The SVD can be performed by the efficient and economic LAPACK subroutines (xGESVD group). The method can also be implemented in MATLAB as well.

3.5.2 Method - 2 : Linearization of Polynomial Eigenvalue Problem (PEP)

In this method if the PEP is linearized to

$$\mathbf{A}z = \lambda \mathbf{B}z\,, \quad \mathbf{A}, \mathbf{B} \in \mathbb{C}^{pN_v \times pN_v} \tag{3.42}$$

where

$$\mathbf{A} = \begin{bmatrix} 0 & \mathbf{I} & 0 & \cdots & 0 \\ 0 & 0 & \mathbf{I} & \cdots & 0 \\ \vdots & \vdots & \ddots & \ddots & \vdots \\ \vdots & \vdots & \ddots & \ddots & \mathbf{I} \\ -\mathbf{A}_0 & -\mathbf{A}_1 & \mathbf{A}_2 & \cdots & -\mathbf{A}_{p-1} \end{bmatrix}$$

$$\mathbf{B} = \begin{bmatrix} \mathbf{I} & & & & \\ & \mathbf{I} & & & \\ & & \ddots & & \\ & & & \mathbf{I} & \\ & & & & -\mathbf{A}_p \end{bmatrix} \tag{3.43}$$

and the relation between x and z is given by $z = (x^T, \lambda x^T, \ldots, \lambda^{p-1}x^T)^T$. $\mathbf{B}^{-1}\mathbf{A}$ is a block companion matrix of the PEP. The generalized eigenvalue problem of Eqn. (3.42) can be solved by the QZ algorithm, iterative method, Jacobi-Davidson method or the rational Krylov method. Each one of these has its own advantages and deficiencies; however, the QZ algorithm is the most powerful method for small to moderate sized problems, which is employed in the subroutines available in LAPACK (xGGEV and xGGES group). MATLAB function *polyeig* is used in many of the codes developed as part of this book for solving the wavenumber for many complex waveguides discussed in this book.

In both of these methods, eigenvalue solver is employed, where for QZ algorithm, the cost of computation is $\sim 30n^3$ and an extra $\sim 16n^3$ for eigenvector computation (n is the order of the matrix). Since, the order of the companion matrix in the second method is 3 times that of first method, the cost is 27 times more, which is significant as this computation is to be performed $N \times M$ times.

The PEP method admits $N_v \times p$ eigenvalues and p eigenvectors. If both $\mathbf{A}_0$ and $\mathbf{A}_p$ are singular, then the problem is potentially ill-posed. Theoretically, the solutions might not exist or might not be unique. Computationally, the computed solutions may be inaccurate. If one, but not both, of $\mathbf{A}_0$ and $\mathbf{A}_p$ is singular, the problem is well posed, but some of the eigenvalues may be zero or infinite, and hence caution should be exercised in rejecting those roots.

There are advantages and disadvantages of both the methods. In the first method, the determinant of the wave matrix needs to be formed, which for large N_v is too difficult to obtain. In this case, resorting to the second method is advantageous as it obviates the necessity of obtaining the lengthy expressions of α_i in Eqn. (3.41). However, in the second method, there is no control over the eigenvalues, as we might be interested sometimes in separating the forward propagating wavenumbers. In this case, the first method is the only option.

Summary

In this chapter, the introductory concepts in wave propagation were presented. First, the concept of phase and group velocities and their basic difference is outlined and from their behavior, the concept of dispersive and non-dispersive waves was explained. Next, some important wave propagation terminologies are explained. This is followed by a detailed section on spectral analysis of motion, wherein a generalized method of determining wavenumbers and wave velocities was presented. The definition of transition frequency and cut-off frequency is given for a general second-order 1-D system. Following this, a discussion is presented on the wave characteristics in anisotropic, inhomogeneous and non-local small scale structures. In the last part of the chapter, the issues in determining the wavenumbers were discussed and following this, two methods, namely the companion matrix method and the method based on linearization of PEP, were presented for determination of wavenumbers and wave amplitudes from characteristic wavenumber polynomial of the system.

Exercises

3.1 A wave is propagating in a medium with an average wavelength λ. If $C_p(\lambda)$ is its phase speed, show that the group speed of the is given by

$$C_g = C_p - \lambda \frac{dC_p}{d\lambda}$$

3.2 The spectrum relation of a certain mechanical waveguide system is given by

$$\omega^2 = C^2k^2 + \omega_0^2, \quad \text{where} \quad C = 5000\,\text{m/s and} \quad \omega_0 = 500\,\text{rad/sec}$$

Plot the wavenumber, phase speed and group speed for this system and discuss its wave behavior

3.3 The water wave in a canal of depth h is governed by following spectrum relation

$$\omega^2 = \left(gk + \frac{\sigma k^3}{\rho}\right) \tanh kh$$

where g is the acceleration due to gravity, σ is the surface tension and ρ is the density of water. The above equation is a transcendental equation having infinite solutions, and it is hard to solve. To solve this, we will make following approximations. For deep water canal where $kh >> 1$, $\tanh kh \approx 1$ and the above spectrum relations reduces to

$$\omega^2 = \left(gk + \frac{\sigma k^3}{\rho}\right)$$

However, for smaller values of $kh\,(kh < \pi/2)$, we can approximate $\tanh kh \approx kh - kh^3/3$. The spectrum relation for this case reduces to

$$\omega^2 = -\frac{\sigma}{3\rho}k^6h^3 + \left(-\frac{g}{3}h^3 + \frac{\sigma h}{\rho}\right)k^4 + ghk^2$$

Determine the wavenumbers, phase and group speeds for both the above cases, plot them and discuss their wave behavior. For the second case, please develop a MATLAB code using *linsolve* function to solve the higher-order spectrum equation. Use $\sigma = 0.073$ N/m, $\rho = 1000$ kg/m^2 and $g = 9.81$ m/s^2.

3.4 The Governing equation, governing the wave propagation in a transmission lines is given by

$$\frac{\partial V}{\partial x} + L\frac{\partial I}{\partial t} + RI = 0$$
$$\frac{\partial I}{\partial x} + C\frac{\partial V}{\partial t} + GV = 0$$

where V is the Voltage and I is the current, both of these are the dependent variables, while L is the inductance, R is the resistance, G is the leakage conductance, and C is the capacitance. Determine the wavenumbers, and speeds for this system and discuss its wave behavior.

3.5 The wavenumbers of a beam having two propagating modes are given by

$$k_1 = \alpha_1\sqrt{\omega}, \qquad k_2 = \alpha_2\left[\omega^{\frac{1}{4}} - 24.7789\right]$$
$$\alpha_1 = 0.005 \qquad \alpha_2 = 0.1$$

This beam is subjected to modulated sine signal of central frequency 50 kHz. (1) Calculate the speeds of propagation of the two modes. Do both modes propagate? If your answer is yes, please justify your answer and plot these wave modes (wavenumber as well as speeds). If your answer is negative, then at what central frequency one should modulate in order to see both these modes propagating. Also, plot all the propagating wave modes (wavenumber and group speeds).

3.6 The wavenumber in a certain infinite beam varies as a function of frequency as given below

$$k = a_0\sqrt{\omega} + b_0\omega$$

where a_0 and b_0 are constants that requires to be determined from wave propagation responses. The beam is subjected to modulated sine signal at a point S_1 in the beam and the wave propagation responses are measured at another point S_2 on the beam, which is 1 meter from S_1. First, a modulated sign signal of central frequency 90 kHz is applied at S_1 and the wave takes 4 millisecs to reach S_2. Next, the central frequency is changed to 160 kHz and for this case the wave takes 2 millisecs. If the central frequency of the modulated signal is next changed to 250 kHz, how long will it take for the wave to reach S_2? For understanding modulated signal, please refer to Chapters 4 and 6.

3.7 The governing equation of a vibrating Carbon nano tube modeled as a rod is given by

$$EA\frac{\partial^2 u}{\partial x^2} + \mu A\frac{\partial^2 u}{\partial x^2 \partial t^2} = \rho A\frac{\partial^2 u}{\partial t^2}$$

where EA is the axial rigidity of the rod, μ is the scale parameter, ρA is the mass of the nano tube. Obtain expressions for wavenumber and group speeds for this system. Do the system exhibit cut-off frequency? Discuss the wave behavior in this system. We will discuss this model in more detail in Chapter 11.

CHAPTER 4

Wave Propagation in One-Dimensional Isotropic Structural Waveguides

Isotropic structures are those where the material has infinite symmetry. That is, the material properties are assumed same in all the directions. In other words, isotropic material system is the simplest of the material systems and it needs only two mechanical properties, namely the Young's modulus and Poisson's ratio, to completely characterize the material system. In this chapter, the wave propagation characteristics in 1-D structures made up of isotropic material is presented. It is to be noted that most metals falls under this category.

We had defined a waveguide in Chapter 3 and the waveguides can be one, two- or three-dimensional depending upon its characteristic lengths in the 3-D coordinate system. The fundamental to analysis of wave propagation analysis is the *Spectral analysis*, which was explained in the last chapter. The nature of governing equations for different waveguides differ and hence their wave characteristics. In this chapter, we will address the wave propagation concepts only for the one-dimensional waveguides. The wave propagation in 2-D isotropic waveguides is addressed in Chapter 7. The main objective here is to determine the nature of waves in these waveguides and this will be presented in terms of the spectrum and the dispersion plots. Another important aspect in the wave propagation studies is the interaction of the waves with the boundaries where, in addition to deriving the necessary equations, the propagation aspect is presented in the form time history plots. This chapter

DOI: 10.1201/9781003120568-4

addresses diverse topics ranging from wave propagation in elementary waveguides, higher-order waveguides, rotating waveguides, and tapered waveguides.

This chapter gives all the basic concepts of wave propagation, which are fundamental to the understanding of topics presented in the later chapters. The chapter begins with wave propagation in 1-D elementary waveguides, namely the rods and beams. Elementary rod supports only longitudinal motion and hence they are called *Longitudinal waveguides*, while the elementary beam (normally referred to as Euler-Bernoulli beam) supports transverse or lateral motion and hence they are normally referred to as *Flexural waveguides.*

Adding certain additional effects to elementary waveguides makes the waveguides higher-order. For example, introducing lateral contraction motion to longitudinal waveguide makes it higher-order longitudinal waveguide and such higher-order longitudinal waveguides is called *Mindlin-Herrmann Rod.* A separate section on wave propagation in higher-order longitudinal and flexural waveguides follows the section on wave propagation in elementary waveguides. Rotating waveguides are found in many engineering and science disciplines and some examples of these include the compressor/turbine blades or helicopter blades. A section on wave propagation in rotating waveguides is presented following higher-order waveguides. This is followed by a section on wave propagation in tapered waveguides.

Deriving governing differential equations is an important step in performing wave propagation studies. For simple waveguides, one can use equilibrium approach, wherein, we can isolate a small section of a waveguide, draw its free body diagram, write all the stress resultants and apply Newton's second law to obtain the governing equations. For more complex waveguides, this method is not feasible. For deriving governing equation for such waveguides, we normally use Hamilton's principle. In this chapter and the chapters that follow, we will be using the Hamilton's principle to derive the governing equation. This principle not only gives the governing differential equation but also the associated essential and natural boundary conditions. The details of this principle are not given in this text. The interested readers can refer to the unified wave propagation text by the author [54] to get more details on this principle.

4.1 WAVE PROPAGATION IN 1-D ELEMENTARY WAVEGUIDES

One-dimensional waveguides are those in which one of the dimensions is very long compared to the other two-dimensions. Some of the 1-D waveguides, although 1-D in terms of its characteristic length, can support 3-D motion. Examples of such waveguides are the framed structures, which are obtained through the assemblage of 1-D waveguides. Wave propagation analysis is an involved process, which requires computation of wavenumbers, reflection coefficients, transmission coefficients and the wave responses and all of these need to be performed in the frequency domain. Hence, one will need two software, one for transforming the variables back and forth from time domain to frequency domain, and the second software to perform the wave analysis for different waveguides. However, if one uses programs like MATLAB, both analysis can be performed using a single code. The complete wave propagation analysis can be performed in four steps as shown in Fig. 4.1.

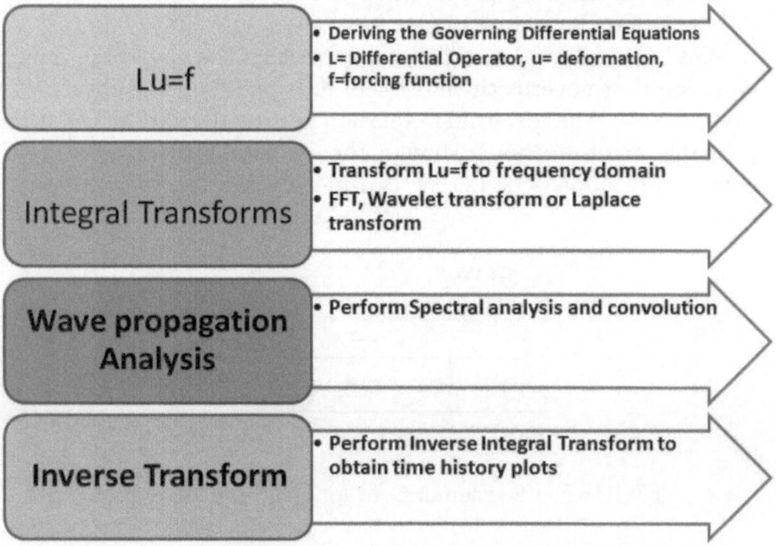

FIGURE 4.1: Steps involved in wave propagation analysis

The first step is to derive the governing equation. After deriving the governing differential equation, we need to transform the field variables (or dependent variables) and the input time history to frequency domain using a suitable integral transforms. The Fourier Transforms was presented in detail in Chapter 2. Although any of the integral transform can used to transform a variable to frequency domain, in this text, we will use only FFT to go back and forth from time and frequency domain. After transforming all the relevant variables to the frequency domain, spectral analysis as explained in Section 3.3 is performed to determine the wavenumbers and wave speeds, which are then used to obtain the frequency domain responses of all the dependent variables. The last step here is to use inverse FFT on the obtained frequency domain responses to get the time domain responses. In this section, we will first present the longitudinal wave propagation in rods, flexural wave propagation in beams and coupled axial and flexural wave propagation in framed structures. As mentioned before, for all these analysis, we will use FFT to transform the variables to frequency domain.

4.2 LONGITUDINAL WAVE PROPAGATION IN RODS

Rod or bar or a strut is the simplest structural element that has one-dimension very long compared to other two-dimensions and they can resist only one type of motion along its longest dimension called the axis of the member. Rod waveguides guides the waves along the axis of the structural member, which is why they are called *Longitudinal waveguides.* In this subsection, we will present the longitudinal wave propagation in rod waveguides and its interaction with different boundary constraints.

We begin this subsection by first writing the governing partial differential equation that governs the motion of rods based on the physics of the deformation in rods. The governing equation can be derived either using Hamilton's principle or otherwise, although they can be derived with least effort using equilibrium approach. Let us consider an 1-D rod waveguide shown in Fig. 4.2.

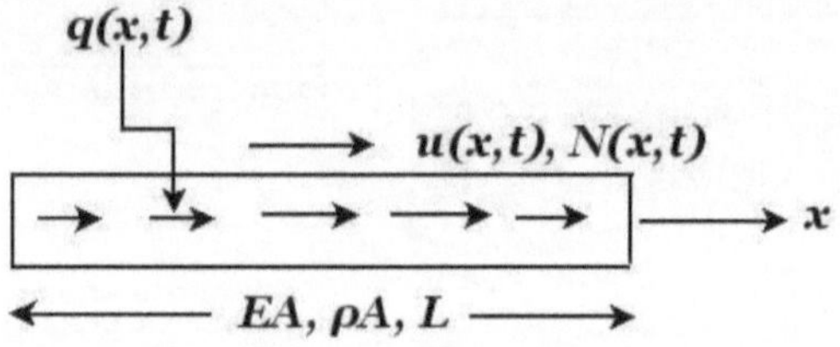

FIGURE 4.2: Elements of longitudinal waveguide with loads

Let the rod be of length L, with axial rigidity EA and mass per unit length ρA. Let u, v and w be the displacements in the three coordinate directions. The displacement field for the rod can be expressed as

$$u(x,y,z,t) = u(x,t), \qquad v(x,y,z,t) = 0, \qquad w(x,y,z,t) = 0 \tag{4.1}$$

Using the above strain fields, the strain energy, kinetic energy and the work done by the external forces can be derived, which are then used in the Hamilton's principle to get the governing partial differential equation for the rod as

$$EA\frac{\partial^2 u}{\partial x^2} + q(x,t)A = \rho A\frac{\partial^2 u}{\partial x^2} \tag{4.2}$$

The Hamilton's principle also gives the natural or force boundary conditions at the two ends of the rod, which is given by

$$N(x,t) = EA\frac{\partial u(x,t)}{\partial x} \tag{4.3}$$

Noting that the speed of the medium C_0 is given by $C_0 = \sqrt{E/\rho}$ and assuming $q(x,t)$ is equal to zero in Eqn. (4.2), we can write the governing equation for an isotropic rod as

$$\frac{\partial^2 u}{\partial x^2} = \frac{1}{C_0^2}\frac{\partial^2 u}{\partial x^2} \tag{4.4}$$

4.2.1 D'Alemberts Solution

Now, let us attempt to solve the PDE given by Eqn. (4.4). In this equation, the dependent variable is a function of x and t. The solution of this equation was found by famous mathematician Jean le Rond d'Alembert [33]. He did so by introducing two new variables $\xi = x - C_0 t$ and $\eta = x + C_0 t$. Using these two new variables, we can now write

$$\frac{\partial^2 u}{\partial x^2} = \frac{\partial^2 u}{\partial \xi^2} + 2\frac{\partial^2 u}{\partial \xi \partial \eta} + \frac{\partial^2 u}{\partial \eta^2}$$
$$\frac{\partial^2 u}{\partial t^2} = C_0^2 \frac{\partial^2 u}{\partial \xi^2} - 2C_0^2 \frac{\partial^2 u}{\partial \xi \partial \eta} + C_0^2 \frac{\partial^2 u}{\partial \eta^2} \tag{4.5}$$

Substituting above equations in Eqn. (4.4), we get

$$\frac{\partial^2 u}{\partial \xi \partial \eta} = 0$$

the solution of which is given by

$$u(x,t) = f(\xi) + g(\eta) = f(x - C_0 t) + g(x + C_0 t) \tag{4.6}$$

where $f(\xi)$ and $g(\eta)$ are some functions of space, and time and represents the forward and backward moving waves traveling with the speed of propagation C_0. We will not pursue with this equation for our study since these solutions are available only for an ideal elementary undamped rod waveguides. We will use spectral analysis route to study their wave behavior.

4.2.2 Spectral Analysis

We will start the wave propagation analysis from Eqn. (4.2). We will now use DFT to transform the dependent variable $u(x,t)$ to frequency domain. That is

$$u(x,t) = \sum_{n=0}^{N} \hat{u_n}(x, \omega_n) e^{i\omega_n t} \tag{4.7}$$

where N is the number FFT points used in the analysis. Using Eqn. (4.7) in Eqn. (4.4), we get

$$\sum_{n=0}^{N} \left[EA \frac{d^2 \hat{u_n}}{dx^2} + \omega_n^2 \rho A \hat{u_n} \right] e^{i\omega_n t} = 0 \tag{4.8}$$

The term inside the bracket is the governing differential equation in the frequency domain, which is given by

$$EA \frac{d^2 \hat{u_n}}{dx^2} + \omega_n^2 \rho A \hat{u_n} = 0 \tag{4.9}$$

The transformation of the governing PDE given by Eqn. (4.4) to frequency domain results in N ordinary differential equation of constant coefficients, whose solution is of the form

$$\hat{u_n}(x, \omega_n) = \mathbf{A_n} e^{-ik_n x}$$

In the above equation, k_n represents the axial wavenumber . In order to determine it, we substitute the above solution in Eqn. (4.9), and the resulting expression for the axial wavenumber is given by

$$k_n^2 = \omega_n^2 \frac{\rho A}{EA} \tag{4.10}$$

We see that that the wavenumber is a linear function of frequency and hence the wave propagation in a rod is *non-dispersive*, that is, the waves do not change its wave profile as it propagates. These aspects were explained in Chapter 3. Next, we will obtain the phase speed C_p and group speeds C_g. They are given by

$$C_p = \frac{\omega}{k} = C_0 = \sqrt{\frac{E}{\rho}}, \qquad C_g = \frac{d\omega}{dk} = C_0 = \sqrt{\frac{E}{\rho}} \tag{4.11}$$

4.2.3 Propagation of Waves in an Infinite Longitudinal Waveguide

Next, in order to obtain the wave responses, we need to obtain the solution for Eqn. (4.9), which is given by

$$u(x,t) = \sum_{n=0}^{N} \left[\mathbf{A_n} e^{-ik_n x} + \mathbf{B_n} e^{ik_n x}\right] e^{i\omega_n t} \tag{4.12}$$

The first term in the brackets of Eqn. (4.12) represents the incident wave with incident wave coefficient $\mathbf{A_n}$, while the second term in the equation represents the reflected wave with reflection coefficient $\mathbf{B_n}$. We will now simulate the propagation of waves in an infinite rod subjected to triangular time input shown in Fig. 4.3. This pulse will be used throughout this text to simulate most of the wave propagation responses. The pulse is transformed into frequency domain using FFT, where a sampling rate of 0.5×10^{-6} with 8192 FFT points. This gives the total time window $N\Delta t = 8192 \times 0.5 \times 10^{-6} = 4092\,\mu$secs.

The FFT of the time input can be considered equal to the incident wave $\mathbf{A_n}$. Since the rod considered here is infinite, the reflected wave can be omitted to get the wave response. The infinite rod wave propagation response of a rod with Young's modulus $E = 70 \times 10^9$ N/m^2, density $\rho = 2700$ kg/m^3, and width and depth of the rod equal to 6.25×10^{-4} m and 0.0254 m, respectively. is shown in Fig. 4.4. Note that, throughout this chapter, we will use the above aluminum properties for all the simulations. The steps to be followed to obtain the responses are given in Fig. 4.1.

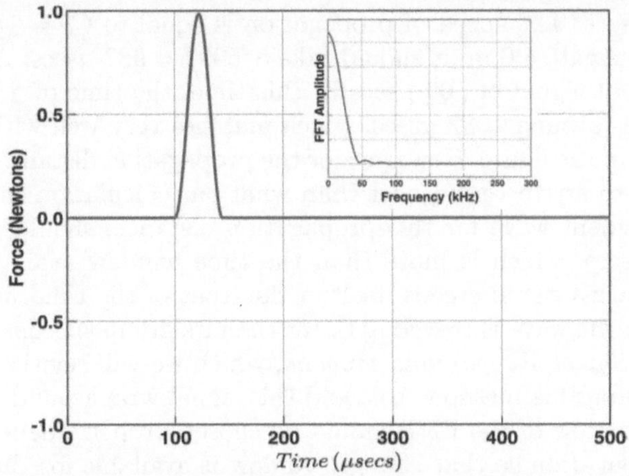

FIGURE 4.3: Input incident signal and its frequency content

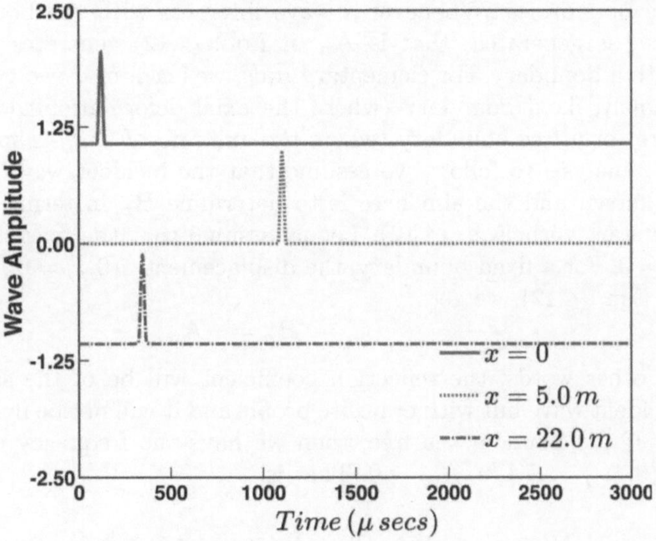

FIGURE 4.4: Wave propagation in an infinite elastic rod at different lengths

We will now analyze the responses shown in Fig. 4.4. We see that as the wave propagates, the shape of the wave profile remains the same due to the fact that the waves in rods are non-dispersive in nature. As mentioned earlier, the plot was generated using a triangular pulse, whose FFT gave a time window of

4092 μsecs. The speed of propagation is equal to $C_0 = \sqrt{E/\rho} = 5092$ m/sec. To propagate 5.0 m, it should take $5/5092 = 982$ μsecs. Adding the header of the input signal of 100 μsecs to this time, the time of arrival of the incident signal is around 1082 μsecs, which matches very well with the time of arrival shown in the figure. However, for the propagation distance of 22 m, the wave is shown to arrive very earlier than what our calculation show. This is because, the incident wave for this propagation distance, should arrive at $22/5092 \approx$ 4320 μsecs, which is more than the time window available for propagation. Hence, this signal creeps itself in the front of the time window portraying as though the wave is traveling faster than its group speeds. This phenomenon is called *Signal Wraparound problem*, which we will revisit in Chapter 6, where we will outline methods to avoid this signal wraparound problem. Hence, the time window of the FFT should be chosen properly depending on the length of propagation so that enough window is available for the wave to propagate its entire length.

4.2.4 Interaction of Waves with Fixed and Free Boundaries

Next, we will see what happens when the longitudinal wave interacts with the boundaries. Whenever a wave interacts with the boundary, a reflected wave is generated, that is $\mathbf{A_n}$ in Eqn. (4.12) generates $\mathbf{B_n}$ on interaction with a boundary. For elementary rods, we encounter two types of boundaries, namely, fixed boundary (where the axial deformation $u(x,t)$ is specified as zero) or a free boundary (where the axil force $F(x,t)$ is specified as zero). In the analysis to follow, we assume that the incident wave $\mathbf{A_n}$ in Eqn. (4.12) is known and the aim here is to determine $\mathbf{B_n}$ in terms of $\mathbf{A_n}$. To do this we start with Eqn. (4.12). Let us assume that the fixed boundary exists at $x = 0$. For a fixed boundary, the displacement $u(0,t) = 0$. Substituting $x = 0$ in Eqn. (4.12), we get

$$\mathbf{B_n} = -\mathbf{A_n}$$

In other words, the reflection coefficient will be of the same magnitude as incident wave but with opposite profile and it will not be frequency dependent.

If the boundary is free, then we have the frequency domain axial force $\hat{N}(0,\omega_n) = EAd\hat{u}_n/\partial x = 0$. That is,

$$\hat{N}(x,\omega_n) = EA\frac{d\hat{u}_n}{dx} = EAik_n\left[-\mathbf{A_n}e^{-ik_nx} + \mathbf{B_n}e^{ik_nx}\right] = 0$$

Substituting $x = 0$ in the above equation, we get

$$\mathbf{B_n} = \mathbf{A_n}$$

That is, if the boundary is free, then the reflected displacement wave will have the same amplitude and time profile as the incident displacement wave.

4.2.5 Reflection from an Elastic Boundary

Elastic boundary in the form of a spring can simulate boundary conditions, that is in between the two extreme boundary conditions, namely the fixed and the free boundary. A rod with elastic boundary containing a spring of constant K is shown in Fig. 4.5.

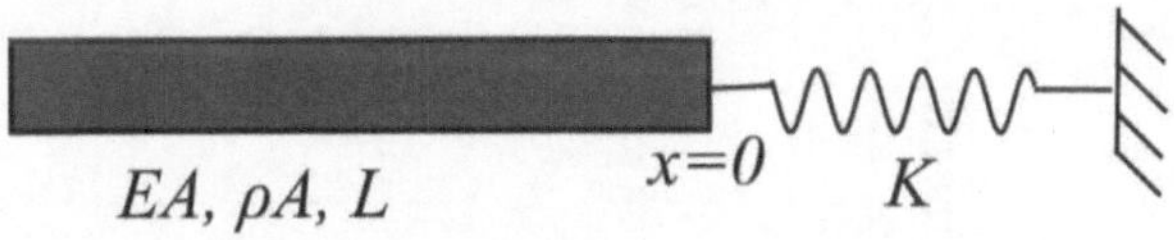

FIGURE 4.5: Longitudinal waveguide with elastic boundary

Here, the waveguide considered is of length L having mass per unit length ρA, and axial rigidity EA. In order to determine the reflected coefficient $\mathbf{B_n}$, we impose the force boundary conditions at $x = 0$, where the elastic boundary is located. That is

$$\hat{N}(0, \omega_n) = EA\frac{d\hat{u}_n}{dx}|_{x=0} = -K\hat{u}_n|_{x=0} \tag{4.13}$$

Substituting Eqn. (4.12) in Eqn. (4.13), we get

$$\begin{aligned} &\left|EAik_n\left[-\mathbf{A_n}e^{-ik_nx} + \mathbf{B_n}e^{ik_nx}\right]\right|_{x=0} \\ &= -K\left|\left[\mathbf{A_n}e^{-ik_nx} + \mathbf{B_n}e^{ik_nx}\right]\right|_{x=0} \end{aligned} \tag{4.14}$$

Substituting $x = 0$ in the equation above and solving for $\mathbf{B_n}$ in terms of the incident wave coefficient $\mathbf{A_n}$, we get

$$\mathbf{B_n} = \frac{ik_nEA - K}{ik_nEA + K}\mathbf{A_n} \tag{4.15}$$

We have seen earlier that for fixed boundary conditions, we have $\mathbf{B_n} = -\mathbf{A_n}$ while for free boundary conditions, we have $\mathbf{B_n} = \mathbf{A_n}$. These values can be obtained if $K = 0$ and $K = \infty$, respectively, in Eqn. (4.15). These can be readily seen in Fig. 4.6.

As the value of the stiffness K increases from a low value to a high value, the profile of the reflected response goes from the fixed boundary response to the free boundary response. For intermediate values of stiffness, the waves show dispersive character.

4.2.6 Reflection and Transmission from a Joint Having Concentrated Mass and Stepped Rod

Normally in structural analysis, it is customary to assume that the structural joints do not have either elasticity or mass, and they are idealized as rigid.

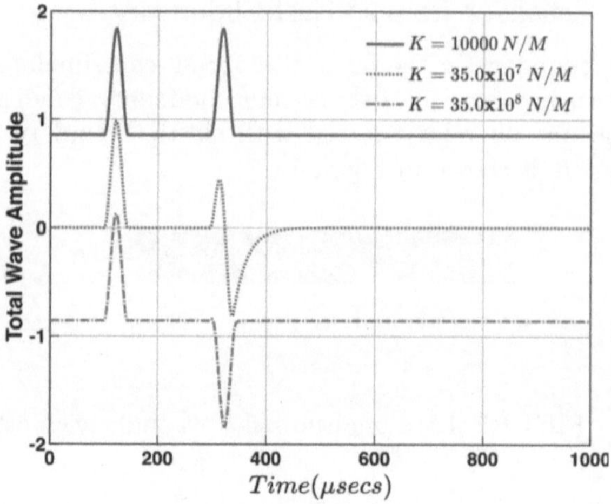

FIGURE 4.6: Reflections from elastic boundary for different values of stiffness coefficient K

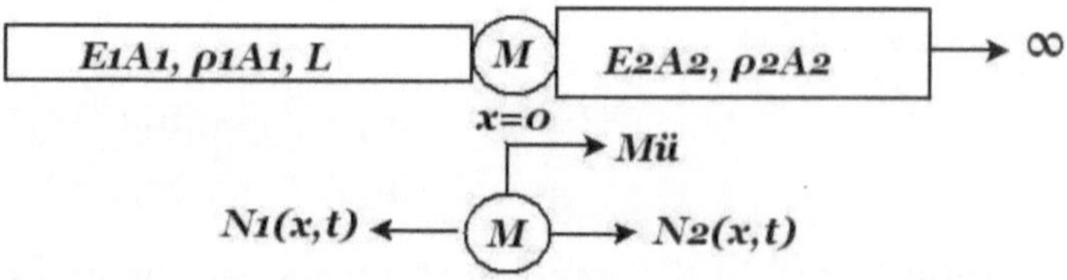

FIGURE 4.7: Longitudinal waveguide with a joint having a mass and free body diagram

Unlike the previous cases, here we have two rod segments, one having property E_1A_1, ρ_1A_1, and L, while the second is an infinite segment having property E_2A_2,and ρ_2A_2 as shown in Fig. 4.7. There will be reflected response in waveguide that is to the left of the joint, while there will be transmitted response in the waveguide that is to the right of joint. Also, the properties of the left waveguide can be higher or lower than the right waveguide, and the wave propagation behavior in these two cases is quite different, irrespective of considering the joint mass.

In this section, the wave propagation analysis procedure is developed considering that the joint connecting two waveguides has a joint mass M. If the joint mass is assumed zero, then it becomes a case of a stepped rod. We begin the analysis by first writing the displacement fields in these two segments as

$$\hat{u}_{1n}(x, \omega_n) = \mathbf{A_{1n}} e^{-ik_{1n}x} + \mathbf{B_{1n}} e^{ik_{1n}} x$$
$$\hat{u}_{2n}(x, \omega_n) = \mathbf{A_{2n}} e^{-ik_{1n}x} \tag{4.16}$$

In writing these displacement fields, it is assumed that the segment to the right of the joint is assumed to extend to infinity. Assuming that the incident wave $\mathbf{A_{1n}}$ is known, there are two unknowns to be determined. Two boundary conditions at the joint location of the rod is necessary to determine the two unknowns, namely the reflection coefficient $\mathbf{B_{1n}}$ and the transmitted coefficient $\mathbf{A_{2n}}$ in terms of the incident signal $\mathbf{A_{1n}}$. This can be done by considering the free body diagram of the joint shown in Fig. 4.7 located at $x = 0$. That is,

$$N_2(0, t) - N_1(0, t) = M\ddot{u} \iff \hat{N}_1(0, \omega_n) - \hat{N}_2(0, \omega_n)$$
$$u_1(0, t) = u_2(0, t) \iff \hat{u}_{1n}(0, \omega_n) = \hat{u}_{2n}(0, \omega_n) \tag{4.17}$$

Using the expressions for $\hat{N}_1 = E_1 A_1 d\hat{u}_1/dx$ and $\hat{N}_2 = E_2 A_2 d\hat{u}_2/dx$, we have

$$E_1 A_1 i k_{1n}[-\mathbf{A_{1n}} + \mathbf{B_{1n}}] = -M\omega_n^2[\mathbf{A_{2n}}] \tag{4.18}$$
$$\mathbf{A_{1n}} + \mathbf{B_{1n}} = \mathbf{A_{2n}} \tag{4.19}$$

Solving the above equations, the reflection coefficient is given by

$$\mathbf{B_{1n}} = \frac{M\omega_n^2 + E_1 A_1 i k_{1n}(1 - S\alpha)}{-M\omega_n^2 + E_1 A_1 i k_{1n}(1 + S\alpha)} \mathbf{A_{1n}} \tag{4.20}$$

where $S = E_2 A_2 / E_1 A_1$, and $\alpha = k_{2n}/k_{1n}$. For large values of mass or at very high frequencies, $\mathbf{B_{1n}} = -\mathbf{A_{1n}}$, which simulates the case of fixed boundary. Similarly at very low frequencies, if both segments have same cross sections, then $\mathbf{B_{1n}} \simeq 0$. The transmitted coefficient is given by

$$\mathbf{A_{2n}} = \frac{2E_1 A_1 i k_{1n}}{-M\omega_n^2 + E_1 A_1 k_{1n}(1 + S\alpha)} \mathbf{A_{1n}} \tag{4.21}$$

For large values of mass or at high frequencies, $\mathbf{A_{2n}} \simeq 0$ and at low frequencies, we have $\mathbf{A_{2n}} \simeq \mathbf{A_{1n}}$. We see that both reflected and transmitted coefficients are frequency dependent.

As mentioned earlier, if the joint is rigid and massless (that is, if $M = 0$), then the formulation is reduced to the case of a *Stepped Rod*. In this case, the reflected and transmitted coefficients become

$$\mathbf{B_{1n}} = \frac{(1 - S\alpha)}{(1 + S\alpha)} \mathbf{A_{1n}}, \qquad \mathbf{A_{2n}} = \frac{2}{(1 + S\alpha)} \mathbf{A_{1n}} \tag{4.22}$$

If we look at these expressions, if $S = 1$ and $\alpha = 1$ (it is, a case with both left and right segments having same properties), the reflected coefficient $\mathbf{B_{1n}} = 0$

and the transmitted coefficient $\mathbf{A_{2n}} = 1$, which is a case of a uniform rod connected by a massless joint. On the other hand, if $S = \infty$ (that is, the right segment having infinite stiffness), then the we have $\mathbf{B_{1n}} = -\mathbf{A_{1n}}$ and $\mathbf{A_{2n}} = 0$, which is the case of fixed end boundary condition. Alternatively, if we have $S = 0$ (that is the case where the left segment having infinite stiffness), then we have $\mathbf{B_{1n}} = \mathbf{A_{1n}}$, which is the case of a free boundary.

To understand these aspects, we will first generate the reflected and transmitted responses for a stepped rod case, that is for the case wherein $M = 0$. Here we generate the reflected and transmitted response for the following two cases;

- **Case-1:** The material properties of the right segments more than the left segment (This is shown in Fig. 4.8).

- **Case-2:** The material properties of the left segments more than the right segment (This is shown in Fig. 4.9).

To generate these plots, we consider $\mathbf{A_{1n}}$ is known and is given as the FFT of the signal shown in Fig. 4.3. This signal is characterized using 8192 FFT points with a sampling time as 0.5 μsecs. The structures is loaded on the left segment and the loading point is at 1.0 m from the joint, where $x = 0$ is located at the joint. The reflected response is measured at the loaded point while the transmitted response on the right segment is measured at 1.0 m from the joints. These responses are for Case-1 is shown in Fig. 4.8, while for Case-2, it is shown in Fig. 4.9. Note that the reflected response also shows the incident pulse.

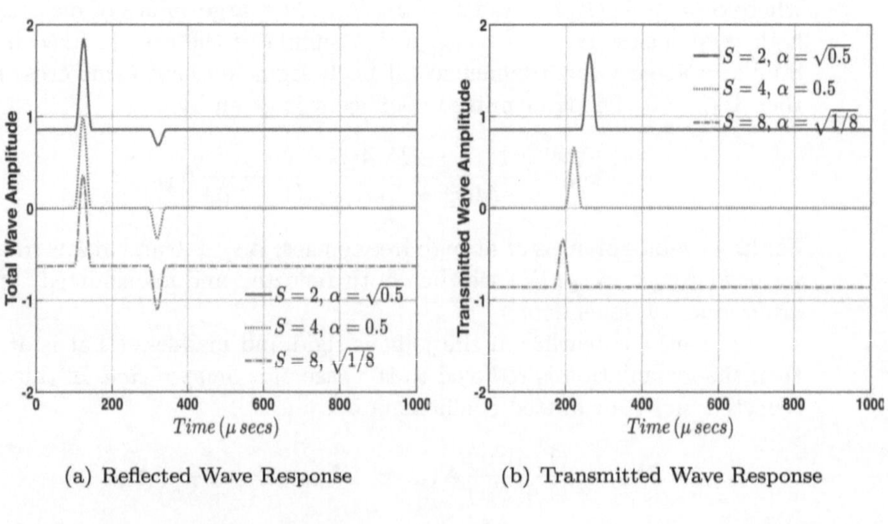

(a) Reflected Wave Response (b) Transmitted Wave Response

FIGURE 4.8: Reflection and transmission for a stepped rod: Case 1

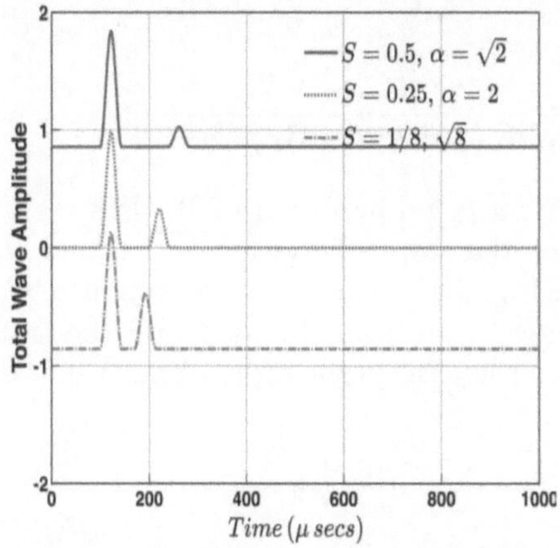

(a) Reflected Wave Response

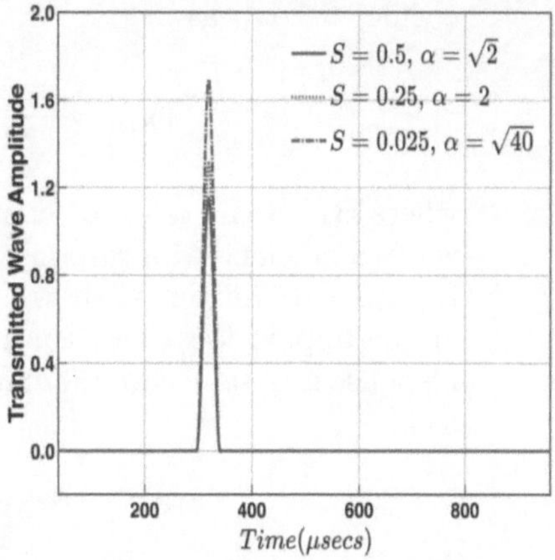

(b) Transmitted Wave Response

FIGURE 4.9: Reflection and transmission for a stepped rod: Case 2

From these figures, we see that the reflected speed does not change for different S and α, while the transmitted wave speeds change with the change in the values of S. When the value of $S \to \infty$, then $\mathbf{A_{1n}} = -\mathbf{B_{1n}}$ and a standing wave will be created with $u\hat{1}_n = 2\cos k_{1n}x$. In other words, the incident and reflected waves add up and propagate as a single wave with higher amplitude.

For Case-2, the reverse happens. That is the transmitted wave travels with the same speeds irrespective of the value of S, while the reflected wave speed changes with the value of S and α.

In the above approach, we assume that the incident wave is given by the triangular pulse shown in Fig. 4.3. This will capture only the first reflection. As the wave traverse through the length of the rod, the incident wave will first hit the joint and generate the reflected wave, which will again traverse through the rod and hit the joint to generate another reflected wave. This process takes place until the wave dies down due to inherent damping. In order to capture this, the incident wave coefficient $\mathbf{A_{1n}}$ need to be accurately determined from the equilibrium at the loading point located at $x = -L$ in Fig. 4.7. That is

$$\hat{N}_1 = E_1 A_1 \frac{d\hat{u}_{1n}}{dx} = \hat{F} \tag{4.23}$$

In the above equation, $\hat{F}$ is the FFT of the externally applied load shown in Fig. 4.3. Substituting for $\hat{u}_{1n}$ from Eqn. (4.16), we can get the incident wave coefficient $\mathbf{A_{1n}}$ as

$$\mathbf{A_{1n}} = \frac{\hat{F}}{ik_{1n}E_1A_1\left(-e^{ik_{1n}L} + \mathbf{B_{1n}}e^{ik_1L}\right)} \tag{4.24}$$

where $\mathbf{B_{1n}}$ is the reflected coefficient given by Eqn. (4.20). Using this incident wave coefficient, the response for the stepped rod for $S = 2$ and $\alpha = \sqrt{0.5}$ is obtained and this is shown in Fig. 4.10. The reflected response is obtained at the impact location situated at 1.0 m from the joint and the transmitted response is obtained at 1.0 m from the joint in the right segment of the stepped rod.

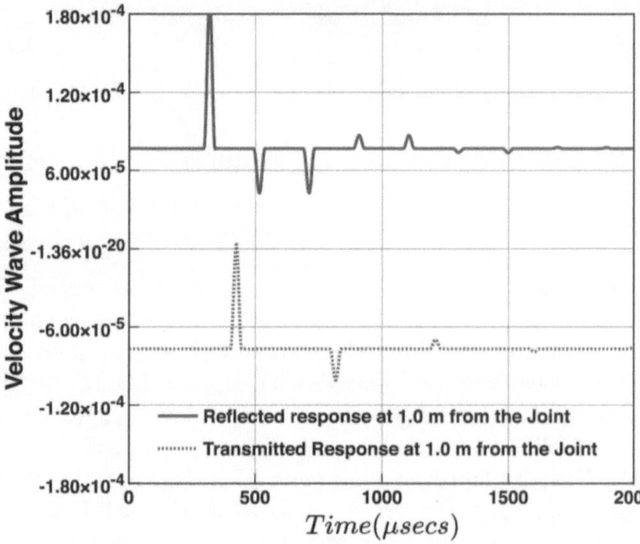

FIGURE 4.10: Reflected and transmitted response in a stepped rod by evaluating incident wave coefficient using Eqn (4.24)

From the above figure, we can clearly see multiple reflection. Note that here, we have plotted velocity wave amplitude instead of displacement wave amplitude as was doing in the earlier plots. This is because, instead of wave coefficient, we have used force as input. From the signal processing aspect (to be dealt in more detail in Chapter 6), force and velocity are comparable quantities. Plotting displacements in this situation will not yield meaningful results due to signal integration issues.

Next, we will study the reflection and transmission characteristics from a joint that has a concentrated mass M. To understand the effect of mass on the reflected and transmitted responses, we will make the material properties of both the left and right segments same. That is $S = 1$ and $\alpha = 1$. To generate these responses that show multiple reflections from the joint, we will use force as input, where the incident and reflected coefficients are given by Eqns. (4.24 & 4.20). These responses are shown in Fig. 4.11.

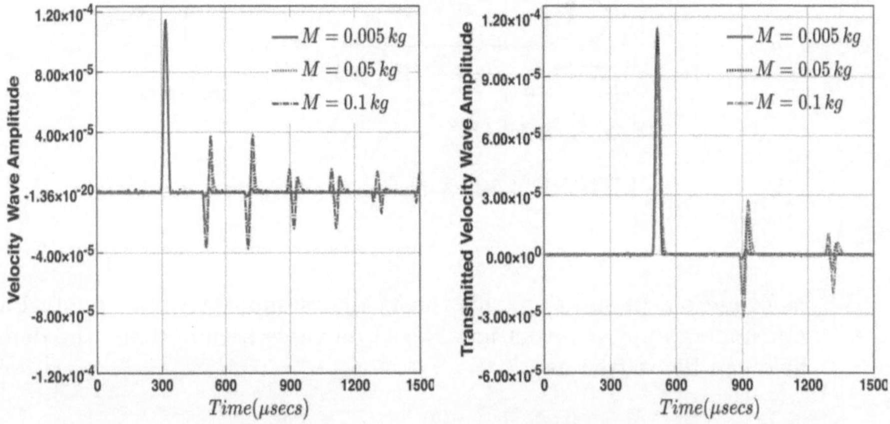

(a) Reflected Wave Response (b) Transmitted Wave Response

FIGURE 4.11: Reflection and transmission from a joint having concentrated mass

As before we plot the velocity wave amplitude as a function of mass. As the value of M increases, we see increase in reflected response and decrease in the transmitted response. Increasing the mass beyond $M = 0.1$ kg will cause the signal to wraparound completely distorting the response. More details on signal wraparound will be dealt in Chapter 6.

4.3 FLEXURAL WAVE PROPAGATION IN BEAMS

Beams are 1-D waveguides that undergoes only traverse or lateral motion. Unlike rods, beams do not have a D-Alembert's solution since they are governed by PDE's that is fourth-order in spatial coordinate and second-order in temporal coordinate. There are several different theories of beams. In this section, we will address wave propagation in elementary beam normally referred to as *Euler-Bernoulli beam.* Higher-order beam and rod theories are addressed in the later part of this chapter. This section is developed on the same lines as the previous section, that is, we will start the analysis from the governing partial differential equation. The beam equation can be derived using Hamilton's principle, the detailed derivation for this case can be found in the unified book of the author [54] and not provided here.

Here, we consider a beam shown in Fig. 4.12. In this theory, we assume that the waveguide is isotropic, homogeneous and they undergo only small

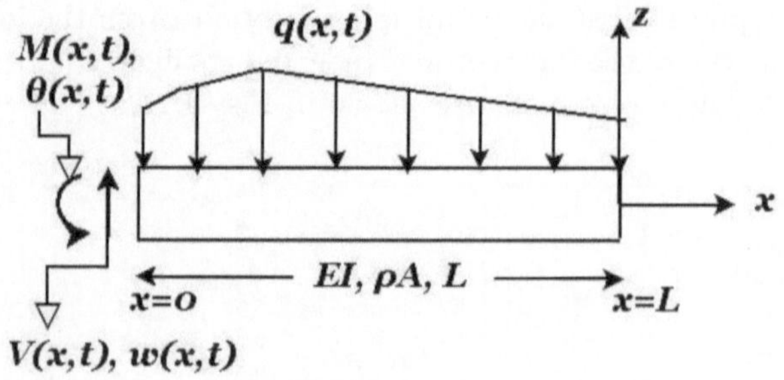

FIGURE 4.12: A elements of flexural waveguide

deformation. In addition, this theory also assumes that the neutral axis does not undergo any deformation. Based on these assumptions, the deformation field can be written as

$$u(x, z, t) = -z\frac{dw}{dx}, \qquad w(x, z, t) = w(x, t) \tag{4.25}$$

where z is the thickness coordinate, $u(x,t)$ is the axial deformation along the x-direction and $w(x,t)$ is the deformation in the z-direction. Due to the flexural loads, the beams bend and the slope or rotation $\theta(x,t)$ are derived from transverse displacement using the expression $\theta(x,t) = dw/dx$. Using these deformation fields, the strain and kinetic energies are evaluated, which are then used in Hamilton's principle to get the equation of motion and the associated boundary conditions. Thus, the governing differential equation for Euler-Bernoulli beam is given by

$$EI\frac{\partial^4 w}{\partial x^4} + q(x,t) = \rho A\frac{\partial^2 w}{\partial t^2} \tag{4.26}$$

The associate forces are the shear force V and the bending moment M, and these are given by

$$V = -EI\frac{\partial^3 w}{\partial x^3}, \qquad M = EI\frac{\partial^2 w}{\partial x^2}$$

4.3.1 Wave Propagation Analysis

We will now perform spectral analysis to determine its parameters, namely the wavenumbers, phase speeds and group speeds. First, the governing differential equation Eqn. (4.26) is transformed to frequency domain using DFT, which is given by

$$w(x,t) = \sum_{n=0}^{N} \hat{w}_n(x, \omega_n)e^{i\omega_n t}$$

where N is the number of FFT points. Substituting this in the homogenous part of Eqn(4.26), we get

$$\sum_{n=0}^{N}\left[EI\frac{d^4\hat{w}_n}{dx^4}+\rho A\omega_n^2\hat{w}_n\right]=0 \tag{4.27}$$

From the above equation, we see that a single PDE, on transformation reduces to N ODEs. From now on, we will drop the subscript n when we perform the computation in the frequency domain. In addition, all variables with "$\hat{()}$" indicates that those variables are frequency dependent. Eqn. (4.27) is an ODE with constant coefficients and it has solutions of the form $\hat{w}=w_0e^{-ikx}$, where k is the wavenumber. On substitution of this solution in Eqn. (4.27), we get

$$\begin{gathered}\left[EIk^4-\rho A\omega^2\right]w_0e^{ikx}=0, \quad k=\beta,\ -\beta,\ i\beta,\ -i\beta n\\ \text{where } \beta=\sqrt{\omega}\left[\frac{\rho A}{EI}\right]^{0.25}\end{gathered} \tag{4.28}$$

That is, there are four possible wave modes corresponding to four values of wavenumber k. The phase and group speeds for Euler-Bernoulli beam are given by

$$C_p=\frac{\omega}{k}=\sqrt{\omega}\left[\frac{EI}{\rho A}\right]^{0.25}, \quad C_g=\frac{d\omega}{dk}=\frac{1}{dk/d\omega}=2\sqrt{\omega}\left[\frac{EI}{\rho A}\right]^{0.25} \tag{4.29}$$

From the above equations, we find the $C_g=2C_p$. From the above equations, following inferences can be drawn:

- The wavenumber is a non-linear function in frequency
- The group and phase speeds are not equal and group speed is twice of its phase speed, and
- The wave speeds change with frequency. Higher frequency components travel faster, while the lower frequency components travel slow.

All the above properties indicate that the waves propagate dispersively in a Euler-Bernoulli beam. In addition, there are a total of four wave modes, two of which are incident wave modes, while the other two are the reflected wave modes. Of the four wave modes, two of them are propagating (since they have real wavenumbers) and the other two are non-propagating, and they are termed as *Evanescent mode* or damping mode. The spectrum plot (plot of wavenumber with frequency) and the dispersion plot of wave speeds with frequency plot for Euler-Bernoulli beam are shown in Fig. 4.13. Both wavenumbers and speeds are plotted in the same figure. Clearly, we see that wave speeds are function of frequency

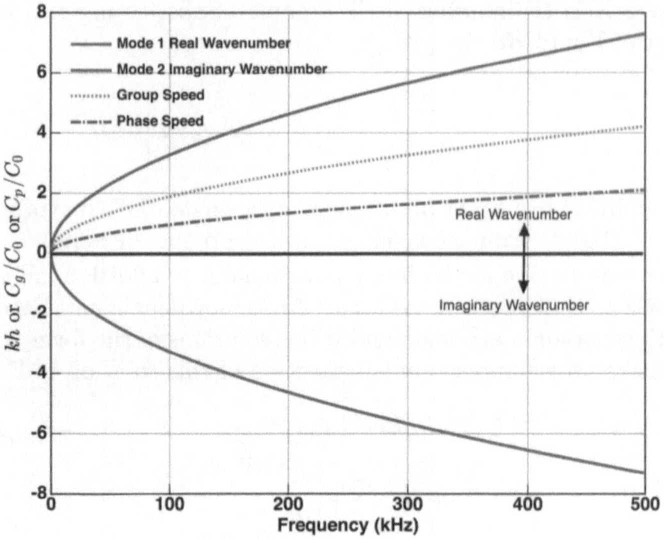

FIGURE 4.13: Spectrum and Dispersion relations for Euler-Bernoulli beam

After the wavenumbers are obtained, the solution of Eqn. (4.27) is given by

$$\hat{w}(x,\omega) = \mathbf{A}e^{-i\beta x} + \mathbf{B}e^{-\beta x} + \mathbf{C}e^{i\beta x} + \mathbf{D}e^{\beta x} \tag{4.30}$$

where **A** and **B** are forward moving modes, while **C** and **D** are the backward moving or reflected modes. Also, **A** and **C** are the forward and backward moving propagating modes, while **B** and **D** are the forward and backward moving evanescent modes.

4.3.2 Propagation of Waves in an Infinite Beam

Next, we will see how the wave propagates in an infinite beam. The main objective here is to visualize the dispersive character of the waves in beams. For simulating the infinite condition, we will omit the terms associated with **C** and **D** in Eqn. (4.30) as they represent reflected wave coefficients and in addition, assume **B**, which is the evanescent component to be zero. Only the incident wave component **A** will be acting, whose temporal variation is shown in Fig. 4.3. It is a smoothed triangular pulse with a pulse width of 50 μsec having a frequency content of near 44 kHz, which is also shown as inset in the same figure. For the purpose of simulations for all the examples now on, unless otherwise stated, following properties of aluminum is used

$$E = 70 \times 10^9\,\mathrm{N/m^2},\ \rho = 2700\,\mathrm{kg/m^3},\ \text{width} = 0.025\,\mathrm{m},\ \text{thickness} = 0.025\,\mathrm{m}$$

First, the frequency domain responses are obtained, which is then fed into inverse FFT to get the time history plots. These time responses are shown in Fig. 4.14.

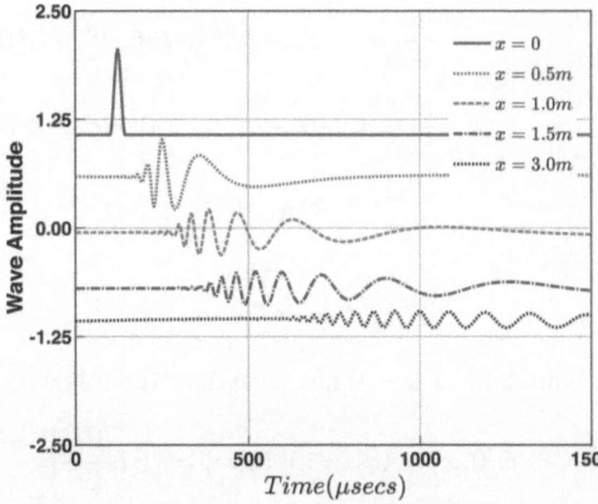

FIGURE 4.14: Wave propagation in an infinite elastic beam at different lengths

This figure clearly demonstrates the level of dispersiveness of waves in a beam as the wave propagates. The initial shape of the wave is completely lost during propagation and the response amplitude reduces with the propagation distance.

4.3.3 Reflection from Boundaries

As seen before, beam is a fourth-order system in space and hence it can accommodate four wave coefficients, which can be determined from the four boundary conditions at the two ends of the beam. The common boundaries are as follows:

- Pinned boundary
- Free boundary
- Fixed boundary

In order to determine the reflected responses for these boundaries, we need to evaluate **C** and **D** in terms of the incident pulse **A** from Eqn. (4.30). It is normally assumed that the evanescent component of incident wave **B** is equal to zero.

The displacement field and their derivatives, which are needed for determining the reflected response from different boundaries, are given by

$$\hat{w}(x,\omega) = \mathbf{A}e^{-i\beta x} + \mathbf{C}e^{i\beta x} + \mathbf{D}e^{\beta x} \tag{4.31}$$

$$\frac{d\hat{w}}{dx} = -i\beta\mathbf{A}e^{-i\beta x} + i\beta\mathbf{C}e^{i\beta x} + \beta\mathbf{D}e^{\beta x} \tag{4.32}$$

$$\frac{d^2\hat{w}}{dx^2} = -\beta^2\mathbf{A}e^{-i\beta x} - \beta^2\mathbf{C}e^{i\beta x} + \beta^2\mathbf{D}e^{\beta x} \tag{4.33}$$

$$\frac{d^3\hat{w}}{dx^3} = i\beta^3\mathbf{A}e^{-i\beta x} - i\beta^3\mathbf{C}e^{i\beta x} + \beta^3\mathbf{D}e^{\beta x} \tag{4.34}$$

$$\beta = \sqrt{\omega}\left[\frac{\rho A}{EI}\right]^{0.25}$$

The pin boundary at $x = 0$ has to satisfy the following boundary conditions

$$\hat{w}(0,\omega) = 0, \qquad \hat{M}(0,\omega) = \left| EI\frac{d^2\hat{w}}{dx^2}\right|_{x=0} = 0 \tag{4.35}$$

Using Eqns. (4.31 & 4.32) in Eqn. (4.35), we get

$$\begin{aligned} &\mathbf{A} + \mathbf{C} + \mathbf{D} = 0 \\ &-\beta^2\mathbf{A} - \beta^2\mathbf{B} + \beta^2\mathbf{D} = 0 \end{aligned} \tag{4.36}$$

Solving the above equation, we get

$$\mathbf{C} = -\mathbf{A}, \qquad \mathbf{D} = 0$$

Hence, the propagating reflected wave is same as inverted incident pulse. Also, the reflected evanescent mode does not get generated if the incident evanescent mode is zero.

Next, we will derive the reflected wave coefficients for a beam with free boundary at $x = 0$. For this case, the boundary conditions that beam has to satisfy are the following.

$$\hat{V}(0,\omega) = \left| EI\frac{d^3\hat{w}}{dx^3}\right|_{x=0} = 0, \qquad \hat{M}(0,\omega) = \left| EI\frac{d^2\hat{w}}{dx^2}\right|_{x=0} = 0 \tag{4.37}$$

where $\hat{V}$ and $\hat{M}$ are the frequency domain shear force and moment resultants, respectively. Now using Eqns. (4.33 & 4.34) in Eqn. (4.37), we get

$$-\beta^2\mathbf{A} - \beta^2\mathbf{C} + \beta^2\mathbf{D} = 0, \quad i\beta^3\mathbf{A} - i\beta^3\mathbf{C} + \beta^3\mathbf{D} = 0 \tag{4.38}$$

Solving the above equation, we get

$$\mathbf{C} = -i\mathbf{A}, \qquad \mathbf{D} = (1-i)\mathbf{A}$$

The above result indicates that even a fully propagating incident pulse generates an evanescent reflected component. The presence of complex number in the reflected coefficient suggests that there is a 90° phase shift between the incident and reflected waves and in addition, one can expect amplitude decrease when the length of propagation increases.

Next, we will derive the reflected wave response for a beam that has fixed boundary at $x = 0$. Here, the boundary conditions that the waveguide needs to satisfy are the following:

$$\hat{w}(0, \omega) = \left| \frac{d\hat{w}}{dx} \right|_{x=0} = 0 \tag{4.39}$$

Substituting for $\hat{w}$ and $d\hat{w}/dx$ from Eqns. (4.31) and (4.32) in Eqn. (4.39), we get

$$\mathbf{A} + \mathbf{C} + \mathbf{D} = 0, \qquad -i\mathbf{A} + i\mathbf{C} + \mathbf{D} = 0 \tag{4.40}$$

Solving the above equation, we get

$$\mathbf{C} = -i\mathbf{A}, \qquad \mathbf{D} = (1 - i)\mathbf{A}$$

These wave coefficients are the same as obtained for the free boundary case. A generalized model for elastic boundary is derived in the next subsection, wherein it is showed that the free and the fixed boundary cases are subset of the general elastic boundary conditions and the reflected wave responses for these cases are derived in the next subsection.

4.3.4 Reflection from Elastic Boundary

A beam with elastic restraints is shown in Fig. 4.15. These elastic restraints, represented in the form of translational spring K and rotational spring K_t, can simulate different boundary conditions. For example, if $K = K_t = \infty$, then it

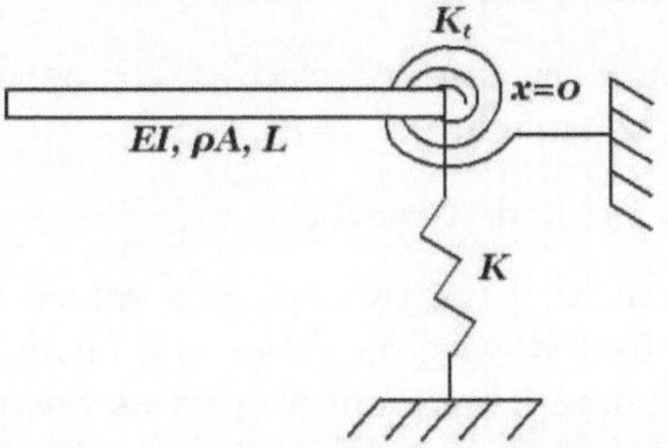

FIGURE 4.15: A beam with elastic boundary

simulates a fixity condition. On the other hand, if $K = K_t = 0$, then it will simulate a free boundary condition. Between these two extreme values, the intermediate values of K and K_t simulates intermediate boundary conditions.

It is of interest to see how the wave interacts with the elastic boundary and study the wave scattering at the boundary for different numerical values of K and K_t.

We will again begin with the solution to the governing differential equation given by Eqn. (4.30), where the aim here is to determine the coefficients **C** and **D** given that the incident pulse **A** is known and $\mathbf{B} = 0$. The boundary conditions, at $x = 0$ that the beam has to satisfy are the following (see Fig. 4.15)

$$\hat{M}(0,\omega) = \left|EI\frac{d^2\hat{w}}{dx^2}\right|_{x=0} = \left|\bar{K}_t\frac{d\hat{w}}{dx}\right|_{x=0}$$
$$\hat{V}(0,\omega) = \left|EI\frac{d^3\hat{w}}{dx^3}\right|_{x=0} = \left|-\bar{K}\hat{w}\right|_{x=0} \tag{4.41}$$

where $\bar{K} = K/(EI\beta^3)$ and $\bar{K}_t = K_t/(EI\beta)$. Substituting Eqns. (4.33 & 4.34) in the above equation, we get the following matrix equation, which requires to be solved to get the wave coefficients

$$\begin{bmatrix} (i-\bar{K}) & -(1+\bar{K}) \\ (-1+i\bar{K}_t) & (1+\bar{K}_t) \end{bmatrix} \begin{Bmatrix} \mathbf{C} \\ \mathbf{D} \end{Bmatrix} = \begin{Bmatrix} (-1+\bar{K}) \\ (1+i\bar{K}_t) \end{Bmatrix} \mathbf{A} \tag{4.42}$$

In the limit of $\bar{K} \to \infty$ and $\bar{K}_t \to \infty$, we get the wave coefficient as $\mathbf{C} = -i\mathbf{A}$ and $\mathbf{D} = (i-1)\mathbf{A}$, which are the wave coefficients for a beam with fixed boundary, which we derived earlier in the previous section. Similarly, in the limit $\bar{K} \to 0$ and $\bar{K}_t \to 0$, then we recover the wave coefficients for a beam with free boundary conditions, which are same as that of the fixed boundary case, and they are given by $\mathbf{C} = -i\mathbf{A}$ and $\mathbf{D} = (i-1)\mathbf{A}$. In order to understand the wave scattering behavior, we consider a beam with elastic boundary situated at $x = 0$, as shown in Fig. 4.15. The responses are generated for the following three cases:

- Free boundary case: $K = 0 \,\&K_t = 0$
- Fixed boundary case: $K = \infty \,\&\, K_t = \infty$ (Here, high value of stiffnesses was used to simulate $K = K_t = \infty$).
- $K = 70 \,\mathbf{x}\, 10^9$ and $K_t = 700$

That is, we have used the two extreme values and an intermediate value to simulate the reflected wave response on a beam at $x = 1.5$ m from the elastic boundary. As before, aluminum properties was used and the incident wave is the triangular pulse shown in Fig. 4.3 sampled with a sampling rate of 0.5 μsecs with 8192 FFT points. The reflected wave response is shown in Fig. 4.16.

From the figure we see that the reflected response for the fixed and the free case is same as they have same wave coefficients in the limiting cases of $K \,\&\, K_t$. For intermediate value of $K \,\&K_t$, we see that the wave amplitude does not change significantly. However, we see that there is significant phase changes.

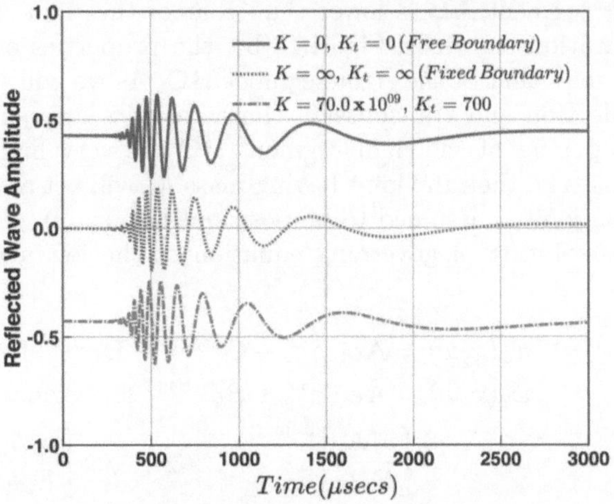

FIGURE 4.16: Reflected response at $x = 1.5$ m from the elastic boundary

4.3.5 Reflection and Transmission from a Stepped Beam and a Joint with Concentrated Mass

In this subsection, the reflection and transmission of waves from the step of a stepped beam and the beam with joint having a concentrated mass M are studied. Here, the equations developed in this section can be used to study separately the wave propagation in these two cases. The configurations of stepped beams with joint mass M are shown in Fig. 4.17. Note that the second segment extends to infinity and the joint is assumed to have a mass M.

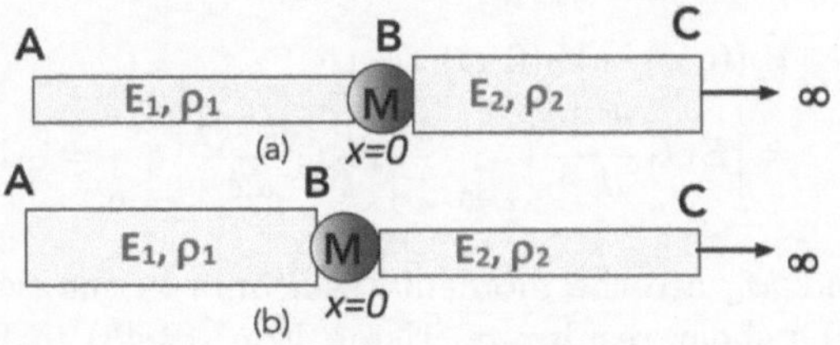

FIGURE 4.17: A stepped beam with a joint mass

In this figure, the segment **AB** has the material and sectional properties as E_1, ρ_1, and I_1, while the segment **BC** has the corresponding properties as E_2, ρ_2, and I_2. The structure is subjected to the incident transverse wave at the point **A**. Here, Fig. 4.17(a) shows a configuration where the properties of the left segment **AB** is lower than that of the right segment **BC**, while the configuration shown in Fig. 4.17(b), the properties of left segment **AB** are higher than that of the right segment **BC**. As we will see later in this section, the reflection and transmission characteristics are quite different. In fact, if the properties of the right segment **BC** are very high compared to the left segment **AB**, then the joint having mass M will act as a fixed boundary. Note that the joint is assumed to be positioned at $x = 0$.

The solution of governing equation of the left and right segments is as follows

$$\hat{w}_1(x,\omega) = \mathbf{A}e^{-i\beta_1 x} + \mathbf{C}e^{i\beta_1 x} + \mathbf{D}e^{\beta_1 x}, \text{ for segment } \mathbf{AB}$$
$$\hat{w}_2(x,\omega) = \bar{\mathbf{A}}e^{-i\beta_2 x} + \bar{\mathbf{B}}e^{-\beta_2 x}, \text{ for segment } \mathbf{BC}.$$
$$\beta_1 = \sqrt{\omega}\left[\frac{\rho_1 A_1}{E_1 I_1}\right]^{0.25}, \qquad \beta_2 = \sqrt{\omega}\left[\frac{\rho_2 A_2}{E_2 I_2}\right]^{0.25} \tag{4.43}$$

where β_1 and β_2 are the transverse wavenumbers in two segments. The incident evanescent wave **B** is assumed zero for the left segment, while for the right segment, evanescent transmitted wave $\bar{\mathbf{B}}$ will get generated and hence included in the analysis. In order to determine the reflected and transmitted responses, we need to determine **C**, **D**, $\bar{\mathbf{A}}$ and $\bar{\mathbf{B}}$ in terms of **A**, which means we need four conditions at the interface or the joint (assumed to be at $x = 0$), which are given by

$$\hat{w}_1(0,\omega) = \hat{w}_2(0,\omega), \tag{4.44}$$
$$\theta_1(0,\omega) = \theta_2(0,\omega) = \left|\frac{d\hat{w}_1}{dx}\right|_{x=0} = \left|\frac{d\hat{w}_2}{dx}\right|_{x=0} \tag{4.45}$$
$$\hat{M}_1(0,\omega) - \hat{M}_2(0,\omega) = \left|E_1I_1\frac{d^2\hat{w}_1}{dx^2}\right|_{x=0} - \left|E_2I_2\frac{d^2\hat{w}_2}{dx^2}\right|_{x=0} = 0 \tag{4.46}$$
$$\hat{V}_1(0,\omega) - \hat{V}_2(0,\omega) = \left|E_1I_1\frac{d^3\hat{w}_1}{dx^3}\right|_{x=0} - \left|E_2I_2\frac{d^3\hat{w}_2}{dx^3}\right|_{x=0} = -M\omega^2\hat{w}(0,\omega) \tag{4.47}$$

where $\hat{M}_1$ and $\hat{M}_2$ are the moment resultants in the two segments, while $\hat{V}_1$ and $\hat{V}_2$ are the shear resultants. Using Eqn. (4.43) in Eqns. (4.45–4.47), we can write the system of equation in matrix form as

$$\begin{bmatrix} 1 & 1 & -1 & 1 \\ i & i & i\alpha & \alpha \\ -1 & 1 & S\alpha^2 & -S\alpha^2 \\ -i & 1 & (\bar{M} - iS\alpha^3) & (\bar{M} + S\alpha^3) \end{bmatrix} \begin{Bmatrix} \mathbf{C} \\ \mathbf{D} \\ \bar{\mathbf{A}} \\ \bar{\mathbf{B}} \end{Bmatrix} = \begin{Bmatrix} -1 \\ i \\ 1 \\ -i \end{Bmatrix} \mathbf{A} \tag{4.48}$$

where $M^* = M\omega^2/(E_1I_1\beta_1^3)$, $\alpha = \beta_2/\beta_1$ and $S = E_2I_2/E_1I_1$. In the above generalized matrix equation, substituting $M^* = 0$ will yield equations for the analysis of the stepped beam. The analytical or closed form solution is a very involved process, and it is preferable to solve this equation numerically. We have used *roots* function in MATLAB to solve for the wave coeffcients. The solution of the above equations will give the reflected coefficients for segment **AB** (**C** and **D**) and transmitted coefficient for right segment **BC** ($\bar{\mathbf{A}}$ and $\bar{\mathbf{B}}$).

We will first simulate the reflected and transmitted wave for a stepped beam. That is, for the simulation, we substitute $M^* = 0$ in Eqn. (4.48). For the simulation, we will analyze both the configurations shown in Figs. 4.17 (a) and (b) and show the difference in the reflected and transmitted wave characteristics. For this purpose, we will again use a triangular pulse shown in Fig. 4.3 as the input wave given by wave coefficient **A**, where the pulse is transformed to the frequency domain by sampling the signal with a sampling rate of 0.5 μsecs and 8192 FFT points. The response in all the cases is measured at 1.0 m from the joint. First, we will plot the reflected and transmitted wave for the case shown in Fig. 4.17(a). If $S = 1$, and $I = 1$, then there is no step and all the wave will be transmitted to infinity. Here, the simulation is performed for three cases, namely, $S = 9$, $I = 1$, $S = 8$, $I = 2$ and $S = 9$, $I = 3$, respectively. The reflected and transmitted wave responses are shown in Figs. 4.18(a) and (b).

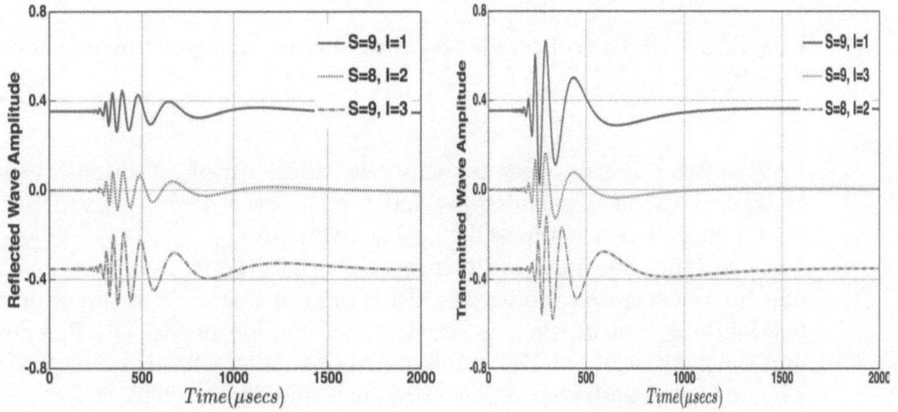

(a) Reflected Wave Response (b) Transmitted Wave Response

FIGURE 4.18: Reflection and transmission for a stepped beam shown in Fig. 4.17(a)

Both the reflected and transmitted responses are plotted to the same scale so that the level of reflection or transmission can be visually estimated. From the figure, we can see that, for all the three cases, there is more transmission than reflection, and the reflection amplitude increases with increase in the value of S and I.

Next, the responses are plotted for the configuration shown in Fig. 4.17(b) and these are shown in Fig. 4.19. Here, the responses are plotted for three different values of parameters S and I, namely $S = 1/9$, $I = 1$, $S = 1/8$, $I = 1/2$ and $S = 1/9$, $I = 1/3$.

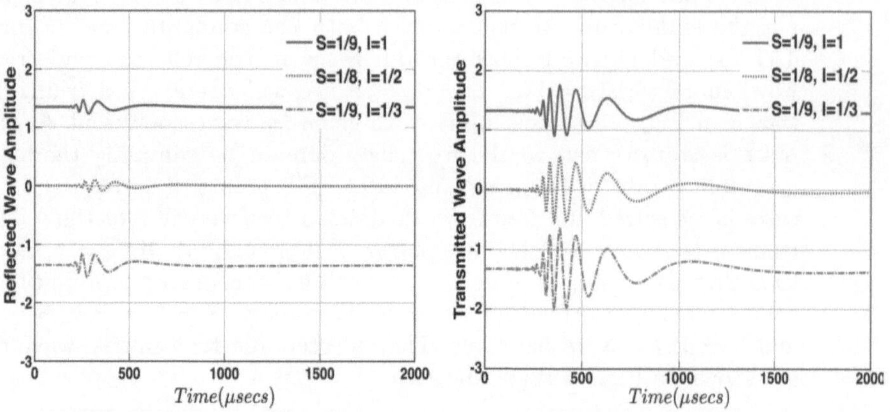

(a) Reflected Wave Response (b) Transmitted Wave Response

FIGURE 4.19: Reflection and transmission for a stepped beam shown in Fig. 4.17(b)

From the figure we see for this case, the level of reflections is very small compared to the transmissions and for the case of $S = 1/9$, $I = 1/3$, the reflection is higher compared to other two cases.

Next, the reflection and transmission characteristics of waves are studied due to concentrated joint mass. To bring out the effect of mass on wave reflection and transmission, we assume that the properties of left segment **AB** and the right segment **BC** are assumed the same. That is we have $S = I = 1$. The reflected and transmitted response due to the joint mass is shown in Fig. 4.20.

From Fig. 4.20, we see that for very small value of mass, the transmitted response is higher compared to reflected response. However, as the value of the joint mass increases, the reflected and the transmission response amplitude does not change significantly. In other words, contrary to the intuition, increase in joint mass does not lead to increase in the reflected response. However, the mass is seen to change the phase of the response, indicating the dependence of frequency ω on the reflected and transmitted responses.

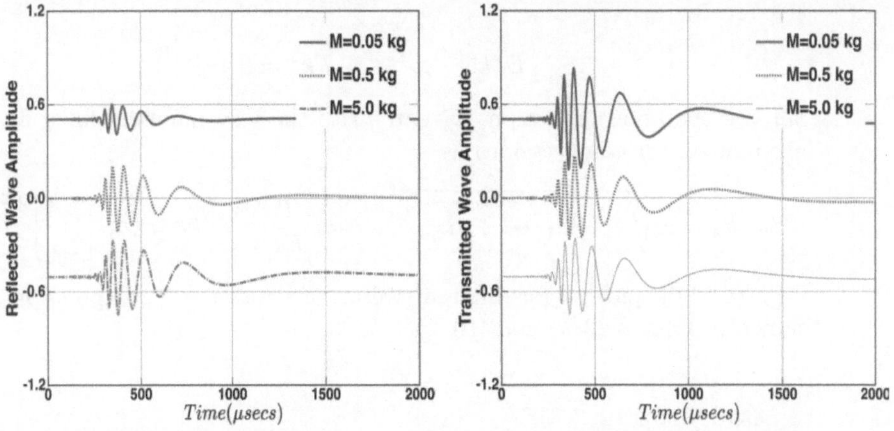

(a) Reflected Wave Response

(b) Transmitted Wave Response

FIGURE 4.20: Reflection and transmission for a beam with concentrated joint mass

4.3.6 Wave Propagation in Beams with Pre-Tension or Pre-Compression

In this section, we will study the effect of pre-tension or pre-compression on the wave propagation characteristics of a beam. Fig. 4.21 shows a beam with pre-tension load with pretension being applied at $x = 0$.

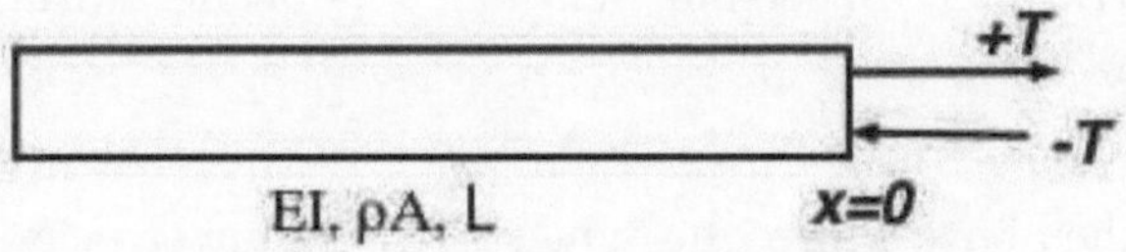

FIGURE 4.21: A beam with pre-tension

The governing differential equation for a beam subjected to pretension (or compression) in the absence of external forces is given by

$$EI\frac{\partial^4 w}{\partial x^4} - T\frac{\partial^2 w}{\partial x^2} = \rho A\frac{\partial^2 w}{\partial t^2} \tag{4.49}$$

where T is the pre-tension. If the beam is loaded with pre-compressive load, the variable T in the above equation should be replaced by $-T$. Using the spectral form of solution of the form $w(x,t) = \sum \hat{w} e^{i(\omega t - kx)}$ in Eqn. (4.49), we get the characteristic equation for determination of wavenumbers, which is given by

$$EIk^4 - \rho A\omega^2 + Tk^2 = 0 \tag{4.50}$$

Solving the above equation, we can write the wavenumbers for a beam with pre-tension (or compression) as

$$k_1,\, k_2 = \pm\sqrt{-\alpha \pm \sqrt{\alpha^2 + 4k_b^4}}, \qquad \alpha = \frac{T}{EI}, \qquad k_b = \sqrt{\omega}\left[\frac{\rho A}{EI}\right]^{0.25} \tag{4.51}$$

Eqn. (4.51) is plotted for different values of tension and compression and it is shown in Figs. 4.22(a) and (b).

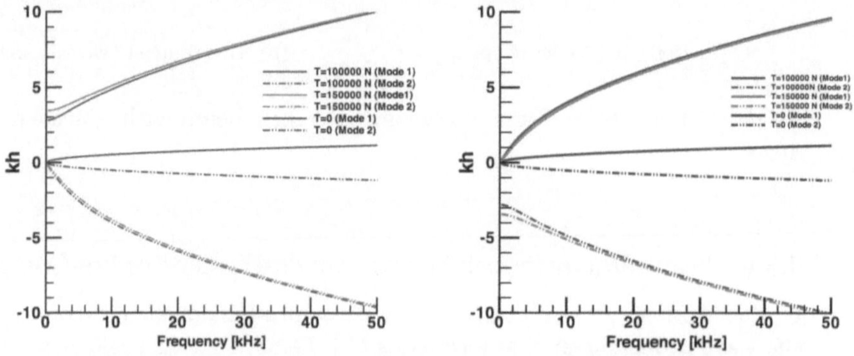

(a) Pre-compression case

(b) Pre-tension case

FIGURE 4.22: Spectrum relation for beams with pre-tension or pre-compression

In the above figure, the values of wavenumber below the zero line are the imaginary values, which means that these modes do not propagate and hence called *evanescent* modes. Fig. 4.22(a) shows the wavenumber variation for different values of T in compression mode. As the value of T increases, the real or propagating part of the wavenumber shifts away from zero, while the imaginary part hardly varies with T and they always start at zero. Shifting of real part of the wavenumber away from zero will cause the wave speeds to increase, which will result in repeated reflections, especially in the case of short waveguides causing failure. Fig. 4.22(b) shows the wavenumber variation for a beam subjected to pre-tension T. The behavior here is quite opposite to the behavior of the pre-compression case. That is, the real part of the wavenumber does not change significantly with the increase in T and follows the normal beam, while the imaginary part shifts away from zero at the lower frequency implying that the waves are going to be damped more faster in the case of pre-tensioned beam as apposed to normal beam.

Next, we will look at the wave speeds. The phase speed and group speed, given by definitions $C_p = \omega/k$ and $C_g = d\omega/dk$, where these two variables are derived from Eqn. (4.51), are shown in Fig. 4.23. The expression of group speed derived from Eqn. (4.51) is given by

$$C_g = \frac{d\omega}{dk} = Real\left(EI\frac{(2k^3+\alpha k)}{\omega(\rho A)}\right), \qquad C_p = Real\left(\frac{\omega}{k}\right) \tag{4.52}$$

where appropriate values of k are substituted to get the speeds of the required mode

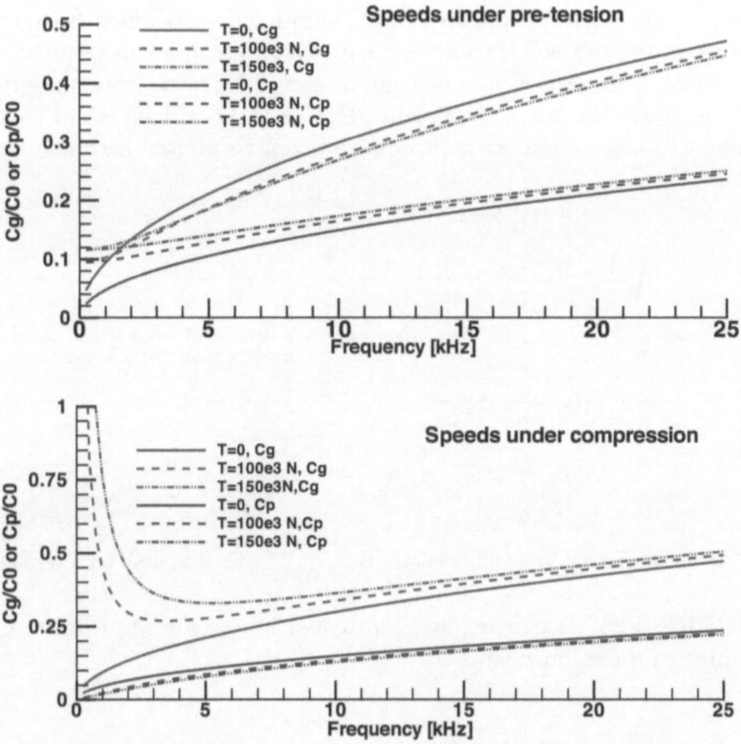

FIGURE 4.23: Phase speed and group speed variation for a beam with pre-tension or compression

The top figure in Fig. 4.23 shows the phase and group speeds under pre-tension. When $T = 0$, it follows the elementary beam values. As T increases, we see that the value of the group speed slightly decreases, while the phase speed increases marginally. This is expected since the wavenumber plot shows increased damping component in the second mode, which contributes to the decreased speed. The bottom figure shows the dispersion plots for pre-compression case. In this case, the phase speeds do not show much change compared to the zero compression case. The group speeds are very high at low frequency, which can cause the propagating mode to travel very fast and this will result in repeated reflections in a short waveguide causing instability.

Next, we study the wave propagation responses in a pre-tensioned (or compression) beam. That is, we first write the solutions to the governing equation given by Eqn. (4.49), which is given by

$$\hat{w}(x,\omega) = \mathbf{A}e^{-ik_1x} + \mathbf{B}e^{-ik_2x} + \mathbf{C}e^{ik_1x} + \mathbf{D}e^{ik_2x} \tag{4.53}$$

where the wavenumbers k_1 and k_2 are determined from Eqn. (4.51). We will see how the wave propagates in an infinite pre-tensioned (or compressioned) beam that is subjected to a pre-compression (or pre-tension)at $x = 0$ shown in Fig. 4.21. The response is measured at $x = 0.375$ m from the input location. In this case, the wave coefficients $\mathbf{B}, \mathbf{C}$ and $\mathbf{D}$ will be equal to zero. All the responses are compared with infinite beam responses without pre-tension.

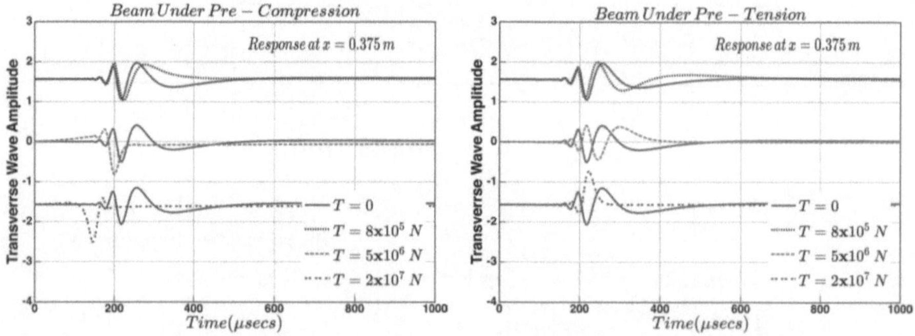

(a) Response for Pre-compression case

(b) Response for Pre-tension case

FIGURE 4.24: Transverse wave propagation response on infinite pre-tension or pre-compression beams

From Fig. 4.24(a), we see that as the value of compression increases, the level of dispersion decreases. For very high value of T, we see substantial increase in the group speeds, which translates into early arrival of incident wave. In addition, we see substantial change in the phase of the response compared to the normal beam.

Next, we will look at the response of a beam subjected to different pre-tension. Even in this case, we see that the beam seems to loose its dispersiveness for higher value of T. In addition, pre-tension introduces substantial changes in the phase of the response compared to normal beam.

4.3.7 Wave Propagation in a Beam on Elastic Foundation

A beam resting on elastic foundation with foundation modulus K is shown in Fig. 4.25. The governing equation that dictates the motion of this beam is given by

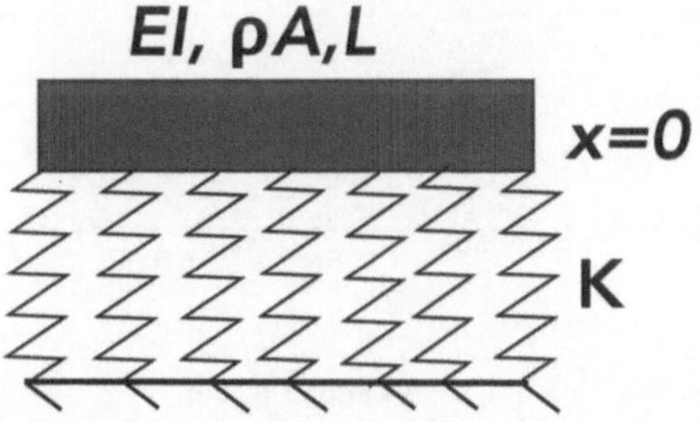

FIGURE 4.25: A beam resting on elastic foundation

$$EI\frac{\partial^4 w}{\partial x^4} + Kw + \rho A\frac{\partial^2 w}{\partial t^2} = 0 \tag{4.54}$$

where $w(x,t)$ is the transverse deformation, K is the foundation modulus, EI and ρA are the flexural rigidity and mass per unit length of the beam, respectively. Transforming this equation to the frequency domain using the spectral transformation $w(x,t) = \sum \hat{w}(x,\omega)e^{i(kx-\omega t)}$, we get the characteristic equation for computation of wavenumbers, which is given by

$$EIk^4 - \rho A\omega^2 + K = k^4 + \frac{K - \rho A\omega^2}{EI} = 0 \tag{4.55}$$

Looking at the above equation, we see that the wavenumber is zero at the frequency

$$\omega_c = \sqrt{\frac{K}{EI}} \tag{4.56}$$

Here, ω_c is called the *translational* or *cut-off frequency* at which the wavenumber changes its wave behavior. At this frequency, either the propagating mode will become evanescent or the evanescent can become propagating. The expression for two wavenumbers is given by

$$k_1 = \pm\left[\frac{\rho A}{EI}\omega^2 - \frac{K}{EI}\right]^{\frac{1}{4}}, \qquad k_2 = \pm i\left[\frac{\rho A}{EI}\omega^2 - \frac{K}{EI}\right]^{\frac{1}{4}} \tag{4.57}$$

The behavior of wavenumber with frequency is shown in Fig. 4.26.

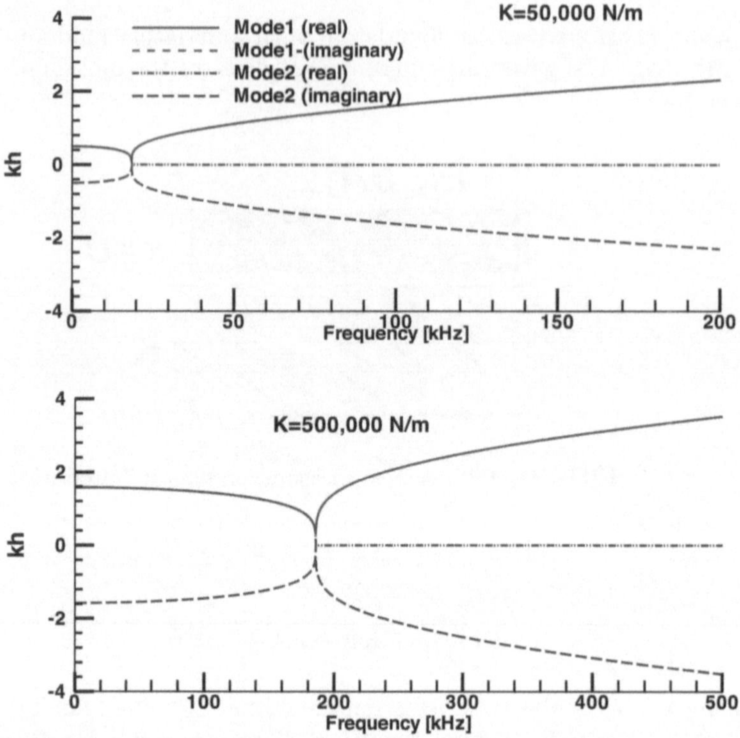

FIGURE 4.26: Spectrum relation for a beam resting on elastic foundation

From the figure, we see that at the cut-off frequency, propagating mode becomes evanescent and evanescent mode becomes propagating. In addition, the cut-off frequency can be shifted by increasing the foundation modulus. Note that we need very high value of foundation modulus to induce this behavior. However, at this value of foundation modulus, the structure will be very stiff and hence wave speeds will attain zero value very quickly. Hence, wave speeds have no meaning and hence not plotted here.

4.3.8 Wave Propagation in a Framed Structure

One can construct a frame structure if two or more beam waveguides that are arbitrarily oriented are combined to make a single structure. An angled joint is an example of a simple frame structure. It is shown in Fig. 4.27(a).

Unlike the beam or rod motions, wherein the bending and axial motions are uncoupled, the framed structure exhibits *bending-axial coupling.* That is, an axial excitation creates a bending motion and vice versa in all connected segments. In wave propagation terminology bending-axial coupling creates

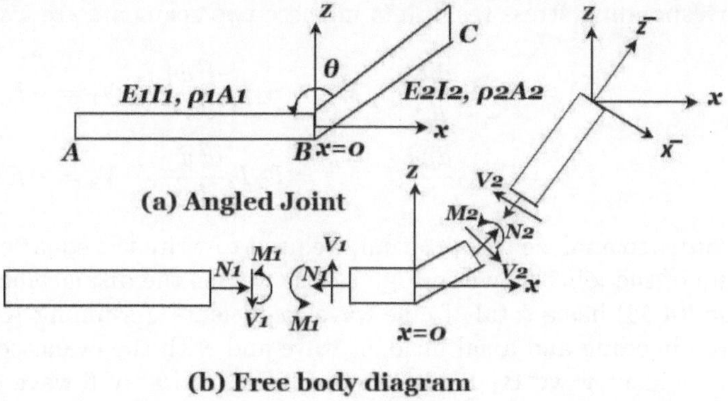

FIGURE 4.27: Framed structure: (a) A simple angled joint and (b) Free body diagram

mode conversion. That is, a bending wave creates an axial wave for non-axial loading. We will now explain the methodology to analyze the wave behavior in an angled joint.

The joint shown in Fig. 4.27(a) has two segments **AB** and **BC**. Both are beam segments. Segment **AB** has the Young's modulus E_1, area moment of inertia I_1, density ρ_1 and area of cross section A_1. The corresponding properties for segment **BC** are E_2, I_2, ρ_2 and A_2. Note that the Segment **BC** is assumed to extend to infinity. Both follow the same governing differential equations pertaining to isotropic rod and beam. The frequency domain solution of wave equation in segments **AB** and **BC** is given by

$$\begin{aligned}
\hat{w}_1(x,\omega) = &\quad \mathbf{A}_1 e^{-ik_1x} + \mathbf{B}_1 e^{-k_1x} \\
&\quad + \mathbf{C}_1 e^{ik_1x} + \mathbf{D}_1 e^{k_1x} \Rightarrow For AB. \\
\hat{w}_2(\bar{x},\omega) = &\quad \mathbf{A}_2 e^{-ik_2\bar{x}} + \mathbf{B}_2 e^{-k_2\bar{x}} \quad \Rightarrow For BC \\
\hat{u}_1(x,\omega) = &\quad \mathbf{P}_1 e^{-ik_{l1}x} + \mathbf{Q}_1 e^{ik_{l1}x} \quad \Rightarrow For AB. \\
\hat{u}_2(\bar{x},\omega) = &\quad \mathbf{P}_2 e^{-ik_{l2}\bar{x}} \quad \Rightarrow For BC
\end{aligned} \tag{4.58}$$

where $\hat{w}_1$ and $\hat{w}_2$ are the frequency domain transverse displacements in segments **AB** and **BC**, while $\hat{u}_1$ and $\hat{u}_2$ are the corresponding axial displacements and x and $\bar{x}$ are the respective coordinate axis along the member axis as shown in Fig. 4.27(b). In addition, k_1 and k_2 are the bending wavenumbers in these two segments, and they are given by $k_1 = \sqrt{\omega}[\rho_1 A_1/E_1 I_1]^{0.25}$ and $k_2 = \sqrt{\omega}[\rho_2 A_2/E_2 I_2]^{0.25}$, while the axial wavenumbers k_{l1} and k_{l2} are given by $k_{l1} = \omega\sqrt{\rho_1 A_1/E_1 A_1}$ and $k_{l2} = \omega\sqrt{\rho_2 A_2/E_2 A_2}$. The corresponding stress resultants in these two segments are given by

$$\hat{N}_1 = E_1 A_1 \frac{d\hat{u}_1}{dx}, \quad \hat{M}_1 = E_1 I_1 \frac{d^2\hat{w}_1}{dx^2}, \quad \hat{V}_1 = -E_1 I_1 \frac{d^3\hat{w}_1}{dx^3}.$$

$$\hat{N}_2 = E_2 A_2 \frac{d\hat{u}_2}{d\bar{x}}, \quad \hat{M}_2 = E_2 I_2 \frac{d^2\hat{w}_1}{d\bar{x}^2}, \quad \hat{V}_2 = -E_2 I_2 \frac{d^3\hat{w}_1}{d\bar{x}^3} \tag{4.59}$$

In order to analyze this problem, we need to write the equations of the equilibrium of the joint shown in Fig. 4.27 as well as the displacement compatibility. Eqn. (4.58) has a total of nine wave coefficients. Assuming $\mathbf{A_1}$ and $\mathbf{P_1}$ are the given bending and axial incident wave and with the evanescent component of the incident wave $\mathbf{B_1}$ being zero, we have a total of 6 wave coefficients to be determined, which requires 6 joint conditions for its determination. Three of these joint conditions come from joint equilibrium conditions, while the other three conditions come from compatibility of the deformations. The joint equilibrium conditions are obtained by resolving the forces in the two segments in the global $x - y$ coordinates and applying Newton's II law, these equations are given by

$$\begin{aligned} -N_1 + N_2 \cos\theta - V_2 \sin\theta &= M\ddot{u}_1 \\ -V_1 + N_2 \sin\theta + V_2 \cos\theta &= M\ddot{w}_1 \\ -M_1 + M_2 &= I\ddot{\phi} \end{aligned} \tag{4.60}$$

where $\phi = dw_1/dx$ is the slope of the beam connecting at the joint. These equations in the frequency domain become

$$\begin{aligned} -\hat{N}_1 + \hat{N}_2 \cos\theta - \hat{V}_2 \sin\theta + M\omega^2\hat{u}_1 &= 0. \\ -\hat{V}_1 + \hat{N}_2 \sin\theta + \hat{V}_2 \cos\theta + M\omega^2\hat{w}_1 &= 0. \\ -\hat{M}_1 + \hat{M}_2 + I\omega^2 \frac{d\hat{w}_1}{dx} &= 0 \end{aligned} \tag{4.61}$$

Next, we write the displacement compatibility at the joint in frequency domain as

$$\hat{u}_1 = \hat{u}_2 \cos\theta - \hat{w}_2 \sin\theta, \; \hat{w}_1 = \hat{u}_2 \sin\theta + \hat{w}_2 \cos\theta, \; \frac{d\hat{w}_1}{dx} = \frac{d\hat{w}_2}{dx} = \hat{\phi} \tag{4.62}$$

Assuming both segments **AB** and **BC** are of same material and cross section, and the joint has zero mass and rotational inertia, then we have $E_1I_1 = E_2I_2 = EI$, $\rho_1 A_1 = \rho_2 A_2 = \rho A$, $k_1 = k_2 = k$ and $k_{l1} = k_{l2} = k_l$. Substituting Eqn. (4.58) in Eqn. (4.59), which is in turn used in Eqns. (4.61) and (4.62) and simplifying, we can write the resulting equations in matrix form as

$$\begin{bmatrix} 1 & 1 & -\cos\theta & -\cos\theta & 0 & -\sin\theta \\ i & 1 & i & 1 & 0 & 0 \\ 0 & 0 & \sin\theta & \sin\theta & 1 & -\cos\theta \\ -K & K & -iK\cos\theta & K\cos\theta & 0 & -i\sin\theta \\ 0 & 0 & iK\sin\theta & -K\sin\theta & -i & -i\cos\theta \\ 1 & -1 & -1 & 1 & 0 & 0 \end{bmatrix} \begin{Bmatrix} \mathbf{C}_1 \\ \mathbf{A}_2 \\ \mathbf{D}_1 \\ \mathbf{Q}_1 \\ \mathbf{B}_2 \\ \mathbf{P}_2 \end{Bmatrix}$$

$$= \begin{Bmatrix} -1 \\ i \\ 0 \\ -iK \\ 0 \\ -1 \end{Bmatrix} \mathbf{A}_1 - \begin{Bmatrix} 0 \\ 0 \\ -1 \\ 0 \\ -i \\ 0 \end{Bmatrix} \mathbf{P}_1 \tag{4.63}$$

where $K = k^3 EI/k_l EA$ with k being the bending wavenumber and k_l being the axial wavenumber. Eqn. (4.63) is solved in MATLAB using *lsolve* function and the code to solve the above equation is provided with this book. Solving Eqn. (4.63), we can obtain the reflected and transmitted coefficients, from which the respective responses can be extracted to a given input. In addition, will can also demonstrate the existence of mode coupling in such structures. First, the angled joint is subjected to axial loading in the member **AB**, where the angled joint is located at a distance of 3.0 m. In the first study, we will load the member **AB** first axially and then transversely and compute the reflection coefficient (RC), wherein the RC for axial and flexural loadings are defined as

$$RC_{axial} = \frac{\mathbf{P_2} e^{ik_l(x-L)}}{\mathbf{P_1} e^{-ik_l x}}, \qquad RC_{transverse} = \frac{\mathbf{C_1} e^{ik(x-L)}}{\mathbf{A_1} e^{-ikx}} \tag{4.64}$$

where L is the distance between the loading point and the joint, the positive direction of x is shown in Fig. 4.27. Here, $\mathbf{P_1}$ and $\mathbf{A_1}$ are the incident triangular pulse, as shown in Fig. 4.3 and $\mathbf{P_2}$ and $\mathbf{C_1}$ are the wave coefficient obtained by solving Eqn. (4.63). Here, aluminum properties are assumed in the simulation for both segments of angled joint. The reflected coefficient in this case is computed at a distance of 1.0 m from the joint and plotted as a function of frequency and the joint angle θ in Fig. 4.28(a).

From the figure we see that most reflection happens at lower frequency and the reflection decreases with increase in frequency. In addition, we see that for higher joint angle θ, the reflection coefficient is also high. This is to be expected since for lower joint angle, substantial energy of the response will get transmitted to the member **BC** leaving the small amount of reflection in member **AB**. If the joint is assumed to have negligible mass and stiffness (as in the present case) and with $\theta = 0°$ (that is, both members **AB** and **BC**

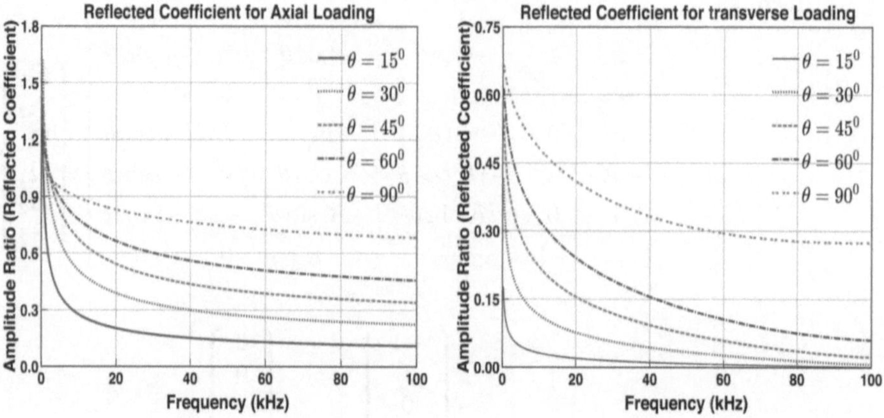

(a) Axial Loading (b) Transverse Loading

FIGURE 4.28: Reflection coefficient for an angled joint

being collinear), we can expect all the energy coming from member **AB** to be transmitted to member **BC**.

Next, we will load the member **AB** transversely at the same point for the same configuration of the angled joint. The reflected coefficient is computed at a distance of 1.0 m from the joint as before, where the transverse reflected coefficient is given by Eqn. (4.65). The transverse reflected coefficient is shown in Fig. 4.28(b). The transverse reflected coefficient follows the same pattern as the case of reflected coefficient for axial loading. However, the magnitude of the reflected coefficient is significantly less compared to the axial load case

Next, we will see how much of energy gets transmitted to the member **BC** due to the incident wave coming from the member **AB** of the angled joint. Here, again the joint is assumed to be located at 3.0 m from the impact location and the transmitted coefficient is computed at a point located at 2.0 m from the joint in member **BC** for different angle θ. The expression for transmitted coefficient (TC) computed by loading the member **AB** both axially and transversely is given by

$$TC_{axial} = \frac{\mathbf{Q_2} e^{-ik_l x}}{\mathbf{P_1} e^{-ik_l x}}, \qquad TC_{transverse} = \frac{\mathbf{A_2} e^{-ikx}}{\mathbf{A_1} e^{-ikx}} \tag{4.65}$$

The transmitted coefficient for both axial and transverse loading is shown in Fig. 4.29.

From the figure we see that the frequency seems to have pronounced effect on the behavior of the joint. In addition, we also see that higher frequencies does not seem to have that significant effect on the transmission for lower angled joint. When the joint is $\theta = 90°$, the joint seems to have more transmission at higher frequencies.

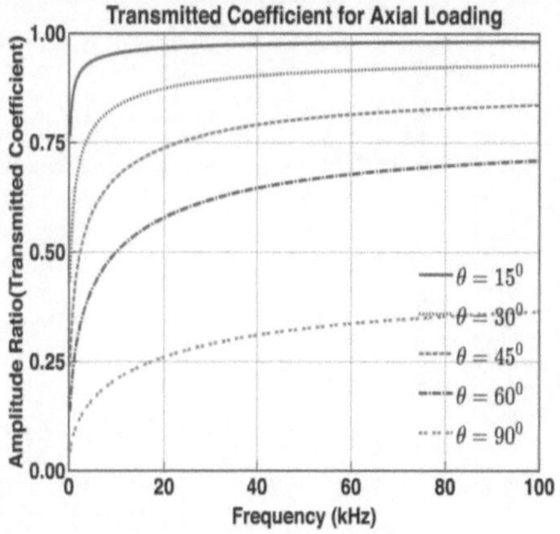

(a) Axial Loading

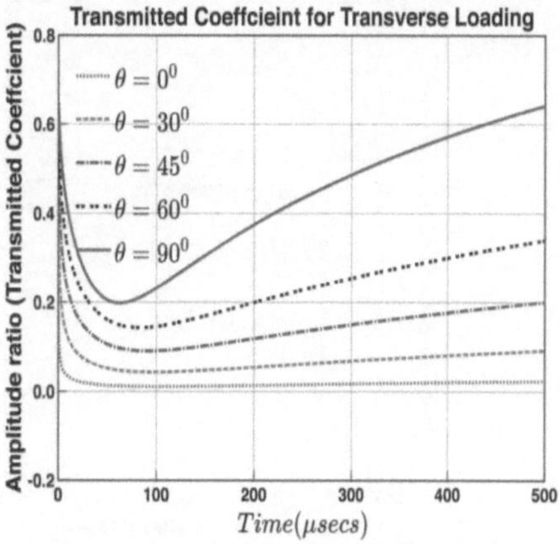

(b) Transverse Loading

FIGURE 4.29: Transmission coefficient for an angled joint

Next, we will study the response amplitudes in the time domain. As mentioned earlier, the angled joint introduces mode coupling, that is, an axial loading will cause axial as well as transverse displacements in both members. Similarly, a transverse loading cause both displacements in both the members. In order to see the extent of these responses, we first load the angled joint shown in Fig. 4.27 axially and compute the response at the loaded point. The joint is assumed to be located 3.0 m from the loading point. The reflected axial response is plotted in Fig. 4.30(a). Next, the member **AB** is loaded transversely at the same point and the reflected axial response is plotted due to this transverse loading at the same point. This response is shown in Fig. 4.30(b). Both these responses are plotted on a same scale to see the differences in responses. It is seen that the magnitude of the reflected axial response due to transverse response is about 60% lower and the magnitude of the reflected response is maximum in both cases when the joint angle θ is 90°.

In the next study, we will study the reflected transverse response to both axial and transverse loading. This is shown in Figs. 4.31(a) and (b), respectively.

In order to bring out the dispersiveness of the transverse waves, the response is plotted at 1.0 from the loading point. Some interesting features can be observed from these figures. As in the case of reflected axial response, both these responses are plotted on a same scale. The waves can be seen to be highly dispersive. In addition, from Fig. 4.31(b), we see that the reflected response for transverse loading increases with increase in the joint angle θ, without much change in the phase. However, from Fig. 4.31(a), we see that for the joint angles $\theta = 15°$ and $\theta = 30°$, the reflected amplitude is much higher than its counterpart in Fig. 4.31(a). Further, when the joint angle is

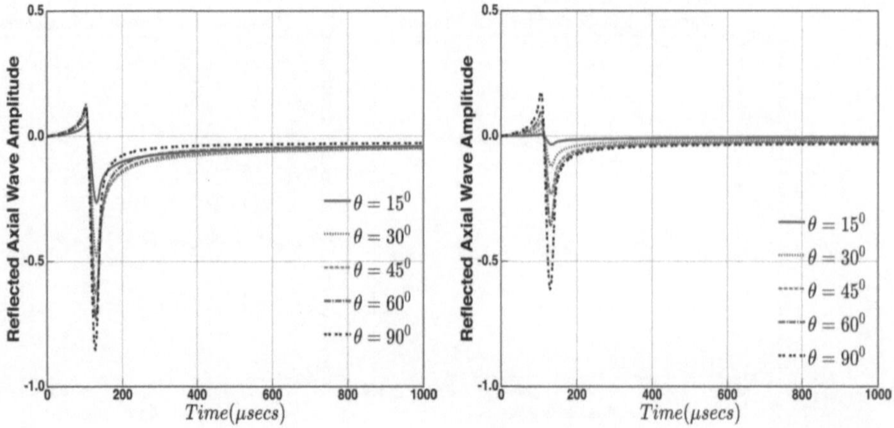

(a) Reflected axial response due to axial loading (b) Reflected axial response due to transverse loading

FIGURE 4.30: Reflected axial response for an angled joint

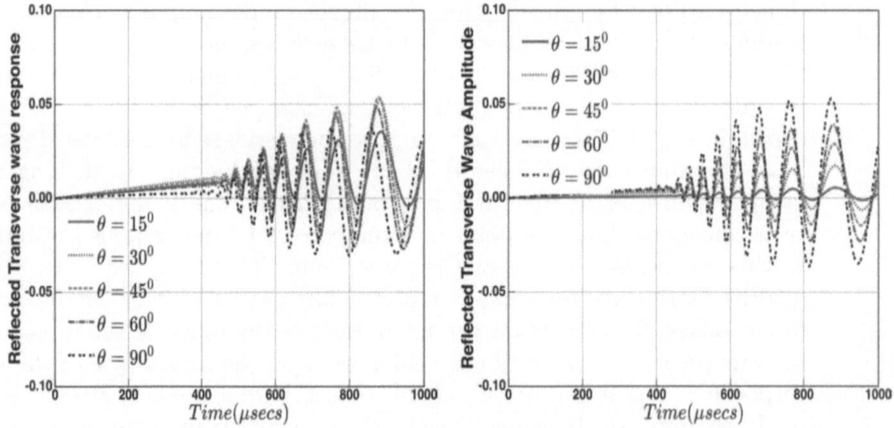

(a) Reflected transverse response due to axial loading (b) Reflected transverse response due to transverse loading

FIGURE 4.31: Reflected transverse response for an angled joint

increased, the response amplitude stagnates to the highest value at $\theta = 60°$. Further increase in the joint angle marginally reduces the response amplitude. In addition, we see that the joint angle θ changes the phase of the response indicating the frequency effect that the joint introduces in the responses. This behavior is unlike the behavior we have in Figs. 4.30(a) and (b).

This section demonstrates the ease with which spectral analysis can be employed to perform wave propagation analysis in multiply connected structures.

4.4 WAVE PROPAGATION IN HIGHER-ORDER WAVEGUIDES

Waveguide structural models are derived from 3-D elasticity equations after making certain assumptions on kinematics. For example, for deriving the elementary beam models, it is assumed that traverse displacement is always the function of only one spatial coordinate, namely the axial coordinate x and the plane sections remain plane before and after bending. Implication of this assumption is that the neutral plane is always the mid-plane and in addition, the line drawn through the cross section will always be normal to the cross section during the process of undergoing bending. Similarly, in elementary longitudinal waveguide, it is assumed that the axial deformation is a function of only the axial coordinate and the waveguide undergoes negligible lateral deformation. Some of these assumptions are relaxed in higher-order waveguides through addition of additional motions that more realistically captures the two-dimensional behavior in one-dimensional models. In this section, the wave propagation behavior of higher-order flexural waveguide, namely the *Timoshenko beam waveguide* and higher-order longitudinal waveguide normally referred to as *Mindlin-Herrmann rod waveguide*, is presented. As before, we first derive the governing equations using Hamilton principle, perform spectral analysis and obtain the wavenumbers and wave speeds. One of the characteristic of higher-order waveguides is the presence of second frequency spectrum and the presence of this mode is demonstrated for Mindlin-Herrmann rod waveguide.

4.4.1 Wave Propagation in Timoshenko Beam

This theory is also referred to in the literature as *First-Order Shear Deformation Theory* (FSDT), and it was proposed by one of the doyens in the area of mechanics, namely Prof. Stephen Timoshenko in early twentieth century [115, 135]. According to this theory, unlike the elementary theory, the shear stress also contributes to the total transverse deformation and this assumption relaxes the need for plane section remaining plane before and after bending and as a result, the beams slopes are not derived from transverse deformation (as was done in the Euler-Bernoulli beam theory), and they are an independent motion and is smaller than the derivative dw/dx, which is the slope in the elementary theory, as shown in Fig. 4.32.

In elementary beam, although shear force at any cross section exist, shear strains are zero. This anomaly is rectified by introducing slope of the beam θ as an independent motion, which introduces a shear strain that contributes to the total deformation. It is for this reason, the deformation field for this

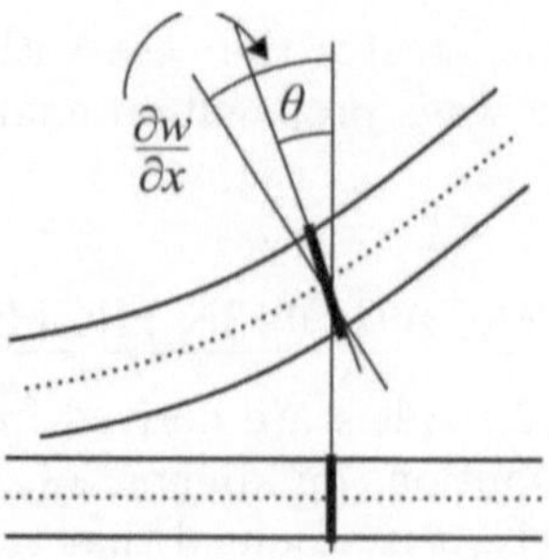

FIGURE 4.32: Timoshenko beam theory: Deformation behavior

theory is given by

$$u(x, z, t) = -z\theta(x, t), \qquad w(x, z, t) = w(x, t) \tag{4.66}$$

The corresponding strains would become

$$\varepsilon_{xx} = -z\frac{\partial \theta}{\partial x}, \qquad \varepsilon_{zz} = 0, \qquad \gamma_{xy} = \frac{\partial w}{\partial x} - \theta \tag{4.67}$$

The relevant stresses are given by

$$\sigma_{xx} = -Ez\frac{\partial \theta}{\partial x}, \qquad \sigma_{zz} = 0, \qquad \tau_{xy} = G\left(\frac{\partial w}{\partial x} - \theta\right) \tag{4.68}$$

where E and G are the Young's modulus and Shear modulus of the material. Using the expressions of the stresses and strains, we can express the strain energy U as

$$\begin{aligned} U &= \frac{1}{2}\int_V [\sigma_{xx}\varepsilon_{xx} + \tau_{xy}\gamma_{xy}]\, dV \\ &= \frac{1}{2}\int_0^L \int_A \left[-Ez^2\left(\frac{\partial \theta}{\partial x}\right)^2 + G\left(\frac{\partial w}{\partial x} - \theta\right)^2\right] dxdA \end{aligned} \tag{4.69}$$

where V and A are the volume and area of cross section of the beam, respectively. The above equation is obtained by substituting Eqns. (4.67) & (4.68) in the strain energy expression. Noting that $\int_A z^2 dA = I$, the area moment of inertia of the cross section of the beam, the first term in Eqn. (4.69) can be written as $1/2\int_0^L EI(\partial\theta/\partial x)^2 dx$. We will look into detail of the second term of this equation. It has to do with the shear stress variation. Since the terms here are constant across every cross section, we can write the second term as $\int_0^L GA(\partial w/\partial x - \theta)^2 dx$.

As mentioned earlier, the elementary or Euler-Bernoulli beam theory neglects shear deformations by assuming that plane sections remain plane and perpendicular to the neutral axis during bending. As a result, shear strains

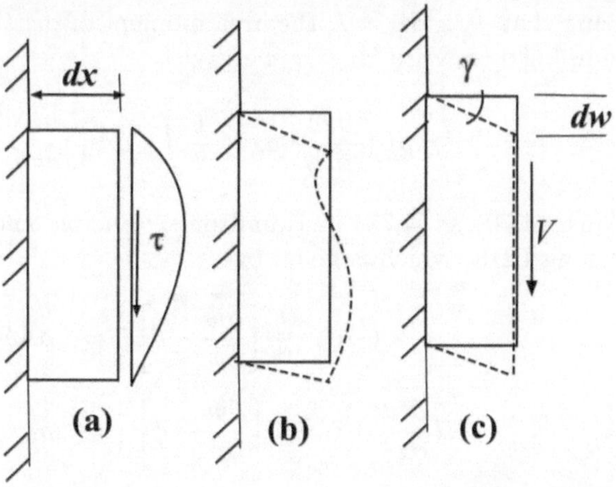

FIGURE 4.33: Shear stress distribution in beam. (a) Elementary beam, (b) Timoshenko beam-actual distribution and (c) Timoshenko beam-assumed distribution

and stresses do not figure in the theory and the shear forces are recovered from equilibrium equation $V = dM/dx$. Using this, the shear stress distribution in elementary beam can be plotted across the beam depth, as shown in Fig. 4.33(a). In reality, the beam cross-section deforms somewhat like what is shown in Fig. 4.33(b). This is particularly the case for deep beams, that is, those beams with relatively thicker cross-sections compared to the beam length, when they are subjected to significant shear forces. Usually the shear stresses are highest around the neutral axis, which is where the largest shear deformation takes place. Instead of modeling this curved shape of the cross-section, the Timoshenko beam theory models the shear stress variation as shown in Fig. 4.33(c). To account for this anomaly, we multiply the second term by a factor K, normally referred to as *shear correction factor* and hence the second term in Eqn. (4.69) becomes $\int_0^L GAK(\partial w/\partial x - \theta)^2 dx$. Now, the Eqn. (4.69) can be written as

$$U = \frac{1}{2}\int_0^L EI\left(\frac{\partial \theta}{\partial x}\right)^2 dx + \frac{1}{2}\int_0^L GAK\left(\frac{\partial w}{\partial x} - \theta\right)^2 dx \tag{4.70}$$

The kinetic energy T can be written as

$$T = \frac{1}{2}\int_V \left[\rho\left(\frac{du}{dt}\right)^2 + \rho\left(\frac{dw}{dt}\right)^2\right] dV \tag{4.71}$$

Substituting Eqn. (4.66) in the above equation, we get

$$T = \frac{1}{2}\int_0^L \int_A \rho z^2 \left(\frac{\partial \theta}{\partial t}\right)^2 dx dA + \frac{1}{2}\int_0^L \int_A \rho \left(\frac{dw}{dt}\right)^2 dx dA \tag{4.72}$$

Recognizing that $\int_A z^2 dA = I$, the area moment of inertia, the above integral can be simplified to write kinetic energy

$$T = \frac{1}{2}\int_0^L \rho I \left(\frac{\partial\theta}{\partial t}\right)^2 dx + \frac{1}{2}\int_0^L \rho A \left(\frac{dw}{dt}\right)^2 dx \tag{4.73}$$

Using Eqns. (4.70) & (4.73) in Hamilton's principle and simplifying, we get the governing PDE, which is given by

$$\begin{aligned} GAK\frac{\partial}{\partial x}\left[\frac{\partial w}{\partial x} - \theta\right] &= \rho A\ddot{w}\,, \\ EI\frac{\partial^2\theta}{\partial x^2} + GAK\left[\frac{\partial w}{\partial x} - \phi\right] &= \rho I\ddot{\theta}\,, \end{aligned} \tag{4.74}$$

where $\ddot{()}$ indicates second derivative of time and represents acceleration of the variable. The Hamilton's principle, in addition to providing the differential equations, also provides the associated boundary conditions, which are given by

$$\text{Specify } w \text{ or } V = GAK\left[\frac{\partial w}{\partial x} - \theta\right] \quad \text{and} \quad \text{Specify } \theta \text{ or } M = EI\frac{\partial\theta}{\partial x} \tag{4.75}$$

where V is the shear force and M is the bending moment.

We will now use spectral analysis to determine the wavenumbers and group speeds. Unlike the previous techniques of using the solution of characteristic equation for the wavenumber determination, we will use techniques outlined in Section 3.5. Since the governing equation given in Eqn. (4.74) is of constant coefficient type, we assume the solution of the form

$$w(x,t) = w_\circ e^{i(kx-\omega_n t)}\,, \theta = \theta_\circ e^{i(kx-\omega_n t)} \tag{4.76}$$

Substituting this in the governing equation Eqn. (4.74), the resulting equation can be cast in the form *Polynomial Eigenvalue Problem* (PEP) given in Section 3.5.2. as

$$\left\{ k^2 \underbrace{\begin{bmatrix} GAK & 0 \\ 0 & EI \end{bmatrix}}_{\mathbf{A}_2} + k \underbrace{\begin{bmatrix} 0 & -iGAK \\ iGAK & 0 \end{bmatrix}}_{\mathbf{A}_1} \right\}.$$

$$+ \left\{ \underbrace{\begin{bmatrix} -\rho A\omega_n{}^2 & 0 \\ 0 & GAK - \rho I\omega_n{}^2 \end{bmatrix}}_{\mathbf{A}_\circ} \right\} \left\{ \begin{matrix} w_\circ \\ \theta_\circ \end{matrix} \right\} = 0 \tag{4.77}$$

where the unknowns are k, $w_{\circ}$ and $\theta_{\circ}$. Note that MATLAB function *polyeig* can be used to solve a PEP problem. Thus, in this case, the order of the matrix polynomial p is 2 and $N_v = 2$. Hence, there are four eigenvalues (k) and eigenvectors $\{w_{\circ}\ \theta_{\circ}\}$. The determinant of the matrix polynomial suggests that the roots are complex conjugate. Solving this equation, the eigenvectors are arranged in a matrix $\mathbf{R}$, so that

$$\{{k_p}^2\mathbf{A}_2 + k_p\mathbf{A}_1 + \mathbf{A}_{\circ}\}\left\{\begin{array}{c} R_{1p} \\ R_{2p} \end{array}\right\} = 0 \tag{4.78}$$

The eigenvalues of PEP matrix give the wavenumber, which are plotted in Fig. 4.34(a).

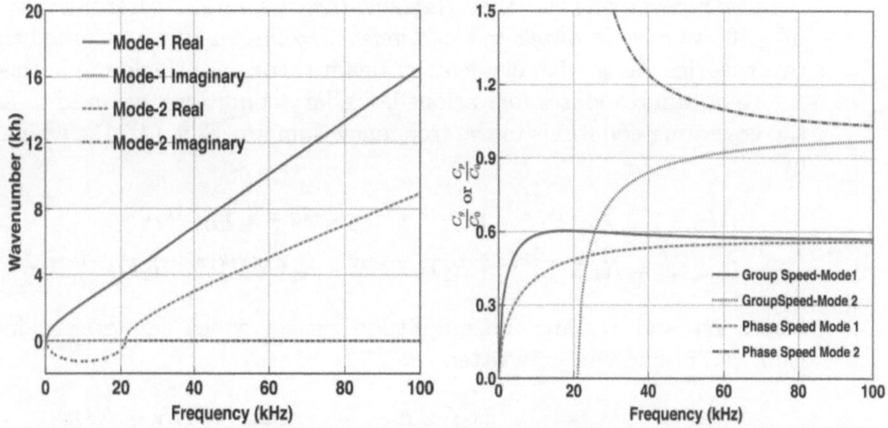

(a) Spectrum Relations (b) Dispersion Relations

FIGURE 4.34: Spectrum and Dispersion relations for a Timoshenko beam

This figure was generated using aluminum properties with the shape factor $K = 0.85$ and the depth and width of the beam being 75 mm. From the figure we see that the second mode shown in the dotted line in Fig. 4.34(a) begins as an evanescent or damping mode and becomes propagating at certain frequency called the *cut-off frequency*. The frequency at which this happens can be determined by looking at the matrix $\mathbf{A}_{\circ}$ in Eqn. (4.77). The frequency at which the determinant of matrix $\mathbf{A}_{\circ}$ becomes zero gives the cut-off frequency. Hence, the cut-off frequency is given by

$$\omega_{Cut-off} = \sqrt{\frac{GAK}{\rho I}} \tag{4.79}$$

We can recover elementary or Euler-Bernoulli beam solutions by setting $GAK \to \infty$ and $\rho I \to 0$. Thus, from Eqn (4.79) we see that for the elementary beam, no cut-off frequency exists as the cut-off frequency escapes to infinity.

Next, we will plot the phase and group speeds, which is shown in Fig. 4.34(b). These can be obtained as before using the relations $C_p = \omega/Real(k)$ and $C_g = Real(d\omega/dk)$.

From these figures, we see that the speeds predicted by elementary beams are very high compared to Timoshenko beam and hence this theory should be used with suspicion especially at higher frequencies. In addition, in thick beams, the shear effects are so pronounced that if not modeled as Timoshenko beams, one would have completely misrepresented the dynamics of the model. There are two other variants of Timoshenko beam theory. If we set $GAK \to \infty$, then we recover what is called *Rayleigh theory* solution. Alternatively, if we set $\rho I \to 0$, we recover *Single mode Timoshenko theory.* It can be noted that these two theories along with elementary beam theory are single mode theories.

To obtain responses for various boundary conditions, we need to first solve the governing equations in the frequency domain (Eqn. (4.74)), which is given by

$$\hat{w}(x,\omega) = \mathbf{A}e^{-ik_1x} + \mathbf{B}e^{-ik_2x} + \mathbf{C}e^{+ik_1x} + \mathbf{D}e^{+ik_2x}.$$
$$\hat{\theta}(x,\omega) = R_1\mathbf{A}e^{-ik_1x} + R_2\mathbf{B}e^{-ik_2x} - R_1\mathbf{C}e^{+ik_1x} - R_2\mathbf{D}e^{+ik_2x} \tag{4.80}$$

where R_1 and R_2 are the amplitude ratios, which can be obtained from Eqn. (4.77) and can be written as

$$R_1 = \frac{ik_1GAK}{GAKk_1^2 - \rho A\omega^2} = \frac{\rho I\omega^2 - GAK - EIk_1^2}{ik_1GAK}$$
$$R_2 = \frac{ik_2GAK}{GAKk_2^2 - \rho A\omega^2} = \frac{\rho I\omega^2 - GAK - EIk_2^2}{ik_2GAK} \tag{4.81}$$

Even though the solution for the variables $\hat{w}$ and $\hat{\theta}$ is written separately, they are expressed only in terms of four wave coefficients, namely **A**, **B**, **C** and **D**, as was the case with Euler-Bernoulli beam. The wave responses for different boundary conditions can be obtained by following the same procedure that was adopted for Euler-Bernoulli beam case and hence not repeated here. The same equations can be used to graphically view the two propagating spectrum of waves beyond the cut-off frequency. We have seen that the waves in Timoshenko beam are highly dispersive. In order to capture the second frequency spectrum, we need to have an incident wave that travels non-dispersively even in a dispersive medium. This can happen only when the frequency spectrum of the input signal has all its energy concentrated only over a very narrow frequency band. *Tone Burst Signal* is one such signal that can satisfy this condition. It is basically a windowed signal, wherein a sine or a cosine input is windowed usually with a *Hanning function.* The properties of this signal are detailed more in Chapter 6. More details of usage of such signals in the context of Structural Health Monitoring can be found in [52]. The usage of this signal in the context of graphically obtaining the second frequency spectrum is highlighted in the next subsection (Section 4.4.2) on higher-order rod waveguide. Similar procedure can be adopted here not only to obtain the second frequency spectrum but also graphically view both propagating spectrums.

4.4.2 Wave Propagation in Mindlin-Herrmann Rod

Elementary rod discussed earlier is called *One Mode Theory* as it allows only the longitudinal displacement. In the Mindlin-Herrmann theory, one additional motion in the lateral direction is allowed and hence this theory is normally referred to as *Two Mode Theory.* This theory was first proposed by Mindlin and Herrmann [98] for cylindrical rod of circular cross section. It was later extended to rectangular cross section in [93]. The assumed displacement field for Mindlin-Herrmann rod is given by

$$u(x, z, t) = u(x, t), \qquad w(x, z, t) = z\psi(x, t) \tag{4.82}$$

where $\psi(x, t)$ is the additional motion in the lateral direction, which arises due to Poisson's ratio. The cross section of the Mindlin-Herrmann rod along with ψ motion is shown in Fig. 4.35(a).

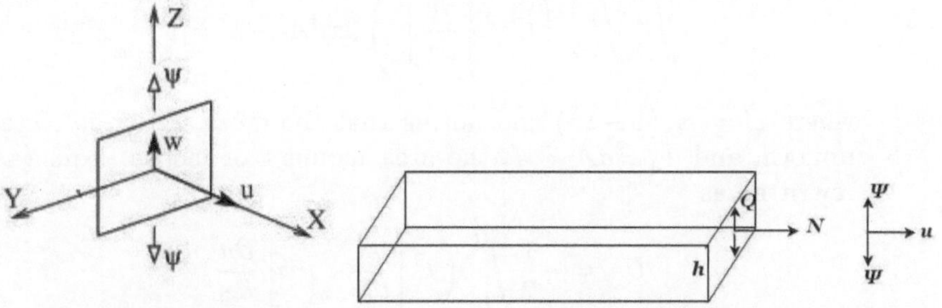

(a) Lateral Contraction (b) Force resultants

FIGURE 4.35: Lateral motion in a Mindlin-Herrmann rod

The governing differential equation is derived first using Hamilton's principle. From the assumed displacement field given by Eqn. (4.82), we can write the strain fields as

$$\varepsilon_{xx} = \frac{\partial u}{\partial x}, \qquad \varepsilon_{zz} = \psi, \qquad \gamma_{xz} = -z\frac{\partial \psi}{\partial x} \tag{4.83}$$

We have 2-D state of stress in this rod, and it is given by

$$\sigma_{xx} = \frac{E}{1-\nu^2}\left[\varepsilon_{xx} + \nu\varepsilon_{zz}\right],$$
$$\sigma_{zz} = \frac{E}{1-\nu^2}\left[\varepsilon_{zz} + \nu\varepsilon_{xx}\right], \quad \tau xz = G\gamma_{xz}$$

or

$$\sigma_{xx} = \frac{E}{1-\nu^2}\left[\frac{\partial u}{\partial x} + \nu\frac{\partial w}{\partial z}\right]$$
$$\sigma_{zz} = \frac{E}{1-\nu^2}\left[\frac{\partial w}{\partial z} + \nu\frac{\partial u}{\partial x}\right], \quad \tau_{xz} = G\left[\frac{\partial w}{\partial x} + \frac{\partial u}{\partial z}\right] \tag{4.84}$$

Substituting for strains from Eqn. (4.83), Eqn. (4.84) becomes

$$\sigma_{xx} = \frac{E}{1-\nu^2}\left[\frac{\partial u}{\partial x} + \nu\psi\right], \quad \sigma_{zz} = \frac{E}{1-\nu^2}\left[\psi + \nu\frac{\partial u}{\partial x}\right], \quad \tau_{xz} = Gz\frac{\partial \psi}{\partial x}$$

where E, ν and G are the Young's modulus, Poisson's ratio and Shear modulus. z is the thickness coordinate. We can write the strain energy expression as

$$U = \int_0^L \int_A \frac{1}{2}\left(\bar{E}\left[\frac{\partial u}{\partial x} + \nu\psi\right]\frac{\partial u}{\partial x} + \bar{E}\left[\psi + \nu\frac{\partial u}{\partial x}\right]\psi\right) dx dA$$
$$+ \int_0^L \int_A \frac{1}{2}\left(Gz^2\left[\frac{\partial \psi}{\partial x}\right]^2\right) dx dA, \quad \bar{E} = \frac{E}{1-\nu^2} \tag{4.85}$$

where $\bar{E} = E/(1-\nu^2)$ and noting that the cross section is rectangular and uniform and $\int_A z^2 dA = I$, the area moment of inertia, Eqn. (4.85) can be rewritten as

$$\begin{aligned} U &= \frac{1}{2}\int_0^L \left(\bar{E}\left[\frac{\partial u}{\partial x} + \nu\psi\right]\frac{\partial u}{\partial x}\right. \\ &+ \left.\bar{E}\left[\psi + \nu\frac{\partial u}{\partial x}\right]\psi + GIK_1\left[\frac{\partial \psi}{\partial x}\right]^2\right) dx \end{aligned} \tag{4.86}$$

In the above equation, K_1 is the shape factor taken to account for approximate shear stress distribution. The kinetic energy can be written as

$$\begin{aligned} T = &\ \frac{1}{2}\int_0^L \int_A \rho\left(\frac{\partial u}{\partial t}\right)^2 + \rho\left(\frac{\partial w}{\partial t}\right)^2 dx dA. \\ = &\ \frac{1}{2}\int_0^L \rho A\left(\frac{\partial u}{\partial t}\right)^2 + \rho I K_2\left(\frac{\partial \psi}{\partial t}\right) dx \end{aligned} \tag{4.87}$$

In the above equation K_2 is the factor introduced to take care of the approximations in lateral inertia variations. Using Eqns. (4.86 & 4.87) in the

Hamilton's principle, we get the following governing equation

$$\bar{E}A\left[\frac{\partial^2 u}{\partial x^2}+\nu\frac{\partial \psi}{\partial x}\right]=\rho A\frac{\partial^2 u}{\partial t^2}$$

$$GIK_1\frac{\partial^2 \psi}{\partial x^2}-\bar{E}A\left[\psi+\nu\frac{\partial u}{\partial x}\right]=\rho IK_2\frac{\partial^2 \psi}{\partial t^2} \tag{4.88}$$

The Hamilton's principle also gives the boundary conditions, which can be written as

$$\text{Specify } u \text{ or } N=\bar{E}A\left[\frac{\partial u}{\partial x}+\nu\psi\right], \qquad \text{Specify } \psi \text{ or } Q=GIK_1\frac{\partial \psi}{\partial x} \tag{4.89}$$

These force resultants acting on a cross section is shown in Fig. 4.35(b). Note that by substituting $GIK_1 \to 0$ and $\rho Ik_2 \to 0$ in the above equations, we get $\psi = -\nu\partial u/\partial x$ and hence the elementary rod equation can be recovered.

In order to determine the wave characteristics of the Mindlin-Herrmann rod, we need to transform the governing equation to frequency domain using Discrete Fourier Transform of the field variable $u(x,t)$ and $\psi(x,t)$. That is,

$$u(x,t)=\sum_{n=0}^{N}\hat{u}(x,\omega)e^{i\omega t}, \qquad \psi(x,t)=\sum_{n=0}^{N}\hat{\psi}(x,\omega)e^{i\omega t} \tag{4.90}$$

Substituting the above equation in Eqn. (4.88), we get the following coupled ordinary differential equation with constant coefficients

$$\bar{E}A\left[\frac{d^2\hat{u}}{dx^2}+\nu\frac{d\hat{\psi}}{dx}\right]+\rho A\omega^2\hat{u}=0.$$

$$GIK_1\frac{d^2\hat{\psi}}{dx^2}-\bar{E}A\left[\hat{\psi}+\nu\frac{d\hat{u}}{dx}\right]+\rho IK_2\omega^2\hat{\psi}=0 \tag{4.91}$$

The above equations are of constant coefficients type and have solutions of the type $\hat{u} = u_0 e^{ikx}$ and $\hat{\psi} = \psi_0 e^{ikx}$. Substituting these in Eqn. (4.91), we can write the resulting equation in matrix form as

$$\begin{bmatrix}(\rho A\omega^2-\bar{E}Ak^2) & i\bar{E}A\nu k\\ -i\bar{E}A\nu k & (\rho IK_2\omega^2-\bar{E}A-GIK_1k^2\end{bmatrix}\begin{Bmatrix}u_0\\ \psi_0\end{Bmatrix}=\begin{Bmatrix}0\\0\end{Bmatrix} \tag{4.92}$$

By setting the determinant of the matrix to zero, we get the fourth-order characteristic equation for determination of wavenumber k, which is given by

$$k^4-k^2[k_s^2r^2+(1-\nu^2)(k_l^2-\alpha)]+k_l^2(1-\nu^2)(k_s^2r^2-\alpha)=0 \tag{4.93}$$

where

$$\alpha=\frac{\bar{E}A}{GIK_1}, \qquad k_l=\omega\sqrt{\frac{\rho}{E}}, \qquad k_s=\omega\sqrt{\frac{\rho}{G}}, \qquad r=\sqrt{\frac{K_2}{K_1}}$$

A careful examination of Eqn. (4.93), we can see that the constant term of this fourth-order polynomial has an explicit function of frequency indicating that one of the wave modes will exhibit a *Cut-off Frequency* ω_c and this frequency can be obtained by equating the constant term in Eqn. (4.93) to zero. That is,

$$k_l^2(1-\nu^2)(k_s^2r^2-\alpha) = 0 \,\text{or}\, k_s^2r^2 = \alpha$$
$$= \omega_c^2 \frac{\rho}{G}\frac{K_2}{K_1} = \alpha \Rightarrow \omega_c = \sqrt{\frac{\bar{E}A}{\rho I K_2}} \tag{4.94}$$

The next big task is to determine the values of correction factors K_1 and K_2 to be used for the wave analysis. It is to be noted that these parameters are introduced as correction factors intended to compensate for the approximations introduced in the displacement fields. It is also to be noted that higher-order effects are introduced in the 1-D waveguide model in order to capture the 2-D responses as accurately as possible using 1-D models. Hence, these correction factors need to be chosen based on 2-D waveguide analysis, which is dealt with in Chapter 7. However, here a brief guideline is provided on the selection of these parameters based on 2-D wave speeds.

We have seen from Eqn. (4.94) that the cut-off frequency occurs at

$$\omega_c = \sqrt{\frac{12EA}{(1-\nu^2)\rho b h^3 K_2}} = \frac{C_p}{h}\sqrt{\frac{12}{(1-\nu^2 K_2}} \tag{4.95}$$

where $C_p = \sqrt{E/\rho}$ is the phase speed of the elementary rod and h is the depth of the rod. From the above equation, we see that the cut-off frequency is inversely proportional to the depth h. If the rod is very thin, then the cut-off frequency will escape to infinity, which means that the lateral contraction mode does not exist. However, for thick rods, the mode transition will take place at very high-frequency. We also see that the parameter K_2 also scales the cut-off frequency inversely. From the exact 2-D waveguide analysis (to be presented in Chapter 7), we can write the cut-off frequency of the $P-Wave$ as

$$\omega_c = \frac{C_p}{\sqrt{1-\nu^2}}\frac{\pi}{h} \tag{4.96}$$

Comparing Eqn. (4.95 & 4.96), we get $K_2 = 1.216$.

Selecting the value of K_1 is more involved. Its value is normally chosen by comparing the wave speeds at very high frequencies with the exact limiting speeds. In the absence of exact analysis methods, Finite Element method with very fine mesh can be employed. More details on the methodology to choose K_1 is given in [93] and they have estimated a value of $K_1 = 1.2$ by comparing higher-order rod solutions with exact 2-D analysis. However, these values of correction factors cannot be constants and varies with the problem that is being solved. In this chapter, we have chosen $K_1 = 0.7$ and $K_2 = 1.5$ for obtaining wavenumber and group speed plots.

Next, by using the properties assumed earlier, we plot the spectrum relation and dispersion relation using Eqn. (4.93). This shown in Figs. 4.36(a) and (b) respectively. The fourth-order characteristic equation for wavenumber means that this rod has two forward moving and two backward moving wave modes. Fig. 4.36 shows the real and imaginary part of both of these modes along with the elementary rod wave mode. Note that the imaginary part is plotted below the zero axis. The first mode is fully propagating and it starts as non-dispersive wave and becomes dispersive at very high frequencies. One can clearly see that both elementary and Mindlin-Herrmann rod have similar behavior at low frequencies and start deviating to become dispersive at high frequencies. The second mode is evanescent to start with and starts propagating at very high frequencies. The transition from evanescent to propagating happens at cut-off frequency, whose expression is given by Eqn. (4.94).

The group speed is obtained by differentiating Eqn. (4.93) with respect to ω and simplifying, we get the following expression

$$C_g = \frac{4k^3 - 2Pk}{k^2 \frac{dP}{d\omega} - \frac{dQ}{d\omega}} \tag{4.97}$$

where

$$P = k_s^2 r^2 + k_l^2(1-\nu^2) - \alpha(1 - nu^2), \qquad Q = k_l^2(1-\nu^2)(k_s^2 r^2 - \alpha)$$

In Eqn. (4.97), appropriate values of the wavenumbers are substituted to obtain the respective group speeds of the two wave modes. The phase speed variation is also shown in Fig 4.36(b), where the phase speeds are obtained as before, that is $C_{p1} = \omega/k_1$ and $C_{p2} = \omega/k_2$ for the two modes of this higher-order rod.

From the figure, we see that the phase speed escape to infinity at the cut-off frequency. The group speed of the longitudinal mode reduces over 40 percent at higher frequencies and beyond the cut-off frequency, compared to the elementary rod, the mode 2 travels at a higher speed compared to mode 1.

Next, we will graphically show the existence of second frequency spectrum in the Mindlin-Herrmann rod. In addition, we will also show the method of deriving the *System Transfer Function* for a system. This can be done by first writing the exact solutions to the governing differential equation, Eqn. (4.91), which is given by

$$\hat{u}(x,\omega) = R_1\mathbf{A}e^{-ik_1x} + R_2\mathbf{B}e^{-ik_2x}$$
$$+ - (R_1)\mathbf{C}e^{+ik_1x} + (-R_2)\mathbf{D}e^{+ik_2x}$$
$$\hat{\psi}(x,\omega) = \mathbf{A}e^{-ik_1x} + \mathbf{B}e^{-ik_2x} + \mathbf{C}e^{+ik_1x} + \mathbf{D}e^{+ik_2x} \tag{4.98}$$

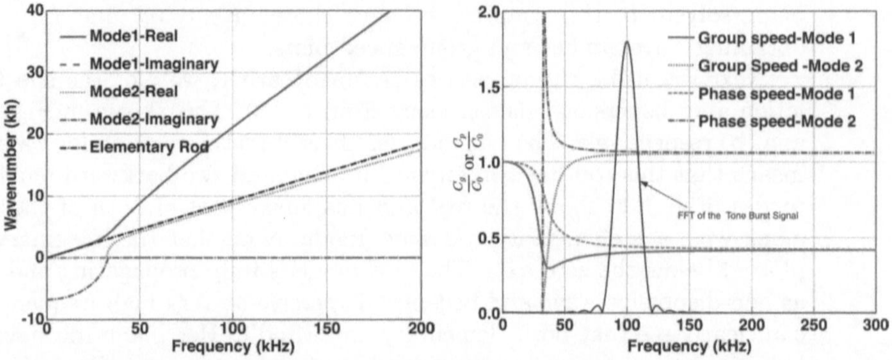

(a) Spectrum Relations (b) Dispersion Relations

FIGURE 4.36: Spectrum and Dispersion relations for a Mindlin-Herrmann Rod

where k_1 and k_2 are wavenumbers obtained by solving Eqn. (4.93). R_1 and R_2 are the amplitude ratios, which are obtained by Eqn. (4.92) and which is given by

$$R_1 = \frac{i\bar{E}A\nu k_1}{\bar{E}Ak_1^2 - \rho A\omega^2} = \frac{i\bar{E}A\nu k_1}{\rho IK_2\omega^2 - \bar{E}A - GIK_1k_1^2}.$$
$$R_2 = \frac{i\bar{E}A\nu k_2}{\bar{E}Ak_2^2 - \rho A\omega^2} = \frac{i\bar{E}A\nu k_2}{\rho IK_2\omega^2 - \bar{E}A - GIK_1k_2^2} \qquad (4.99)$$

Now let us consider an infinite Mindlin-Herrmann subjected to an axial impact at $x = 0$, which is shown in the inset of Fig. 4.37. Since the rod is infinite, the reflected coefficients in Eqn. (4.98) are omitted from the solution and the resulting equation can be written as

$$\hat{u}(x,\omega) = \mathbf{A}e^{-ik_1x} + \mathbf{B}e^{-ik_2x}, \qquad \hat{\psi}(x,\omega) = R_1\mathbf{A}e^{-ik_1x} + R_2\mathbf{B}e^{-ik_2x} \quad (4.100)$$

where the amplitude ratios R_1 and R_2 are given in Eqn. (4.99). We now impose the condition that the transverse force resultant $Q = 0$ at $x = 0$, where Eqn. (4.89) gives the expression for Q in terms of lateral contraction ψ. imposing this condition, we can write the deformation field as

$$\hat{u}(x,\omega) = \mathbf{A}\left(R_1e^{-ik_1x} - R_2\frac{k_1}{k_2}e^{-ik_2x}\right)$$
$$\hat{\psi}(x,\omega) = \mathbf{A}\left(e^{-ik_1x} - \frac{k_1}{k_2}e^{-ik_2x}\right) \qquad (4.101)$$

To determine **A** we impose the force boundary condition at $x = 0$, that is

$$\hat{\mathbf{F}} = \bar{E}A\left(\frac{d\hat{u}}{dx} + \nu\psi\right) = \hat{P}/2$$

imposing this condition, we get the value of **A** as

$$\mathbf{A} = \frac{\hat{P}k_2}{2(ik_1k_2(R_2 - R_1) + \nu(k_2 - k_1))}$$

Using this value of **A** in Eqn. (4.101), we can write the expression for lateral contraction as

$$\hat{\psi}(x,\omega) = \hat{P}\left[\frac{k_2}{2(ik_1k_2(R_2 - R_1) + \nu(k_2 - k_1))}\left(e^{-ik_1x} - \frac{k_1}{k_2}e^{-ik_2x}\right)\right] \tag{4.102}$$

The Eqn. (4.102) is of the form

Output = [Transfer Function] Input

The term in the big square bracket in Eqn. (4.102) is the system transfer function of the system, which is a function of both the frequency ω and the spatial variable x. A plot of this function with frequency will show the position of the resonances. It is a very important parameter that determines how the structure responds to dynamic loads. In other words, system transfer function is a direct by-product of the spectral analysis.

To graphically view the second propagating mode in a dispersive system, we need a modulated sine pulse of 100 kHz central frequency is considered and is shown in Fig. 4.37(a). Such a pulse, as mentioned before, will travel non-dispersively even in a dispersive medium. This contractional displacement wave amplitude is also plotted for different distances in the same figure (Fig. 4.37(b)).

Fig. 4.37(b) clearly shows the two propagating modes. The FFT of the modulated sine signal is shown in Fig. 4.36 (b), which we clearly peaks at 100 kHz. The group speeds of the two wave modes corresponding to 100 kHz are $0.47C_0$ and $1.05C_0$, respectively, where $C_0 = \sqrt{E/\rho}$ is 5091 m/sec for aluminum properties. That is, at 100 kHz, first mode propagates lot slower than the second mode. Let us now compute the time of arrival of the two propagating modes, say at $x = 0.5$ m for which we can look at Fig. 4.37(b). The time taken for the two wave modes to travel 0.5 m are $0.5/0.47C_0 = 210\,\mu$secs and $0.5/1.05C_0 = 94\,\mu$secs The initial pulse shown in Fig. 4.37(a) has a zero header of 150 μsecs. If we add this header to the computed time of arrival, the actual time of arrival for two modes are respectively equal to 360 μsecs and 244 μsecs. These times for the two modes clearly matches with the time of arrival predicted from the dispersion relations.

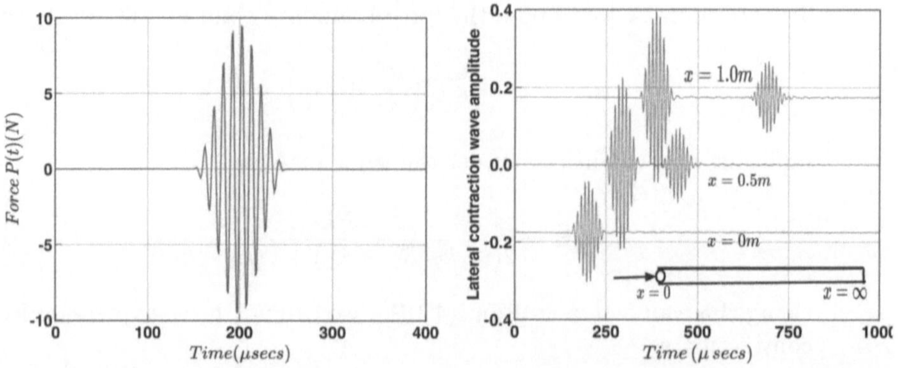

(a) 100 kHz Modulated sine signal

(b) Lateral displacement wave amplitude

FIGURE 4.37: Second frequency spectrum in a Mindlin-Herrmann rod: (a) Modulated signal and (b) Lateral contraction wave amplitude

4.5 WAVE PROPAGATION IN ROTATING BEAMS

Rotating helicopter blades are normally idealized as rotating beams. The analysis of these structural elements leads to differential equations with variable coefficients, which are introduced due to the variation of centrifugal force and geometry along the beam length. In general, rotation complicates the analysis of engineering structures. Here, the governing partial differential equation for a uniform rotating beam is derived using Hamilton's principle and the variable coefficients associated with the centrifugal term are replaced by the maximum centrifugal force to make it constant coefficient governing equation. With this assumption, the rotating beam problem is now transformed to a case of beam subjected to an axial force. Even though this averaging seems to be a crude approximation, one can use this as a powerful model in analyzing the wave propagation characteristics of the rotating structure.

The rotating beam of flexural rigidity EI and mass/unit length ρA is shown in Fig. 4.38(a) and the corresponding stress resultants are shown in Fig:4.38(b)

As before, we will first derive the governing equation of motion for a rotating beam, which we will use it to obtain the wave parameters. Note that the rotation acts like a pretension on the beam, with pretension replaced by the maximum force due to centrifugal force caused by rotation Ω. Hence, the governing equation is derived assuming that the beam is uniform and in this case Euler-Bernoulli beam theory assumptions is used, which is given by

$$EI\frac{\partial^4 w}{\partial x^4} - T_{max}\frac{\partial^2 w}{\partial x^2} + \rho A\frac{\partial^2 w}{\partial t^2} = 0 \tag{4.103}$$

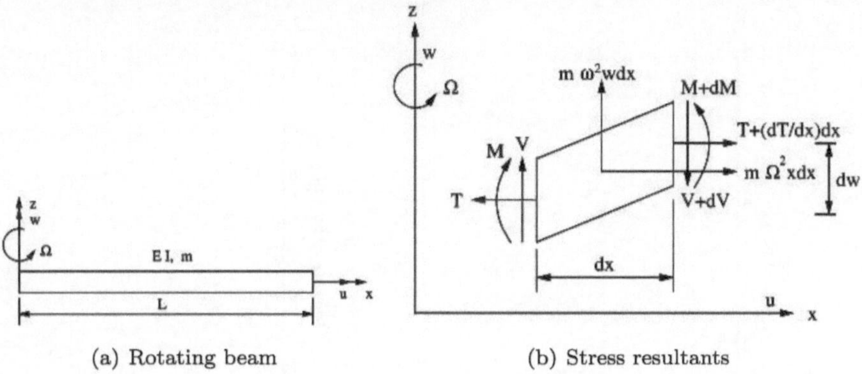

FIGURE 4.38: Configuration and stress resultants in a rotating beam

where

$$T_{max} = \int_0^L \rho A(x)\Omega^2 x dx = \frac{\rho A \Omega^2 L}{2}$$

The above equation governs the transverse displacement of a beam with axial load undergoing out of plane motion. In this particular problem, the axial load is proportional to the rotation speed Ω. Though approximate, this mathematical model can give significant insight into the study of spectrum relations of a rotating beam.

In order to determine its wave characteristics, we need to transform Eqn. (4.103) to frequency domain using DFT, which is given by

$$w(x,t) = \sum_{n=0}^{N} \hat{w}(x,\omega)e^{i(kx-\omega t)}$$

Substituting this in the governing equation, we get fourth-order characteristic equation for computation of wavenumbers, which is given by

$$k^4 + \alpha^2 k^2 - k_b^4 = 0, \qquad \alpha = \sqrt{\frac{T_{max}}{EI}}, \qquad k_b^4 = \frac{\rho A \omega^2}{EI} \tag{4.104}$$

Solving the above equation, we get

$$\sqrt{2}k_1, k_2 = \pm\sqrt{-\alpha^2 \pm \sqrt{\alpha^4 + 4k_b^4}} \tag{4.105}$$

where k_b is the wavenumber of the non-rotating beam. The phase speed is obtained using the expression $C_p = \omega/Real(k)$, while the group speed can be obtained by definition $C_g = Real(d\omega/dk)$. This can be computed by differentiating Eqn. (4.104) with respect to ω and simplifying. In doing so, we

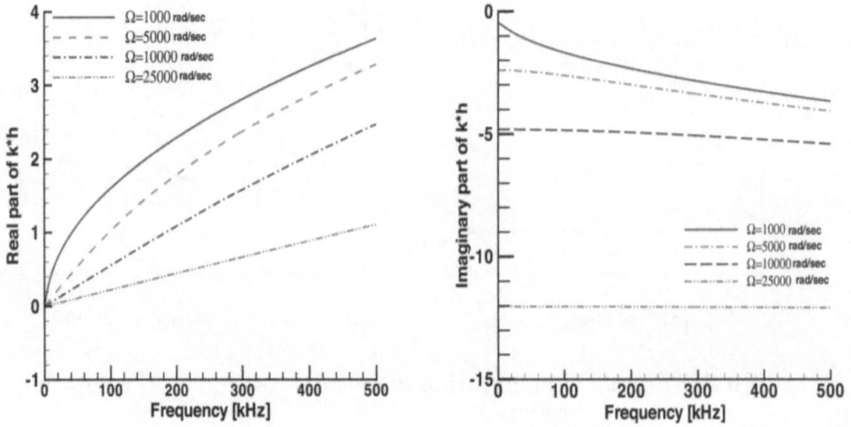

(a) Propagating wavenumber

(b) Evanacent wavenumber

FIGURE 4.39: Spectrum relations for a rotating beam

get

$$C_g = \frac{d\omega}{dk} = \frac{(2k^3 + \alpha^2 k)\omega}{kb^4} \tag{4.106}$$

First, the spectrum relations are plotted in Fig. 4.39. Fig. 4.39(a) shows the real part of the propagating wavenumber (k_1), while Fig. 4.39(b) shows the evanescent wavenumber k_2. One can see some interesting features. When the rotational speed is low, the waves are dispersive as in the case of elementary beam. When the rotational speed increases, the slope of the curve changes, the values of the wavenumber decrease, and the level of dispersiveness reduces and at very high speeds, the slope tends to be 45° to the *x-axis* indicating that the waves are non-dispersive. In other words, at very high speeds, the wave speeds will be very high due to reduction in the slope of the wavenumber curve and waves tend to become non-dispersive. These aspects can be confirmed by plotting the dispersion relation, which is shown in Fig. 4.40. At low frequency, the speeds follow elementary beam wherein the group speeds are nearly twice the phase speeds. At very high-frequency, the speeds are very high and group speeds are nearly equal to phase speeds indicating that the propagating wave will be non-dispersive. At high-frequency, if the rotating waveguide is short and rotating at very high-frequency, then repeated reflections caused by high speeds can eventually lead to failure. Also, the evanescent component of the wavenumber is very high at very high-frequency.

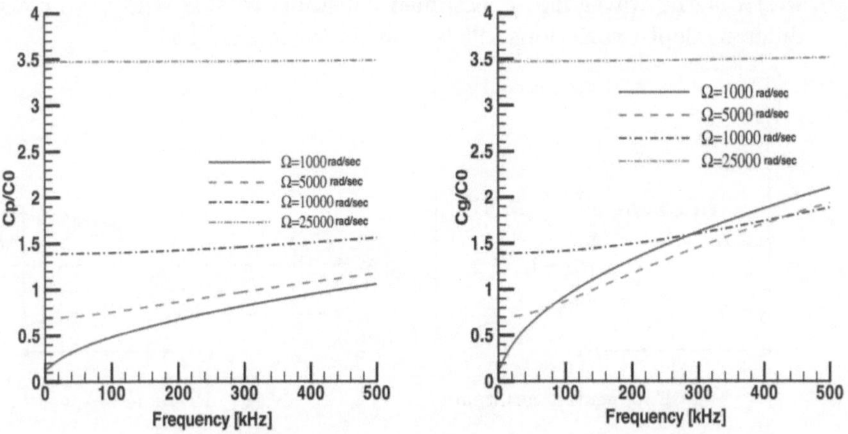

(a) Phase Speed (b) Group Speed

FIGURE 4.40: Dispersion relations for a rotating beam

4.6 WAVE PROPAGATION IN TAPERED WAVEGUIDES

Tapered waveguides are those in which any of the dimensions, say length or depth or thickness, vary across the structures. This changes the cross-sectional properties, namely the area or area moment of inertia of the cross section, which in turn changes the governing equations. This variation in cross-sectional properties leads to variable coefficient differential equations.

Wavenumbers and wave speeds can be computed only to those waveguides whose motions are defined by a governing differential equation that are of constant coefficient type. In the case of waveguides, which are defined by variable coefficient differential equation, determination of wave parameters is challenging, quite involved and quite a few argue that the concept of wavenumber does not exist for such waveguides. However, if one solves the governing equations either analytically or numerically, then visualizing the response, one can infer on its wave behavior. There are certain classes of variable coefficient differential equation that can be converted to constant coefficient equation using suitable substitution. In most cases, depending upon the type of variable coefficient equation, exact solution to such equation of motion may not be available. In such cases, numerical solutions such as finite element methods, spectral element methods, etc. are the approaches that has to be adopted to obtain the solutions.

As mentioned earlier, tapered structures are inhomogeneous structures wherein their dimensions vary across their respective dimensions. Here, we are dealing with one-dimensional waveguides and hence we assume only the depth of the waveguide vary along the length of the waveguide, while the width of the waveguide is assumed constant. In this study, we consider two different depth variations. These are shown in Fig. 4.41.

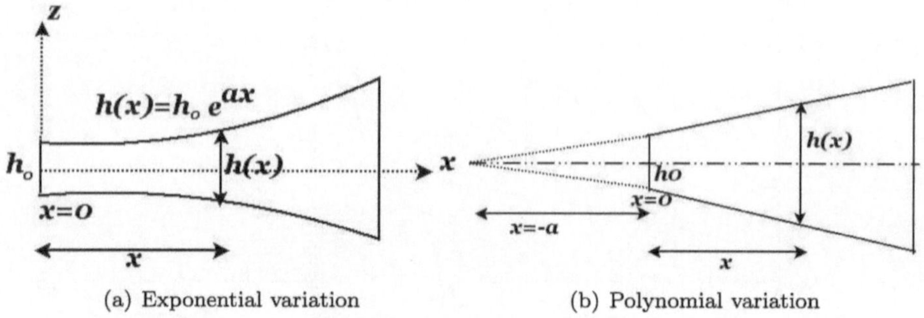

FIGURE 4.41: Thickness variation in a tapered 1-D waveguide

Fig. 4.41(a) shows the exponential variation of the depth, where the depth varies as

$$h(x) = h_0 e^{\alpha x} \tag{4.107}$$

where h_0 is the depth at $x = 0$ and α is the taper parameter and its value normally lies between 0 and 1. Fig. 4.41(b) shows the polynomial variation of the depth and its variation is given by

$$h(x) = h_0 \left(1 + \frac{x}{a}\right)^m \tag{4.108}$$

where again h_0 is the depth of the waveguide at $x = 0$, a is the taper parameter and m is the constant, which determines the polynomial variation of the depth with respect to x coordinate.

As mentioned earlier, solving governing equations of motion of tapered waveguide to study its wave propagation is a challenging one. Hence, in this section, we will look at the propagation characteristics of some simple tapered longitudinal and flexural waveguides. The aim here is to see how the taper influences the wave behavior in 1-D isotropic waveguides.

4.6.1 Wave Propagation in Tapered Rod Having Exponential Depth Variation

Assumption of exponential variation of depth is very useful for certain types of variable coefficient differential equations, wherein it converts such equations to constant coefficient type, which are more amenable for exact solutions. One such situation where this assumption is applicable is in the case of the governing equation for the tapered rod. The governing equation of a rod whose depth varies along its length can be easily derived using Hamilton's principle or otherwise and is given by

$$\frac{\partial}{\partial x}\left(EA(x)\frac{\partial u}{\partial x}\right) = \rho A(x)\frac{\partial^2 u}{\partial t^2} \tag{4.109}$$

where $u(x,t)$ is the axial deformation and E is the Young's modulus, which is assumed to be constant everywhere. Transforming the above equation to the frequency domain, we get

$$\frac{d}{dx}\left(EA(x)\frac{d\hat{u}}{dx}\right) + \rho A(x)\omega^2\hat{u} = 0 \tag{4.110}$$

Assuming the breadth variation as constant, and using Eqn. (4.107), we can write the area of cross section as

$$A(x) = bh(x) = bh_0 e^{\alpha x} = A_0 e^{\alpha x} \tag{4.111}$$

where h_0 is the depth at $x = 0$. Substituting Eqn. (4.111) in Eqn. (4.110) and simplifying, we get

$$\frac{d^2\hat{u}}{dx^2} + \alpha\frac{d\hat{u}}{dx} + k_L^2\hat{u} = 0 \tag{4.112}$$

where $k_L = \omega\sqrt{\rho A_0/EA_0}$ is the wavenumber of the uniform cross section rod evaluated at $x = 0$. The above equation is a constant coefficient differential equation, whose solution is of the form $\hat{u}(x,\omega) = Ae^{ikx}$. Substituting this in Eqn. (4.112), we get the following quadratic characteristic equation for determination of wavenumbers, which is given by

$$k^2 - i\alpha k - k_L^2 = 0 \tag{4.113}$$

Solving the above equation, we get

$$2(k_1, k_2) = i\alpha + \sqrt{4k_L^2 - \alpha^2} \tag{4.114}$$

The examination of this equation reveals the following:

- The first term in Eqn. (4.114) represents the damping component. In most cases $\alpha < 1$ and higher value represents steeper taper. In other words, steeper the taper, faster will the wave attenuate.
- With $\alpha < 1$, the second term is a small perturbation to uniform wavenumber. Hence, one cannot see perceivable change in the wavenumber variation compared to uniform rod.

Since the wavenumber can be explicitly determined, we can compute both the phase and group speeds. The phase speed of this tapered rod is given by

$$C_p = Real\left(\frac{\omega}{k_1}\right) = Real\left(\frac{2\omega}{i\alpha + \sqrt{4k_L^2 - \alpha^2}}\right) = \frac{\omega\sqrt{4k_L^2 - \alpha^2}}{2k_L^2} \tag{4.115}$$

and the group speed of this rod is given by

$$C_g = Real\left(\frac{d\omega}{dk}\right) = Real\left(\frac{1}{\frac{dk}{d\omega}}\right) \tag{4.116}$$

Differentiating Eqn. (4.114) with respect to ω and substituting in Eqn. (4.116), we can write the group speed of the tapered rod with exponential variation of depth as

$$C_g = \frac{\omega\sqrt{4k_L^2 - \alpha^2}}{2k_L^2} \tag{4.117}$$

From Eqns. (4.115 & 4.117), it is interesting to see that both these speeds are equal. Does this indicate that the waves in this waveguide are non-dispersive? The response analysis presented next will reveal answer to this question.

Hence, in order to understand the wave attenuation and also study other wave characteristics, the response needs to be plotted. Since the governing equation is reduced to constant coefficient type, we can write the solution to Eqn. (4.110) as

$$\hat{u}(x,\omega) = \mathbf{A}e^{-ik_1x} + \mathbf{B}e^{-ik_2x} \tag{4.118}$$

where k_1 and k_2 are the wavenumbers given by Eqn. (4.114). If an infinite rod configuration is assumed, then $\mathbf{B} = 0$ and the solution $\hat{u}(x,\omega) = \mathbf{A}e^{-ik_1x}$ is plotted with $\mathbf{A}$ being the incident triangular pulse used in the earlier studies. Wave responses are plotted at $x = 0.5$ m from the impact site for different values of α. The response plot is generated using aluminum properties used in the earlier examples and this is shown in Fig. 4.42.

From the above figure, we see the propagation is similar to uniform rod. As the value of α increases, the peak amplitude decreases. The waves are non-dispersive in nature and the speeds of wave do not change much compared to uniform rod.

4.6.2 Wave Propagation in Tapered Rod Having Polynomial Depth Variation

The governing differential equation is given by Eqn. (4.109) and its frequency domain representation is given by Eqn. (4.110). Using Eqn. (4.108), we can

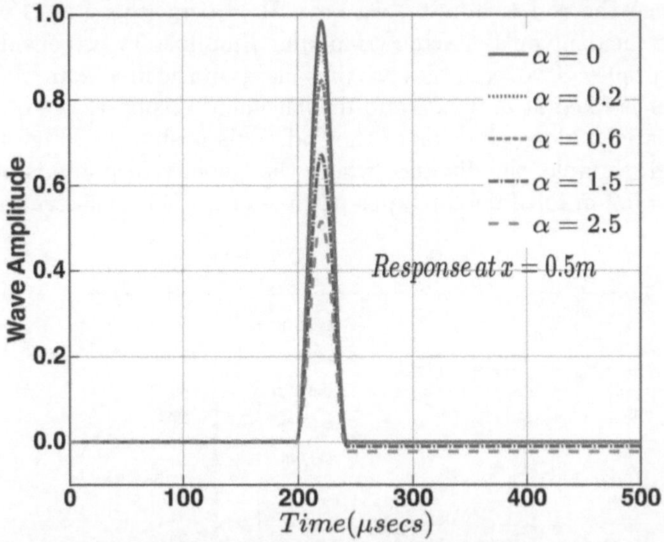

FIGURE 4.42: Response of an infinite tapered rod with exponentially varying depth

write the varying area as

$$A(x) = bh(x) = bh_0\left(1 + \frac{x}{a}\right)^m = A_0\left(1 + \frac{x}{a}\right)^m \tag{4.119}$$

Using this in the governing equation (Eqn. (4.110)), we get the following variable coefficient differential equation.

$$\frac{d^2\hat{u}}{dx^2} + \frac{m}{a+x}\frac{d\hat{u}}{dx} + k_L^2\hat{u} = 0 \tag{4.120}$$

where k_L is the uniform rod wavenumber. The solution of the above equation is quite involved and is not straightforward. Some details of the solution of this type of equations are given in [3, 118, 137]. The solution of Eqn. (4.120) is in terms of Bessel functions and is given by

$$\hat{u}(x,\omega) = \mathbf{A}s^pJ_p(s) + \mathbf{B}s^pY_p(s), \;\; p = \frac{1}{2}(1-m), \;\; s = k_L(a+x) \tag{4.121}$$

In order to understand the wave propagation in such tapered rod, we again consider an infinite aluminum rod with properties of the rod as considered earlier. We consider for the case $m = 1$, which yields $p = 0$. Thus, the solution for this case reduces to

$$\hat{u}(x,\omega) = \mathbf{A}J_0(s) + \mathbf{B}Y_0(s) \tag{4.122}$$

Since the rod is infinite, we have $\mathbf{B} = 0$ in Eqn. (4.121). Here, $\mathbf{A}$ is again the incident pulse having triangular time history as considered in the earlier examples. Two examples are presented; in the first example, the taper parameter is fixed at $a = 0.05$ and displacement responses are obtained at different locations along the axis of the rod. This is shown in Fig. 4.43(a). In the second example, the distance where the velocity response is obtained is fixed at $x = 0.2$ m for different taper parameter a. This is shown in Fig. 4.43(b).

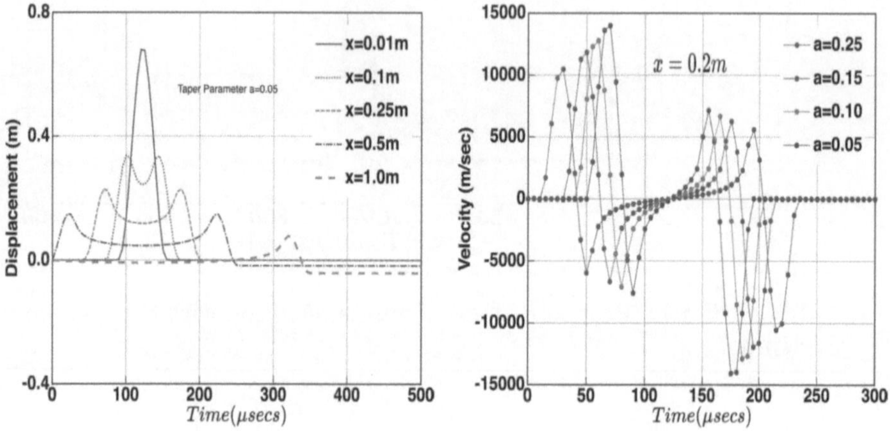

(a) Displacement response at different distances for a given taper $a = 0.05$

(b) Velocity responses at $x = 0.2$ m for different taper value a

FIGURE 4.43: Response of an infinite tapered rod ($p = 1$) with polynomial depth variation

From the figure we can conclude the following:

1. From the displacement response shown in Fig. 4.43(a), we see that as the wave propagates in the positive x-*direction*, the original triangular pulse splits into two pulses of smaller amplitudes, one of which propagates to the left and does not appear after the wave propagates over $x = 0.5$ m, while the second pulse propagates with diminishing amplitudes.

2. At $x = 0$, the responses shoots up to a very high value and attenuates significantly over a very short distance of $x = 0.3$ m

3. The velocity response shown in Fig. 4.43(b) clearly shows that the increase in the taper increases the group speed of the longitudinal wave. The waves are clearly dispersive in nature.

The MATLAB script is provided with this book, which can solve this problem. The script uses inbuilt MATLAB function to compute the Bessel functions used in the solution. Solution of Eqn. (4.120) for other values m is very difficult to obtain and in such cases we need to apply approximate techniques to solve the equations.

4.6.3 Wave Propagation in Tapered Beam

Again for the beam structure, it is assumed that only the depth varies along the axis of the beam as shown in Fig. 4.41(b). The exponential variation of depth is not considered here as it does not help reducing the variable coefficient equation to constant coefficient differential equation. Thus, we assume that the depth varies as given in Eqn. (4.108). Thus, the area and area moment of inertia variation become

$$A(x) = A_0 \left(1 + \frac{x}{a}\right)^m, \qquad I(x) = I_0 \left(1 + \frac{x}{a}\right)^{3m} \tag{4.123}$$

The governing differential equation for a tapered beam with polynomial variation of depth is given by

$$\frac{\partial^2}{\partial x^2}\left(EI(x)\frac{\partial^2 w}{\partial x^2}\right) + \rho A(x)\frac{\partial^2 w}{\partial t^2} = 0 \tag{4.124}$$

Transforming the above equation to the frequency domain, we can write the resulting equation as

$$\frac{d^2}{dx^2}\left(EI(x)\frac{d^2 \hat{w}}{dx^2}\right) - \rho A(x)\omega^2 \hat{w} = 0 \tag{4.125}$$

Substituting the area and area moment of inertia variation in the above equation and simplifying, we get

$$\begin{aligned}(a+x)^2 m\frac{d^4\hat{w}}{dx^4} + 6m(a+x)^{2m-1}\frac{d^2\hat{w}}{dx^2}\cdot \\ +3m(3m-1)(a+x)^{2m-2}\frac{d^2\hat{w}}{dx^2} - a^{2m}k_b^4\hat{w} = 0\end{aligned} \tag{4.126}$$

where $k_b^4 = \omega^2\rho A_0/EI_0$, A_0 is the area at $x = 0$ and I_0 is the area moment of inertia at $x = 0$. The solution of this equation for any arbitrary value of m is indeed very difficult. However, for $m = 1$, we can write the governing equation as

$$\left((a+x)\frac{d^2}{dx^2} + 2\frac{d}{dx} + ak_b^2\right)\left((a+x)\frac{d^2}{dx^2} + 2\frac{d}{dx} - ak_b^2\right)\hat{w} = 0$$

The above equation can be written as product of two separate equations, each of these is in the form of Bessel's equations, one of which has a Bessel solution of first kind as its solution, while the second equation has Bessel solution of second kind as its solution. The details again can be found in [3]. The complete solution is given by

$$\hat{w}(x,\omega) = \frac{1}{\sqrt{s}}\left[\mathbf{A}J_2(s) + \mathbf{B}Y_2(s) + \mathbf{c}I_2(s) + \mathbf{D}K_2(s)\right]$$
$$\text{where} \qquad s = 2k_b\sqrt{a}\sqrt{a+x} \tag{4.127}$$

where $J_2(s)$ and $Y_2(s)$ are the second-order Bessel functions of first kind and $I_2(s)$ and $K_2(s)$ are the Bessel functions of second kind.

In order to study its wave response, we again consider an infinite beam, which means that **C** and **D** are equal to zero. Assuming **B** also equal to zero, we have

$$\hat{w}(x,\omega) = \frac{\mathbf{A}J_2(s)}{\sqrt{s}}$$

where **A** is the input triangular pulse and $J_2(s) = 2J_1(s)/s - J_0(s)$. Again we use the same triangular pulse with peak force equal to 1 N is used here to obtain the flexural response. To understand the wave responses, two different studies are performed. In the first case, the value of the taper parameter $a = 0.02$ is fixed and the flexural displacement responses is taken at a number of location along the axis of the beam. In the second study, the responses are obtained for different taper values at a given distance of $x = 0.2$ m. Unlike in the earlier cases, here, the displacement responses are plotted as some special features of the tapered beam waveguide can be extracted from these responses. These responses are shown in Fig. 4.44.

From the figure, following inferences can be drawn:

- The response is maximum at the impact location and the response attenuates significantly over a very small distance.
- The responses at larger distances are highly dispersive with significant reduction in amplitude.
- The wave, as it propagates, it splits is symmetrically into smaller waves. This aspect is similar to the tapered rod behavior.
- The waves show significant phase changes for longer distance propagation and for higher values of taper.

This section shows how spectral analysis can be used to understand the physics of waves in a tapered waveguide. We have only shown for simple waveguides where the exact solutions are possible. Wave propagation in different tapered waveguides are still an open area of research and much needs to be done in terms of understanding physics behind propagating waves in such waveguides.

For the tapered rod and beam cases with polynomial variations, the response are obtained using the MATLAB script, where the inbuilt Bessel function in MATLAB is used

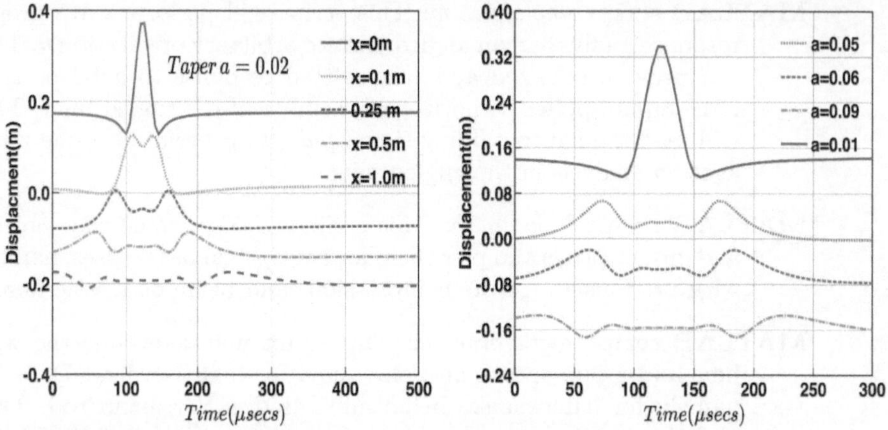

(a) Response at different distances for a given taper $a = 0.02$

(b) Responses at $x = 0.2$ m for different taper value a

FIGURE 4.44: Response of an infinite tapered beam ($m = 1$) with polynomial depth variation

Note on MATLAB® scripts provided in this chapter

In this chapter, following MATLAB scripts are provided for performing wave propagation analysis in 1-D isotropic waveguides, namely rods and beams. The given programs perform wave propagation analysis on waveguides with different boundaries, wave propagation in higher-order Mindlin-Herrmann rod and Timoshenko beam waveguides, special beam cases such as beam on elastic foundations and beams with pretension, wave propagation in angled joint and tapered waveguides. By running these simple MATLAB codes and generating time response plots, the reader will be able to appreciate the content of this book much better.

MATLAB script *rod.m*: This script will perform wave propagation response analysis in 1-D rod waveguides using broad band signals for the following cases (a) simple transmission, (b) wave interaction with elastic boundary and (c) wave interaction with a concentrated mass as well as stepped rod. The script will read the file containing signal *pul.txt* for performing the wave propagation analysis.

MATLAB script *beam.m*: This script will perform wave propagation response analysis in 1-D beam waveguides using broad band signals for the following cases (a) simple transmission, (b) wave interaction with elastic boundary and (c) wave interaction with a concentrated mass as well as stepped beam. The script will read the file containing signal *pul.txt* for performing the wave propagation analysis.

MATLAB script *anglejoint.m*: This script will perform wave propagation response analysis in an angled joint of arbitrary orientation θ. The script can provide reflected and transmitted response to axial or transverse force input applied to horizontal member of the angled joint. The script will read the file containing signal *pul.txt* for performing the wave propagation analysis in the angled joint.

MATLAB script *kcgbeam.m*: This script will determine the wavenumbers and group speeds and plot them for two special beam cases, namely beam with pre-tension (or pre-compression) and beam on elastic foundations

MATLAB script *higherorder.m*: This script will determine the wavenumbers and group speeds and plot them for higher-order-1-D waveguides, namely for Timoshenko beam and Mindlin-Herrmann rod. The script can also obtain wave propagation response on an infinite Timoshenko beam or Mindlin-Herrmann rod. The second propagating spectrum can be captured in these higher-order waveguides using this MATLAB script. This script uses two function scripts namely *timok1k2.m* and *mhk1k2.m*, which are used to compute the wavenumbers for these two higher-order waveguides. The script can also perform wave propagation analysis using broad band pulse, in which case it will externally read the Gaussian signal from the file *pul.txt*

MATLAB script *tapered.m*: This script will perform wave propagation response analysis in tapered rod and beam. This script considers both broad band and tone burst signal. For the tapered rod case, both exponential variation and polynomial variation of the depth are considered, while in the case of beam, only linear polynomial variation of the depth is considered.

Summary

This chapter comprehensively presents the fundamental aspects on the wave propagation in 1-D isotropic structures, the understanding of which is essential for the material to be presented at the later chapters. The basic objective here is to bring out the physics of wave propagation in simple 1-D isotropic waveguides and understand the interaction of the waves with boundaries and constraints. The wave propagation in both longitudinal and flexural waveguides and their interaction with the boundaries such as fixed, free and hinged boundaries are presented in this chapter. In addition, the effects of additional constraints such as elastic constraints, elastic foundations and pretension on the wave propagation are presented. Next, the method of coupling the axial and bending mode through framed structure is presented, and the propagation of waves in such structures is discussed in detail. Following this, through spectral analysis, wave propagation in higher-order waveguides such as Timoshenko beam and Mindlin-Herrmann rod is presented. In particular, the effect of introducing higher-order effects in the elementary waveguides on wave propagation is highlighted.

Next, the wave propagation in rotating flexural waveguides is addressed, and it is shown the effect of high rotating speeds on the wave parameters, namely the wavenumbers and the wave speeds is that it makes the waves non-dispersive in a beam structure. The last topic in this chapter is on the wave propagation in waveguides defined by variable coefficient differential equations such as tapered waveguides. In this section, the wave propagation in some simple cases of tapered rod and beam is presented and the physics of wave propagation in these waveguides are highlighted. In summary, this chapter presents a grand overview of the power of spectral analysis in the analysis of wave propagation in structures and materials. Wave propagation in viscoelastic waveguides was also the part of this chapter in the unified book [54]. However, it was felt that a dedicated chapter for studying viscoelastic waveguides is essential since several viscoelastic models are available and the physics of wave propagation in these structures are vastly different. Hence in the next chapter detailed viscoelastic wave propagation in 1-D waveguides is presented.

Exercises

4.1 Plot the spectrum and dispersion relations for a rod that has a resistive force that is displacement dependent. That is

$$F_R = Cu(x,t)$$

where C is the damping constant.

4.2 Two rods are connected by a linear spring of spring constant K as shown below. The left rod segment is of length L, while the right rod segment extents to infinity. Assume both the rod segments is made of Aluminum. The left end of the left rod segment is subjected to an impact $F(t)$, which is the signal used in all the examples in this chapter. Determine both the reflected velocity response (in the left rod segment) and transmitted velocity response (in the right rod segment)

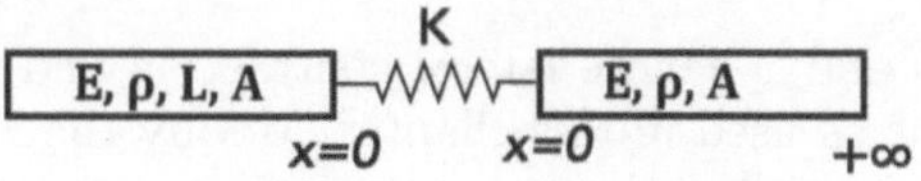

Figure for Problem 4.2

4.3 In this chapter, we studied the elementary and Mindlin-Hermann (MH) rod theories. Elementary rod theory is a single mode theory, while the MH theory is a two mode theory. An improvement of the elementary theory is a *One Mode Theory*. This theory was first formulated by [87]. The governing partial differential equation governing the axial motion $u(x,t)$ of rod and the internal force $F(x,t)$ are given by

$$EA\frac{\partial^2 u}{\partial x^2} + \nu^2\rho J\frac{\partial^4 u}{\partial x^2\partial t^2} = \rho A\frac{\partial^2 u}{\partial t^2}$$

$$F(x,t) = EA\frac{\partial u}{\partial x} + \nu^2\rho J\frac{\partial^3 u}{\partial x\partial t^2}$$

where EA is the axial rigidity of the rod, ν is the Poisson's ratio, ρ is the density and J is the polar moment of inertia.

1. Discuss the spectrum and dispersion relations for this rod. Does this system exhibit an *Escape frequency*, where the wavenumber escapes to infinity?
2. Using a tone burst signal sampled at a frequency beyond the escape frequency, show that no propagation takes place on an infinite rod. Use Aluminum properties.

4.4 Next, we want to study the wave modes in the three mode rod theory. Such higher-order theories will approximate more closely to the 2-D Lamb wave modes, which is discussed in Chapter 7. The Governing equation for the three mode theory is given by

$$(\lambda+2\mu)A\frac{\partial^2 u}{\partial x^2} + \lambda A\frac{\partial \psi}{\partial x} = \rho A\frac{\partial^2 u}{\partial t^2}$$

$$\mu I\frac{\partial^2 \psi}{\partial x^2} - (\lambda+2\mu)A\psi - \lambda A\frac{\partial u}{\partial x} - 2\mu Ah\frac{\partial \phi}{\partial x} = \rho I\frac{\partial^2 \psi}{\partial t^2}$$

$$(\lambda+2\mu)I\frac{\partial^2 \phi}{\partial x^2} - 5\mu A\phi + \frac{10}{48}\mu Ah\frac{\partial \phi}{\partial x} = \rho I\frac{\partial^2 \phi}{\partial t^2}$$

where λ and μ are Lame's constants and they are related to Young's modulus and Poisson's ratio as

$$\lambda = \frac{\nu E}{(1+\nu)(1-2\nu)}, \qquad \mu = G = \frac{E}{2(1+\nu)}$$

Here, ψ and ϕ denote lateral contraction and rotation. Other variables are same as used in this chapter. Modify the MATLAB code developed in this chapter to determine the spectrum and dispersion relations for this model and plot the wavenumbers group speeds obtained from all the rod theories and discuss the result. Use aluminum beam properties.

4.5 The beam shown below has one end rotationally restrained, while translationally it is free as shown below. Setup the equations that relates the incident and reflected coefficients.

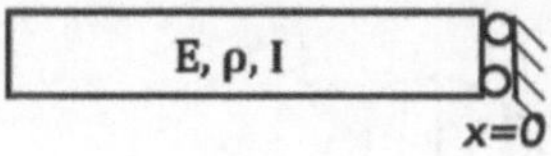

Figure for Problem 4.5

4.6 Two beams are connected by a rotational springs as shown in the figure below. The left beam segment is of length L, while the right beam segment extents to infinity. Assume both the beam segments are made of aluminum. The left end of the left beam segment is subjected to an impact $F(t)$, which is the signal used in all the examples in this chapter. Determine both the reflected velocity response (in the left beam segment) and transmitted velocity response (in the right beam segment).

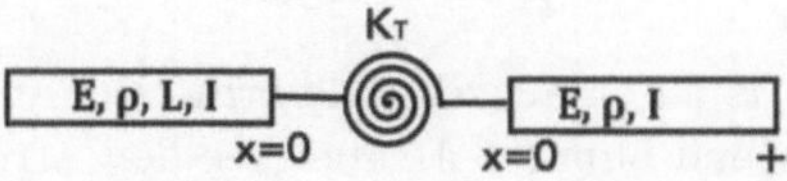

Figure for Problem 4.6

4.7 Till now we have solved only wave propagation problems due to point impact. If the load is distributed, it will also have a spatial component. On a 1-D waveguide, the distributed load distribution can be written as

$$q(x,t) = p(x)F(t), \qquad \text{where} \qquad p(x) = C_0 + C_x + C_2x^2 + C_3x^3$$

where C_0, C_1, C_2 and C_3 are some constants and x is the axial coordinate of the beam.

1. First, establish the frequency domain solution for transverse displacement $\hat{w}(\omega, x)$ as

$$\hat{w}(\omega, x) = \mathbf{A}e^{-ikx} + \mathbf{B}e^{-kx} + \mathbf{C}e^{ikx} + \mathbf{D}e^{kx} - \frac{\hat{F}(\omega)}{\rho\omega^2}(C_0 + C_1x + C_x^2 + C_3x^3)$$

2. Next, solve the problem of the impact of an infinite beam with parabolically distributed load as shown in the below figure. Here,

$$p(x) = p_0\left[1 - \frac{x^2}{a^2}\right] \qquad |x| =\leq a, \qquad p_0 = \frac{3}{4}\frac{F}{a}$$

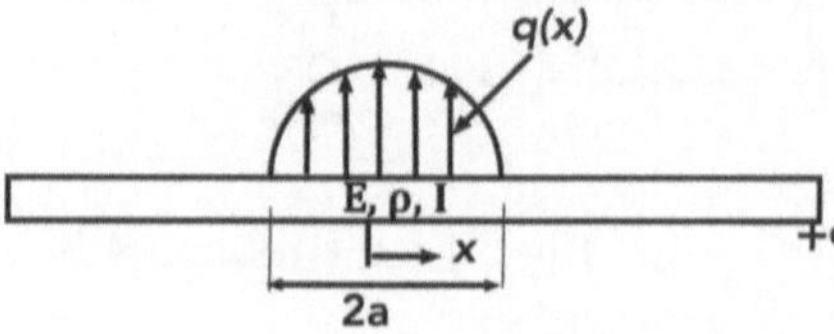

Figure for Problem 4.7

3. Using the above solution determines the responses at a number points along the axis of the beam

4.8 The flutter in a beam-like structure is governed by the following PDE

$$EI\frac{\partial^4 w}{\partial x^4} + \rho A\frac{\partial^2 w}{\partial t^2} + \alpha\frac{\partial w}{\partial x} + \beta\frac{\partial w}{\partial t}$$

where $w(x,t)$ is the transverse deformation and constants α and β are functions of Mach number M and free flow stream pressure p, and they are given by

$$\alpha = \frac{2p}{\sqrt{M^2-1}}, \qquad \beta = \frac{2p(M^2-2)}{(M^2-1)^{1.5}}$$

Obtain the spectrum and dispersion relation on this model and plot them as a function of Mach number M for some free stream pressure p

4.9 Two beams are connected to each other by a distributed spring of spring constant K as shown below. Both beams are assumed to have same properties. If $w_1(x,t)$ and $w_2(x,t)$ are the displacement of the top and bottom beams, the governing equation of this elastically coupled double beam is given by

$$EI\frac{\partial^4 w_1}{\partial x^4} - \rho A\frac{\partial^2 w_1}{\partial t^2} + K(w_1 - w_2) = 0$$
$$EI\frac{\partial^4 w_2}{\partial x^4} - \rho A\frac{\partial^2 w_2}{\partial t^2} - K(w_1 - w_2) = 0$$

Determine (a) the wavenumbers and group speeds for this elastically coupled double beam as a function of spring constant K, (b) Using the Gaussian pulse of certain frequency bandwidth and applied at some point on the top beam of an infinite double beam determine time responses at a few locations both on the top and bottom beams as a function of spring constant K.

4.10 The figure below shows a curved beam section with its stress resultants. Assuming the beam is isotropic with axial and bending rigidity of EA

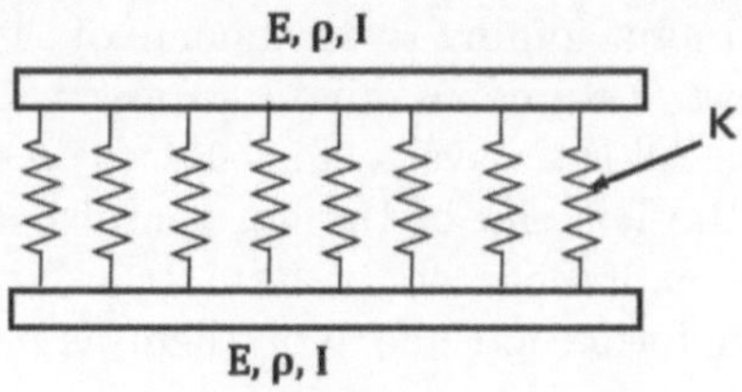

Figure for Problem 4.9

and EI, respectively, the governing PDE for this beam is given by

$$EA\frac{\partial^2 u}{\partial s^2} + \frac{1}{R^2}\left[EI\frac{\partial^2 u}{\partial s^2} - EAR\frac{\partial w}{\partial s} + EIR\frac{\partial^3 w}{\partial s^3}\right] = \rho A\frac{\partial^2 u}{\partial t^2}$$
$$EI\frac{\partial^4 w}{\partial s^4} + \frac{1}{R^2}\left[EAw - EAR\frac{\partial u}{\partial s} + EIR\frac{\partial^3 u}{\partial s^3}\right] = -\rho A\frac{\partial^2 w}{\partial t^2}$$

The stress resultants for this beam are given by

$$EA\left[\frac{\partial u}{\partial s} - \frac{w}{R}\right], \qquad M = EI\left[\frac{\partial w^2}{\partial s^2} + \frac{1}{R}\frac{\partial u}{\partial s}\right]$$
$$V = EI\left[\frac{\partial w^2}{\partial s^2} + \frac{1}{R}\frac{\partial u}{\partial s}\right]$$

where $u(s,t)$ and $w(s,t)$ are the axial and transverse displacement fields and R is the radius of curvature. Assuming aluminum properties deter-

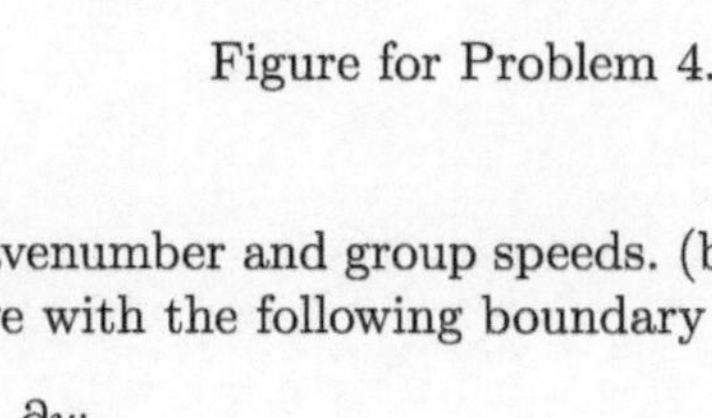

Figure for Problem 4.10

mine (a) wavenumber and group speeds. (b) Assuming only the forward moving wave with the following boundary conditions at $s = 0$

$$\frac{\partial w}{\partial s} = 0, \qquad u(0,t) = 0, \qquad V(0,t) = -\frac{1}{2}P(t)$$

construct the responses at $s = 0$, $s = R$, $s = 2R$ and $s = 3R$. See [34] for more details on this problem

4.11 A right angled joint joining two segments of same properties is shown in the figure below. Assume aluminum properties. (a) set up the equations for determining all the wave coefficients. (a) Assuming a Gaussian pulse is applied at the left end of the horizontal beam segment of length L, determine the expression for system transfer function for both axial and bending deformation and plot them as a function of frequency at $x = 0$, $y = 0$

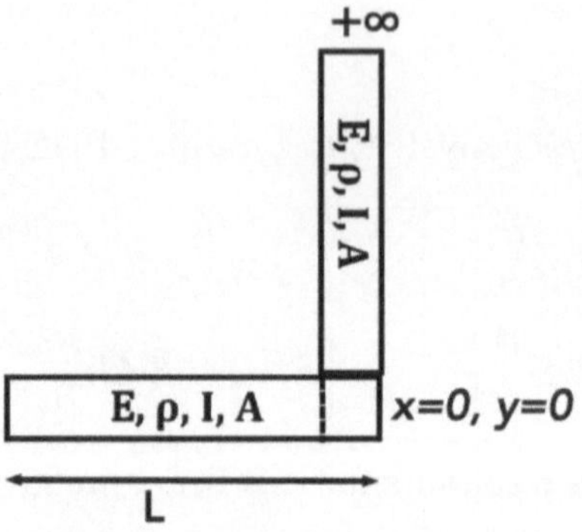

Figure for Problem 4.11

CHAPTER 5

Wave Propagation in Viscoelastic Waveguides

The distinction between elastic and viscoelastic materials is not very clear. However, viscoelasticity is the property of materials that exhibit both viscous (dash pot-like) and elastic (spring-like) characteristics when undergoing deformation. Viscoelastic materials are those for which the relationship between stress and strain depends on time. That is, their constitutive relations depend on strain rate, stress rate or both. A purely viscous materials like water when stressed resist the shear flow and the strain linearly. On the other hand, metallic structures, when stressed, deform and return back to its original position when the stress is removed. A viscoelastic material is the one that exhibits both these properties. Materials such as food, synthetic polymers, wood, soil and biological soft tissue as well as metals at high temperature display significant viscoelastic effects.

The viscoelastic possess the following four important properties [147]:

- When the material is stressed, at gets strained. However, when the stress is removed, the material does not spring back to original position. This process causes what is termed as *hysteresis.* The area under the stress-strain curve is the amount of energy available for dissipation. This is the energy that is normally absorbed in the case the structure that is subjected to impact loading.
- **Stress relaxation** (Here, constant strain results in decreasing stress)
- **Creep** (Here, constant stress results in increasing strain), and
- The constitutive model of the viscoelastic material will depend on stress σ, strain ϵ, the strain rate $\dot{\epsilon}$, stress rate $\dot{\sigma}$ and even higher derivatives of stress or strain depending on the model used. The constitutive models are discussed in the later part of this chapter.

DOI: 10.1201/9781003120568-5

A good discussion on the theory of viscoelasticity can be found in [147].

Creep and stress relaxation tests are convenient for studying material response at long times (minutes to days), but less important at shorter times (seconds and less). Dynamic tests in which the stress (or strain) resulting from a sinusoidal strain (or stress), are often well suited for filling out the short time range of viscoelastic response. We illustrate these ideas with a discussion of the stress resulting from a sinusoidal strain. In a typical dynamic test carried out at a constant temperature, one programs a loading machine to prescribe a cyclic history of strain to a material sample rod given by

$$\epsilon(t) = \epsilon_0 \sin(\omega t) \tag{5.1}$$

where ϵ_0 is the amplitude and ω is the frequency. The response of stress as a function of time t depends on the characteristics of the material, which can be separated into several categories, some of which are given below:

- **A Purely Elastic or Hookean Solid**: Here, for the stress is proportional to strain at all times, that is, $\sigma(t) = E\epsilon(t)$. Substituting Eqn. (5.1) in this equation, we get $\sigma(t) = E\epsilon_0 \sin(\omega t)$. Here the stress amplitude $\sigma_0 = E\epsilon_0$ is linear with strain amplitude ϵ_0 and the response of stress caused by strain is immediate. That is, the stress is in *phase* with the strain

- **A Purely Viscous Material**: For this material, the stress is proportional to the strain rate, that is, $\sigma(t) = \eta d\epsilon/dt$, where η is the coefficient of viscosity with a unit of *Pascal – second*. Hence, the strain given in Eqn(5.1) becomes

  $$\sigma(t) = \eta\epsilon_0\omega \cos(\omega t) = \eta\epsilon_0\omega \sin\left(\omega t + \frac{\pi}{2}\right) \tag{5.2}$$

  From the above equation, we see that stress amplitude $\eta\epsilon_0\omega$ is linear with strain amplitude and they depend on frequency ω. In addition, the stress response is out-of-phase with the strain response by 90°.

Viscoelastic materials find great use in attenuating the waves in structures. We will see in this chapter how the spectral analysis can elegantly be employed for materials exhibiting linear viscoelasticity property for studying its wave propagation characteristics, especially the wavenumbers, phase speeds and the group speeds. The fundamental to spectral analysis is the characterization of constitutive models for viscoelastic waveguides. In the next section, we will see how different constitutive models can be developed for the viscoelastic waveguides using the concept of springs and dashpots. In the development of the constitutive models, we will deal with only linear viscoelastic material, wherein the characteristic feature is that the stress is linearly proportional to the strain or its strain history, and it is important to note that the property of linearity of response does not refer to the shape of any material response curve. Linear viscoelasticity is usually applicable only for small deformations and/or for linear materials. Thus, infinitesimal strain theory needs to be employed for their analysis.

5.1 CONSTITUTIVE MODELS FOR VISCOELASTIC WAVEGUIDES

Springs and dashpots are the basic building blocks in developing the constitutive models for viscoelasticity. These can be connected in various forms to construct different constitutive models. As mentioned earlier, springs account for elastic behavior, while the dashpots are used to describe the viscous or (fluid) behavior. The behavior of these elements is shown in Fig. 5.1.

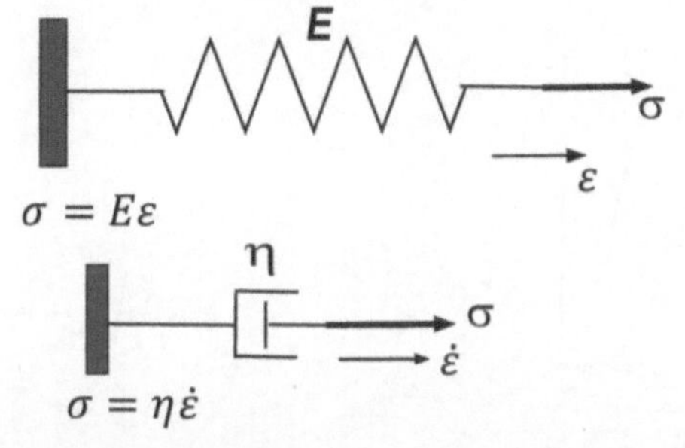

FIGURE 5.1: Elastic and viscous representation in a viscoelastic material

Note that when the spring is subjected to constant stress, it produces a constant deformation, while when the dashpot is subjected to constant stress, it produces constant rate of deformation. Also, in linear viscoelasticity assumption, the deformation of the spring is completely recoverable upon the release of stress, while in the case of dashpot, the deformation is permanent.

Using the spring and dashpot as basic elements, we will develop constitutive models for viscoelastic material by different arrangements of these elements. Here, we will discuss six different viscoelastic models, two of them are two-parameter models, while the other four are three-parameter models. Higher parameter models can also be derived using the concepts used in deriving two and three-parameter models, which is left to the readers to derive themselves. Since the constitutive models are time dependent, Fourier Transform of these models will yield the equivalent constitutive relations in the frequency domain of the form $\hat{\sigma} = \hat{E}(\omega)\hat{\epsilon}$, where $\hat{\sigma}$ is the frequency domain stress, $\hat{\epsilon}$ is the frequency domain strain with $\hat{E}(\omega)$ being the equivalent frequency domain Young's modulus, which will be complex and will be of the form $a + ib$, where the real part a is called the *Storage Modulus* while the imaginary part b is called the *Loss Modulus*. The storage modulus is due to elastic part while the loss modulus is due to viscous part of the constitutive model. It is this loss modulus that is responsible for hysteresis and hence energy dissipation in high impact situations.

We will first derive the constitutive model and hence $\hat{E}(\omega)$ for two different two-parameter models, followed by four different three-parameter models. In these models, some are viscous dominated models, which we call *fluid models*, while others are spring dominated models, which we call it as *solid models*.

5.1.1 Two-Parameter Models

There are two different two-parameter viscoelastic models, the arrangement of the spring and dashpot for these models are shown Fig. 5.2. The first is the *Kelvin-Voigt* model shown in Fig. 5.2(a) and the second is the *Maxwell* model.

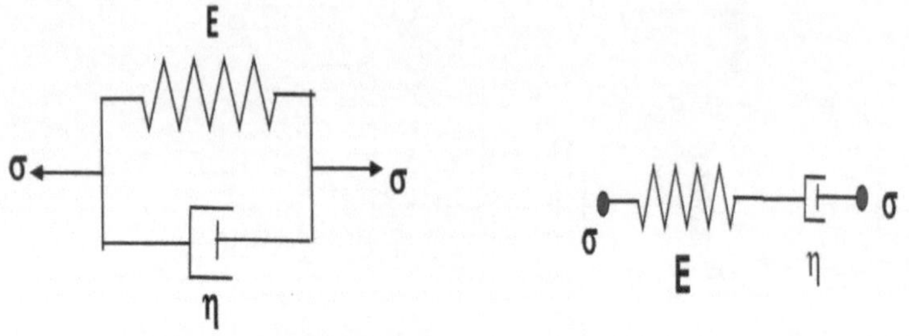

(a) Kelvin-Voigt Model (b) Maxwell Model

FIGURE 5.2: Two-parameter viscoelasticity models

For the Kelvin-Voigt model, the spring of stiffness E and dashpot η are connected in parallel. Here, the total stress is σ, while the stress in spring is σ_{spring} and the stress in the dashpot is $\sigma_{dashpot}$. The total stress σ is equal to

$$\sigma = \sigma_{spring} = \sigma_{dashpot} \tag{5.3}$$

and the strain ϵ will follow the relation

$$\epsilon = \epsilon_{spring} + \epsilon_{dashpot} \tag{5.4}$$

where

$$\sigma_{spring} = E\epsilon, \qquad \sigma_{dashpot} = \eta\dot{\epsilon} \tag{5.5}$$

Using Eqns. (5.4 & 5.5) in Eqn. (5.3), we can write the *Constitutive equation of the Kelvin-Voigt model* as

$$\sigma(x,t) = E\epsilon(x,t) + \eta\dot{\epsilon}(x,t) = E\epsilon + \eta\frac{d\epsilon}{dt} \tag{5.6}$$

Taking Fourier Transform on Eqn. (5.6), we get

$$\begin{aligned}\hat{\sigma} &= E\hat{\epsilon} + i\eta\omega\hat{\epsilon}\\ \hat{\sigma}(\omega) &= \hat{E}(\omega)\hat{\epsilon}(\omega)\end{aligned} \tag{5.7}$$

where the frequency domain modulus $\hat{E}(\omega)$ for Kelvin-Voigt model is given by

$$\hat{E}(\omega) = E + i\eta\omega \tag{5.8}$$

The value of $\hat{E}(\omega)$ is very critical for wave propagation analysis. The loss modulus provides necessary damping for wave attenuation

Next, we will derive the constitutive relations for the Maxwell model. In this model, the spring and the dashpot are connected in series, as shown in Fig. 5.2(b). In this case, the total strain is equal to

$$\epsilon = \epsilon_{spring} + \epsilon_{dashpot}$$

where

$$\epsilon_{spring} = \frac{\sigma_{spring}}{E}, \quad \dot{\epsilon}_{dashpot} = \frac{\sigma_{dashpot}}{\eta} \tag{5.9}$$

In this model, we have the total stress σ will follow

$$\sigma = \sigma_{spring} = \sigma_{dashpot} \tag{5.10}$$

Combining Eqns. (5.9 & 5.10), we can write the *constitutive equation for Maxwell model* as

$$\dot{\sigma} + \frac{E}{\eta}\sigma = E\dot{\epsilon} \tag{5.11}$$

Taking Fourier transform on Eqn. (5.11), we can write the constitutive equations for the Maxwell model as

$$\hat{\sigma} = \left(\frac{i\eta E\omega}{E + i\omega\eta}\right)\hat{\epsilon} \tag{5.12}$$

The frequency domain modulus for Maxwell model is given by

$$\hat{E}(\omega) = \left(\frac{i\eta E\omega}{E + i\omega\eta}\right) \tag{5.13}$$

Among the two-parameter models, Maxwell model is considered more fluid-oriented model, while the Kelvin-Voigt model is solid-oriented model.

5.1.2 Three-Parameter Models

Here, constitutive models of four viscoelastic models having three-parameters are discussed in this section. The arrangements of spring and dashpots for these four viscoelastic models are shown in Fig. 5.3.

These are

- Standard Linear Solid-I Model (SLS-I) shown in Fig. 5.3(a)
- Standard Linear Solid-II Model (SLS-II) shown in Fig. 5.3(b)
- Standard Linear Fluid-I Model (SLF-I) shown in Fig. 5.3(c)

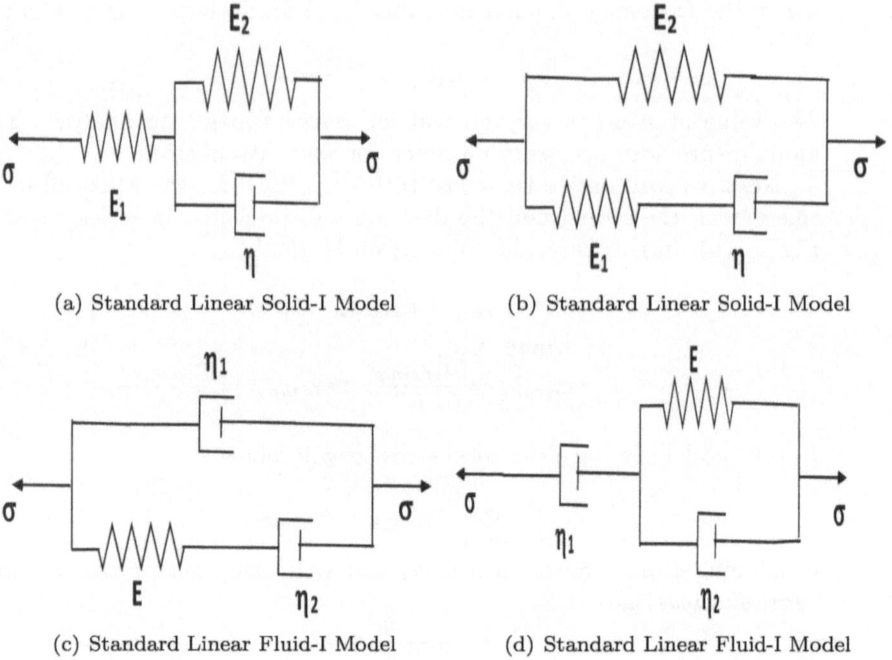

(a) Standard Linear Solid-I Model (b) Standard Linear Solid-I Model

(c) Standard Linear Fluid-I Model (d) Standard Linear Fluid-I Model

FIGURE 5.3: Four parameter viscoelasticity models

- Standard Linear Fluid-II Model (SLF-II) shown in Fig. 5.3(d)

Let us now consider SLS-I model, where two springs of constants E_1 and E_2 and a dashpot of constant η are connected, as shown in Fig. 5.3(a). It is a spring and the Kelvin-Voigt model connected in series.

Let $\sigma(t)$ and $\epsilon_1(t)$ be the stress and strain in the element with spring constant E_1. Also, let $\sigma_1(t)$ and $\sigma_2(t)$ be the stresses in element corresponding to element having spring constant E_2 and dashpot constant η, respectively. Note that the strain in these two elements are same and is equal to $\epsilon_2(t)$. In the element having spring constant E_1, we have

$$\sigma(t) = E_1\epsilon_1(t) \tag{5.14}$$

Similarly, in the segment 2 that contains elements E_2 and η, the following equations hold good

$$\sigma_1(t) = E_1\epsilon_2(t), \qquad \sigma_2(t) = \eta\dot{\epsilon}_2(t) \tag{5.15}$$

In addition, the complete viscoelastic material system shown in Fig. 5.3(a) has to also satisfy the following stress equilibrium and strain compatibility conditions.

$$\sigma(t) = \sigma_1(t) + \sigma_2(t), \qquad \epsilon(t) = \epsilon_1(t) + \epsilon_2(t) \tag{5.16}$$

where $\epsilon(t)$ is the strain of the whole systems. The aim here is to determine the relation between σ and its time derivatives with ϵ and its time derivatives using Eqns. (5.14–5.16). To do this, using the above equations is not straight forward. We can come up with the required constitutive relation by transforming the above time domain equations to frequency domain using Fourier or Laplace transform and later converting them back to time domain using the inverse transforms. The procedure is illustrated below. The Fourier Transform of the first of Eqn. (5.16) can be written as

$$\hat{\sigma} = E_2\hat{\epsilon}_2 + i\omega\eta\hat{\epsilon}_2 = \hat{\epsilon}_2(E_2 + i\omega\eta) \tag{5.17}$$

The frequency domain counterpart of the second of Eqn. (5.16) can be written using the frequency domain representation of Eqns. (5.14 & 5.15) and Eqn. (5.17) as

$$\hat{\epsilon} = \frac{\sigma}{E_1} + \frac{\sigma}{E_2 + i\omega\eta} \tag{5.18}$$

Simplifying Eqn. (5.18), we get

$$\hat{\epsilon}E_1E_2 + \hat{\epsilon}i\omega E_1\eta = \hat{\sigma}E_2 + \hat{\sigma}i\omega\eta + \hat{\sigma}E_1 \tag{5.19}$$

Taking inverse Fourier Transform on Eqn. (5.19) and noting that inverse transforms of $\hat{\epsilon}i\omega = \dot{\epsilon}$ and $\hat{\sigma}i\omega = \dot{\sigma}$, Eqn. (5.19) directly gives the constitutive model for the SLS-I viscoelastic model, which is given by

$$\sigma + \left(\frac{\eta}{E_1 + E_2}\right)\dot{\sigma} = \epsilon\left(\frac{E_1E_2}{E_1 + E_2}\right) + \dot{\epsilon}\left(\frac{E_1\eta}{E_1 + E_2}\right) \tag{5.20}$$

We have shown here how Fourier Transform can be used elegantly to derive the constitutive models for viscoelastic solids. The utility of this method will be more in the case of multi-parameter viscoelastic medium. Now the frequency domain modulus $\hat{E}(\omega)$ can be obtained as before by taking the Fourier Transform of the constitutive model (Eqn. (5.20)). That is,

$$\begin{aligned} ((E_1 + E_2) + i\omega\eta)\hat{\sigma} &= (E_1E_2 + iE_1\omega\eta)\hat{\epsilon} \\ \hat{\sigma} = \hat{E}(\omega)\hat{\epsilon}, \qquad \hat{E}(\omega) &= E_1\frac{E_2 + i\omega\eta}{E_1 + E_2 + i\omega\eta} \end{aligned} \tag{5.21}$$

Unlike the two-parameter model, the static Young's modulus at $\omega = 0$ is equal to

$$\hat{E}(0) = \frac{E_1E_2}{E_1 + E_2}$$

Next, the constitutive model for SLS-II model is derived. The arrangement of springs and dashpot for this model is shown in Fig. 5.3(b). In this model, the spring of constant E_1 is connected in parallel with a Maxwell solid.

The stress and strain in the segment having spring constant E_1 are $\epsilon_1(t)$ and $\sigma_1(t)$. Similarly, the stress and strain in the second segment having both spring and dashpot are $\epsilon_2(t)$ and $\sigma_2(t)$, respectively. The corresponding constitutive model in these two segments is

$$\epsilon_1(t) = \frac{\sigma_1(t)}{E_1}, \qquad \sigma_2(t) + \frac{\eta}{E_2}\dot{\sigma}_2 = \eta\dot{\epsilon}_2 \tag{5.22}$$

If $\sigma(t)$ and $\epsilon(t)$ are the overall stress and strain for this arrangement, they have to satisfy the following conditions

$$\epsilon(t) = \epsilon_1(t) = \epsilon_2(t) \qquad \text{and} \qquad \sigma(t) = \sigma_1(t) + \sigma_2(t) \tag{5.23}$$

Following the same procedure that was adopted for the SLS-I model, we can obtain the constitutive relations for SLS-II as

$$\sigma + \frac{\eta}{E_2}\dot{\sigma} = E_1\epsilon + \frac{\eta(E_1 + E_2)}{E_2}\dot{\epsilon} \tag{5.24}$$

As before, taking Fourier Transform on Eqn. (5.24), we can obtain $\hat{E}(\omega)$ as

$$\hat{E}(\omega) = \frac{E_1E_2 + i\omega\eta(E_1 + E_2)}{E_2 + i\omega\eta} \tag{5.25}$$

There are two more models (SLF-I & SLF-II) shown in Figs. 5.3(c) and (d). The arrangement of the spring and dashpot elements is very similar to SLS-I & SLS-II models shown in Figs. 5.3(a) and (b) except the spring and dashpot elements are interchanged, that is SLF-I & SLF-II models will have two dashpot elements η_1 & η_2 and a spring element of constant E. Since the arrangement of the dashpots and spring is very similar to SLS-I & SLS-II models, the derivation of the constitutive relations is omitted here. Following the same procedure, we can write the constitutive equation for SLF-I and SLF-II models as given below

Standard Linear Fluid-I Model:

$$\sigma(t) + \frac{\eta_2}{E}\dot{\sigma}(t) = (\eta_1 + \eta_2)\dot{\epsilon}(t) + \frac{\eta_1\eta_2}{E}\ddot{\epsilon}(t) \tag{5.26}$$

Standard Linear Fluid-II Model:

$$\sigma(t) + \frac{(\eta_1 + \eta_2)}{E}\dot{\sigma}(t) = \eta_1\dot{\epsilon}(t) + \frac{\eta_1\eta_2}{E}\ddot{\epsilon}(t) \tag{5.27}$$

As before, taking Fourier Transform on the above equation, we can obtain frequency domain modulus $\hat{E}(\omega)$, which for these two models are given by

Standard Linear Fluid-I Model:

$$\hat{E}(\omega) = \frac{i\omega E(\eta_1 + \eta_2) - \omega^2\eta_1\eta_2}{E + i\omega\eta_2} \tag{5.28}$$

Standard Linear Fluid-I Model:

$$\hat{E}(\omega) = \frac{iE\eta_1\omega - \omega^2\eta_1\eta_2}{E + i\omega(\eta_1 + \eta_2)} \tag{5.29}$$

Unlike the SLS models, the SLF models have second time derivative of strains. Hence, the constitutive model for a general linear viscoelastic solid will be a n parameter solid and its equation can be written as

$$a_0\sigma + a_1\frac{d\sigma}{dt} + a_2\frac{d^2\sigma}{dt^2} + ...a_n\frac{d^n\sigma}{dt^2}.$$
$$= b_0\epsilon + b_1\frac{d\epsilon}{dt} + b_2\frac{d^2\epsilon}{dt^2} + ...b_n\frac{d^n\epsilon}{dt^n} \tag{5.30}$$

Following the procedure outlined for two and three-parameter models, we can determine the complex frequency domain Young's modulus $\hat{E}(\omega)$ for this n parameter model.

5.2 WAVE PROPAGATION IN VISCOELASTIC ROD

The propagation of waves in viscoelastic rod is governed by the same partial differential equation except that the Young's modulus E is a function of time. That is, if $u(x,t)$ is the axial deformation of the rod having density ρ and area of cross section A, the governing partial differential equation of the viscoelastic rod is given by

$$\frac{\partial}{\partial x}\left(E(t)A\frac{\partial u}{\partial x}\right) = \rho A\frac{\partial^2 u}{\partial t^2} \tag{5.31}$$

Transforming Eqn. (5.31) to the frequency domain using DFT, and assuming the beam to be of same cross section throughout the length of the rod, we have

$$\hat{E}(\omega)A\frac{d^2\hat{u}}{dx^2} + \rho\omega^2\hat{u} = 0$$
$$\frac{d^2\hat{u}}{dx^2} + k_L^2\hat{u} = 0 \tag{5.32}$$

where $\hat{u}(x,\omega)$ is the frequency domain axial deformation and k_L is the axial wavenumber of the viscoelastic rod. It has the same form as the uniform rod. However, the presence of $\hat{E}(\omega)$ in the expression makes k_L fully complex. Hence the waves as it propagates will attenuate. The level of attenuation will be different for different viscoelastic models discussed in the previous section. We will plot the spectrum and dispersion relation for some of these models and see how the physics of wave propagation changes in comparison with uniform rod. As before, the solution of Eqn. (5.32) is given by

$$\hat{u}(x,\omega) = \mathbf{A}e^{-ik_Lx} + \mathbf{B}e^{ik_Lx}, \qquad k_L = \omega\sqrt{\frac{\rho A}{\hat{E}(\omega)A}} \tag{5.33}$$

The wavenumber given in Eqn. (5.33) is complex due to the presence of complex modulus $\hat{E}(\omega)$. That is, they will be of the form $a+ib$, where the real part is the propagating part primarily contributed by the elastic spring constant

E and the imaginary part b is primarily contributed by the dashpots and this part enables response reduction and wave attenuation.

We will now study the wave characteristics of a viscoelastic rod modeled by these six viscoelastic models. Using the wavenumber expression, we can determine both the phase speeds C_p (given by $Real[\omega/k_L]$) and the group speed C_g (given by $d\omega/dk_L$). The generalized group speed expression for any viscoelastic model can be derived by differentiating the wavenumber k_L given by Eqn. (5.33), which is given by

$$C_g = Real\left[\frac{d\omega}{dK_L} = \frac{2k_L\hat{E}(\omega)^2}{\rho\left(2\omega\hat{E}(\omega) - \omega^2\frac{d\hat{E}(\omega)}{d\omega}\right)}\right] \tag{5.34}$$

Given $\hat{E}(\omega)$, $d\hat{E}(\omega)/d\omega$ can be obtained for all the six viscoelastic models, which are tabulated in Table 5.1.

TABLE 5.1: Viscoelastic models and their complex modulus and their derivative

Viscoelastic model	$\hat{E}(\omega)$	$\frac{d\hat{E}(\omega)}{d\omega}$
Kelvin-Voigt	$E + i\eta\omega$	$i\eta$
Maxwell	$\frac{i\eta E\omega}{E+i\eta\omega}$	$\frac{iE^2\eta}{(E+i\eta\omega)^2}$
SLS-I	$E_1\frac{E_2+i\omega\eta}{E_1+E_2+i\omega\eta}$	$\frac{iE_1^2\eta}{(E_1+E_2+i\omega\eta)^2}$
SLS-II	$\frac{E_1E_2+i\omega\eta(E_1+E_2)}{E_2+i\omega\eta}$	$\frac{iE_2^2\eta}{(E_2+i\omega\eta)^2}$
SLF-I	$\frac{i\omega E(\eta_1+\eta_2)-\omega^2\eta_1\eta_2}{E+i\omega\eta_2}$	$\frac{iE^2(\eta_1+\eta_2)-2\omega E\eta_1\eta_2-i\omega^2\eta_1\eta_2^2}{(E+i\omega\eta_2)^2}$
SLF-II	$\frac{iE\eta_1\omega-\omega^2\eta_1\eta_2}{E+i\omega(\eta_1+\eta_2)}$	$\frac{iE^2\eta_1-2\omega\eta_1\eta_2E-i\omega^2\eta_1\eta_2(\eta_1+\eta_2)}{(E+i\omega(\eta_1+\eta_2))^2}$

5.2.1 Wave Propagation in Two-Parameter Viscoelastic Rod

We will first study the wave propagation characteristics of two-parameter models, that is the Kelvin-Voigt and Maxwell models. The wavenumber and group speeds of all the models depend on the value of the complex modulus $\hat{E}(\omega)$. The real and imaginary part of the wavenumber is largely contributed by this complex modulus. That is, the propagating component is mostly contributed by the spring constant E, while the damping component is always contributed by the dashpot constant η. It is necessary to find the values of these two-parameters to obtain the limiting behavior of no damping or full damping.

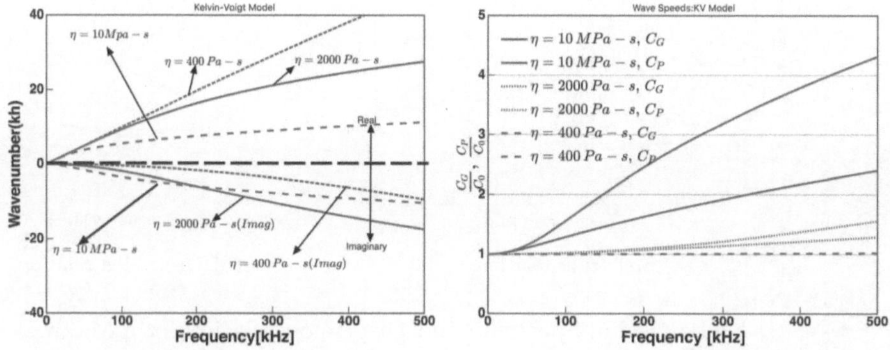

(a) Spectrum relation (b) Dispersion relation

FIGURE 5.4: Spectrum and Dispersion relation for a Kelvin-Voigt rod

Let us first consider the case of Kelvin-Voigt model. The value of the complex modulus is given in Table 5.1. Here, $\eta = 0$ recovers the elastic beam wavenumber, that is, the waves will propagate without any attenuation. If the value of the η is going to be large such that the numerical values of real and imaginary are comparable, then there will hardly be any propagation. Fig. 5.4 shows the spectrum and dispersion relation for a Kelvin-Voigt rod. Here, in getting these plots, the spring constant E is assumed as 3.0 GPa with the value of density equal to 1200 kg/m^3. From the plots we see that the wavenumber is complex, with the imaginary part increasing with increase in value of the dashpot constant η, which means that the waves will be heavily damped. It is also seen that higher value of η results in very high and unrealistic speeds. However, the waves in these cases will not propagate due to the presence of very high imaginary part in the wavenumber. From the spectrum and dispersion relations, we can infer that the wave propagation in Kelvin-Voigt rod is dispersive.

Next, the dispersion characteristic of Maxwell viscoelastic rod is discussed. Again we will look at the limiting behavior. The expression for complex modulus is given in Table 5.1. Here, $\eta = 0$ makes $\hat{E}(\omega) = 0$, which means that the wavenumber $k_L = \infty$ and hence the group speeds $C_g = 0$. This give us the hint that low values of η will result in no propagation. On the other hand, if $\eta \to \infty$, then $\hat{E}(\omega) = E$, which recovers elastic rod solution. Hence, in the case of Maxwell model, it is necessary to have high value of η for the wave to propagate with attenuation. Low value of η will completely damp out the waves.

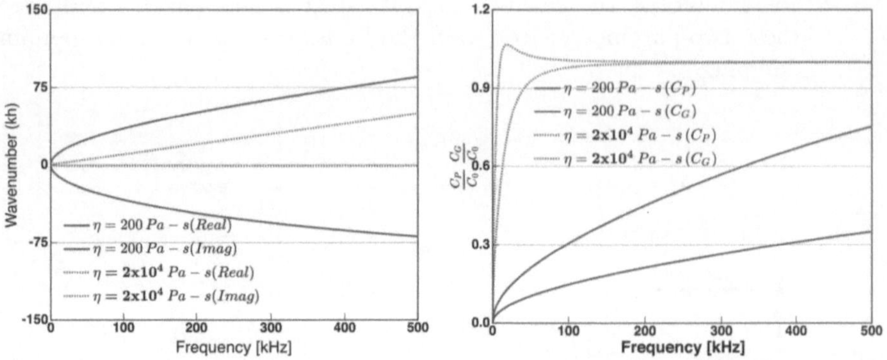

(a) Spectrum relation

(b) Dispersion relation

FIGURE 5.5: Spectrum and Dispersion relation for a Maxwell rod

Fig. 5.5 shows the spectral and dispersion relations for the Maxwell rod. As in the previous case, we have fixed the value E and ρ as 3.0 GPa and 1200 kg/m^3. The spectrum and dispersion relation were obtained for different values of η. For small values of η, the group and phase speeds are different, and they are less than that of the elastic rod and these vary with frequency, indicating that the waves are dispersive. At these values, the imaginary part of the wavenumber is very large and their numerical values are comparable to that of real part and hence there will hardly be any propagation. However, when the value of η is high, then, although speeds starts from zero frequency, it quickly attains the speed of elastic rod within over a very small frequency range and continues to propagate at this speed at all other frequencies. In summary, high value of η tends to make the rod behave elastically.

We will now look at the wave response for the Kelvin-Voigt and Maxwell infinite rod. The approach is very similar to the elastic rod presented in the previous chapter. That is, we will use the solution given in Eqn. (5.33) where we will ignore the reflected coefficient since we are considering a long infinite rod. That is, we have

$$\hat{u}(x, \omega) = \mathbf{A}e^{-ik_L x} \tag{5.35}$$

Here, the incident wave coefficient is $\mathbf{A}$, which is assumed as triangular pulse shown in Fig. 4.3, which is sampled with a sampling rate of 0.5 μsecs with

8192 FFT points. Fig. 5.6 shows the wave propagation in the Kelvin-Voigt rod. For generating these plots, as before we have used $E = 3.0$ GPa with $\rho = 1200$ kg/m^3.

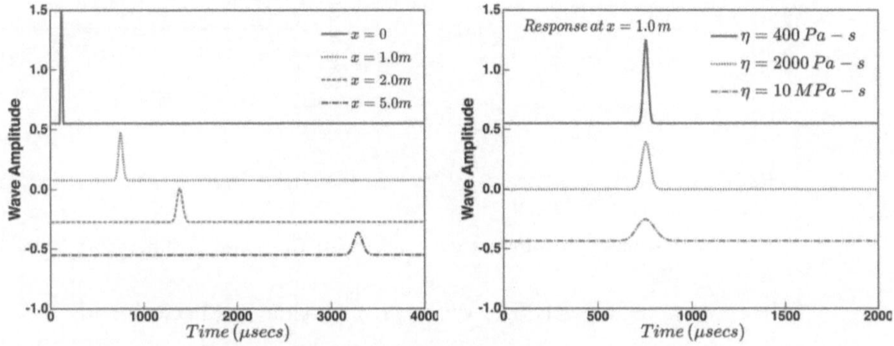

(a) Responses at different x (b) Response at $x = 1.0m$ for different η

FIGURE 5.6: Wave propagation in Kelvin-Voigt rod

Fig. 5.6(a) shows the wave propagation in a Kelvin-Voigt rod having $\eta = 2000\,Pa\text{-}s$ at different position x. From the plot we see that as the propagation distance increases, the amplitude reduction happens with broadening of the pulse width of the triangular pulse. The response amplitude reduces by more than 80% over a propagation distance of 5.0 m. This response reduction will be higher for higher value of η. This is seen in Fig. 5.6(b), where the wave propagation response is captured at $x = 1.0$ m for different values of η.

Using the same infinite rod configuration, we will plot the wave responses for a Maxwell rod, which is shown in Fig. 5.7. As before, we will use the same value of E and ρ as used in the previous example. Fig. 5.7(a) shows the wave responses for different x for a very high value of $\eta = 19000$ MPa-s. The plot shows that by the time wave propagates 2.0 m, 95% of the response amplitude is lost. Note that for the Maxwell rod, the value of the η should be high in order to have some propagation. Fig. 5.7(b) shows the propagation of the longitudinal wave when the value of η is moderate. Here we have plotted the response for $\eta = 9000$ Pa-s. For this moderate value of η, we see that the response drops to zero over a very small distance of 0.5 m.

5.2.2 Wave Propagation in Three-Parameter Viscoelastic Rod

We will first study propagation of waves in the SLS-I model . As before, looking at the expression of the complex modulus (given in Table 5.1), we will first look at the limiting condition for wave propagation (or no propagation). There are three-parameters to look at, namely E_1, E_2 and η. Let us now derive

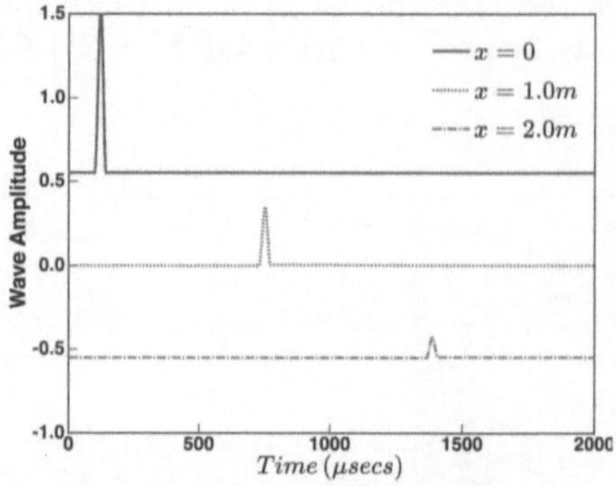

(a) Responses at different x

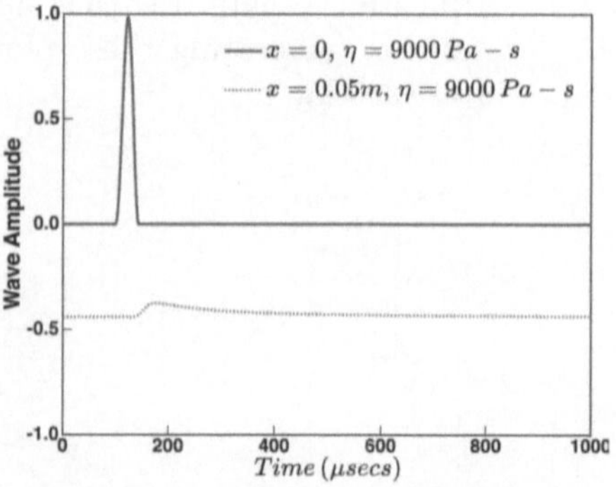

(b) Response at different x for moderate η

FIGURE 5.7: Wave propagation in Maxwell rod

the limiting conditions.

$$\hat{E}(\omega) = E_1 \left[\frac{E_2 + i\omega\eta}{E_1 + E_2 + i\omega\eta} \right]$$

$$\textbf{for} \quad \eta \to 0, \quad \hat{E}(\omega) = \hat{E}(0) = \frac{E_1 E_2}{E_1 + E_2}$$

$$\textbf{for} \quad \eta \to \infty, \quad \hat{E}(\omega) = \hat{E}(\infty) = E_1$$

$$\textbf{for} \quad E_2 \to 0, \quad \hat{E}(\omega) = \frac{iE_1\omega\eta}{E_1 + i\omega\eta}$$

which is Maxwell complex modulus

$$\textbf{for} \quad E_2 \to \infty, \quad \hat{E}(\omega) = \hat{E}(\infty) = E_1$$

In other words, this model will provide low to high wave attenuation with increase in the value of η from low to high. In addition, low value of E_2 will give the Maxwell model like responses with high attenuation.

We will first plot the wavenumbers and wave speeds for this model. Fig. 5.8 shows the wavenumber and speeds as a function of η and E_2. In generating these plots, the value of E_1 is fixed at 3.0 GPa and the value of ρ is fixed at 1200 kg/m^3.

Fig. 5.8(a) and (b) shows the wavenumber and group speed plots for different values of η. Here, the value of E_2 is fixed at 3.0 GPa for generating the plot. From these two plots, we can conclude that the level of dispersion and magnitude of the group speeds increase with increase in the value of η. At higher frequencies, the wave tends to attain the speeds of the elastic rod. Next, we will plot the speeds as well as the wavenumber as a function of the modulus E_2. These plots are shown in Fig. 5.8(c) and (d). In generating these plots, the value of η is fixed at 2000 Pa-s. From these plots, we can clearly

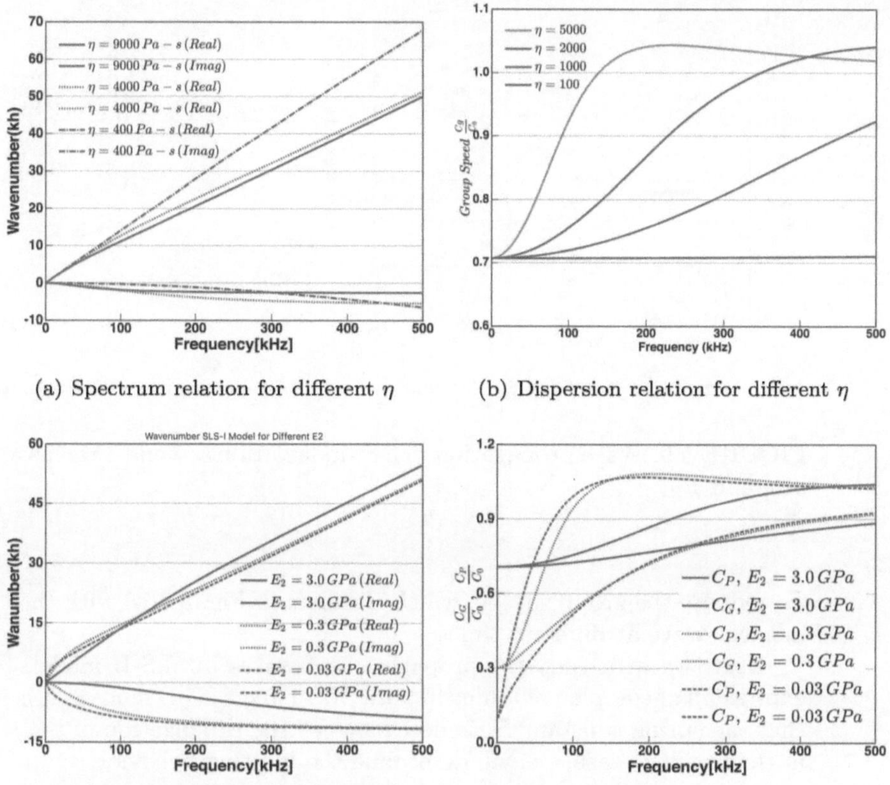

(a) Spectrum relation for different η

(b) Dispersion relation for different η

(c) Spectrum relation for different E_2

(d) Dispersion relation for different E_2

FIGURE 5.8: Spectrum and Dispersion relation for a Standard Linear Solid-I rod

see for low value of E_2, the spectrum and dispersion relations tend toward to that of Maxwell rod.

Next, the nature of wave propagation in SLS-I rod is presented as a function of η for a given E_2 and as a function of E_2 for a given η. These two plots are shown in Fig. 5.9

Fig. 5.9(a) shows the wave propagation responses using the triangular pulse used previously at $x = 1.0$ m for a fixed value of $E_2 = 3.0$ GPa with the value of η varying. The figure clearly shows that the amplitude decreases with increase in value of η and in addition shows widening of the pulse width with decreasing amplitude, which is one of the fundamental characteristics of viscoelastic responses. Fig. 5.9(b) shows the wave propagation responses for different values of E_2. Here, the value of η is fixed at 2000 Pa-s and the response is measured at 2.0 m from the impact site. Decrease in the value of

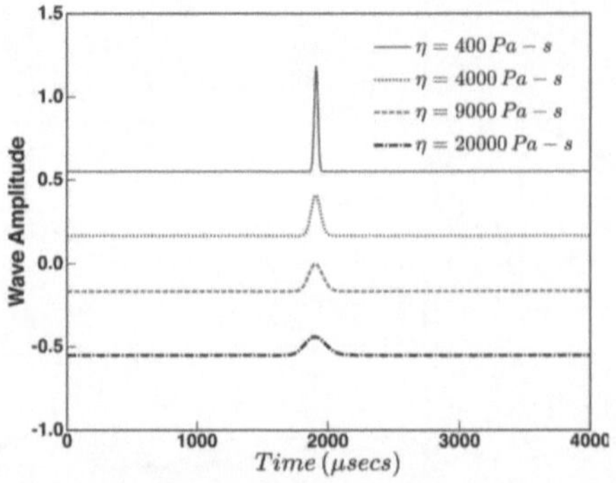

(a) Response at $x = 1.0m$ for different η

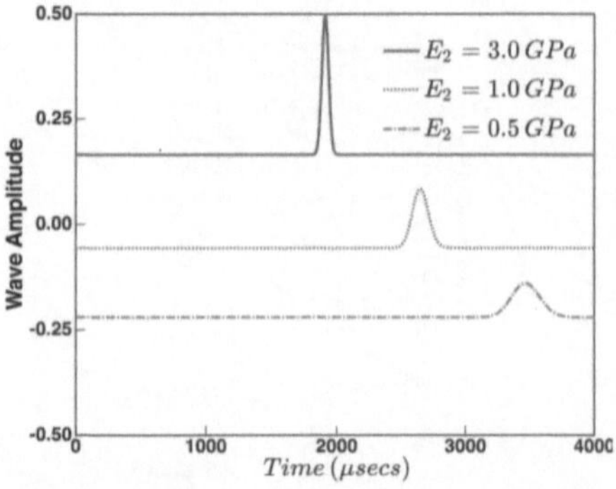

(b) Response ar $x = 2.0m$ for different E_2

FIGURE 5.9: Wave propagation in a Standard Linear Solid-I rod as a function of η and E_2

E_2 reduces the group speeds, which is shown in Fig. 5.9(b) with the arrival of incident wave at different times.

Next, we will study the propagation of waves in SLS-II model. From the expression of complex modulus in both SLS-I and SLS-II models, it is apparent that the spring constant E_1 is necessary for the propagation of the waves. As in the previous case, we will determine the limiting behavior.

$$\hat{E}(\omega) = \left[\frac{E_1E_2 + i\omega\eta(E_1+E_2)}{E_2 + i\omega\eta}\right]$$

$$\textbf{for} \quad \eta \to 0, \quad \hat{E}(\omega) = \hat{E}(0) = E_1 \qquad \textbf{Elastic rod}$$

$$\textbf{for} \quad \eta \to \infty, \quad \hat{E}(\omega) = \hat{E}(\infty) = E_1 + E_2$$

$$\textbf{for} \quad E_1 \to 0, \quad \hat{E}(\omega) = \frac{iE_2\omega\eta}{E_2 + i\omega\eta}$$

$$\textbf{which is Maxwell complex modulus}$$

$$\textbf{for} \quad E_1 \to \infty, \quad \hat{E}(\omega) = \infty \qquad \textbf{Hence no propagation}$$

$$\textbf{for} \quad E_2 \to 0, \quad \hat{E}(\omega) = E_1 \qquad \textbf{Elastic rod}$$

$$\textbf{for} \quad E_2 \to \infty, \quad \hat{E}(\omega) = E_1 + i\omega\eta$$

$$\textbf{which is Kelvin} - \textbf{Voigt complex modulus}$$

From the above relation, we can recover both Kelvin-Voigt and Maxwell model solution by proper choice of spring constants E_1 and E_2. Here, we are not plotting the wavenumber and group speed for the SLS-II viscoelastic rod as the limiting conditions itself throw lot of light on how the waves will behave in this rod. Fig. 5.10 shows the propagation of waves in this model considering the same triangular pulse considered in earlier examples.

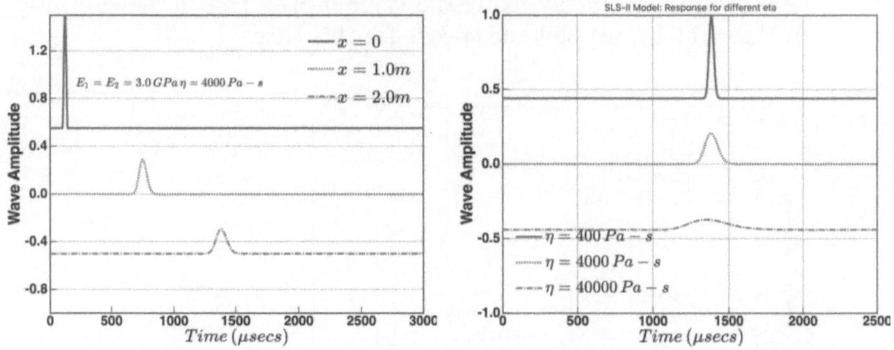

(a) Response at different x for $\eta = 4000\, Pa-s$

(b) Response ar $x = 2.0m$ for different E_2

FIGURE 5.10: Wave propagation in a Standard Linear Solid-II rod

In Fig. 5.10(a), the viscoelastic response is plotted for different values of x. To generate this plot, both $E_1 = E_2 = 3.0$ GPa, $\eta = 4000$ Pa-s and $\rho = 1200$ kg/m^3 are assumed. In this plot, we can clearly see that the amplitude decreases with the distance of propagation. In Fig. 5.10(b), the responses for different values of η are shown. Here, both $E_1 = E_2 = 3.0$ GPa is assumed. The responses are on the expected lines. It is to be noted that both SLS-I and SLS-II models are more solids oriented viscoelasticity model.

Next, we will study the SLF-I and SLF-II models, which are more fluid-oriented viscoelasticity models. As before, we will first establish the limiting behavior of the various parameters of a SLF-I model. In this model, we have three-parameters, namely E, η_1 and η_2. The limiting behavior for various parameters for this model is given below

$$
\begin{aligned}
\hat{E}(\omega) &= \left[\frac{i\omega E(\eta_1+\eta_2) - \omega^2\eta_1\eta_2}{E + i\omega\eta_2}\right] \\
\text{for} \quad & \eta_2 \to 0, \quad \hat{E}(\omega) = i\omega\eta_1 \qquad \textbf{No Propagation} \\
\text{for} \quad & \eta_1 \to 0, \quad \hat{E}(\omega) = \frac{iE\omega\eta_2}{E+i\omega\eta_2} \qquad \textbf{Maxwell model behavior} \\
\text{for} \quad & \eta_1 \to \infty, \quad \hat{E}(\omega) = \infty \qquad \textbf{Hence no propagation} \\
\text{for} \quad & \eta_2 \to \infty, \quad \hat{E}(\omega) = E - i\omega\eta_1 \qquad \mathbf{E >> \omega\eta_1} \textbf{ for Propagation}
\end{aligned} \tag{5.36}
$$

From the above equation, we can expect that the propagation in this model will be more complex compared to other earlier models studied. From the above table, we can conclude that the values of η_1 and η_2 can neither be very high nor very low. Here, we plot wavenumber as a function of frequency for different value of η_1 by fixing the value of η_2,. This is shown in Fig. 5.11(a). In Fig. 5.11(b), we plot the speeds for this case.

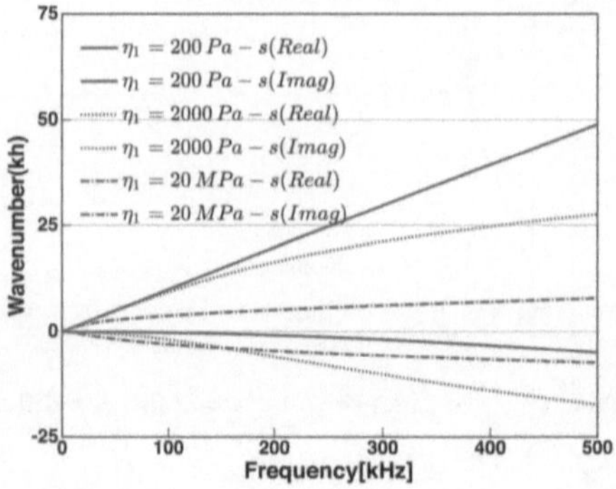

(a) Spectrum Relation for different η_1

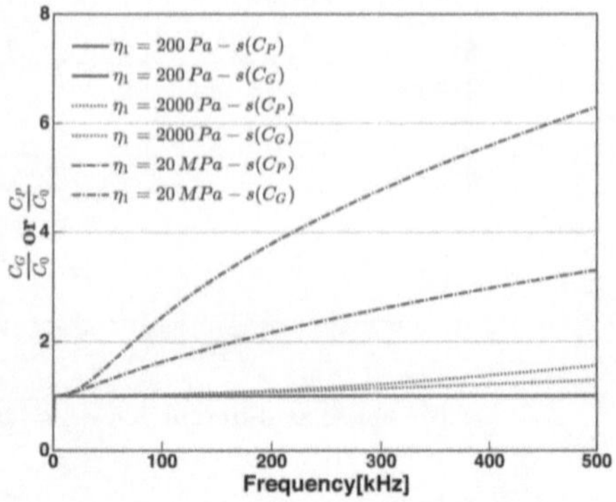

(b) Dispersion Relation for different η

FIGURE 5.11: Spectrum and Dispersion relations for a Standard Linear Fluid-I rod

From these plots, we see that for low values of η_1, the imaginary part of the wavenumber is substantial, while, the speed increases significantly. Next, we will plot the responses, where the same triangular pulse is used as incident wave. These plots are shown in Fig. 5.12. Here $E = 3.0$ GPa is used to generate the plots. Fig. 5.12(a) shows the viscoelastic wave propagation for low η_2 and high η_1. As expected, the response reduces rapidly over a propagating distance of 2.0 m. In Fig. 5.12(b), wave propagation is shown for low values of η_1 and η_2. The figures shows that the response dies town to zero over a very small propagating distance of 0.05 m.

Next, we will study the propagation of the waves in the last of the three-parameter model, namely the SLF-II Model . As before, we will establish the limiting behavior, which is given below:

$$
\begin{aligned}
&\hat{E}(\omega) = \left[\frac{i\omega E\eta_1 - \omega^2\eta_1\eta_2}{E + i\omega(\eta_1 + \eta_2}\right)\right] \\
&\textbf{for} \quad \eta_1 \to 0, \quad \hat{E}(\omega) = 0 \qquad \textbf{No Propagation} \\
&\textbf{for} \quad \eta_1 \to \infty, \quad \hat{E}(\omega) = E + i\omega\eta_2 \quad \textbf{Kelvin - Voigt model behavior} \\
&\textbf{for} \quad \eta_2 \to 0, \quad \hat{E}(\omega) = \frac{iE\omega\eta_1}{E + i\omega\eta_1} \qquad \textbf{Maxwell model behavior} \\
&\textbf{for} \quad \eta_2 \to \infty, \quad \hat{E}(\omega) = i\omega\eta_1 \qquad \textbf{No Propagation}
\end{aligned}
\tag{5.37}
$$

These expression shows that we can recover Maxwell and Kelvin-Voigt behavior by choosing $\eta_2 = 0$ and $\eta_1 = \infty$. In other words, if we need to have

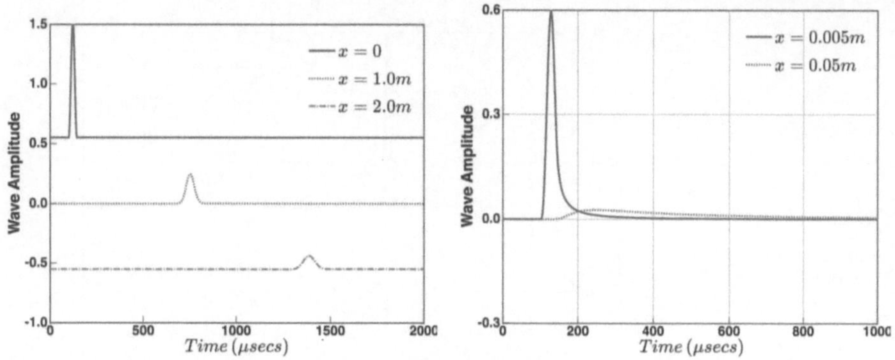

(a) Response at different x for $\eta_1 = 2000\,Pa-s$ and $\eta_2 = 200\,MPa-s$

(b) Response at different x for $\eta_1 = \eta_2 = 200\,Pa-s$

FIGURE 5.12: Wave propagation in a Standard Linear Fluid-I rod

propagation in this rod model, we need to have high value of η_2 and moderate value of η_1. Here, we are not plotting the wavenumbers and group speeds as the behavior of this model is obvious from the above limiting behavior. Fig. 5.13 shows the wave propagation response at = 1.0 m as a function of η_2. Here η_1 value is fixed at 200 MPa-s. The response clearly shows that as the value of η_2 reaches very high, the peak response goes nearly to zero.

5.3 WAVE PROPAGATION IN VISCOELASTIC BEAMS

The elastic beam modeled using Euler-Bernoulli theory has two modes of wave propagation, the propagating mode having real wavenumber and an evanescent mode having purely imaginary wavenumber. The wave propagation characteristic of the elastic Euler-Bernoulli was studied in the last chapter. One of the fundamental characteristics of viscoelastic models is that they make the wavenumber complex and this we have seen in the previous section on viscoelastic rod. Making the elastic Euler-Bernoulli beam viscoelastic, there can exist a possibility of transforming the purely imaginary evanescent to complex wavenumber, that is having both real and imaginary parts. Also, complex wavenumber can also give rise to situation that can produce negative group speeds over certain frequency band. We study these aspects for the both two-parameter viscoelastic models presented in the previous sections. We leave the study of the viscoelastic wave propagation in the four 3 parameter models to the readers. This can be accomplished by simply following the procedures outlined in the present section as well as in the previous section.

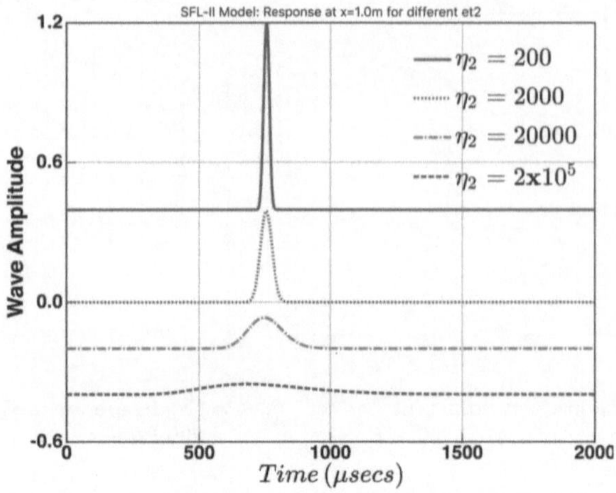

FIGURE 5.13: Wave propagation in Standard Linear Fluid-II rod model

The forward moving and damped wavenumbers for a elastic Euler-Bernoulli beam is given by

$$k_b, \qquad ik_b, \qquad \text{where} \qquad k_b = \sqrt{\omega}\left[\frac{\rho A}{EI}\right]^{0.25}$$

The phase and group speeds for this model are given by

$$C_P = \frac{\omega}{k_b}, \qquad C_G = \frac{d\omega}{dk_b}$$

where E, I, A and ρ have their usual meaning. In the case of viscoelastic beams, Young's modulus E will be complex and is represented as $\hat{E}(\omega)$, and the derivation of the complex modulus was presented in Section 5.1. The wavenumber for a viscoelastic Euler-Bernoulli beam is given by

$$k_b, \qquad ik_b, \qquad \text{where} \qquad k_b = \sqrt{\omega}\left[\frac{\rho A}{\hat{E}(\omega)I}\right]^{0.25}$$

and the corresponding phase and group speeds for this model are given by

$$C_p = Real\left(\frac{\omega}{k_b}\right), \qquad C_g = Real\left[\frac{4k_b^3\hat{E}(\omega)^2 I}{2\omega\rho A\hat{E}(\omega) - \omega^2 I k_b^4 \frac{d\hat{E}}{d\omega}}\right]$$

The details of the complex modulus $\hat{E}(\omega)$ are given in Table 5.1. The limiting conditions for all these six different models were derived in the previous section and these conditions also hold good for viscoelastic beams as well. Hence, we will directly present the propagation of waves in viscoelastic beams using the two 2-parameter viscoelastic models.

We will now study the wave propagation characteristics in the two-parameter Kelvin-Voigt and Maxwell models. Based on the limiting conditions, the range of parameters to be chosen is similar to the case of viscoelastic rods. Fig. 5.14 shows the spectrum and dispersion relations. In generating these plots, we have used the value of the spring constant $E = 30.0$ GPa, $\eta = 2000$ Pa-s and the density $\rho = 2000$ kg/m^3.

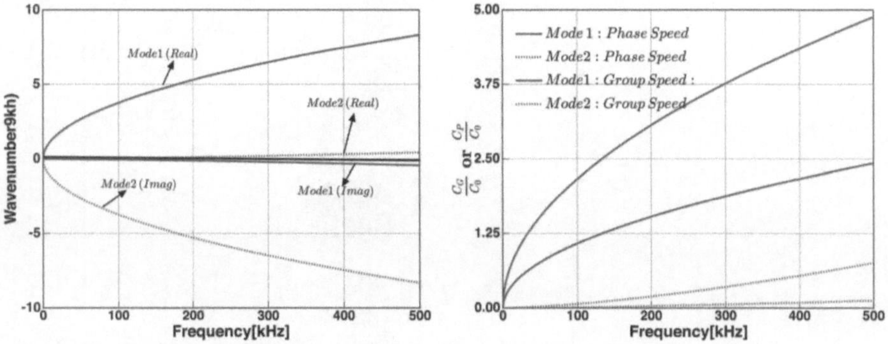

(a) Spectrum relation

(b) Dispersion relation

FIGURE 5.14: Spectrum and Dispersion relation for a Kelvin-Voigt beam

From Fig. 5.14(a), we can see that the second mode (wavenumber), although have substantial imaginary component, a small amount of real part is also introduced due to complex modulus. In Fig. 5.14(b), the speeds of this propagating second mode can also be seen. The complex modulus also seems to substantially increase speeds of the primary propagating mode. Although there is a second propagating mode, this mode has imaginary part that is many orders higher than the real part, which will damp out this mode completely.

Next, we will show the wave propagation in a Kelvin-Voigt beam. Here again we will consider an infinite beam. The solution for the governing beam equation in the frequency domain will have only incident coefficient and is given by

$$\hat{w}(x,\omega) = \mathbf{A}e^{-ik_b x} + \mathbf{B}e^{-k_b x}, \qquad k_b = \sqrt{\omega}\left[\frac{\rho A}{\hat{E}(\omega) I}\right]^{0.25} \tag{5.38}$$

where $\hat{w}$ is the transverse displacement. Unlike the response analysis in elastic beam in the previous chapter, where we assumed $\mathbf{B}$ to be zero, here we cannot assume this wave coefficient to be zero since there is a possibility of the second wavenumber, which is always imaginary in the elastic beam case, to have some real part. Hence, we have to account the both the wave coefficients to determine the response. Hence, to evaluate this, we need two conditions at $x = 0$, where the transverse force $F(t)$ is assumed to act. In the response estimation in this section, two different excitations are considered, one is the triangular pulse shown in Fig. 4.3 and second is the 60 cycle burst signal shown in Fig. 5.15. Now to determine the wave coefficients **A** and **B** in Eqn. (5.38),

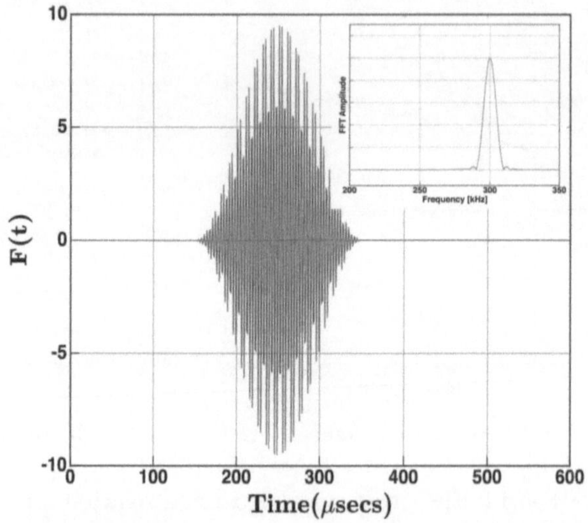

FIGURE 5.15: 300 kHz tone burst signal

the two conditions at $x = 0$ are given by

$$\begin{aligned} &\frac{d\hat{w}}{dx} = 0 \qquad \textit{Due to symmetry condition} \\ &-\hat{F} + 2\hat{V} = 0 \qquad \textit{Due to force equilibrium} \\ &\textit{where} \qquad \hat{V} = -\hat{E}(\omega) I \frac{d^3\hat{w}}{dx^3} \end{aligned} \tag{5.39}$$

where $\hat{F}$ is the FFT of the force $F(t)$. Using Eqn. (5.38) in Eqn. (5.39) and simplifying, we can write the solution as

$$\hat{w}(x, \omega) = \frac{i\hat{F}}{4\hat{E}(\omega) I k_b^3} \left(e^{-ik_b x} - i e^{-k_b x}\right) \tag{5.40}$$

Note that the above equation is derived assuming the second mode as imaginary. When the mode becomes complex, the expression will not change except that the second term exponent will have both the real and imaginary components.

Using Eqn. (5.40), we will plot the wave propagation response for a Kelvin-Voigt model. Here we will use the burst signal shown in Fig. 5.15 sampled using a sampling rate of 1.0 μsecs with 8192 points to plot the wave response. Here, the velocity wave amplitude is plotted since we are using the force $F(t)$ as input rather than the wave coefficient **A**. The responses are shown in Fig. 5.16.

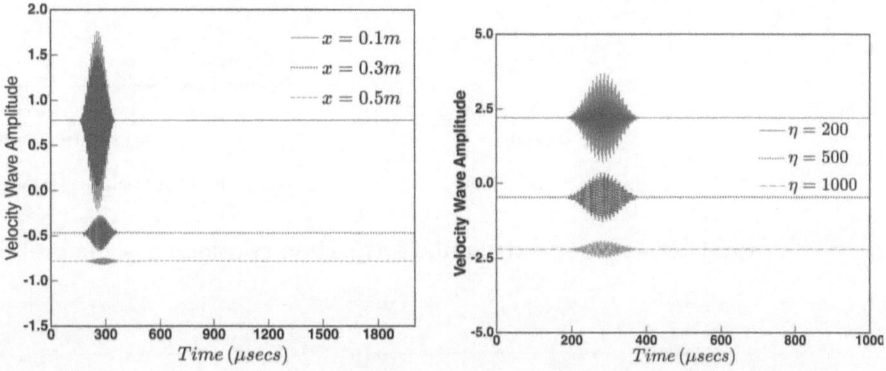

(a) Response at different position x for $\eta =$ 2000 $Pa - s$

(b) Response at different η at $x = 0.5m$

FIGURE 5.16: Wave propagation responses for a Kelvin-Voigt beam

In Fig. 5.16(a), the wave propagation response at different x from the impact site is plotted for a fixed value of η = 2000 Pa-s. We see that by the time the wave propagates a distance of 0.5 m, the amplitude reduces by more than 90%. In Fig. 5.16(b), the wave propagation response is plotted at $x =$ 0.5 m for different values of η. As expected, increase in the value of η brings down the wave amplitude, which is on expected lines.

Next, the wave propagation characteristics for a Maxwell viscoelastic beam are presented in Figs. 5.17 and 5.18.

Based on the limiting conditions for this model, the values of η should be high for propagation of waves. Fig. 5.17(a) shows the spectrum relation for different values of η. For lower values of η, the real part of the second wavenumber becomes very small causing the speeds to become high. This can be reflected in Fig. 5.17(b), where the speeds are plotted for the same values of η. One interesting aspect is that the second wave mode exhibits a small frequency band over which the group speed becomes negative. However, due to high values of the imaginary part of the second wavenumber over this frequency band, this wave will not propagate.

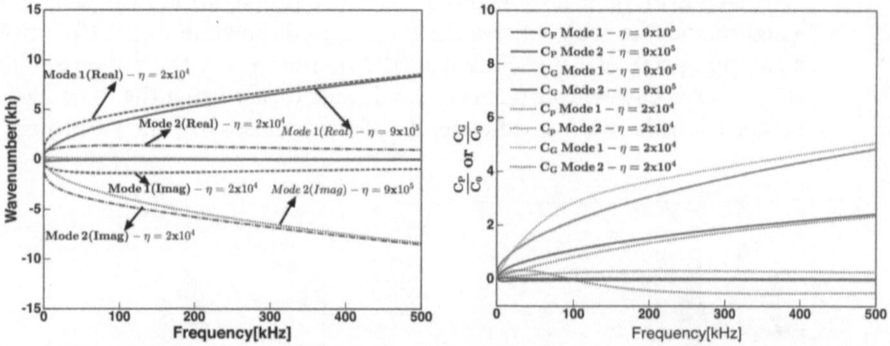

(a) Spectrum relation

(b) Dispersion relation

FIGURE 5.17: Spectrum and Dispersion relation for a Maxwell beam

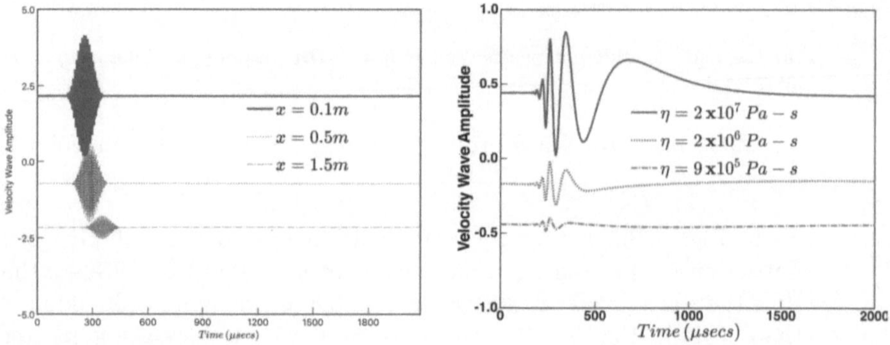

(a) Response at different position x for $\eta =$ 900x10^5 $Pa - s$

(b) Response at different η at $x = 0.5m$

FIGURE 5.18: Wave propagation responses for a Maxwell beam

Next, the wave propagation responses are plotted. Here, we have used the tone burst excitation (Fig. 5.15) as well as the broad band signal (Fig. 4.3). The responses are shown in Fig. 5.18. The idea of using burst signal is to see if at all the second mode, which is having negative speeds beyond 100 kHz, can capture the wave moving backwards. It is for this reason that the central frequency of the burst signal was created at 300 kHz. However, from the plots we see that this mode is not propagating due to high value of the imaginary part of the wavenumber. The plot also shows all the features of viscoelastic wave propagation, that is reduction of responses with decrease in the values of η and the response reduction can be seen as a function of distance of propagation.

Similar approaches can be followed to obtain beam viscoelastic responses in the three-parameter models, which is not presented here. The approach presented here is so general, that it can be extended to models having more than three-parameters. The provided MATLAB scripts can be used to generate the rod and beam responses for all the six viscoelastic models.

Note on MATLAB® scripts provided in this chapter

In this chapter, the following two MATLAB scripts are provided for performing wave propagation analysis in viscoelastic waveguides. The following two major MATLAB scripts are provided in this chapter.

MATLAB script [*viscoparam.m*] This script computes the wavenumber, group and phase speeds for both rods and beams for all the six different viscoelastic modes. The program is written in modular fashion, wherein computation of wave parameter for each of these six models is written as separate MATLAB function scripts, whose details are given below:

- **MATLAB script function** *kelvinrod.m* Computes the wavenumber and speeds for *Kelvin-Voigt* rods
- **MATLAB script function** *kelvinbeam.m* Computes the wavenumber and speeds for *Kelvin-Voigt* beams
- **MATLAB script function** *maxwellrod.m* Computes the wavenumber and speeds for *Maxwell* rods
- **MATLAB script function** *maxwellbeam.m* Computes the wavenumber and speeds for *Maxwell* beams
- **MATLAB script function** *SLSIrod.m* Computes the wavenumber and speeds for *SLS-I* rods
- **MATLAB script function** *SLSIbeam.m* Computes the wavenumber and speeds for *SLS-I* beams
- **MATLAB script function** *SLSIIrod.m* Computes the wavenumber and speeds for *SLS-II* rods
- **MATLAB script function** *SLSIIbeam.m* Computes the wavenumber and speeds for *SLS-II* beams
- **MATLAB script function** *SLFIIrod.m* Computes the wavenumber and speeds for *SLF-I* rods
- **MATLAB script function** *SLFIbeam.m* Computes the wavenumber and speeds for *SLF-I* beams

MATLAB script function *SLFIIrod.m* Computes the wavenumber and speeds for *SLF-II* rods

MATLAB script function *SLFIIbeam.m* Computes the wavenumber and speeds for *SLF-II* beams

MATLAB Script [*visco.m*] This script computes and plots the following:

MATLAB script function *triangular.m* Reads the time domain gaussian signal *pul.txt* and computes its FFT. The FFT of this signal is used by the program *visco.m* to compute to wave propagation responses.

MATLAB script function *tb.m* Generates the tone burst signal and computes its FFT. The FFT of this signal is used by the program *visco.m* to compute to wave propagation responses.

Summary

In this chapter, propagation of linear viscoelastic propagation of waves in one-dimensional waveguides is presented. Viscoelastic waveguides are those which have constitutive models that are dependent on not only stresses and strains but also their time derivatives. The constitutive models for a viscoelastic material are derived using the elastic component modeled as a springs and the viscous component modeled as dashpots. The result of these models is that the material will have Young's modulus, which is complex, and it is a function of frequency ω. The real part of modulus is the *storage modulus*, while the imaginary part is the *loss modulus*. It is this complex modulus that makes the waves to attenuate in the viscoelastic waveguides.

The frequency domain representation of the constitutive relations for all the six viscoelastic models is obtained and their limiting behavior is derived for each one of the models. This process helped us to identify the range of constitutive model parameters (E, E_1, E_2, η_1, etc.) to be chosen in each one of these models in order that the wave either propagates or gets damped out with minimal propagation. Wave propagation parameters such as phase speeds, group speeds and wavenumbers are derived as a function of constitutive model parameters and presented for all of these models. In some of the models, the evidence of negative speeds seems to be present. However, a close examination of these models reveals that these wave modes will not propagate due to high value of imaginary number in the corresponding wavenumber. The response attenuation is always associated with broadening of the input signal, especially when the waveguide is subjected to gaussian signal. Also, several response studies on these models are performed to show the effectiveness of the viscoelastic models in attenuation of the waves. In summary, this chapter shows the power of spectral analysis in handling complex wave propagation in 1-D viscoelastic waveguides.

Exercises

5.1 The reflected and transmission coefficients for a stepped isotropic rod was derived in Chapter 4 (see Section 4.2.6) and it was shown that these coefficients were fully real and were not frequency dependent. Following the same procedure as was followed in Chapter 4, derive these coefficients for a stepped viscoelastic rod. Obviously, these coefficients will be complex as well as frequency dependent. Plot these reflected and transmitted coefficients as a function frequency for different complex modulus $\hat{E}_2$ of the right rod segment. Perform the analysis for the following viscoelastic models (a) Kelvin-Voight, (b) SLS-I model and (c) SLS-II model.

5.2 A viscoelastic material insert is sandwiched between the isotropic aluminum rod system as shown in the figure below. The viscoelastic insert

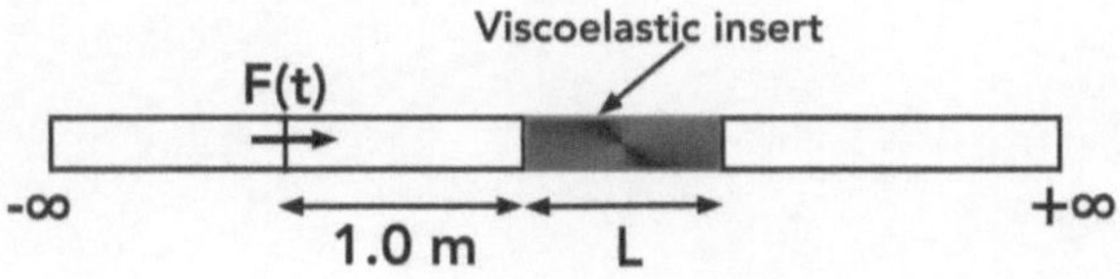

Figure for Problem 5.2

can be made from any of the six viscoelastic models discussed in the chapter. The aim of this example is to show the reduction in response amplitudes in the downstream of the viscoelastic insert and this reduction in the response will depend on the length of the insert L. First, develop the equations needed to solve for the wave coefficients. For different lengths L, plot the axial velocity response at 0.5 m from the insert for an input force $F(t)$, which is assumed as a Gaussian pulse. Plot these responses individually for the insert made from six viscoelastic models discussed in this chapter and discuss the results obtained.

5.3 Plot the wavenumber and group speeds for a viscoelastic beam on an elastic (Winkler) foundation of stiffness K for all the six viscoelastic models discussed in this chapter.

5.4 Plot the wavenumber and group speeds for a viscoelastic beam subjected to pre-tension and pre-compression of value $\pm T$ for all the six viscoelastic models discussed in this chapter.

5.5 Generate the wavenumber and group speeds plots for a *Timoshenko* viscoelastic beam for (a) Kelvin-Voight (b) SLS-I model and (c) SLS-II model. Plot these wave parameters for different values of visco-elastic parameters. Following the same procedure as was explained in Chapter

4, generate second frequency spectrum for an infinite beam made from all of the models. Compare the solutions with the isotropic Aluminum beam.

5.6 Perform the same analysis as given in Problem 5.5; however, for this case consider higher-order *Mindlin-Herrmann* rod.

CHAPTER 6

Signal Processing Aspects in Wave Propagation

Wave propagation is all about time series signals and their representation in the frequency domain. In the last two chapters, we have used different time signals and their DFT counterparts to perform wave propagation in 1-D waveguides. The central to all the analysis in the previous two chapters and to the chapters that will follow this chapter is the transformation of the signal back-and-forth to frequency domain and time domain. This was made possible by Discrete Fourier Transform enforced through Fast Fourier Transform. DFT is the discrete frequency domain representation of a continuous signal, which we call it as *sampled wave forms.* Due to induced periodicity in both the time and the frequency domain representation of the signal, there are hosts of signal processing issues, which pollutes the spectral content of the sampled wave form. In this chapter, we will discuss these issues associated with signal sampling and also outline the methods to overcome these problems.

6.1 SIGNAL PROCESSING ISSUES OF SAMPLED WAVEFORMS

This section discusses some common problems encountered in handling experimentally measured signals. Experimentally obtained signals are truncated and the Fourier Transform of the truncated signal is quite different compared to the signal with full trace. This is due to periodicity assumption of the signal. The quality of the sampled signal depends on two factors, that is, the signal itself and it's time sampling rate. If the time sampling rate is not high enough then all high-frequency wave components will appear as low frequency waves as a result of *Signal Aliasing.* If ΔT is the time sampling rate, the highest detectable frequency is given by $f = 1/2\Delta T$. That is, if the highest fre-

DOI: 10.1201/9781003120568-6

quency component of the signal is known, then the sample rate is chosen from the above criteria. We will discuss signal aliasing in little more detail in this section.

Another problem normally encountered while sampling signals is the *Spectral Leakage* of the response. This normally occurs due to non-availability of the full trace of the signal. The spectral energy of the signal that is associated with truncated signal will leak into neighboring window distorting the spectral estimates of the signal. These leakages will manifest in the form of side lobes in the frequency domain representation of the signal. This problem is mainly due to periodicity assumption of the signal both in time and frequency domain in DFT. One of the ways to minimize the spectral leaks is to window the signals. In this section we will discuss in detail all the issues concerning with time signals that are relevant to study wave propagation in structures and materials.

6.1.1 Signal Aliasing

The signal does not alias when they are represented by continuous time sinusoids. However, when the same signal is represented using discrete sinusoids, one may expect signal aliasing as per the *Nyquist Sampling Theorem* [40]. We will explain the reasons for signal aliasing and the remedy to overcome it.

It is well known that the discrete-time sinusoids are not unique. The value of sine or cosine function at frequency say ω is same as the value of the sine or cosine function at frequency $\omega + 2n\pi$ for all integer values n. This is the fundamental reason for the signal to alias. For example, we have

$$\cos\left(\frac{\pi}{4}n\right) = \cos\left(\frac{9\pi}{4}n\right) = \cos\left(\frac{\pi}{4}n + 2\pi n\right) = \cos\left(\frac{\pi}{4}n\right) \tag{6.1}$$

The above property does not apply to continuous sinusoids because the variable *time* is not limited to integers and signal functions are continuous functions of time. From Eqn. (6.1), we see that discrete functions are unique only between the frequency interval of 0 and 2π. Aliasing of the signal will always occur when we convert a continuous time signal to a discrete time signal unless there is a unique one-to-one mapping between continuous and discrete time frequency. That is, the continuous sinusoids with distinct frequencies are always unique. When the same is sampled with discrete sinusoids, it does not fully inherit the uniqueness of discrete sinusoids. To demonstrate signal aliasing, let us consider the following three continuous functions having frequencies 60 Hz, 340 Hz and 460 Hz

$$x_1(t) = \cos(2\pi^* 60t + \pi/3)\ x_2(t) = \cos(2\pi^* 340t - \pi/3)\ x_3(t) = \cos(2\pi^* 340t + \pi/3) \tag{6.2}$$

Let the above function be discretely sampled at 400 Hz or at 1/400 sec = 0.025 secs. The continuous function given in Eqn. (6.2) is sampled at a very fine interval of $t = 0.00025$ sec in order simulate the continuous time signal. The discrete and the continuous representation of the three signals $x_1(t)$, $x_2(t)$ and $x_3(t)$ is shown in Figs. 6.1 (a), (b) & (c).

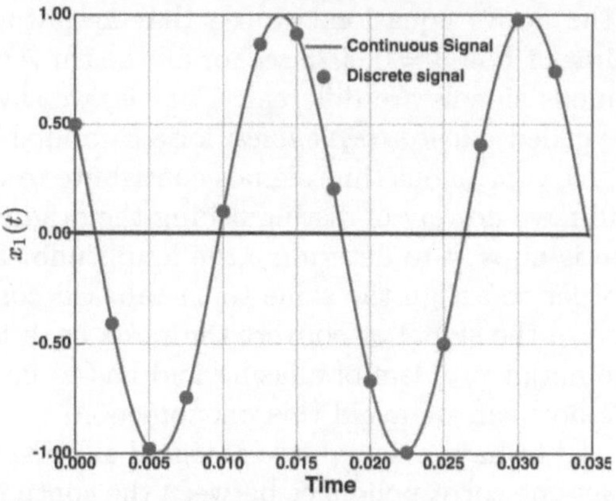

(a) Frequency 60 Hz

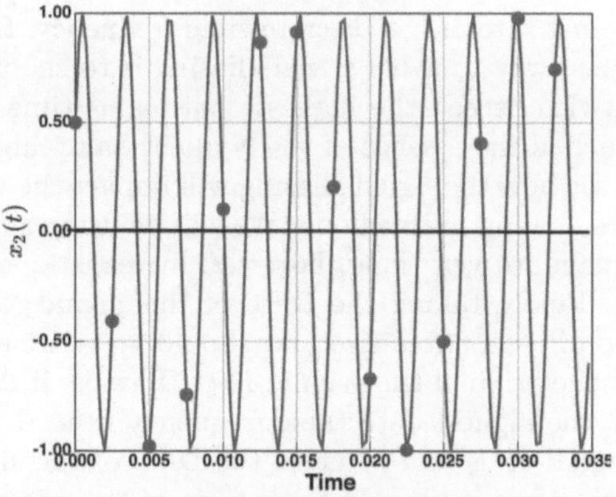

(b) Frequency 340 Hz

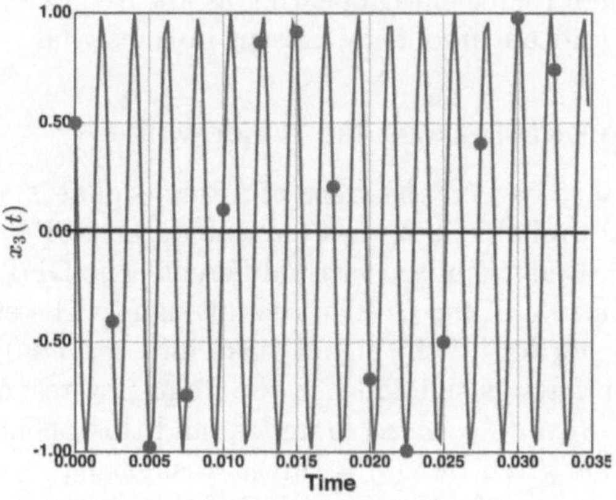

(c) Frequency 460 Hz

FIGURE 6.1: Aliasing in signal processing

From the above equation, we see that amplitudes of the signal at each sampled time of $1/400 = 0.0025$ sec for all the three signals are same, although the continuous signals are different. That is, we say that 60 Hz, 340 Hz and 460 Hz are called *aliased frequencies* for a sampling rate of 400 Hz. Hence, we see that three continuous time signals contribute to the same discrete sinusoid and we will have no way of distinguishing the original continuous time signal. Also, there is no way to determine the amplitude as well as the phase of the signal in order to assign the same to the aliased continuous time signal if we try to analyze the signal or convert the same back to continuous time signal. This is the major problem of aliasing and has to be avoided at any cost. The question is how can we avoid this phenomenon?

If we need to have a signal free of signal aliasing problem, then there need to be one-on-one correspondence between the continuous function and the discrete time function. This can happen if we limit the continuous time function frequency range to say $-f_s/2 < f < f_s/2$, where $f_s = 1/\Delta T$ and ΔT is the time sampling rate of the discrete-time frequency. In the case wave propagation, the best way to avoid signal aliasing is to choose the time sampling rate ΔT that twice exceed the highest continuous time frequency present in the signal, which is the essence of the Nyquist Sampling Theorem.

Let us see how the signal aliasing will impact the wave propagation analysis that we are dealing in this book. We will be using a variety of the signals. The actual signals are very long, however, in realistic situation, they need to be truncated. Before taking the DFT of this signal, one must ensure that the signals are of reasonable size and should to great extent retain the spectral content of the original time signal $F(t)$. That is, if ΔT is the sampling rate of the signal, the highest detectable frequency needed is the Nyquist Frequency, which is equal to $1/2\Delta T$. Hence, the DFT of the signal will depend on $F(t)$ and the way it is sampled. In order to meet the criteria as outlined by Nyquist Sampling theorem, the sampling rate of the signal $F(t)$ should be high enough to avoid high frequencies appearing as low frequency, which is dictated by the highest significant frequency present in the signal.

6.1.2 Spectral Leakage and Windowing

As mentioned earlier, the trace of a time signal is always not fully available. The signal available is always truncated. The CFT of the full signal and the DFT truncated signal are vastly different. The DFT of a truncated signal is a smeared version of the CFT of the full signal. The energy associated with the truncated portion of the signal leaks into the DFT of the truncated signal, which manifests as side lobes in the frequency response. This phenomenon is called the *Signal Leakage*. To understand this phenomenon, let us consider a sine function given by $y(t) = \sin \omega_0 t = \sin 2\pi f_0 t$, where $0 < t < \infty$, where f_0 is assumed as 50 kHz. The CFT of this signal is given by

$$\hat{y}(\omega) = \int_{t=0}^{t=\infty} y(t)\, e^{i\omega t} dt = \int_{t=0}^{t=\infty} \sin \omega_0 t e^{i\omega t} = -\pi\delta(\omega - \omega_0) \tag{6.3}$$

The signal $y(t)$ and its CFT of the signal are shown (in the insert on the right) in Fig. 6.2.

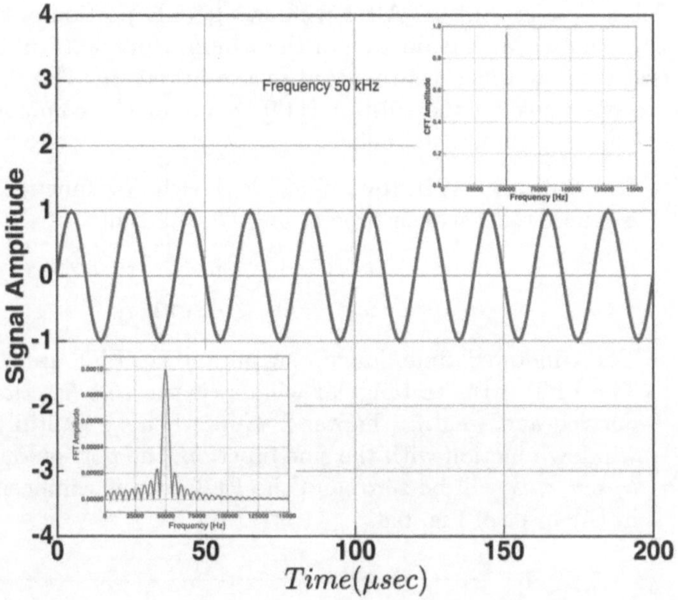

FIGURE 6.2: Sine signal, its CFT and its DFT sampled at frequency $\omega = 50$ Hz

The amplitude of the CFT of $y(t) = \sin \omega_0 t$, where $\omega_0 = 50$ kHz, is the dirac delta function evaluated at $\omega = \omega_0$, which is shown in the right inset of Fig. 6.2. If this signal is truncated at 150 kHz as shown in the above figure and if the DFT is taken on this truncated sine signal, the transform is what is shown on the left insert in Fig. 6.2. The DFT of the signal, although peaks exactly at $\omega = \omega_0 = 50$ kHz, there are a number of side lobes close to $\omega = \omega_0$. These lobes are due to signal leakage caused by the truncation of the sine function. These side lobes in the frequency domain appear even when the sine function are complete by itself, that is, the signal is truncated after completing a period due to the fact that the number of points to be chosen in FFT are in the powers of 2^n, for any integer n. The leakage will be much more if the signal is truncated somewhere in between before it completes the full period.

If we need to reasonably estimate the spectral content of the signal, we need to eliminate these side lobes. This can be done through a technique called *Signal Windowing.* Windowing reduces the amplitude of discontinuities at the boundaries of the signal. Windowing consists of a function called the *Window functions*, which when multiplied with the original signal, makes the end points of the waveform meet thus making it devoid of any sharp transitions.

In this process, it will ensure that the original function will be zero valued outside the window domain. The window functions are chosen such that its amplitude varies smoothly and reduces to zero at the edges somewhat similar to Poisson's distribution. After windowing, only portion of the original signal overlapping within the domain of the window function will be left. There are several window functions reported in the literature. The reader can refer to the recent book on the subject [114]. Some of the commonly used window functions are summarized below:

- **Rectangular Window:** If $w(t)$ is the window function, and if the signal is windowed between time t_0 and t_1 such that

$$\begin{aligned} w(t) &= 1 \qquad \text{for} \qquad t0 < t < t_1 \\ w(t) &= 0 \qquad \text{elsewhere} \end{aligned}$$

  The windowed time domain signal and its FFT are shown in Fig. 6.3. The FFT of the rectangular window is the *sinc function*, which we have derived and seen in Chapter 2. Now when we multiply the rectangular window function with the sine function, the portion outside the domain $t_0 < t < t_1$ will be zero, and the FFT of this composite signal is given in the inset of Fig. 6.3.

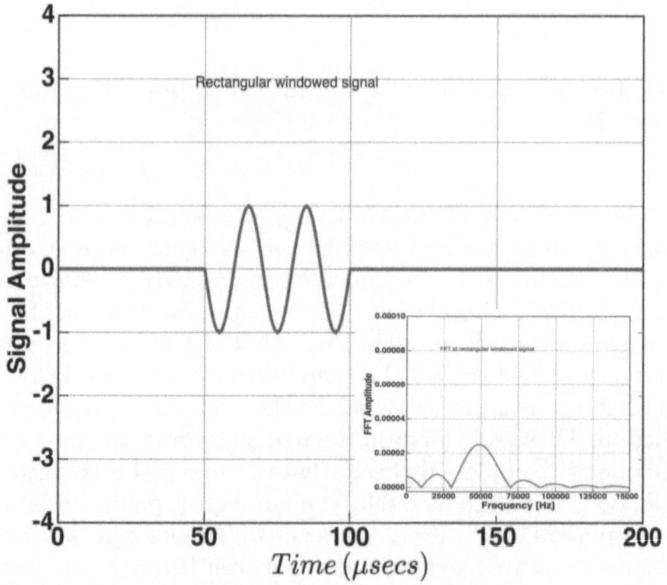

FIGURE 6.3: Rectangular windowed sine signal, its CFT and FFT sampled at frequency $\omega = 50$ Hz

From the figure we see that the rectangular window certainly reduces the side lobes but does not eliminate it completely.

- **Triangular Window:** Here again if $w(t)$ is the window function and if the signal is windowed between time t_0 and t_1, where the length of the window is denoted by $L = t_1 - t_0$ and if N is the number of points within L, where $N = f_s/L$ and f_s is the sampling frequency expressed in Hertz. The triangular window function is given by

$$w(t) = 1 - \left| \frac{t - \frac{N}{2}}{\frac{L}{2}} \right| \qquad \text{for} \qquad t_0 < t < t_1$$
$$w(t) = 0 \qquad \text{elsewhere}$$

- **Hanning Window:** If $w(t)$ is the Hanning window function, and if the signal is windowed between time t_0 and t_1, where the length of the window is denoted by $L = t_1 - t_0$ and if N is the number of points within L, where, $N = f_s/L$ and f_s is the sampling frequency expressed in Hertz, the expression for Hanning window is given by

$$w(t) = 0.5 \left[1 - \cos\left(\frac{2\pi t}{N}\right) \right] \qquad \text{for} \qquad t0 < t < t_1$$
$$w(t) = 0 \qquad \text{elsewhere}$$

The profile of Hanning window function is shown in Fig. 6.4(a). The

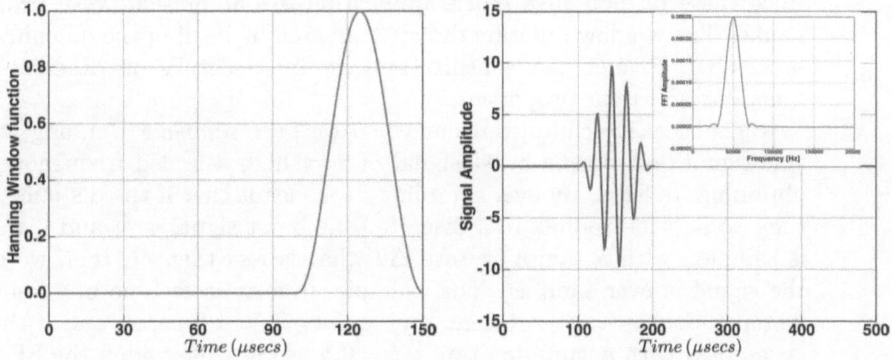

(a) Hanning window profile for sampled frequency 50 kHz

(b) Hanning Windowed sine signal and its FFT

FIGURE 6.4: Hanning window profile, Hanning windowed signal and its FFT

profile of Hanning function is similar to smoothed triangular pulse we used in our earlier wave propagation studies. When this Hanning window function is applied to the sine signal, the resulting time signal and the corresponding FFT of this windowed time signal are shown in Fig. 6.4(b). From the above figure, the side lobes, initially seen in the FFT of the pure sine signal, are 99% eliminated. Hanning window is one of the very effective window functions used to reduce spectral leakage. The Hanning windowed sine signal is nothing but the *tone burst signal* or *modulated sine signal*, which we used in the previous two chapters. This signal is a single frequency dominated signal and has great utility in studying wave propagation in dispersive waveguides such as beam waveguides studied in the earlier chapters. This signal propagates non-dispersively even in a dispersive waveguides. This is because, all the energy in this signal is dominated around a small band of the sampled frequency f_s.

- **Hamming Window::** This window function $w(t)$ for the Hamming window is given by

$$w(t) = 0.54\left[1 - \cos\left(\frac{2\pi t}{N}\right)\right] \qquad \text{for} \qquad t0 < t < t_1$$
$$w(t) = 0 \qquad \text{elsewhere}$$

  This choice of factor 0.54 will place a zero-crossing at frequency $5\pi(N-1)$, which cancels the first side lobe of the Hanning window, giving it a height of about one-fifth that of the Hanning window.

There are several other window functions, which we are not mentioning here since these do not have much application in our present wave propagation studies.The window function design is an area by itself in the domain of signal processing. Reader can consult [114] for more details on different window functions and their properties.

Signal leakage can also occur when we *over sample* a Hanning windowed sine signal or the tone burst signal of very high sampled frequency f_s while obtaining its FFT. By over sampling, it is meant that if the distance between the two adjacent points in a discrete tone burst signal is δt and if this signal is sampled with a sampling rate ΔT that is less than δt, then we call that the signal is over sampled. For example, in a discrete tone burst signal, the distance between two adjacent time points $\delta t = 1.0$ μsecs, and if this signal is sampled with a sampling rate $\Delta t = 0.5$ μsecs while taking the FFT of this discrete tone burst signal, we call this case as over sampled. This kind leakage can happen when the sampling frequency f_s is very high.

To demonstrate the signal leakage due to over sampling of the signal, let us consider a 200 kHz tone burst signal shown in Fig. 6.5(a) and its FFT shown in the inset. This signal is sampled with a sampling rate of 1 μsecs. The FFT plot clearly shows a single peak corresponding to 200 kHz. Now the same signal is sampled with a sampling rate, smaller than 1 μsecs, and the FFT plot for this case is shown in Fig. 6.5(b).

If the sampling rate is reduced, it shifts the Nyquist frequency further down in the frequency scale. If the signal is over sampled, that is, if the sampling rate is reduced, we see additional peaks in the FFT plots (which should not be

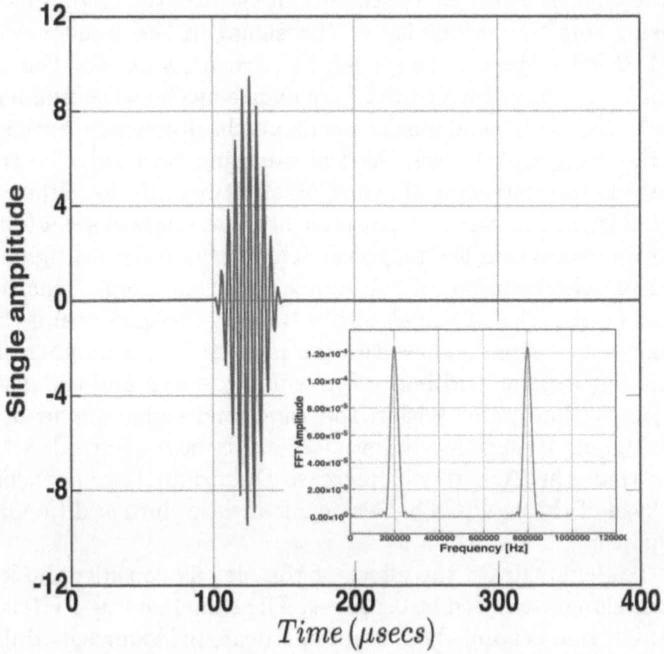

(a) 200 kHz tone burst signal and its FFT

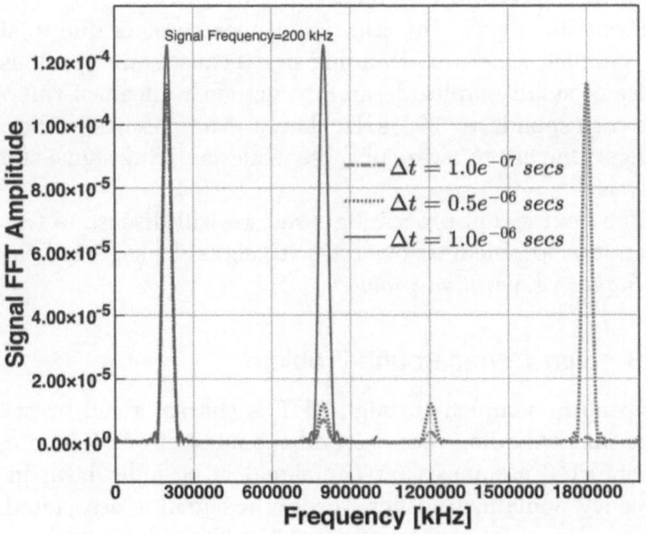

(b) FFT of 200 kHz signal for different sampling rate

FIGURE 6.5: Signal leakage in high-frequency tone burst signal there). For the case of 1 μsecs sampling rate, the second peak beyond Nyquist arising due to periodicity of the signal in the frequency domain shows up at 800 kHz. We call this peak as *periodic peak.* For the over sampled case, which increases the Nyquist frequency as well as the frequency of the periodic peak, the additional peaks occur at the frequency location of the periodic of the original 1.0 μsecs. As the sampling rate reduces, the number of such peaks keeps increasing showing up additional peaks. When such over sampled signal is used in wave propagation studies, one will see additional non-existent spurious modes in the response. This is again due to signal leakage occurring due to over sampling of the signal. The tone burst signal is characterized by number of cycles of signal within the chosen window and the frequency of the original sine signal. These two-parameters fix the number of points within the Hanning window and hence the window length and the distance between two adjacent time point within the windowed signal (or in other words ΔT) of the signal. If such a windowed signal is now sampled with a sampling rate less than the ΔT, it will increase the original window length thus causing leakage of the signal. This leakage translates into additional peaks in the FFT response.

To demonstrate the effect of this, let us consider the same 200 kHz tone burst signal, sampled at 0.1 μsecs. The signal and its FFT is shown in Fig. 6.5. If this signal is applied on an infinite beam of Young's modulus 70 GPa, density $\rho = 2700$ kg/m^3, with length and width of 0.025 m. If this signal is allowed to propagate a distance of 7.0 m of an infinite beam, the transverse wave response is shown in Fig. 6.6.

From this figure, the additional peaks arising due to signal leakage of an over sampled windowed Hanning signal cause some spurious non-existing wave modes of small amplitude and occur much ahead of the original wave mode that corresponds to 200 kHz. These examples show the ill effects of signal leakage. One has to aware of these while sampling signals for wave propagation analysis.

The next signal processing issue we will discuss in the next subsection is yet another problem associated with signal leakage and this problem is called the *Signal wraparound problem.*

6.1.3 Signal Wraparound Problem

A waveform sampled through FFT is characterized by two-parameters, that is the time sampling rate ΔT and the number of FFT point N. As mentioned earlier, FFT assumes that the signal is periodic both in the time and the frequency domain. In other words, the signal is associated with a finite time window. It is quite well known that the dispersive signal traveling in a medium with small attenuation normally does not die down within the chosen window, no matter how long the time window is. The trace of the signal beyond the chosen time window will start leaking in to the initial part of the time history thereby completely distorting the time response. This problem is referred to

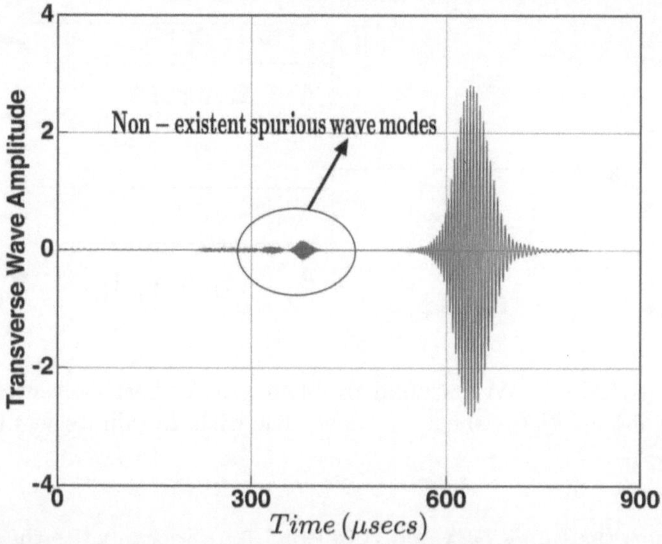

FIGURE 6.6: Transverse response due to oversampled tone burst signal with a sampling frequency of 200 kHz

as *Signal Wraparound problem.* To understand this problem better, we consider a 1-D cantilevered bar undergoing axial motion $u(x,t)$ (Fig. 6.7(a)) and subjected to a time dependent load $P(x,t)$. The governing equation for this problem earlier derived in Chapter 4 is given by

$$EA\frac{\partial^2 u}{\partial x^2} = \rho A\frac{\partial^2 u}{\partial x^2} \tag{6.4}$$

The solution of the above equation in the frequency domain is given by

$$\hat{u}_n(x,\omega) = \mathbf{A}e^{-ikx} + \mathbf{B}e^{ikx} \tag{6.5}$$

where k is the wavenumber, $\mathbf{A}$ is the incident wave coefficient, and $\mathbf{B}$ is the reflected wave coefficient, which are to be determined from the two boundary conditions. $\hat{u}_n$ is the FFT of the axial displacement $u(x,t)$ and x is the axial coordinate. The boundary conditions for the cantilever problem are $\hat{u}_n = 0$ at $x = 0$ and $EA\frac{d\hat{u}_n}{dx} = \hat{P}$, at $x = L$, where $\hat{P}$ is the FFT of input force $P(x,t)$ and L is the length of the cantilever bar. Substituting these boundary conditions in Eqn. (6.5), we get

$$\mathbf{A} = -\mathbf{B}, \qquad \mathbf{A} = \frac{\hat{P}}{EA}\frac{Li}{2kL\cos kL} \tag{6.6}$$

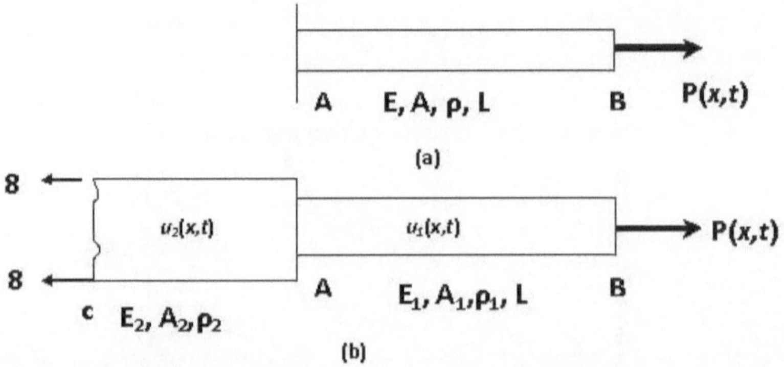

FIGURE 6.7: Wraparound problem: (a) A short cantilever bar subjected to axial load. (b) A short cantilever bar with an infinite segment attached

Using the values of **A** and **B** in Eqn. (6.5), we can write the axial displacement in the transformed Fourier domain as

$$\hat{u}_n(x, \omega) = \mathcal{H}(\omega)\hat{P} \tag{6.7}$$

where $\mathcal{H}(\omega)$ is the system transfer function, and it is given by

$$\mathcal{H}(\omega) = \frac{L \sin kx}{EAkL \cos kL} \tag{6.8}$$

The axial response for this case is obtained by following the usual procedure of convoluting the transfer function with the input load. For the response to die down within the chosen time window, it is necessary that the transfer function be complex. In the present case, the transfer function is real only as it has *sine* and *cosine* functions of ω, which has a finite value for all ω. Hence, no matter how long the rod member is, the response will never die down within the chosen time window. This is one of the severe limitations of FFT in analyzing finite structures.

The total time window $T = N\Delta T$, where N is the number of FFT points, and ΔT is the time sampling rate. Hence, the key to avoid wraparound problem is to increase the time window. This can be done either by increasing the number of FFT points, increasing the time sampling rate or a combination of these. Note that, increasing the sampling rate sometime leads to *aliasing* problems, the consequences of which was explained in the previous subsection. Alternatively, one can add a small amount damping to the wavenumber to make it complex as $k = k(1 - i\eta)$, where η is a small damping constant. The above methods may still not work for those systems such as the cantilever rod problem, which gives real transfer function. For such problems, the signal wraparound is eliminated by using a different modeling philosophy.

In the finite structure, the energy gets trapped due to repeated reflections from the fixed boundaries, which causes the signal to wraparound. By allowing some leakage of the responses from the boundary, one can add some artificial damping so that good resolution in the time response can be obtained. The modeling philosophy is shown in Fig. 6.7(b), wherein, the fixed boundary is replaced by an infinite rod having axial rigidity EA many times higher than that of the regular rod segment AB. We will now derive the transfer function for this new system.

Let $u_1(x,t)$ be the solution for the actual cantilever rod AB and let $u_2(x,t)$ be the solution for the infinite segment AC. Let $\hat{u}_{1n}$ and $\hat{u}_{2n}$ be their respective Fourier Transform. As per Eqn. (6.5), the solutions in these two segments can be written as

$$\begin{aligned} \hat{u}_{1n}(x,\omega) &= \mathbf{A}e^{-ikx} + \mathbf{B}e^{ikx}, \text{ For Segment AB of Length } L \\ \text{where} \qquad k^2 &= \omega\frac{\rho_1 A_1}{E_1 A_1} \end{aligned} \tag{6.9}$$

$$\begin{aligned} \hat{u}_{2n}(x,\omega) &= \bar{\mathbf{A}}e^{-i\bar{k}x}, \qquad \text{For Segment AC} \\ \text{where} \qquad \bar{k}^2 &= \omega\frac{\rho_2 A_2}{E_2 A_2} \end{aligned} \tag{6.10}$$

Note that the solution of the infinite bar (Segment AC) does not have any expression corresponding to the reflected wave. Hence, there are three wave coefficients that need to be determined. The three conditions that are necessary for their determination are obtained as follows. Considering point A as the origin, we have the following three conditions:

- At $x = 0$, we have, $\hat{u}_{1n} = \hat{u}_{2n}$
- At $x = 0$, we have the total force, that is $E_1A_1\frac{d\hat{u}_{1n}}{dx} + E_2A_2\frac{d\hat{u}_{2n}}{dx} = 0$, and
- At $x = L$, $E_2A_2\frac{d\hat{u}_{2n}}{dx} = \hat{P}$, where $\hat{P}$ is the FFT of the axial force.

Using these conditions in Eqns. (6.10) and (6.10), after simplification, we get the solution for the finite cantilever bar as

$$\hat{u}_{1n}(x,\omega) = \hat{P}\left[\frac{-iL((1-\beta)e^{-ikx} + (1+\beta)e^{ikx})}{2E_1A_1kL(\beta\cos kL + i\sin kl)}\right], \ \beta = \frac{A_2}{A_1}\sqrt{\frac{E_2\rho_2}{E_1\rho_1}} \tag{6.11}$$

The term in the brackets in Eqn. (6.11) is the transfer function, which has the real part as well as the imaginary part, indicating that the wave as it propagates, it also attenuates. That is, if the time window is large enough, the wraparound problems can be prevented. The level of attenuation can be manipulated by appropriately choosing β, or in other words the axial rigidity E_2A_2 is chosen such that the response dies out within the chosen window. If $\beta = \infty$, we recover back the fixed bar solution, which was derived in Eqn.

(6.7). Also, if we substitute $\beta = 0$, we simulate a free-free bar, whose solution is given by

$$\hat{u}_{1n}x, \omega) = \hat{P}\left[\frac{L\cos kx}{2E_1A_1kL\sin kL}\right] \tag{6.12}$$

In this equation, the term inside the brackets is the transfer function, which is again real, indicating that severe wraparound problems will be encountered if one uses FFT to solve the problem. From the above discussion, it is clear that in order to avoid signal wraparound and have good time resolution, it is necessary to attach an infinite segment of appropriate material properties to the short finite segment.

We will now show the effect of signal wraparound on the velocity responses of a finite cantilevered rod of length $L = 5.0$ m, width $b = 0.025$ m, and thickness $h = 0.025$ m. The rod is made of aluminum having a Young's modulus of $E = 70$ GPa and density $\rho = 2.7 \times 10^3$ kg/m^3. The rod is loaded at the tip by a force $P(x, t)$, whose time history and its FFT is shown in Fig. 4.3. When this rod is subjected to this impact load, the axial wave will be setup and it will propagate with a speed of 5092 m/sec. We will measure the velocity response to this load at the impact site ($x = 0$ m). This wave will propagate 10 m (5 m forward and 5 m backward after reflection) and hence the response will contain incident wave and a reflected wave appearing at a time $t = 10/5092 = 1963\,\mu$secs. If we add the zero heads of 100 μsec of the input signal, we should see reflections appearing at 2063 μsecs.

Here, two studies are performed. In the first, we will see how the signal warparound can be controlled by introducing damping in the wavenumber of the form $k = k(1 - i\eta)$, where $\eta = 0.01$ is assumed. As mentioned earlier, this damping is needed to make the system transfer function complex. Here, we will first perform this study for a sampling rate $\Delta T = 1\,\mu$sec and for three different numbers of FFT points. This plot is shown in Fig. 6.8(a). For the case of $N = 2048$, the total time window available is $N\Delta T = 2048\,\mu$secs. However, the reflection due to fixed boundary will appear around 2063 μsecs. Hence, the signal will not die down within the chosen time window and the balance signal beyond 2048 μsecs will leak and start appearing in front, which is clearly seen in Fig. 6.8(a). In addition, the response shows severe undulations of the response caused by signal wraparound and increasing the time window beyond 2048 μsecs is not able to completely eliminate the signal wrap around problems.

In the next study, we increase the time sampling rate ΔT to 5.0 μsecs, which increase the time window 5 times to the previous case. The response plots for this case are shown in Fig. 6.8(b). The increase in time window is to make sure that the response dies down within the chosen time window. The figure shows that for $N = 2048$ and for $N = 4096$, the signal wraparound still exists. However, for $N = 8192$, the signal wrap around problem is completely eliminated.

Next, we model the same problem, by adding an infinite rod segment of properties much higher than the original rod, as shown in Fig. 6.7(b). Here,

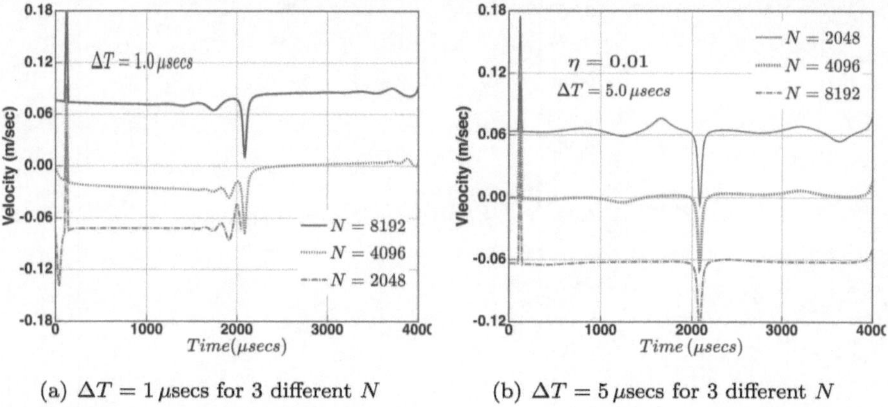

FIGURE 6.8: Control of signal warp around problem through damping

the Young's modulus of the infinite segment (E_2) is varied and the response is obtained for two cases, namely $E_2 = 1.2E_1$ and $E_2 = 2E_1$. These responses are shown in Fig. 6.9.

The idea here is, by addition the infinite rod segment, a small amount of leakage of response is allowed, so that the response dies down within the chosen time window. Note that if $E_2 = \infty$, it simulates the perfect fixity, which is the actual boundary condition in the problem. The aim here is to get a wraparound free responses for the case of $\Delta T = 1.0\,\mu$sec and $N = 2048$. The previous example for this case showed significant signal wraparound. The figure shows that very clean response, free of signal wrap around, is obtained by this procedure. However, the magnitude of the reflected response is very small, and it is seen to grow with the increase in the value of E_2. Note that, too high a value of E_2 will again cause signal wrap around problem.

This problem shows that, one of the methods to avoid signal wraparound due to reflections coming from the boundary is by adding an infinite segment to allow some signal leakage so that the response dies down within the chosen window. This can also be demonstrated for higher-order waveguides such as beams, frames and 2-D waveguides.

6.2 PROPAGATION AND RECONSTRUCTION OF SIGNALS

Signal processing aspect is very important in wave propagation studies. The two important parameters that characterize the spectral content in the signal are the sampling rate ΔT and the number of FFT points N. In the previous sections, some of the common issues of signal processing that are relevant for wave propagation analysis was presented. In this section, some of the

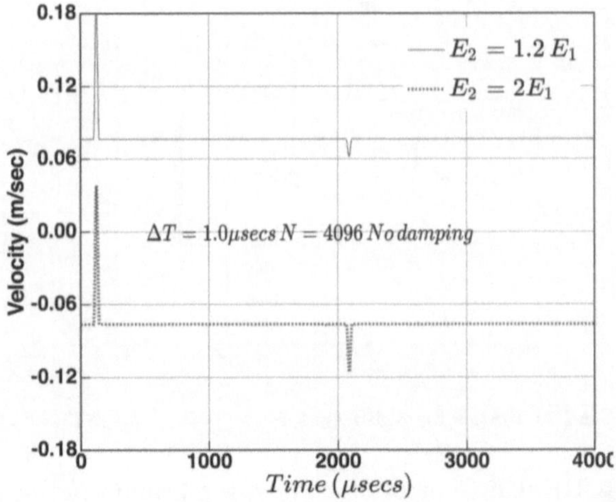

FIGURE 6.9: Control of signal warp around problem through infinite segment attachment

important concept of signal processing required for propagating signals in non-dispersive and dispersive waveguides is highlighted.

Wave propagating in a non-dispersive waveguide preserves its shape since the wave propagates at same speeds at all frequencies. However, when the propagation distance is long, if the chosen time window ($N\Delta T$) is not sufficient to accommodate this long length of propagation, the wave will propagate out of the chosen window. Because of the inherent periodicity of the FFT in the frequency domain, the waves that have propagated out in the neighboring window will propagate into view from the left. This is actually shown in Chapter 4 (Fig. 4.4) in the context of wave propagating in an infinite rod. This is again a form of signal wraparound problem. Hence, one should choose the time window such that enough room for propagation of the signal to the required distance is provided. Hence, the time window not only depends on ΔT and N but also the propagating distance.

Propagating signals in dispersive media are more complicated since the wave speed is a function of frequency. That is, at different frequencies, the speeds are different. Hence, the time window required for the analysis is dictated by the slower frequency components as they will take longer time to propagate. The dispersive nature changes the complete profile (or shape) of the wave as they propagate. This is due to non-linear relationship between the wavenumber k and frequency ω. Most dispersive signals have long tail. This is due to low frequency components that travel with lower speeds and as a result they take long time to arrive. Also, these low frequency components having longer tails does not die down within the chosen time window, irrespective of how large the time window is. In summary, these low frequency components are of nuisance value and require to be filtered out. In addition, the trace of the signal gets distorted in the initial time periods due to low frequency components as well as the signal wraparound. To identify these effects and also the effect of signal integration(to be discussed in the next sub section), a zero header is always provided to the signal. These effects can be reduced by adopting the following methods:

- **Band-pass Filtering**: In this method, all the frequency components below the specified frequencies can be blocked. Essentially, the signal is manipulated in the frequency domain to reduce the initial distortion. Fig. 6.10 shows the original signal and also the band-passed filtered signal

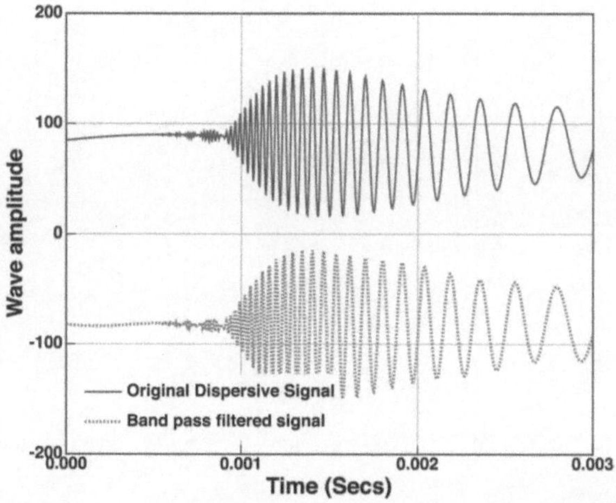

FIGURE 6.10: Band pass filtered dispersive signal

The original FFT of the signal created using 2048 FFT points is sent to a band pass filter, where all the frequencies below 1300 Hz are blocked. This process significantly reduces the initial signal distortion. One can use MATLAB to perform this operation.

- **n-point Moving Average**: This method is based on the obtained time domain response. The equation for simple n point moving average filter is given by

$$y(t) = \frac{1}{n}\sum_{i=0}^{n-1} x(t-i) \tag{6.13}$$

In this equation, $y(t)$ is the modified input, $x(t)$ is the current input, and $x(t-1)$ is the previous input and so on. Here, n is the length of the signal over which the average is taken or the length of the average.

- The third method is to simply extend the time window by increasing the time sampling rate or number of FFT points or the both.

6.2.1 Integration of the Signals

In many wave experiments, strains are measured instead of displacements and the displacements are extracted by integrating strains. In doing so, if the constant of integration is not considered, then the results obtained will be erroneous. To demonstrate the effect of improper integration of signals, we will consider an infinite rod and will plot the displacement response, as shown in Fig. 6.11 for a triangular input shown in Fig. 4.3.

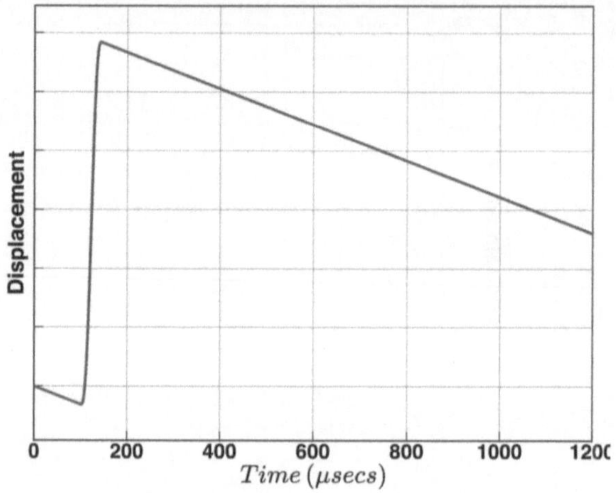

FIGURE 6.11: Displacement response at 3.0 m from impact site in an infinite rod

Typically, the displacement response will be similar to a Heaviside function that rises at certain time and sustain the deformation level over a long period of time. However, the periodicity of the signal forces the signal go to zero at the end of the time window. This distorts the signal as shown in Fig. 6.11. The signal trace in this case is completely continuous from the end of the window and beginning of the next.

To minimize this problem, we should have output at the same differentiation level as the input. The input here is the force, which is essentially time dependent. However, the output is both space and time dependent and the space dependence comes from the first or the second derivative in space depending on the type of waveguide in which the wave is propagating. The output can also involve higher time derivatives. To determine the differentiation level of output and the input, we should establish equivalence of space and time derivatives. This equivalence for rod and beam waveguides can be derived by considering their respective displacement fields. If $u(x,t)$ is the axial deformation of the rod, and $w(x,t)$ is the transverse displacement in a beam, and their respective wavenumbers for these two waveguides are given by

$$\begin{aligned} &\text{Rod Waveguide} \quad u(x,t) = \sum \mathbf{A}_n \, e^{-i(k_L x - \omega t)} \\ &\text{Beam Waveguide} \quad w(x,t) = \sum \mathbf{A}_n \, e^{-i(k_b x - \omega t)} \\ &k_L = \omega\sqrt{\rho A/EA} \quad k_b = \sqrt{\omega}(\rho A/EI)^{0.25} \end{aligned} \tag{6.14}$$

Using the above expressions, we derive the space and time derivative equivalence as follows. Let us consider the equivalence of first derivative of the axial displacement and its corresponding temporal derivative. Let us consider du/dx. Using Eqn. (6.14), we can write the first derivative as

$$\frac{du}{dx} = - - ik_L \sum A_n \, e^{-i(k_L x - \omega t)} \rightarrow k_L u$$

Since the wavenumber is directly proportional to frequency, we have

$$k_L u \rightarrow \omega u$$

From Eqn. (6.14), the time derivative of $u(x,t)$ is given by

$$\frac{du}{dt} = -i\omega \sum A_n \, e^{-i(k_L x - \omega t)} \rightarrow \omega u$$

Hence, $\frac{du}{dx}$ is at the same differentiation level as $\frac{du}{dt}$. The rest of the derivatives for the beam and rod waveguides is given in Table 6.1.

TABLE 6.1: Equivalence of temporal and spatial derivatives for rod and beam waveguides

Rod Waveguide	**Variable**	**Beam Waveguide**
	Slope	$\frac{dw}{dx} \rightarrow k_b w$
$\frac{du}{dx} \rightarrow k_L u \rightarrow \omega u \rightarrow \frac{du}{dt}$	Axial Strain	
	Bending Strain	$\frac{d^2w}{dx^2} \rightarrow k_b^2 w \rightarrow \omega w \rightarrow \frac{dw}{dt}$
$\frac{d^2u}{dx^2} \rightarrow k_L^2 u \rightarrow \omega^2 u \rightarrow \frac{d^2u}{dt^2}$	Axial Loading	
	Shear Force	$\frac{d^3w}{dx^3} \rightarrow k_b^3 w$
	Bending Loading	$\frac{d^4w}{dx^4} \rightarrow k_b^4 w \rightarrow \omega^2 w \rightarrow \frac{d^w}{dt^2}$

For example, in the case of rod, the axial force $F = EA\frac{du}{dx}$. Hence, the applied force is directly proportional to the axial strain and as per Table 6.1, the equivalent temporal derivative is du/dt, which is the axial velocity. Hence, by plotting the velocity instead of displacement will avoid the problems associated with signal integration, which is shown in Fig. 6.11. The best way to obtain displacements is to first obtain the velocity response and then time integrate to obtain the wave response.

Summary

In this chapter, some of the signal processing issues associated with the spectral analysis were discussed. The main issue in handling time signals is the signal truncation, which is the source of many signal processing problems. The main signal processing issues discussed in this chapter include *Signal Aliasing*, *Signal Leakage* and *Signal wraparound.* These problems were discussed using a number of examples. A special signal leakage issue arising out of over sampling the signal in some time signals such as modulated sine signal was also discussed. The methods of eliminating these problems were also highlighted. One of the methods to eliminate signal leakage is to window the signal. Two different window functions, namely the *Rectangular* and *Hanning*, were discussed in detail, and the wave propagation examples were presented to show how these window functions were signal leakage. A full section on signal wraparound problem and its remedy were presented using examples. The last part of the chapter dealt with issues concerning propagation and reconstruction of the signals. Here, the signal processing aspects of dispersive and non-dispersive signals were discussed in detail. In particular, the usage of techniques such as *band pass filters* and *n-point moving average method* was discussed in the context of obtaining better signal quality. Finally, a subsection on the issues related to *Signal Integration* brought forth the problem associated with choosing proper output time variable (displacement, velocity or acceleration) for the time domain responses. Here, the importance of equivalence of the spatial and temporal derivatives were discussed and established. Note that no separate MATLAB scripts particular to this chapter are given. The scripts provided in the earlier and the following chapters can be used to understand the materials presented in this chapter.

Exercises

6.1 In this chapter, we had shown that for a rod with one end fixed, the transfer function is real and hence the response reconstruction will give meaningless result having severe signal wraparound problems. However, if a stiff infinite rod segment is added to this finite rod, we showed that the system transfer function will become complex and signal wraparound can be eliminated. Alternatively, we can use a dashpot as shown below to make system transfer function complex. For this

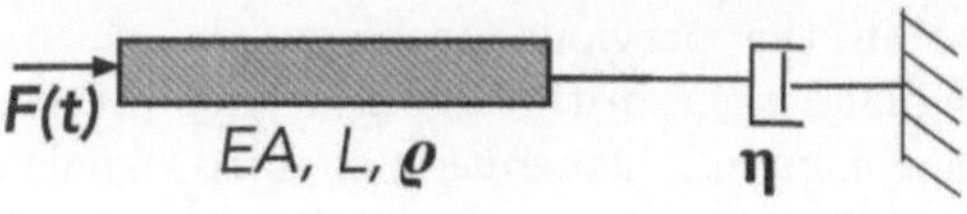

Figure for Problem 6.1

configuration, determine the system transfer function and show it is fully complex.

6.2 Determine the FFT of the following Heaviside function

$$\begin{aligned} F(t) &= 0 \qquad t < t_0 \\ &= 1 \qquad t > t_0 \end{aligned}$$

where $t_0 = 300\,\mu$secs. Use 8192 FFT points with a sampling rate of 1.0 μsecs. Can you propagate this signal? What problems will you encounter?. Alternatively, if the signal is modified as

$$\begin{aligned} F(t) &= 0 \qquad t < t_0 \\ &= e^{-\alpha t}, \qquad t > t_0 \end{aligned}$$

Where, α is a constant. Can you propagate this signal? If so, use this signal and obtain time domain responses on an (a) infinite aluminum rod and (b) infinite aluminum beam at a number of locations along the axis of the beam.

6.3 A linear chirp signal is normally used in many applications as one can control the amount of signal energy over a specified bandwidth. However, its use in propagating signal in structural waveguides is limited. You will see that this signal is a non-stationary exhibiting significant signal leakage in the frequency domain. In order to understand the chirp signal propagation in a structural waveguides, propagate a chirp signal of following two different properties on an infinite aluminum waveguide (rod as well as beam waveguide)

$$\begin{aligned} &T_0 = 0\,\mu\text{secs}, \qquad T_1 = 300\,\mu\text{secs}, f_0 = 20\text{ kHz}, \qquad f_1 = 300\text{ kHz} \\ &T_0 = 0\,\mu\text{secs}, \qquad T_1 = 300\,\mu\text{secs}, f_0 = 50\text{ kHz}, \qquad f_1 = 55\text{ kHz} \end{aligned}$$

Plot the velocity responses at a number of different locations for both rod and beam waveguides. Comment on the suitability of chirp signals for the structural wave propagation problems. Use MATLAB function *chirp* to generate the linear input chirp signal.

6.4 As mentioned in the previous problem, the chirp signals are non-stationary signals, wherein the signal energy within a specified bandwidth can be controlled. Generation of chirp signals are also relatively simple compared to tone burst signal. Responses obtained using chirp signals are very useful to obtain responses due to other types of signals, for example tone burst signal of a given central frequency ω_c. This is done by using the following relation in the frequency domain

$$\begin{aligned} \hat{G}(\omega, x) &= \frac{\hat{u}_c(\omega, x)}{\hat{F}_c(\omega)} = \frac{\hat{u}_t(\omega, x)}{\hat{F}_t(\omega)} \\ &= \hat{u}_t(\omega, x) = \frac{\hat{F}_t(\omega)\hat{u}_c(\omega, x)}{\hat{F}_c(\omega)} \end{aligned}$$

where $\hat{G}$ is the system transfer function, $\hat{u}_c$ is the response of the structure at an location x due to an input chirp signal $\hat{F}_c$, and $\hat{u}_t$ is the required response of the tone burst signal $\hat{F}_t$ sampled at a central frequency of ω_c. Using the above relations, determine the tone burst signal responses $\hat{u}_t$ using the chirp signal generated responses $\hat{u}_c$ obtained using the input chirp signal $\hat{F}_c$ given in the previous problem (Problem 6.3) on (a) aluminum infinite rod waveguide, (b) aluminum infinite beam waveguide. Vary the central frequency ω_c and compare the accuracy of group speeds obtained from the time response of the reconstructed force with the dispersion relations.

6.5 Generate a chirp signal with following properties

$$T_0 = 0\,\mu\text{secs}, \qquad T_1 = 300\,\mu\text{secs}, \qquad f_0 = 50 \text{ kHz}, \qquad f_1 = 55 \text{ kHz}$$

Window the above signal using Hanning window. Propagate this windowed chirp signal in an infinite Timoshenko infinite beam and see if two propagating wave modes can be captured using this signal.

CHAPTER 7

Wave Propagation in Two-Dimensional Isotropic Waveguides

In Chapter 4, wave propagation analysis was presented for 1-D isotropic longitudinal and flexural waveguides using the spectral analysis. In particular, the interaction of incident waves with boundaries and constraints was presented. In this chapter, we will again use spectral analysis and apply these to two-dimensional isotropic waveguides. Introduction of an additional dimension introduces additional complexity in wave propagation. That is, it introduces a second wavenumber in the second dimension and these two wavenumbers are normally coupled. Additional dimensions also introduce additional waves. In 1-D waveguides, we saw the existence of a number of wave modes, each of them identified by different wavenumbers and these were normally called temporal modes. Introduction of additional dimension and hence the wavenumber, introduces spatial modes, which are additional waves arising out of different boundary conditions at the different waveguide edges of the 2-D waveguides.

In this chapter, we will address the wave propagation in 2-D waveguides subjected to both in-plane and out-of-plane excitations. These two cases are the 2-D equivalent of rods and beams studied in Chapter 4. Here, we will first derive the governing equation for two-dimensional waveguides and detail the procedure to solve the equation using *Helmholtz Decomposition*. This process decomposes the complicated coupled governing equations into two independent partial differential equations involving *Scalar Potential* and *Vector Potential*. Displacements are derived from these potentials. In order to perform spectral analysis, first a Fourier Transform in time is performed. This removes the variable time from the equation, however, the resulting equation continues to be partial differential equation in two spatial directions. To reduce it further, an additional transform is enforced in one of the spatial

DOI: 10.1201/9781003120568-7

direction by introducing an additional spatial wavenumber. This will reduce the partial differential equation to ordinary differential equation on which we can perform spectral analysis.

Spectral analysis can now be applied to these independent reduced scalar and vector potential equations and the wave propagation characteristics of different waves, namely the *P*-waves, the *SV-waves* and the *SH-waves* in a semi-infinite 2-D media will be studied. Then for a doubly bounded media, where different boundary conditions can generate new type of waves, wave propagation will be studied for certain specified boundary conditions. The last part of this chapter will deal on wave propagation in plate type structure that are subjected out-of-plane loads.

7.1 GOVERNING EQUATIONS OF MOTION

Using the principles of elasticity (see [134], [54]), we will now derive the partial differential equation governing the motion for 2-D waveguides. Such equation is called the *Navier's Equation.* A 2-D waveguide having material properties defined in terms of Lame's constants λ and μ and density ρ is shown in Fig. 7.1.

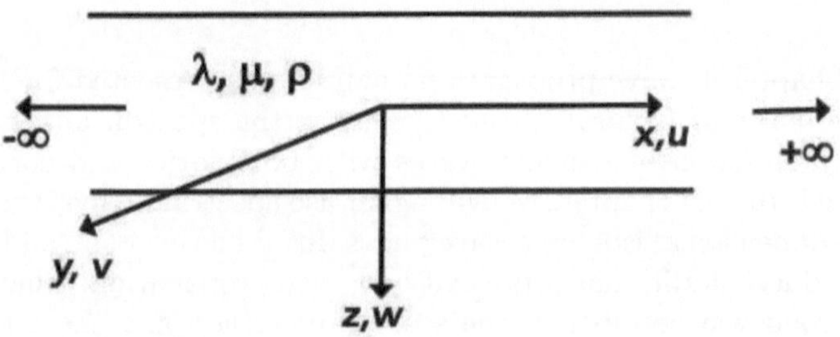

FIGURE 7.1: The coordinate axes of a 2-D waveguide

where the Lamé constants are related to Young's modulus and Poisson's ratio through the relations

$$\lambda = \frac{\nu E}{(1+\nu)(1-2\nu)}, \qquad \mu = G = \frac{E}{2(1+\nu)}$$

The displacement components in the three coordinate directions are u, v, and w. Here, we consider only a 2-D solid in x-z plane and hence the relevant stresses, strains and displacements are the following:

$$\begin{aligned}
&\text{Stress Components}: \sigma_{xx},\ \sigma_{zz},\ \tau xz\\
&\text{Strain Components}: \epsilon_{xx},\ \epsilon_{zz},\ \gamma_{xz}\\
&\text{Displacement Components}: u,\ w
\end{aligned}$$

We begin with the strain displacement relationship (see [54] for more details), which for 2-D solid is given by

$$\epsilon_{xx} = \frac{\partial u}{\partial x}, \qquad \epsilon_{zz} = \frac{\partial w}{\partial z}, \qquad \gamma_{xz} = \frac{\partial u}{\partial z} + \frac{\partial w}{\partial x} \tag{7.1}$$

The stress-strain relationship in terms of Lame's constants for an isotropic solid is given by (see [54] for details).

$$\begin{aligned} &\sigma_{xx} = 2\mu\epsilon_{xx} + \lambda(\epsilon_{xx} + \epsilon_{zz}), \quad \sigma_{zz} = 2\mu\epsilon_{zz} + \lambda(\epsilon_{xx} + \epsilon_{zz}) \\ &\tau_{xz} = \mu\gamma_{xz} \end{aligned} \tag{7.2}$$

Substituting Eqn. (7.1) in Eqn. (7.2), we get

$$\begin{aligned} &\sigma_{xx} = 2\mu\frac{\partial u}{\partial x} + \lambda\left(\frac{\partial u}{\partial x} + \frac{\partial w}{\partial z}\right), \; \sigma_{zz} = 2\mu\frac{\partial w}{\partial z} + \lambda\left(\frac{\partial u}{\partial x} + \frac{\partial w}{\partial z}\right), \\ &\tau_{xz} = \mu\left(\frac{\partial u}{\partial z} + \frac{\partial w}{\partial x}\right) \end{aligned} \tag{7.3}$$

Next, we write the 2-D equilibrium equation in the absence of body force (see [54] for more details) as

$$\frac{\partial \sigma_{xx}}{\partial x} + \frac{\tau_{xz}}{\partial z} = \rho\frac{\partial^2 u}{\partial t^2}, \qquad \frac{\partial \tau_{xz}}{\partial x} + \frac{\sigma_{zz}}{\partial z} = \rho\frac{\partial^2 w}{\partial t^2} \tag{7.4}$$

Differentiating Eqn. (7.3) and substituting the resulting equation in Eqn. (7.4) and simplifying, we can write the *Navier's equation* as

$$\begin{aligned} \mu\nabla^2 u + (\lambda + \mu)\left[\frac{\partial^2 u}{\partial x^2} + \frac{\partial^2 w}{\partial x \partial z}\right] &= \rho\frac{\partial^2 u}{\partial t^2} \\ \mu\nabla^2 w + (\lambda + \mu)\left[\frac{\partial^2 w}{\partial z^2} + \frac{\partial^2 u}{\partial x \partial z}\right] &= \rho\frac{\partial^2 w}{\partial t^2} \\ \text{where} \qquad \nabla^2 = \frac{\partial^2}{\partial x^2} + \frac{\partial^2}{\partial z^2} \end{aligned} \tag{7.5}$$

Here, Eqn. (7.5) represents the governing equation that dictates the motion of 2-D solids. The 3-D equivalent of this equation is quite complicated to write in the expanded form and can be written in tensorial notation

$$(\lambda + \mu)\frac{\partial^2 u_p}{\partial x_p \partial x_i} + \mu\frac{\partial^2 u_i}{\partial x_p^2} = \rho\frac{\partial^2 u_i}{\partial t^2} \qquad i = x, y, z \tag{7.6}$$

In the above equation, p is the *dummy index* and hence the index needs to be summed over x, y and z.

7.2 SOLUTION OF NAVIER'S EQUATION

Solving wave propagation in 2-D isotropic solid requires solution of Eqn. (7.5), which is a set of coupled partial differential equations u and w as dependent variable and x, z and t as independent variable. Direct solution of this equation is almost not possible for a host of boundary conditions that the 2-D waveguide can accommodate. *Helmholtz Decomposition* is the suggested method that can be used to exactly solve Eqn. (7.5) for certain waveguide configuration. This method expresses the main field variables, namely the deformation u and w in terms of a *Scalar Potential* Φ and a *Vector Potential* H_y. In the 3-D case, we will express the displacement field u, v and w in terms of a scalar potential and three vector potentials H_x, H_y and H_z. For solving the 2-D Navier's equation Eqn. (7.5), we will use the following Helmholtz decomposition

$$u(x,t) = \frac{\partial \Phi(x,t)}{\partial x} + \frac{\partial H_y(x,t)}{\partial z}, \quad w(x,t) = \frac{\partial \Phi(x,t)}{\partial z} - \frac{\partial H_y(x,t)}{\partial x}$$

$$\text{with condition} \quad \frac{\partial H_y}{\partial y} = 0 \tag{7.7}$$

The 3-D counterpart of this equation is again quite complicated and can be written in tensor notation as

$$u_i = \frac{\partial \Phi}{\partial x_i} + \epsilon_{ilm}\frac{\partial H_m}{\partial x_l}, \qquad \text{with condition} \; \frac{\partial H_k}{\partial x_k} = 0 \tag{7.8}$$

where ϵ_{ilm} is a permutation symbol, which is equal to zero if any of the two subscripts are same, and equal to one if i, l, m are cyclic or equal to -1 if they are non-cyclic.

Substituting Eqn. (7.7) in Eqn. (7.5) and simplifying, we get two uncoupled equations in terms of the scalar potential Φ and vector potential H_y, and they are given by

$$(\lambda + \mu)\nabla^2\Phi = \rho\frac{\partial^2 \Phi}{\partial t^2}, \quad \mu\nabla^2 H_y = \rho\frac{\partial^2 H_y}{\partial t^2} \tag{7.9}$$

The first equation involving the scalar potential Φ is associated with the *P*-wave. They are also called by other names such as *Dilatational waves, Irrotational waves, Longitudinal waves* or more commonly as *Primary waves or P*-waves. They travel with a speed given by

$$C_p^2 = \frac{\lambda + \mu}{\rho} = \frac{E(1-\nu)}{\rho(1+\nu)(1-2\nu)} = c_0^2\frac{1-\nu}{(1+\nu)(1-2\nu)}$$

where c_0 is the longitudinal speed of elementary rod. From the above equation, we see $C_p > c_0$. Let us examine this for a few values of Poison's ratio ν.

If	$\nu = 0.0$	then	$C_p = c_0$
If	$\nu = 0.2$	then	$C_p = 1.11c_0$

Similarly, the second equation of Eqn. (7.9) is in terms of vector potential H_y, which is associated with S-wave and it travels with the speed given by

$$C_S^2 = \frac{\mu}{\rho} = \frac{E}{2\rho(1+\nu)} = c_0^2 \frac{1}{2(1+\nu)}$$

From the above equation, we see that C_s is always less that c_0.

The propagation of P-waves and S-waves has great relevance to those engaged in earthquake engineering research. The P-wave motion is same as that of sound wave in air, that is, it alternately pushes (compresses) and pulls (dilates) the structure. The motion of the particles is always in the direction of propagation. This wave, like sound, travels both in solids and fluids. It may be mentioned that, because of sound like nature, when P-wave emerges from deep in the Earth to the surface, a fraction of it is transmitted into atmosphere as sound waves. Such sounds, if frequency is greater than 15 cycles per second, are audible to animals or human beings. These are known as *Earthquake Sound.* The pictorially, the propagation of P-waves and S-waves is shown in Fig. 7.2.

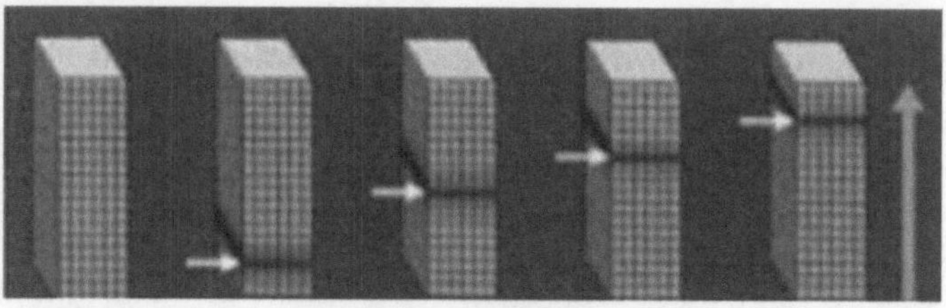

(a) P-wave Propagation

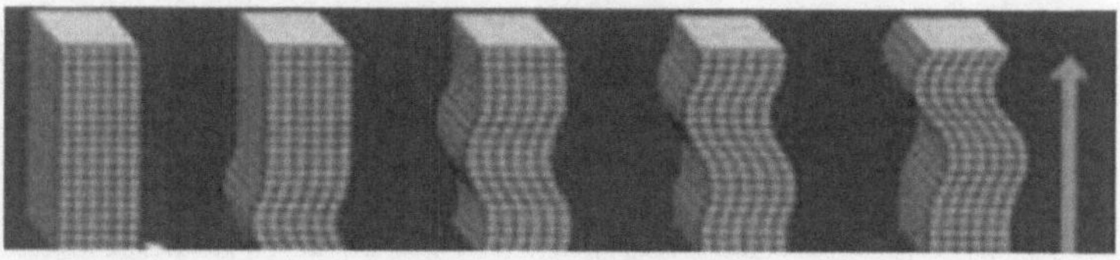

(b) S-wave propagation

FIGURE 7.2: Propagation of P- and S-waves in a 2-D solid

7.3 PROPAGATION OF WAVES IN INFINITE 2-D MEDIA

In this section, an infinite 2-D medium is considered, wherein the waves are propagating in only one plane and in the present case, we consider x-z plane. Here, we will provide the necessary equations and conditions that exist for propagation of P-waves alone, S_V-waves alone or S_H-waves alone.

The S-wave is also called the *Secondary wave* or *Rotational wave* or *Shear wave*. These waves travel slower than the P-wave and hence they cannot propagate in liquids. The particle motion of the S-wave is perpendicular (transverse) to the propagation. There are two variants of S-wave. If the particle motion of the S-wave is up and down in the vertical plane, then it is named as S_V-*wave*. However, if the S-wave oscillates in the horizontal plane, then such an S-wave is referred to as S_H *wave*. The direction of propagation of P, S_V and S_H waves is shown in Fig. 7.3.

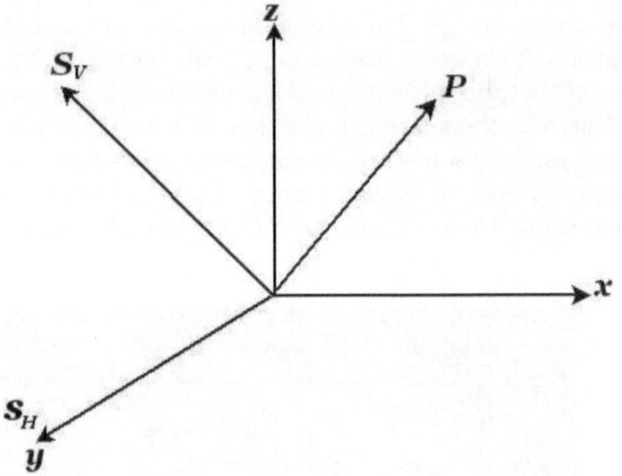

FIGURE 7.3: Propagation directions of P-wave, S_H wave and S_V wave

7.3.1 Propagation of *P*-waves

As mentioned earlier, we consider waves propagating in an infinite medium in the x-z plane, which means that that the y dependence of all the variables does not exist. We assume that the P-wave is propagating in isolation without the existence of either S_V or S_H waves. Since the P-wave is associated with only Φ, we have that these waves satisfy the following differential equation

$$\frac{\partial^2 \Phi}{\partial x^2} + \frac{\partial^2 \Phi}{\partial z^2} = \frac{1}{C_p^2} \frac{\partial^2 \Phi}{\partial t^2} \tag{7.10}$$

where $C_p = \sqrt{(\lambda + \mu)/\rho}$, with the following displacement field variation

$$u = \frac{\partial \Phi}{\partial x}, \qquad w = \frac{\partial \Phi}{\partial z}, \qquad v = 0 \tag{7.11}$$

In 2-D media, we deal with tractions rather than with forces. Determination of stress components is essential for determining the tractions (see the [54]).

Stresses are determined using the Eqns. (7.1 & 7.11) in Eqn. (7.2) and the resulting equations are given by

$$\sigma_{xx} = (\lambda + 2\mu)\nabla^2\Phi - 2\mu\frac{\partial^2\Phi}{\partial z^2}, \; \sigma_{zz} = (\lambda + 2\mu)\nabla^2\Phi - 2\mu\frac{\partial^2\Phi}{\partial x^2}.$$

$$\sigma_{yy} = \lambda\nabla^2\Phi, \qquad \sigma_{xz} = 2\mu\frac{\partial^2\Phi}{\partial x\partial z}, \qquad \sigma_{xy} = \sigma_{yz} = 0 \tag{7.12}$$

We note that the out-of-plane displacement is zero; however, the out-of-plane stresses are non-zero, indicating that the displacement field given by Eqn. (7.11) gives rise to *Plane Strain* type conditions. The waves here are non-dispersive where the waveguide undergoes "push" or "pull" motion in the direction of the propagation and the behavior of waves is similar to longitudinal waves in rods.

7.3.2 Propagation of S_V *Waves*

If we consider a situation where only the waves associated with vector potential H_y are propagating in an infinite medium in the x-z plane, such waves satisfy the differential equation given by

$$\frac{\partial^2 H_y}{\partial x^2} + \frac{\partial^2 H_y}{\partial z^2} = \frac{1}{C_s^2}\frac{\partial^2 H_y}{\partial t^2} \tag{7.13}$$

where $C_s = \sqrt{\mu/\rho}$. The displacement field in this case is given by Eqn. (7.7) after removing the Φ dependencies, this equation becomes

$$u = \frac{\partial H_y}{\partial z}, \qquad v = 0, \qquad w = -\frac{\partial H_y}{\partial x} \tag{7.14}$$

The above displacement field is defined in the x-z plane, and it is in the direction tangential to the surface represented by $H_y(x, y, t)$. The propagation direction is shown in Fig. 7.3. Since the wave is perpendicular to P-wave propagation, this wave is called S_V *Wave*. The stresses are computed as done before. That is, we use Eqns. (7.1 & 7.14) in Eqn. (7.2) and the resulting equations are given by

$$\sigma_{xx} = 2\mu\frac{\partial^2 H_y}{\partial x\partial z}, \qquad \sigma_{zz} = -\sigma_{xx}, \qquad \sigma_{xy} = \sigma_{yz} = \sigma_{yy} = 0$$

$$\sigma_{xz} = \mu\left[-\frac{\partial^2 H_y}{\partial x^2} + \frac{\partial^2 H_y}{\partial z^2}\right] \tag{7.15}$$

If we draw a *Mohr's Circle* using the above stress components, one will see that the state of stress is for the case of a pure shear along the planes oriented at ±45°. The behavior of the S_V *wave* is very similar to the behavior of a beam under shear loading.

7.3.3 Propagation of S_H Waves

The S_H waves are generated when the displacement fields are such that there are no in-plane displacement and the out-of-plane wave propagates along the in-plane directions. In other words, the dependency of variables on the out-of-plane variable y does not exist and hence as per Eqn. (7.6), there will be two vector potentials that will participate in the motion, namely H_x and H_z and hence these waves need to satisfy the following two governing equations and the constraint equation arising due to Eqn. (7.8), and they are given by

$$\frac{\partial^2 H_x}{\partial x^2} + \frac{\partial^2 H_x}{\partial z^2} = \frac{1}{C_s^2}\frac{\partial^2 H_x}{\partial t^2}, \qquad \frac{\partial^2 H_z}{\partial x^2} + \frac{\partial^2 H_z}{\partial z^2} = \frac{1}{C_s^2}\frac{\partial^2 H_z}{\partial t^2}$$

$$\text{where} \qquad \frac{\partial H_x}{\partial x} + \frac{\partial H_z}{\partial z} = 0 \tag{7.16}$$

with the following boundary conditions

$$u = w = 0, \qquad v = -\frac{\partial H_x}{\partial z} + \frac{\partial H_z}{\partial x} \tag{7.17}$$

The direction of propagation is perpendicular to both P-wave and S_V wave as shown in Fig. 7.3. The corresponding stresses can be obtained as before, and they are given by

$$\sigma_{xx} = \sigma_{zz} = \sigma_{zz} = \sigma_{xz} = 0$$

$$\sigma_{xy} = \mu\left[-\frac{\partial^2 H_x}{\partial x \partial z} + \frac{\partial^2 H_z}{\partial x^2}\right] = \mu\nabla^2 H_z$$

$$\sigma_{zy} = \mu\left[-\frac{\partial^2 H_x}{\partial z^2} + \frac{\partial^2 H_z}{\partial x \partial z}\right] = -\mu\nabla^2 H_x \tag{7.18}$$

Thus, S_H wave generates only shear stresses. The above form of equations requires determination of two vector potentials and is cumbersome to handle. Alternatively, since there is only one displacement unknown, namely $v(x, y, t)$ in the out-of-plane direction, it is easy to work with this variable. That is, the governing equation reduces to

$$\frac{\partial^2 v}{\partial x^2} + \frac{\partial^2 v}{\partial z^2} = \frac{1}{C_s^2}\frac{\partial^2 v}{\partial t^2} \tag{7.19}$$

The corresponding normal stresses are zero, while the shear stresses in terms of out-of-displacement are given by

$$\sigma_{xy} = \mu\frac{\partial v}{\partial x}, \qquad \sigma_{zy} = \mu\frac{\partial v}{\partial z} \tag{7.20}$$

It is to be noted that the out-of-plane stress σ_{yy} is zero and hence a condition of *Plane Stress* exists in the case of S_H wave propagation. The solution to

Eqn. (7.19) will be of the form

$$v(x,t) = A_0 e^{-ikx} e^{i\eta z} e^{-i\omega t} \tag{7.21}$$

where η is the wavenumber in the z direction. Substituting Eqn. (7.21) in Eqn. (7.19), we will get the equation to compute its wavenumber, which is given by

$$k = \sqrt{k_s^2 - \eta^2} \tag{7.22}$$

where $k_s = \omega/C_s$, and C_s is the shear wave speed given by $C_s = \sqrt{\mu/\rho}$. A look at Eqn. (7.22) say that when $k_s << \eta$, the wavenumber will be imaginary and the waves can undergo multiple cut-off frequencies and they occur at

$$\omega_{cut-off} = \frac{\eta}{C_s} \tag{7.23}$$

From the above equation, we see that different values of η provide different cut-off frequency. However, the values of η and k should be determined according to the boundary conditions at $x = z = 0$. Fig. 7.4 shows the plot of wavenumber and phase & group speeds with frequency. These plots are generated using aluminum properties. All those aspects discussed above can be clearly seen in these figures.

7.4 WAVE PROPAGATION IN SEMI-INFINITE 2-D MEDIA

In the last section, we discussed the wave propagation in infinite 2-D media where we saw that the two types of waves, namely the P-wave and S-wave propagating non-dispersively. When the media is bounded as shown in Fig. 7.5, then the propagating P-wave and S-wave will interact with the boundary and generate reflections and transmissions. In other words, these two wave couple themselves depending upon the boundary conditions at $z = 0$ and can generate entirely new type of waves. .

The study of wave propagation in 2-D semi-infinite media is of great relevance in the area of seismology, where it is quite well known that during earthquakes, in addition to P-wave and S-wave, a third wave called the *Rayleigh Wave* propagates and it follows the arrival of both the P and S waves. In this section, we will see how this third wave get generated.

Since, both P-wave and S-wave propagate in the semi-infinite media, the governing equations that governs the wave behavior in the 2-D semi-infinite media is same as given in Eqns. (7.10 & 7.13), which are written here in the frequency domain as

$$\nabla^2\hat{\Phi} + k_p^2\hat{\Phi} = 0, \qquad k_p = \omega\sqrt{\frac{\rho}{\lambda+\mu}} \tag{7.24}$$

$$\nabla^2\hat{H}_y + k_s^2\hat{H}_y = 0, \qquad k_s = \omega\sqrt{\frac{\rho}{\mu}} \tag{7.25}$$

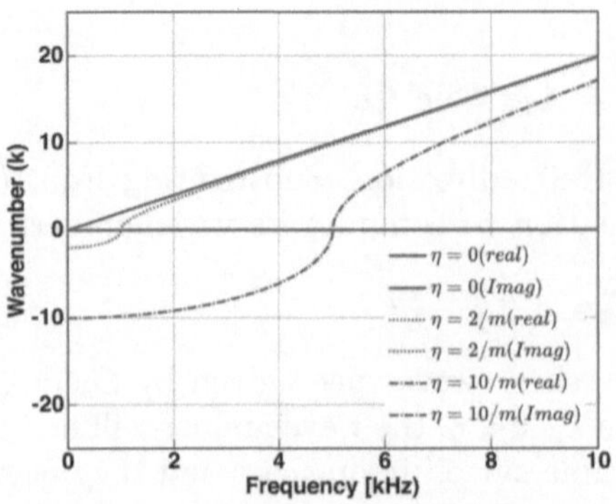

(a) $S_H Wave$ Spectrum Relation

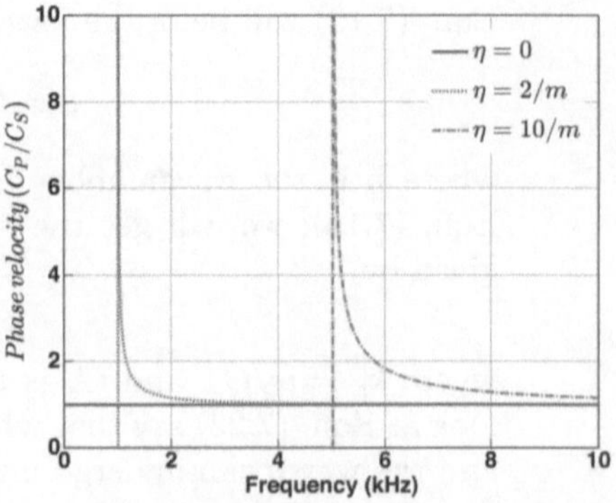

(b) $S_H Wave$ Dispersion Relation-Phase speed

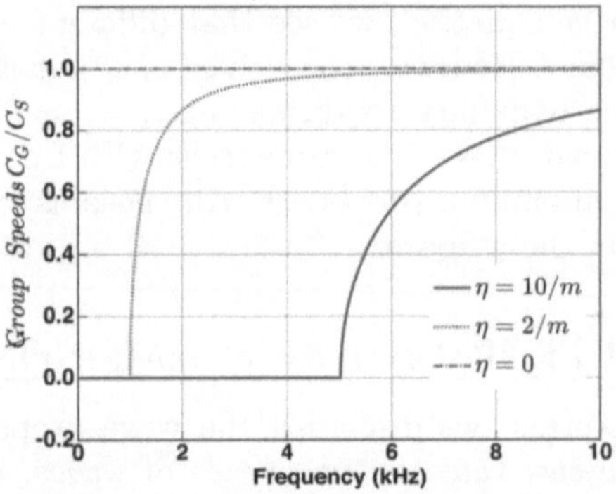

(c) $S_H Wave$ Dispersion Relation-Group speed

FIGURE 7.4: Spectrum & Dispersion relations for S_H waves in a infinite 2-D solid

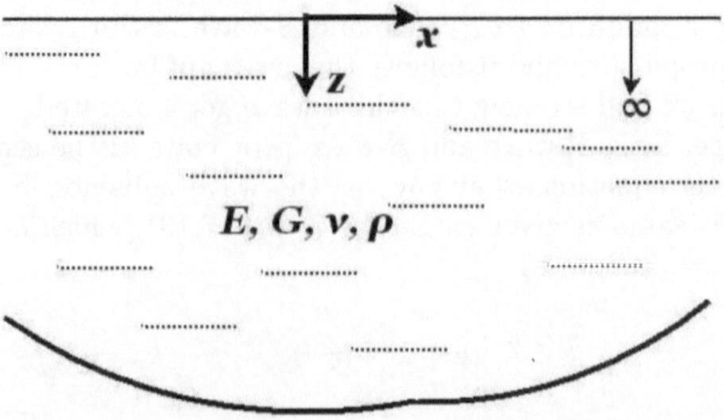

FIGURE 7.5: A semi-infinite media with the coordinate axes

The above partial differential equation needs to be solved on a domain shown in Fig. 7.5, where we can conveniently apply the variable-separable method for their solutions. That is, the solution can be assumed of the form

$$\hat{\Phi}(x,y,\omega) = f(z)e^{-ikx}, \qquad \hat{H}_y(x,y,\omega) = g(z)e^{-ikx}$$

The solution approach is very similar to the solution of the S_H waves in infinite domain (Eqn. (7.19)) presented in the previous section. Note that we have the common term e^{-ikx} between the terms in order to preserve the phase of the response. Substituting the above equation in the governing differential equation Eqn. (7.25), we get the following equations

$$\frac{\partial^2 f}{\partial z^2} - [-k^2 + k_p^2]f = 0. \qquad \frac{\partial^2 g}{\partial z^2} - [-k^2 + k_s^2]g \tag{7.26}$$

The above equations are constant coefficient differential equations, which have exponential solutions. Noting that

$$\eta^2 = -k^2 + k_p^2, \qquad \bar{\eta}^2 = -k^2 + k_s^2 \tag{7.27}$$

and considering that the domain is semi-infinite, wherein no terms associated with reflections are possible either in the x direction or z direction, we can assume the solution of potential given Eqn. (7.25) as

$$\hat{\Phi} = \mathbf{A}e^{i\eta z}e^{-ikx}, \qquad \hat{H}_y = \mathbf{B}e^{i\bar{\eta} z}e^{-ikx} \tag{7.28}$$

In the above equation, η and $\bar{\eta}$ are the wavenumbers in the z direction and they can also be complex or negative. Eqn. (7.27) represents the wavenumber equations of P and S wave propagating in a semi-infinite medium that are coupled to the wavenumber in the z direction. We will now plot the spectrum relation (k vs. frequency) for different values of η and $\bar{\eta}$. Here, η and $\bar{\eta}$ represent spatial modes and they, like their temporal counterpart, have a spatial window L associated with them, and they are chosen as $\eta = \bar{\eta} = 2m\pi/L$, where m represents the spatial modes and can take values $m = 0, 1, 2, ...M$. The spectrum relations for P- and S-waves are shown in Figs. 7.6(a) and (b), respectively. These plots were generated using aluminum properties.

From Fig. 7.6, we can see that both P-wave and S-wave are dispersive initially and tend to become non-dispersive at higher frequencies. Their behavior is very similar to the spectrum relation of the S_H wave propagation presented in the previous section. Depending upon the spatial mode m, new spatial modes are generated, which are evanescent to start with, and become propagating beyond the cut-off frequency as shown in the figure. Higher the value of η, higher is the value of the cut-off frequency.

Next, we will compute the dispersion relations for P and S waves for different values of η and $\bar{\eta}$, where we have assumed $\eta = \bar{\eta}$. The phase speeds are calculated using the usual definitions, which are given by

$$C_p = real\left(\frac{\omega}{k}\right) \qquad k_1^2 = k_p^2 - \eta^2, \qquad k_2^2 = k_s^2 - \bar{\eta}^2 \tag{7.29}$$

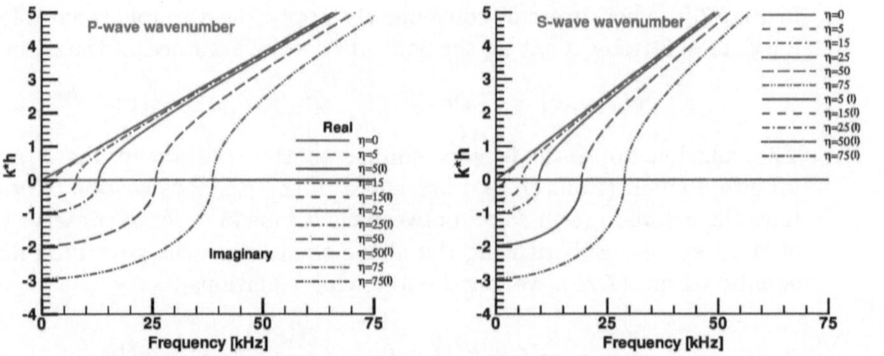

(a) P-wave Spectrum Relation (b) S-wave Spectrum Relation

FIGURE 7.6: Spectrum relation of P and S waves in a semi-infinite 2-D solid

As before, we can obtain the group speeds of these wave by differentiating Eqn. (7.27) with respect to ω and in doing so, we can write the group speeds of P and S waves as

$$\frac{d\omega}{dk_1} = C_g^{p-wave} = \frac{k_1}{k_p}C_p, \qquad \frac{d\omega}{dk_2} = C_g^{s-wave} = \frac{k_2}{k_s}C_s \tag{7.30}$$

where $k_p = \omega/C_p$ and $k_s = \omega/C_s$ are the wavenumber of P and S waves and C_p and C_s are the phase speeds of P and S waves.

Let us next plot the wave speeds for these waves. Fig. 7.7 shows the phase and group speeds of P-wave, while Fig. 7.8 shows the same for S-wave.

From Figs. 7.7 and 7.8, we see that the waves are dispersive initially and tend to become non-dispersive at higher frequencies. The P-wave speeds tend to attain the value of C_p at higher frequencies, while the S-wave speeds tend to reach the value of C_s. In addition, the phase speed escape to infinity at cut-off frequencies.

Next, we will evaluate the displacements, strains and stresses. The aim of deriving these expressions is to enforce boundary conditions at $z = 0$ and see if new waves are generated due to these boundary conditions. Using the potentials, we can write the displacement components using Eqns. (7.7) and (7.28), they can be written for the present case as

$$\begin{aligned}\hat{u} &= -ik\mathbf{A}e^{i\eta z}e^{-ikx} + i\bar{\eta}\mathbf{B}e^{i\bar{\eta}z}e^{-ikx} \\ \hat{v} &= 0 \\ \hat{w} &= i\eta\mathbf{A}e^{i\eta z}e^{-ikx} + ik\mathbf{B}e^{i\bar{\eta}z}e^{-ikx}\end{aligned} \tag{7.31}$$

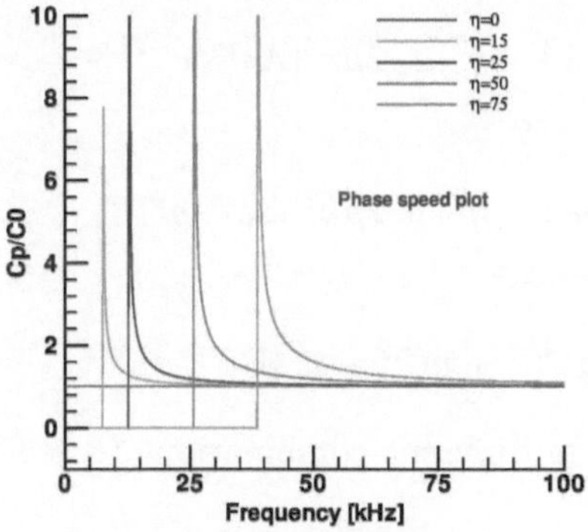

(a) Phase Speed

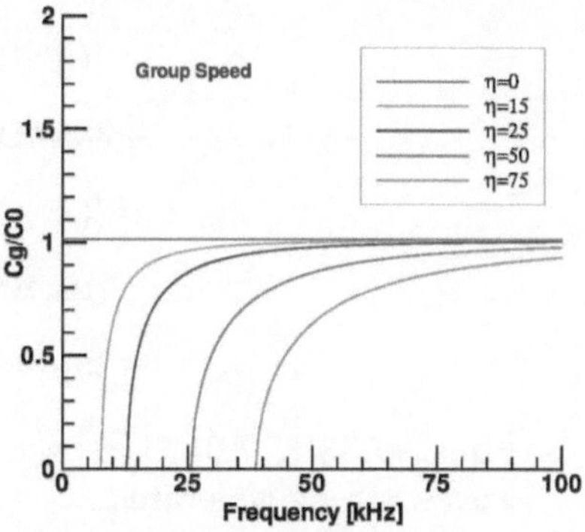

(b) Group Speed

FIGURE 7.7: Phase and Group speed variation for *P*-wave

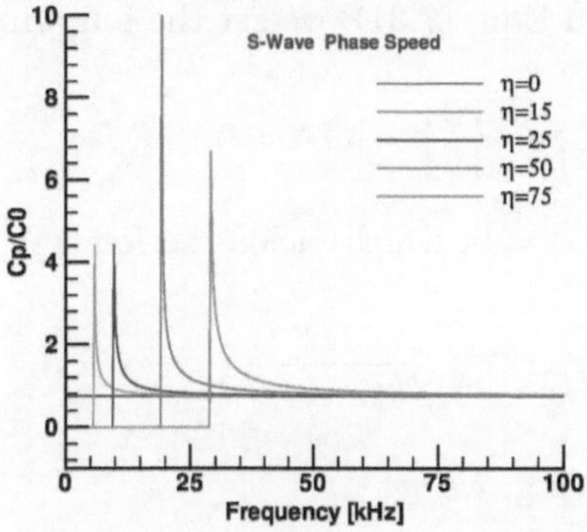

(a) Phase Speed

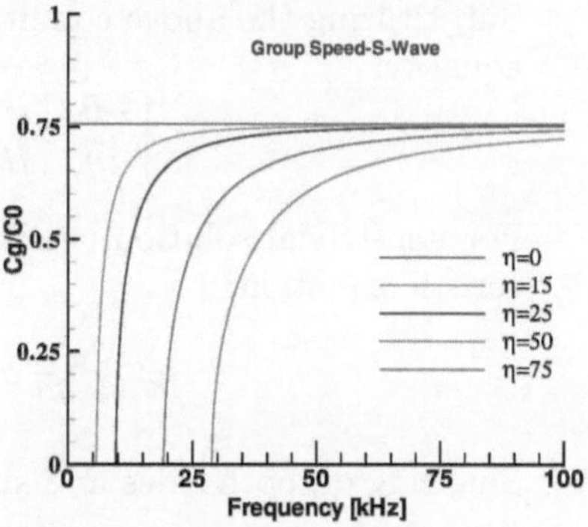

(b) Group Speed

FIGURE 7.8: Phase and Group speed variation for *S*-wave

Using the above equations, we can write the strains as

$$\begin{aligned}
\hat{\epsilon}_{xx} = \frac{\partial \hat{u}}{\partial x} = - &\quad k^2 \mathbf{A} e^{i\eta z} e^{-ikx} + \bar{\eta} k \mathbf{B} e^{i\bar{\eta} z} e^{-ikx} \\
\hat{\epsilon}_{zz} = \frac{\partial \hat{w}}{\partial z} = &\quad -\eta^2 \mathbf{A} e^{i\eta z} e^{-ikx} - k\bar{\eta} \mathbf{B} e^{i\bar{\eta} z} e^{-ikx} \\
\hat{\gamma}_{xz} = \frac{\partial \hat{u}}{\partial z} + \frac{\partial \hat{w}}{\partial x} = &\quad k\eta \mathbf{A} e^{i\eta z} e^{-ikx} - \bar{\eta}^2 \mathbf{B} e^{i\bar{\eta} z} e^{-ikx} \\
&\quad +\eta k \mathbf{A} e^{i\eta z} e^{-ikx} + k^2 \mathbf{B} e^{i\bar{\eta} z} e^{-ikx}
\end{aligned} \tag{7.32}$$

Using the equation above in Eqn. (7.3) and simplifying, we can write down the stresses as

$$
\begin{aligned}
\hat{\sigma}_{xx} &= [(\lambda+2\mu)(-k^2-\eta^2)+2\mu\eta^2]\mathbf{A}e^{i\eta z}e^{-ikx} \\
&+2\mu k\bar{\eta}\mathbf{B}e^{i\bar{\eta}z}e^{-ikx} \\
\hat{\sigma}_{zz} &= [(\lambda+2\mu)(-k^2-\eta^2)+2\mu k^2]\mathbf{A}e^{i\eta z}e^{-ikx} \\
&-2\mu k\bar{\eta}\mathbf{B}e^{i\bar{\eta}z}e^{-ikx} \\
\hat{\tau}_{xz} &= 2\mu k\eta\mathbf{A}e^{i\eta z}e^{-ikx}+\mu(k^2-\bar{\eta}^2)\mathbf{B}e^{i\bar{\eta}z}e^{-ikx}
\end{aligned}
\tag{7.33}
$$

Eqns. (7.33 & 7.31) are used next in different combinations to see how new waves can be generated.

7.4.1 Fixed Boundary Condition

Here, we impose the following boundary conditions:

$$
\hat{u}(x,0,\omega)=0, \qquad \hat{w}(x,0,\omega)=0
$$

Substituting the above conditions in Eqn. (7.31), we get the following matrix equation

$$
\begin{bmatrix} -ik & i\bar{\eta} \\ i\eta & ik \end{bmatrix} \begin{Bmatrix} \mathbf{A} \\ \mathbf{B} \end{Bmatrix} = \begin{Bmatrix} 0 \\ 0 \end{Bmatrix} = \hat{\mathbf{G}}\mathbf{A}=0 \tag{7.34}
$$

For non-trivial solution, we need $|\hat{\mathbf{G}}|=0$, which yields the following characteristic equation

$$
k^2+\eta\bar{\eta}=k^2\sqrt{k_p^2-k^2}\sqrt{k_s^2-k^2}=0
$$

Squaring on both sides and simplifying, we get

$$
k^2(k_p^2+k_s^2)=k_p^2k_s^2
$$

Substituting $k_p=\omega/C_p$ and $k_s=\omega/C_s$ with $C_p=(\lambda+\mu)/\rho$ and $C_s=\mu/\rho$, we can solve for wavenumber as

$$
k_1,k_2=\pm\frac{\omega}{\sqrt{C_p^2+C_s^2}} \tag{7.35}
$$

We see that the fixed boundary at the surface generates a wave that propagates non-dispersively with a speed of $c=\sqrt{C_p^2+C_s^2}$. This example also shows how the boundary condition couples the P and S waves. This is similar to non-dispersive wave propagation of a simple rod but with a speed of C_p+C_s. The spectrum and dispersion relations are similar to that of an axial rod. Using the wavenumber expressions from Eqn. (7.35), we can write the transverse wavenumbers η and $\bar{\eta}$ as

$$\eta = \sqrt{k_p^2 - k^2} = \sqrt{\frac{\omega^2}{C_p^2} - \frac{\omega^2}{C_p^2 + C_s^2}} = k\frac{C_s}{C_p}$$

$$\bar{\eta} = \eta = \sqrt{k_s^2 - k^2} = \sqrt{\frac{\omega^2}{C_s^2} - \frac{\omega^2}{C_p^2 + C_s^2}} = k\frac{C_p}{C_s}$$

where k takes the value of either k_1 or k_2 given in Eqn. (7.35). We can now write the amplitude ratio from Eqn. (7.34) as

$$\frac{\mathbf{B}}{\mathbf{A}} = \frac{k}{\bar{\eta}} \tag{7.36}$$

Substituting for η and $\bar{\eta}$, we can write the reflected coefficient as

$$\frac{\mathbf{B}}{\mathbf{A}} = \frac{C_s}{C_p} \tag{7.37}$$

From this relation, we see that the reflected coefficient is independent of frequency and the amplitude of the reflected pulse $\mathbf{B}$ will always be lower than incident pulse $\mathbf{A}$ since the ratio C_s/C_p is always less than 1.

7.4.2 Mixed Boundary Condition

Unlike the previous case, boundary conditions involving both displacement and stresses can be imposed at the surface $z = 0$ shown in Fig. 7.5. There are two possibilities that exist, which are given below.

$$\hat{w}(x, 0, \omega) = 0, \qquad \text{and} \qquad \hat{\tau}_{xz}(x, 0, \omega) = 0$$
$$\hat{\sigma}_{zz}(x, 0, \omega) = 0, \qquad \text{and} \qquad \hat{u}(x, 0, \omega) = 0$$

We will analyze the wave propagation for both these cases. We will now consider the first case. This is the case where $\hat{w} = 0$ and $\hat{\tau}_{xz} = 0$. Using Eqns. (7.31 & 7.33) at $z = 0$, we get the following matrix equation

$$\begin{bmatrix} i\eta & ik \\ 2k\eta & (k^2 - \bar{\eta}^2 \end{bmatrix} \begin{Bmatrix} \mathbf{A} \\ \mathbf{B} \end{Bmatrix} = \begin{Bmatrix} 0 \\ 0 \end{Bmatrix} = \hat{\mathbf{G}}\mathbf{A} = 0 \tag{7.38}$$

Setting the determinant of the matrix to zero will result in the following characteristic equation

$$i\eta(-k^2 - \bar{\eta}^2) = 0$$

Expressing η and $\bar{\eta}$ in terms of k, the above equation can be written as

$$i\sqrt{k_p^2 - k^2}(-k^2 - k_p^2 + k_s^2) = 0 \Rightarrow \eta = 0, \qquad \text{or} \qquad k = k_p = \omega/C_p$$

That is, for this boundary condition, only P-wave will propagate. The amplitude ratio for this case can be obtained by Eqn. (7.38), which is given by

$$\frac{\mathbf{B}}{\mathbf{A}} = \frac{\eta}{k} \Rightarrow 0$$

This implies that the second condition is automatically satisfied and is not necessary to obtain the wave behavior.

Let us now impose the second set of mixed boundary conditions, that is $\hat{\sigma}_{zz} = 0$ and, $\hat{u} = 0$ at $z = 0$. These boundary conditions are imposed using Eqns. (7.31 & 7.33) and the resulting equations can be written as

$$\begin{bmatrix} -ik & i\bar{\eta} \\ (\lambda+2\mu)(-k^2-\eta^2)+2\mu k^2 & -2\mu k\bar{\eta} \end{bmatrix} \begin{Bmatrix} \mathbf{A} \\ \mathbf{B} \end{Bmatrix} = \begin{Bmatrix} 0 \\ 0 \end{Bmatrix} = \hat{\mathbf{G}}\mathbf{A} = 0 \qquad (7.39)$$

As before, setting the determinant to zero, we have

$$i2\mu k^2\bar{\eta} - i\bar{\eta}(\lambda+2\mu)(-k^2-\eta^2) - i2\mu k^2\bar{\eta} = 0 \Rightarrow i\bar{\eta}(-k^2-\eta^2) = 0 \qquad (7.40)$$

The above characteristic equation looks similar to the previous case but with $\bar{\eta}$ taking the place of η. This means, only the propagation of S-wave will take place. The above two boundary conditions illustrate how one can uncouple P and S waves in a semi-infinite media.

7.4.3 Traction Free Boundary Conditions: A Case of Rayleigh Surface Waves

This case was first investigated by Lord Rayleigh [70], wherein he found a new wave that propagated on a traction free surface. In this subsection, we apply traction free boundary condition at $z = 0$, that is $\hat{\sigma}_{zz} = 0$ and $\hat{\tau}_{xz} = 0$ using Eqn. (7.33), and this results in the following matrix equation

$$\begin{bmatrix} (\lambda+2\mu)(-k^2-\eta^2)+2\mu k^2 & -2\mu k\bar{\eta} \\ 2\mu k\eta & \mu(k^2-\bar{\eta}^2) \end{bmatrix} \begin{Bmatrix} \mathbf{A} \\ \mathbf{B} \end{Bmatrix} = \begin{Bmatrix} 0 \\ 0 \end{Bmatrix} = 0 \qquad (7.41)$$

As before, we can get the characteristic equation by setting the determinant of the matrix to zero and simplifying, which is given by

$$(2k^2 - k_s^2)^2 + 4k^2\eta\bar{\eta} = 0 \qquad (7.42)$$

In the equation above, we substitute for η and $\bar{\eta}$ to get the following equation

$$(2-m)^2 - 4\sqrt{(1-n)(1-m)} = 0, \qquad m = \frac{c^2}{C_s^2}, \qquad n = \frac{c^2}{C_p^2} \qquad (7.43)$$

where c is the speed of the wave, which we need to compute. Multiplying Eqn. (7.43) by $(2-m)^2 + 4\sqrt{(1-n)(1-m)}$ and canceling m, following polynomial equation for solving for m is obtained

$$m^3 - 8m^2 + (24 - 16\Gamma)m - 16(1-\Gamma) = 0, \qquad \Gamma = \frac{C_s^2}{C_p^2} = \frac{1-2\nu}{2(1-\nu)} \qquad (7.44)$$

Eqn. (7.44) is a cubic polynomial in m and can be solved using *roots* function in MATLAB. Alternatively, we can also solve this equation using Cardano's method [17]. In this book, a MATLAB script by name *eqn.m* is provided, which solves Eqn. (7.44) for obtaining wavenumber modes for Rayleigh waves and the code also plots these wavenumbers for different values of Poisson's ratio. The important aspect is that they are frequency independent. The three roots may be real, or one root can be real with other two roots being imaginary. That is, as we saw in the earlier chapters, the nature of roots of the characteristic equation has profound effect on its wave propagation.

Looking at Eqn. (7.44), we see that the character of the roots will change depending upon the value of Poisson's ratio ν. For low values of ν, all the three roots will always be real. As the value of ν increases, for certain Poisson's ratio called the *Critical Poisson's ratio*, two of the modes become complex . Here, we compute the roots m_1, m_2 and m_3 of Eqn. (7.44) using MATLAB script *eqn.m*, and the wavenumber ratio k/k_s is extracted as follows:

$$m = \frac{c^2}{C_s^2} = \frac{\omega^2}{k^2}\frac{k_s^2}{\omega^2} = \frac{k_s^2}{k^2}$$

Here, we plot $\sqrt{1/m_1} = k_1/k_s$, $\sqrt{1/m_2} = k_2/k_s$ and $\sqrt{1/m_2} = k_2/k_s$ and this is shown in Figs. 7.9(a) and (b)

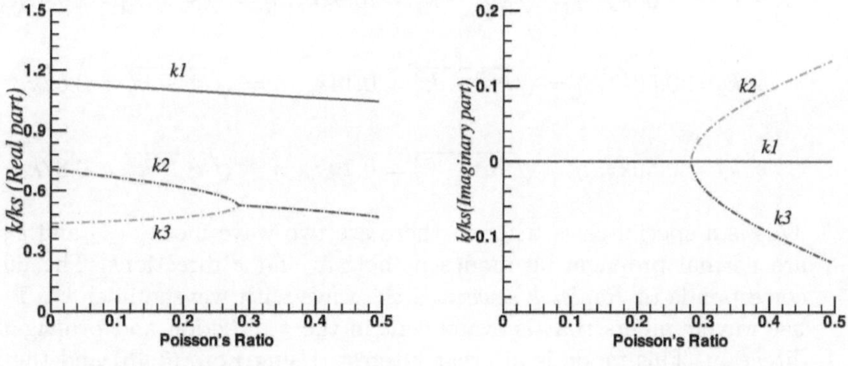

(a) Real Part (b) Imaginary Part

FIGURE 7.9: Spectrum relation for Rayleigh waves

From Fig. 7.9, we see that the real root k_1 decreases with increase in Poisson's ratio, while the other two roots k_2 and k_3 are purely real until the value of $\nu = 0.285$. Beyond this value, the roots k_2 and k_3 are complex and they attenuate very fast as they propagate. Here, $\nu = 0.285$ for aluminum material is the *critical Poisson's ratio* at which Rayleigh wave propagation takes place. This wave is non-dispersive and its phase and group speeds are independent of frequency. Their utility in seismology is well documented.

Once the wavenumbers are determined, we first write η as a function of Rayleigh wavenumber k_r and Rayleigh wave speed C_r as

$$\eta = k\sqrt{1-\Gamma_r^2} = k_r\eta_r, \qquad \Gamma_r = \frac{c}{C_p} = \frac{C_r}{C_p}$$

to obtain the responses using the following expressions:

$$\text{Scalar Potential} \qquad \Phi = \mathbf{A}e^{ik_r\eta_r}e^{ik_r(x-\omega t)}.$$

$$\text{Vector Potential} \qquad H_y = \mathbf{B}e^{ik_r\eta_r}e^{ik_r(x-\omega t)}.$$

$$u = \frac{\partial\Phi}{\partial x} + \frac{\partial H_y}{\partial z}, \qquad w = \frac{\partial\Phi}{\partial z} - \frac{\partial H_y}{\partial x}.$$

$$\text{Amplitude Ratio} \qquad \frac{\mathbf{B}}{\mathbf{A}} = \frac{i\sqrt{1-\Gamma_r^2}}{2-\Gamma_S^2}, \qquad \Gamma_S = \frac{C_r}{C_S} \tag{7.45}$$

Let us consider a special case of $\nu = 0.25$. For this value, the Lame's constants λ and μ are equal and Eqn. (7.44) can be factorized exactly. Also, the three wavenumbers can be obtained from Fig. 7.9. The values of the three wavenumbers and the associated values of η and $\bar{\eta}$ are as follows:

- $k_1 = 1.08k_s$, $\eta = \sqrt{kp^2 - k_1^2} = i0.92k_s$, $\bar{\eta} = \sqrt{k_s^2 - k_1^2} = i0.42k_s$
- $k_2 = 0.56k_s$, $\eta = \sqrt{kp^2 - k_2^2} = 0.14k_s$, $\bar{\eta} = \sqrt{k_s^2 - k_2^2} = 0.82k_s$
- $k_3 = 0.50k_s$, $\eta = \sqrt{kp^2 - k_3^2} = 0.29k_s$, $\bar{\eta} = \sqrt{k_s^2 - k_3^2} = 0.87k_s$

This is a special case, wherein there are two wave modes (k_2 and k_3), which are normal propagating modes in both x and z directions. The mode that corresponds to *Rayleigh's wave* is the wave with wavenumber k_1. This mode has wavenumbers that is evanescent in the z direction and propagating in x direction. This mode is of great interest. Using Eqn. (7.45) and the material properties of aluminum, we will plot the responses on a semi-infinite 2-D media. These responses are shown in Fig. 7.10. These responses were generated using the triangular pulse used in the previous chapters and are shown in Fig. 4.3. Here, the coefficient **B** is assumed to take this triangular shaped pulse and the coefficient **A** is evaluated from the amplitude ratio given in Eqn. (7.45).

Both these figures show that as the propagation distance increases, the response amplitude decreases and this decrease is more rapid along the z direction. Reference [144] has discussed the problem of Rayleigh wave problem

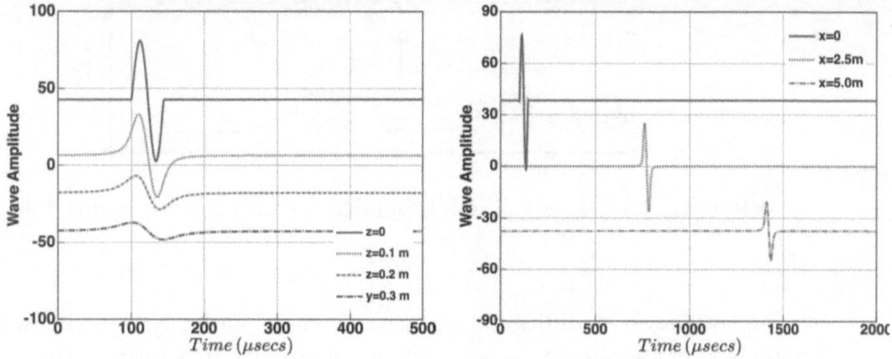

(a) Response at $x = 0$ for different z (b) Response at $z = 0$ for different x

FIGURE 7.10: Rayleigh wave responses for an incident triangular pulse

in more detail. The author of this reference had derived the approximate value of the primary root k_1 for general Poisson's ratio ν as

$$\frac{k_1}{k_s} = \frac{1+\nu}{0.87 + 1.12\nu}$$

From the discussions presented in this and the previous subsections, the following inferences can be drawn:

- There is a need of taking both the scalar potential Φ and vector potential H_y together for satisfying the boundary conditions.
- Different boundary conditions on the surface generate different surface waves and *Rayleigh wave* is one such wave generated due to traction free boundary at the surface.
- The Rayleigh wave equation gives three modes. For low Poisson's ratio, all the modes are real and propagating, which is not of much interest, especially in the area of seismology. When the Poisson's ratio is close to critical Poisson's ratio, then among the three roots, two of them decay rapidly as they propagate. This Rayleigh mode is of great interest, especially to those working in the area of seismology.

7.5 WAVE PROPAGATION IN DOUBLY BOUNDED MEDIA

In a doubly bounded media, an additional surface is introduced at a distance h from the top surface, as shown in Fig. 7.11. Let us assume that the bounded media is isotropic having Young's modulus E, Poisson's ratio ν, rigidity modulus G and density ρ.

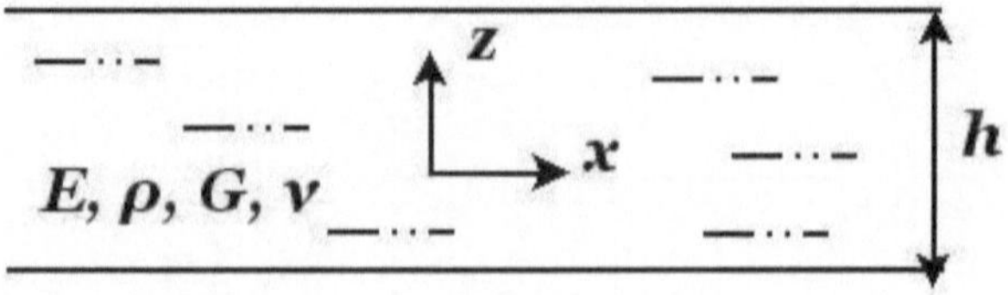

FIGURE 7.11: A 2D doubly bounded media with the coordinate axes

Introduction of additional surface will cause the wave to confine within this bounded region thereby it acts like a waveguide. In addition, the additional bottom surface will start reflecting the incident wave coming from the top surface. As in the case of semi-infinite media, different boundary conditions at the top and bottom surface will generate newer waves. Hence, the same approach to study the dispersion relations of an infinite 2-D media will be again followed here. However, for most of the boundary conditions, the approach leads to complicated transcendental equation, the exact solution of which is very difficult to obtain. These equations need to be solved numerically. The traction free boundary at the top and bottom surface of a thin plate is of great importance in *Structural Health Monitoring*, wherein the waves that get generated due to this boundary condition and normally referred to as *Lamb waves* are very useful in detecting damages in structural members as they can travel long distances with little attenuation.

In this case, two loading cases arise, which are normally referred to as *Symmetric case*, where the wave causes bending or beam-like action, while the second case corresponds to *Anti-symmetric case*, which simulates rod-like motion. Bounding the waves between two surfaces causes standing waves in the z direction, while along x direction, propagation happens. In the next subsection, we will derive the displacement and stresses and look at the generation of newer waves for the symmetric case. A similar procedure can be applied to obtain these quantities for the anti-symmetric case. Based on the theory developed in this section, MATLAB scripts are provided, that will help reader to compute and plot the wave modes for two different boundary conditions, namely *zero displacement conditions* and *zero traction conditions* at both the edges.

7.5.1 Symmetric Loading Case

In this case, we assume that the loading on this bounded media is symmetric about the z direction and hence, we can write the scalar and vector potentials as

$$\Phi(x,z,t) = \mathbf{A}\cos(\eta z)e^{-ik(x-\omega t)}, \quad H_y(x,z,t) = \mathbf{B}\sin(\bar{\eta} z)e^{-ik(x-\omega t)} \tag{7.46}$$

where η and $\bar{\eta}$ are the wavenumbers in the z direction corresponding to scalar and vector potentials, respectively. Using the potentials, we can evaluate the displacements in the frequency using Eqn. (7.7), which are given by

$$\begin{aligned}\hat{u}(x,z,\omega) &= \left(-ik\mathbf{A}\cos(\eta z) + \bar{\eta}\mathbf{B}\cos(\bar{\eta}z)\right)e^{-ikx}\\ \hat{v}(x,z,\omega) &= 0.\\ \hat{w}(x,z,\omega) &= \left(-\eta\mathbf{A}\sin(\eta z) + ik\mathbf{B}\sin(\bar{\eta}z)\right)e^{-ikx}\end{aligned} \tag{7.47}$$

Using the displacement variation and the strain-displacement relations, strains are evaluated and using the isotropic constitutive model, we can obtain the stresses as done previously. The stress variation in this case is given by

$$\begin{aligned}\hat{\sigma}_{xx} &= \left(-(2\mu k^2 + k_p^2)\mathbf{A}\cos(\eta z) - 2\mu ik\bar{\eta}\mathbf{B}\cos(\bar{\eta}z)\right)e^{-kx}\\ \hat{\sigma}_{zz} &= \left(\mu(k^2 - \eta^2)\mathbf{A}\cos(\eta z) + 2\mu ik\bar{\eta}\mathbf{B}\cos(\bar{\eta}z)\right)e^{-kx}\\ \hat{\tau}_{xy} &= \left(\mu ik\eta\mathbf{A}\sin(\eta z) - 0.5\mu(\bar{\eta}^2 - k^2)\mathbf{B}\sin(\bar{\eta}z)\right)e^{-ikx}\end{aligned} \tag{7.48}$$

As done in the semi-infinite case, we will use the displacement and stress expressions to impose different symmetric boundary condition at the top and bottom surfaces located at $z = -h/2$ and $z = +h/2$.

7.5.2 Fixed Boundary Condition

In this case, we apply fixed boundary condition on the two surfaces located symmetrical about $z = -h/2$ and $z = +h/2$. That is,

$$\hat{u}(x, \pm h/2, \omega) = 0, \qquad \hat{w}(x, \pm h/2, \omega) = 0$$

Using the above condition in Eqn. (7.47), we get

$$\begin{bmatrix}-ik\cos(\eta h/2) & \bar{\eta}\cos(\bar{\eta}h/2)\\ -\eta\sin(\eta h/2) & ik\sin(\bar{\eta}h/2)\end{bmatrix}\begin{Bmatrix}\mathbf{A}\\ \mathbf{B}\end{Bmatrix} = \begin{Bmatrix}0\\ 0\end{Bmatrix} \tag{7.49}$$

Setting the determinant of the matrix to zero, we get the following characteristic equation

$$k^2 + \eta\bar{\eta}\frac{\tan(\eta h/2)}{\tan(\bar{\eta}h/2)} = 0 \tag{7.50}$$

A look this equation will tell us that for $h << 1$, that is when the thickness of media is very small, Eqn. (7.50) becomes

$$k^2 + \eta^2 = 0 \rightarrow kp^2 = 0$$

This means that P-wave will not propagate and only the S-wave will propagate in a doubly bounded media with above boundary conditions.

In order to solve the above equation, we will first look at the cut-off frequencies for the symmetric loading case. This is obtained by substituting $k = 0$ in Eqns. (7.71 & 7.72). That is the cut-off frequencies for symmetric and anti-symmetric lamb wave modes occur at

$$\omega_{cut-off} = \frac{2m\pi}{h}C_s, \qquad \frac{(2m-1)\pi}{h}C_p \qquad m = 1, 2, ... \tag{7.51}$$

Any solution method will have wavenumber branches emanating from these cut-off frequencies.

7.5.3 Method of Solution of Dispersion Equations in a Doubly Bounded Media

The complete solution of Eqn. (7.50) gives the spectrum relations. This equation is a transcendental equation and is multiple valued and in addition, the wavenumber could be purely real, purely imaginary or it can be complex (that is, the wavenumber will have both real and imaginary parts). Solving this analytically is very difficult and may not be possible. There are several different numerical techniques that researchers have employed to obtain the dispersion relations. Some of the techniques to solve the above equation could be *Bisection method, Secant method or it is also called Regula Falsi method, Newton-Raphson method*, etc. Most of the numerical methods require assumption of initial guess of the roots, which in most times will end up being tricky. Success of some of the methods depends on how well one chooses the initial guess. The choice of the initial guess depends on the physics of the problem that the given transcendental equation represents and as such there are no definitive rules in choosing them. The algorithm and the Fortran 77 code to solve transcendental equations of the above methods are given in [41]. The detailed theoretical basis of these methods is given in [22]. Most of these methods can only pick the real roots and cannot handle imaginary roots and also the complex roots. Most of these methods are iterative in nature and they make the error associated with the dispersion function minimum for the iteratively determined wavenumber. However, for a root, which is complex in nature, we need to make the error associated with both real and imaginary parts minimum. Doyle, in his book [34], has given an algorithm to solve for wavenumbers in doubly bounded media using the above concept.

There are a number of researchers who have used different methods to determine the wavenumber-frequency (or phase speed-frequency) relations for a doubly bounded media. Some of these methods are reported in the references [43], [60], [100], and [106]. A simple method to find the roots of the dispersion equation is to simply scan the function for each frequency in terms of small wavenumber steps . This process was used in [100] to determine the real, imaginary and complex modes. However, for doing this, it is necessary to take very small wavenumber steps so that the roots are not missed. Many researchers have written the dispersion relations in terms of phase speeds and solved it as the basic unknown.

While the solutions of the doubly bounded media having stress free boundary conditions (Lamb wave case, which is presented in the next subsection) are widely reported in the literature, the solution of the dispersion relation for the doubly bounded media with fixed boundary condition having the dispersion relation given by Eqn. (7.50) is not available in the literature. The dispersion relations for this case (Eqn. (7.50)) resemble very closely with the Lamb wave equations (to be discussed in the later subsection of this section) and hence the techniques adopted in the solution of Lamb wave dispersion relations can be adopted here. There are many general purpose codes such as *Disperse* [64] developed by Imperial College London can be used to solve dispersion relation for the stress free doubly bound media. However, this software cannot be used to solve the present problem of doubly bounded media with fixed boundary conditions. In addition, these software provide only real only wave modes. Another open source software *Dispersion Calculator* [63] developed by German Aerospace Center (DLR) can also provide solution to the dispersion relation for Lamb wave modes in isotropic and anisotropic medium. However, this software again does not provide solutions for the present problem. Many commercial software such as *MATLAB*, *Mathematica* or *Maple* can be used to develop methods to solve dispersion relations in a doubly bounded media. In this chapter, we have developed the MATLAB scripts for the present problem of fixed boundary conditions as well as for the case of stress-free boundary conditions. The provided MATLAB scripts can obtain purely real, purely imaginary as well as complex roots of the transcendental equations.

In this section, we will present a novel method to solve dispersion relation for any general boundary conditions of a doubly bounded media that gives real, imaginary and complex roots of the wavenumber. The method is developed using MATLAB software. It is worth mentioning that this method of finding roots is not robust but it works if MATLAB and bracketing technique are used together.

The methodology for determining the roots of a transcendental equation such as Eqn. (7.50) by bracketing technique is summarized below:

Real roots: In finding the real roots by bracketing technique, approximate forms are not necessary. The exact form, either in terms of tan or sin and cos, can be used. First step involves bracketing the root. This is achieved as follows. Let us say that the dispersion equation is written in the following form.

$$LHS + RHS = 0$$

Then, around the root, it is expected that there can be a change of sign in the term $(LHS + RHS)$. Therefore, the wavenumber k is marched from zero at an interval within a range of real numbers. Whenever there exist a change of sign for the term, that interval bracket of k is stored. Then, as next step, the bracketed interval is subdivided and further bracketed.

This process is repeated until the difference in lower and upper limit interval values is less than 10^{-8}. It is to be mentioned here that, by this technique not all real k were determined for some transcendental equations. Upon examining the behavior of the term with respect to k, it was observed that the term becomes a pure imaginary number at higher k. Further, it was found that there existed some roots at higher k in the sense that magnitude of this imaginary part was going to zero at certain values of k, which were not captured by the above bracketing algorithm. Therefore, in order to capture these missing roots, an additional condition was put in the developed code which converts the pure imaginary numbers to real numbers.

Imaginary roots: In order to find pure imaginary roots, if at all exist, k is substituted as

$$k = ik$$

into the spectrum relations Eqn. (7.50). Then, again, the above bracketing technique has been applied to the term $(LHS + RHS)$. It is to be noted that, after substitution of above equation into the spectrum relation, the term $(LHS+RHS)$ becomes purely real, without any possibility of imaginary or complex values to the term. This process tracks the pure imaginary roots

Complex roots: Finding complex roots by the bracketing technique is a bit tricky. If a complex number $a + ib$ is zero, then both $a = b = 0$. By this, if the term $(LHS + RHS)$ is zero for a complex k, then both real and imaginary parts of the term are zeros. This can be implemented as absolute value of the term, that is, $[real(term)^2 + imag(term)^2] \rightarrow 0$. That is, finding a point in complex k plane such that it is a local minimum for the absolute value of the term as well as it is also zero for the absolute value of term.

Once the wavenumber is obtained, phase speed for the ith mode is obtained using the expression $C_{p_i} = \omega/k_i$. Three different MATLAB codes are provided for computing real, imaginary and complex roots for Eqn. (7.50).

Using the procedure outlined above, the wavenumbers for the fixed boundary conditions are obtained for the purely real, purely imaginary and complex roots by solving Eqn. (7.50). These are shown in Figs. 7.12 (a), (b) & (c). Fig. 7.12(d) shows the variation of phase speed obtained using real roots. These plots are generated using aluminum properties with depth of the doubly bounded media being 25 mm.

From the figure, we see that there are only about three modes up to 250 kHz. In Fig. 7.12(c), it is found that the first complex root was only after 450 kHz. For this mode, the imaginary part of wavenumber is quite high indicating that they may hardly propagate in the medium.

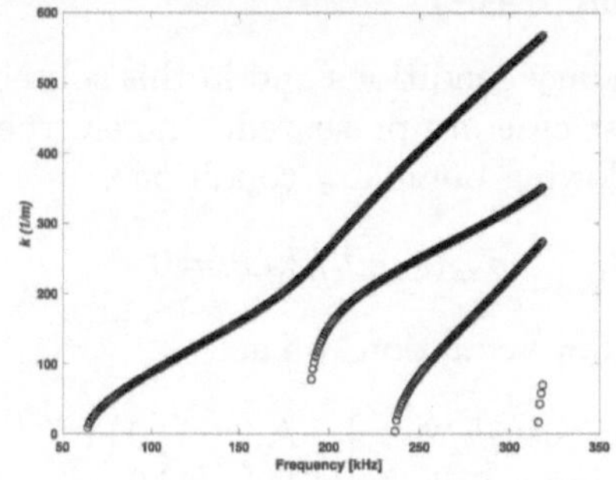

(a) Real roots

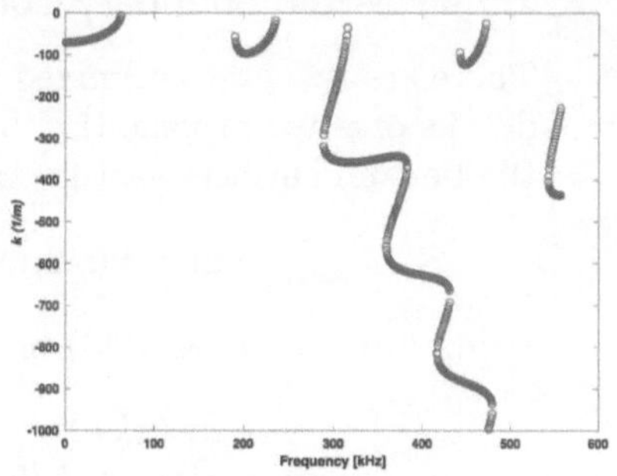

(b) imaginary roots

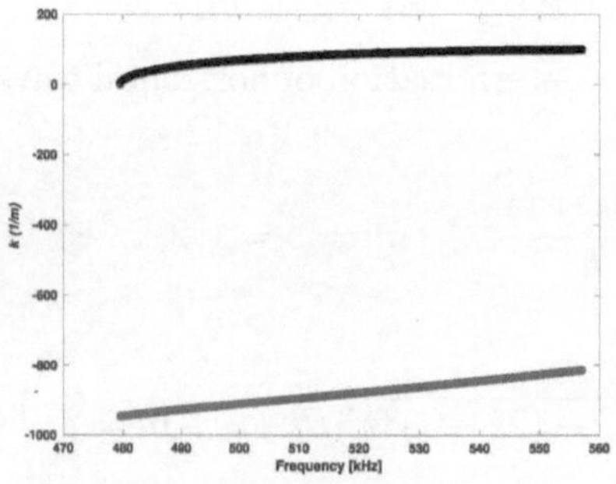

(c) Complex roots

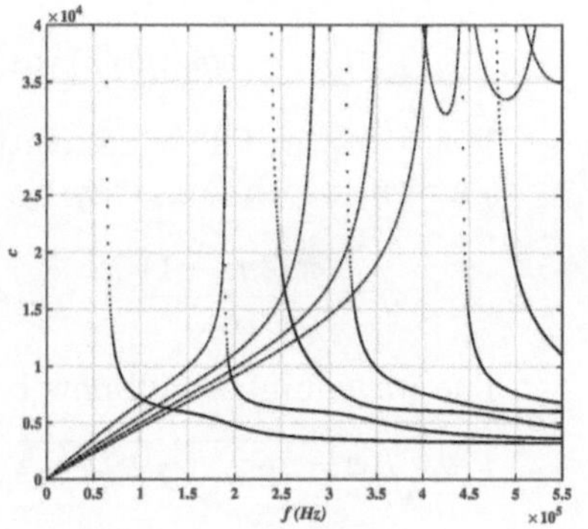

(d) Phase Velocity

FIGURE 7.12: Spectrum and Dispersion relation for a 2-D isotropic media with fixed boundary condition. Red lines indicate the imaginary roots

Next, the amplitude of the reflection coefficient for the case of doubly bounded media with fixed boundary condition is obtained using Eqn. (7.49) and is given by

$$\frac{\mathbf{B}}{\mathbf{A}} = \frac{ik\cos\eta h/2}{\bar{\eta}\cos\bar{\eta}h/2} = \frac{\eta\sin\eta h/2}{ik\sin\bar{\eta}h/2} \tag{7.52}$$

For $h << 1$, we can write Eqn. (7.52) after simplification as

$$\frac{\mathbf{B}}{\mathbf{A}} = \frac{1}{\sqrt{1-c^2/C_s^2}} = \frac{1-c^2/C_p^2}{\sqrt{1-c^2/C_s^2}}$$

From the above equation, we can say that when the phase speed of the medium becomes equal to the shear wave speed or the longitudinal wave speed, no reflection can take place from the fixed boundary. The above equations were derived considering the loading to be symmetric. Similar equations can be derived for anti-symmetric loadings.

7.5.4 Mixed Boundary Condition: Case I

There are two cases of mixed boundary conditions and in this subsection, the details of wave propagation for first case are presented, wherein the top and the bottom surfaces satisfy the following boundary conditions.

$$\hat{u}(x, \pm h/2, \omega) = 0, \qquad \hat{\sigma}_{zz}(x, \pm h/2, \omega) = 0$$

Using Eqns. (7.47 & 7.48) in the above equation, we get

$$\begin{bmatrix} -ik\cos(\eta h/2) & \bar{\eta}\cos(\bar{\eta}h/2) \\ \mu(k^2-\eta^2)\cos(\eta h/2) & 2\mu ik\bar{\eta}\cos(\bar{\eta}h/2) \end{bmatrix} \begin{Bmatrix} \mathbf{A} \\ \mathbf{B} \end{Bmatrix} = \begin{Bmatrix} 0 \\ 0 \end{Bmatrix} \tag{7.53}$$

Setting the determinant of the matrix to zero gives

$$\cos(\eta h/2)\cos(\bar{\eta}h/2)\left[2\mu k^2\bar{\eta} - \mu(k^2-\eta^2)\bar{\eta}\right] = 0 \tag{7.54}$$

This gives the following relations for determination of horizontal wavenumbers η and $\bar{\eta}$, which are given by

$$\eta = \frac{\pi(2m-1)}{h}, \qquad \bar{\eta} = \frac{\pi(2s-1)}{h}, \qquad m = s = 1, 2, ...\infty \tag{7.55}$$

The wavenumber will now become

$$k = \sqrt{k_p^2 - (2m-1)^2\pi^2/h^2} = \sqrt{k_s^2 - (2s-1)^2\pi^2/h^2}, \qquad m = s = 1, 2, ...$$

The wavenumber variation exhibits multiple cut-off frequencies, which is obtained by equating the above equation to zero. The multiple cut-off frequencies are given by

$$\omega_{cut-off} = C_p(2m-1)\pi = C_s(2s-1)\pi, \qquad m = s = 1, 2, 3..... \tag{7.56}$$

The reflection coefficient in this case becomes

$$\frac{\mathbf{B}}{\mathbf{A}} = \frac{ik\cos(\eta h/2)}{\bar{\eta}\cos(\bar{\eta}h/2)} = \frac{i(k^2-\eta^2)\cos(\eta h/2)}{2k\bar{\eta}\cos(\bar{\eta}h/2))} \tag{7.57}$$

The above equations are derived considering the loading to be symmetric. Similar equations can be derived for anti-symmetric loadings.

7.5.5 Mixed Boundary Condition: Case II

We consider here the second of the mixed boundary condition, which is given by

$$\hat{w}(x, \pm h/2, \omega) = 0, \qquad \hat{\tau}_{xz}(x, \pm h/2, \omega) = 0$$

Using Eqns. (7.47 & 7.48) in the above equation, we get

$$\begin{bmatrix} -\eta\sin(\eta h/2) & ik\sin(\bar{\eta}h/2) \\ \mu ik\eta\sin(\eta h/2) & -0.5\mu(\bar{\eta}^2-k^2)\sin(\bar{\eta}h/2) \end{bmatrix} \begin{Bmatrix} \mathbf{A} \\ \mathbf{B} \end{Bmatrix} = \begin{Bmatrix} 0 \\ 0 \end{Bmatrix} \tag{7.58}$$

Setting the determinant of the above matrix equal to zero, we get

$$\sin(\eta h/2)\sin(\bar{\eta}h/2)k_s^2/2 = 0 \tag{7.59}$$

which gives

$$\eta = \frac{2n\pi}{h}, \qquad \bar{\eta} = \frac{2m\pi}{h} \tag{7.60}$$

Using the above equations, we can write the wavenumber expression as

$$k = \sqrt{k_p^2 - 4\pi^2 n^2/h^2} = \sqrt{k_s^2 - 4\pi^2 m^2/h^2}, \qquad n = m = 1, 2, ... \tag{7.61}$$

As before, the wavenumber exhibits a number of cut-off frequencies, and they are obtained by setting the above equation equal to zero, which gives

$$\omega_{cut-off} = \frac{C_P}{h}2n\pi = \frac{C_S}{h}2m\pi, \qquad n = m = 1, 2, ... \tag{7.62}$$

The variation of the wavenumber with frequency is similar to what is shown in Fig. 7.6. The amplitude ratio for wave propagating in this case is given by

$$\frac{\mathbf{B}}{\mathbf{A}} = \frac{-\eta\sin(\eta h/2)}{ik\sin(\bar{\eta}h/2)} = \frac{ik\eta\sin(\eta h/2)}{-0.5(\bar{\eta}^2 - k^2)\sin(\bar{\eta}h/2))} \tag{7.63}$$

As before, the above equations are derived considering the loading to be symmetric. Similar equations can be derived for anti-symmetric loadings.

7.5.6 Traction Free Surfaces: A Case of Lamb Wave Propagation

We will now study the wave characteristics in a 2-D isotropic media that has zero traction in both top and bottom surfaces. This is an important case that has far reaching applications in seismic engineering, geological applications, and more recently in aerospace, civil and nuclear engineering for detection of structural flaws . The boundary conditions specified in this case are given as

$$\hat{\sigma}_{zz}(x, \pm h/2, \omega) = 0, \qquad \hat{\tau}_{xz}((x, \pm h/2, \omega) = 0 \tag{7.64}$$

We will consider both symmetric and anti-symmetric loading. For symmetric loadings, we can assume the scalar and vector potential in the frequency domain as

$$\hat{\Phi} = \mathbf{A}\cos(\eta z)e^{-ikx}, \qquad \hat{H}_y = \mathbf{B}\sin(\bar{\eta}z)e^{-ikx} \tag{7.65}$$

If the loading is anti-symmetric, then the scalar and vector potential is assumed as

$$\hat{\Phi} = \mathbf{A}\sin(\eta z)e^{-ikx}, \qquad \hat{H}_y = \mathbf{B}\cos(\bar{\eta}z)e^{-ikx} \tag{7.66}$$

Using these potentials, we can obtain the displacements $\hat{u}$ and $\hat{w}$ using Eqn. (7.7). For the symmetric loading case, the deformations are given by

Eqn. (7.47). For anti-symmetric loading, the deformations are given by

$$\begin{aligned}\hat{u}(x,z,\omega) &= (-ik\mathbf{A}\sin(\eta z) + \bar{\eta}\mathbf{B}\sin(\bar{\eta}z))\,e^{-ikx}\\ \hat{v}(x,z,\omega) &= 0.\\ \hat{w}(x,z,\omega) &= (-\eta\mathbf{A}\cos(\eta z) + ik\mathbf{B}\cos\bar{\eta}z))\,e^{-ikx}\end{aligned} \tag{7.67}$$

Using the strain-displacement and stress-strain relationship, the variation of the stress components can be obtained. For symmetric loading case, the stress variation is given by Eqn. (7.48). For anti-symmetric loading, the corresponding stresses are given by

$$\begin{aligned}\hat{\sigma}_{xx} &= \left(-(2\mu k^2 + k_p^2)\mathbf{A}\sin(\eta z) - 2\mu ik\bar{\eta}\mathbf{B}\sin(\bar{\eta}z)\right)e^{-kx}\\ \hat{\sigma}_{zz} &= \left(\mu(k^2-\eta^2)\mathbf{A}\sin(\eta z) + 2\mu ik\bar{\eta}\mathbf{B}\sin(\bar{\eta}z)\right)e^{-kx}.\\ \hat{\tau}_{xy} &= \left(\mu ik\eta\mathbf{A}\cos(\eta z) - 0.5\mu(\bar{\eta}^2-k^2)\mathbf{B}\cos(\bar{\eta}z)\right)e^{-ikx}\end{aligned} \tag{7.68}$$

Applying the traction free boundary condition for both the symmetric and the anti-symmetric loading cases given by Eqn. (7.64), we get the following matrix relations for symmetric loading case

$$\begin{bmatrix}\mu(k^2-\eta^2)\cos(\eta h/2) & 2\mu ik\bar{\eta}\cos(\bar{\eta}h/2)\\ \mu ik\eta\sin(\eta h/2) & -0.5\mu(\bar{\eta}^2-k^2)\sin(\bar{\eta}h/2)\end{bmatrix}\begin{Bmatrix}\mathbf{A}\\ \mathbf{B}\end{Bmatrix} = \begin{Bmatrix}0\\ 0\end{Bmatrix} \tag{7.69}$$

For the anti-symmetric case, this relation becomes

$$\begin{bmatrix}\mu(k^2-\eta^2)\sin(\eta h/2) & 2\mu ik\bar{\eta}\sin(\bar{\eta}h/2)\\ \mu ik\eta\cos(\eta h/2) & -0.5\mu(\bar{\eta}^2-k^2)\cos(\bar{\eta}h/2)\end{bmatrix}\begin{Bmatrix}\mathbf{A}\\ \mathbf{B}\end{Bmatrix} = \begin{Bmatrix}0\\ 0\end{Bmatrix} \tag{7.70}$$

Setting the determinant of the matrices in Eqns. (7.69 & 7.70) for symmetric and anti-symmetric loading cases to zero, and simplifying, we get the wavenumber-frequency relation for symmetric and anti-symmetric lamb wave modes, which are given below:

- *Symmetric Lamb wave equation*

$$\frac{\tan(\eta h/2)}{\tan(\bar{\eta}h/2)} = -\frac{4k^2\eta\bar{\eta}}{(\bar{\eta}^2-k^2)^2} \tag{7.71}$$

- *Anti-symmetric Lamb wave equation*

$$\frac{\tan(\bar{\eta}h/2)}{\tan(\eta h/2)} = -\frac{4k^2\eta\bar{\eta}}{(\bar{\eta}^2-k^2)^2} \tag{7.72}$$

Eqns. (7.71 & 7.72) are, as in the case of fixed boundary case, multi-valued function and are very difficult to solve analytically. They look very similar to the dispersion relation for doubly bounded fixed surface, whose dispersion equation is given by Eqn. (7.50). The cut-off frequencies for the symmetric and anti-symmetric cases are given by

- **Symmetric Lamb wave mode:**

$$\omega_{cut-off} = \frac{2m\pi}{h}C_s, \qquad \frac{(2m-1)\pi}{h}C_p \qquad m = 1, 2, ... \tag{7.73}$$

- **Anti-symmetric Lamb wave mode:**

$$\omega_{cut-off} = \frac{2m\pi}{h}C_p, \qquad \frac{(2m-1)\pi}{h}C_s \qquad m = 1, 2, 3...... \tag{7.74}$$

We will adopt the *bracketing technique* to solve the transcendental equations for symmetric and antisymmetric loading cases given by Eqns. (7.71 & 7.72). The bracketing technique was explained in Section 7.5.3.

The Lamb wave modes for symmetric and antisymmetric loadings are shown in Figs. 7.13 and 7.14. The figure shows, purely real, purely imaginary and complex modes for both symmetric and antisymmetric loading cases. The figures also show the phase speed variation for real only roots.

As before, the only one complex root is tracked and plotted. The imaginary component is again very high and it may be possible that these modes may not propagate . All the modes display very high dispersive behavior.

7.6 WAVE PROPAGATION IN THIN PLATES

Plate waveguides are 2-D counterpart of 1-D flexural waveguides or beams and hence can resist out-of-plane loads. The flexural waves in this type of waveguides can be expected to be dispersive. As in the case of beams, if the plates support shear deformation, then the theory on which the mechanics of such plates is based is called the *Mindlin plate theory*, which is the 2-D counterpart of the Timoshenko beam theory. In this section, we will present only the wave propagation in thin plates, the theory of which will be based on Classical Plate Theory, which is the 2-D counterpart of Euler-Bernoulli beam theory .

As we had done for all other waveguides, we need to first derive the governing partial differential equation. For this, we consider a plate waveguide shown in Fig. 7.15.

Let the plate have a length L along y-axis and width W in the direction of x, which is assumed to be propagating direction. The plate be of thickness t. The displacement field governing the mechanics of the plate is similar to that of elementary beam. However, these fields are two dimensional in nature. The displacement fields for a thin plate are given by what is well known as Classical Plate Theory and they can be written as

$$u(x, y, z, t) = -z\frac{\partial w}{\partial x}(x, y, t), \; v(x, y, z, t) = -z\frac{\partial w}{\partial y}(x, y, t).$$
$$w(x, y, z, t) = w(x, y, t) \tag{7.75}$$

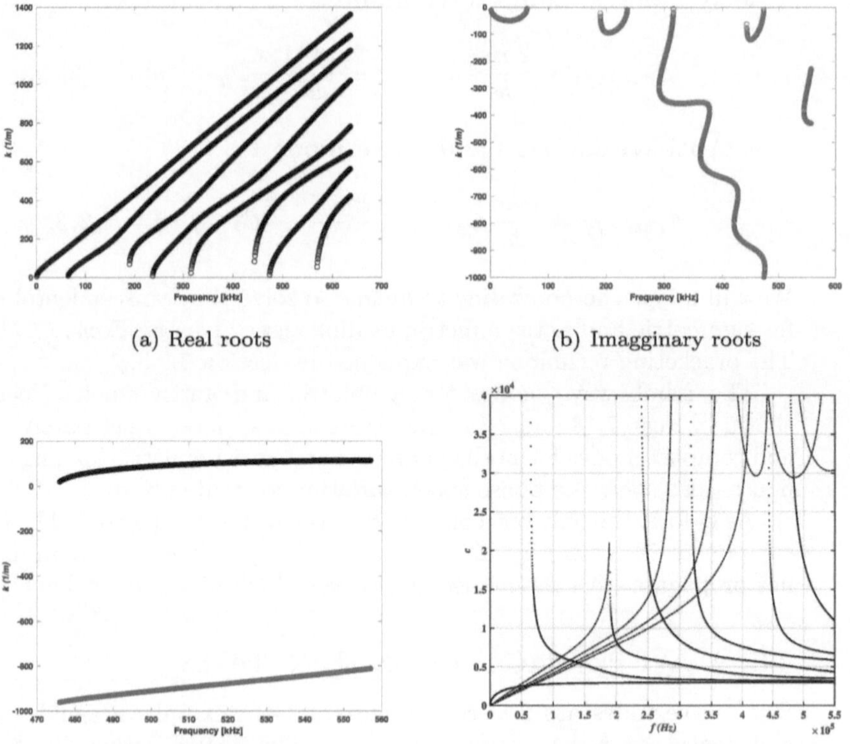

(a) Real roots (b) Imagginary roots

(c) Complex roots (d) Phase Velocity

FIGURE 7.13: Spectrum and Dispersion relation for a 2-D isotropic media with traction free boundary: Symmetric Loading. Red lines indicate the imaginary roots

The corresponding strains are obtained using strain-displacement relations, which are given by

$$\epsilon_{xx} = \frac{\partial u}{\partial x} = -z\frac{\partial^2 w}{\partial x^2}, \qquad \epsilon_{yy} = \frac{\partial v}{\partial y} = -z\frac{\partial^2 w}{\partial y^2}.$$

$$\gamma_{xy} = \frac{\partial u}{\partial y} + \frac{\partial v}{\partial x} = -2z\frac{\partial^2 w}{\partial x \partial y}, \qquad \gamma_{yz} = \gamma_{zx} = 0 \tag{7.76}$$

As in the case of beams, we see that in thin plates, the transverse shear strains are zero although the shear forces exist. The main difference between the plate and the beam deformation is that the in-plane shear stress is a function of the thickness coordinate z and such shear strains are not present in the case of beams.

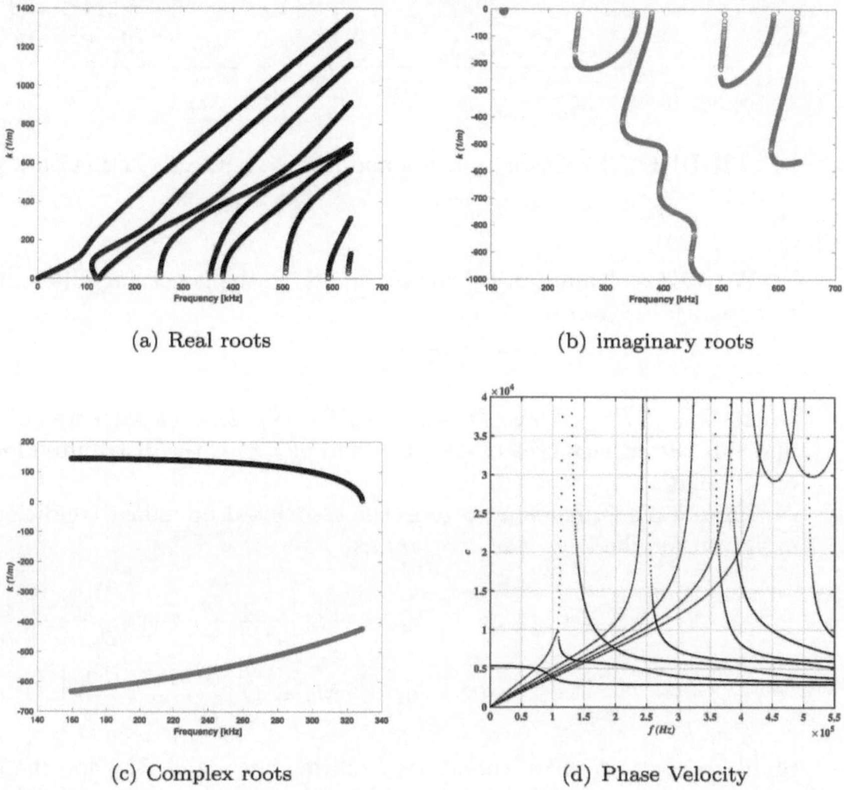

FIGURE 7.14: Spectrum and Dispersion relation for a 2-D isotropic media with traction free boundary: Anti-Symmetric Loading. Red lines indicate the imaginary roots

Next, we write the stress-strain relationship. Here, we assume that the plate is in a state of plane stress and we can write the constitutive model as

$$
\begin{aligned}
\sigma_{xx} &= \frac{E}{1-\nu^2}\left[\epsilon_{xx} + \nu\epsilon_{yy}\right] = -\frac{Ez}{1-\nu^2}\left[\frac{\partial^2 w}{\partial x^2} + \nu\frac{\partial^2 w}{\partial y^2}\right] \\
\sigma_{yy} &= \frac{E}{1-\nu^2}\left[\epsilon_{yy} + \nu\epsilon_{xx}\right] = -\frac{Ez}{1-\nu^2}\left[\frac{\partial^2 w}{\partial y^2} + \nu\frac{\partial^2 w}{\partial x^2}\right] \\
\tau_{xy} &= -G\gamma_{xy} = -2zG\frac{\partial^2 w}{\partial x \partial y} \qquad (7.77)
\end{aligned}
$$

In the above equation, E is the Young's modulus, G is the shear modulus and ν is the Poisson's ratio.

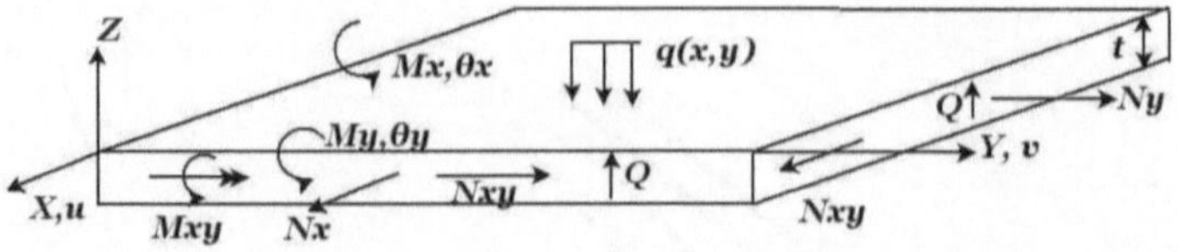

FIGURE 7.15: Displacements and stress resultants acting on a plate

We will use Hamilton's Principle to derive the governing differential equation, which is given by

$$D\nabla^4 w + \rho h \frac{\partial^2 w}{\partial t^2} = q \tag{7.78}$$

where $\nabla = \partial/\partial x + \partial/\partial y$, $D = Et^3/12(1-\nu^2)$, t is the thickness of the plate and A is the surface area of the plate and $q(x,y)$ is the distributed load acting on the plate.

Hamilton's Principle also gives the associated boundary conditions, which are given for the edge x *is constant* as

$$\begin{aligned} w = 0 \quad & \text{or} \quad V_{xz} = -D\left[\frac{\partial^3 w}{\partial x^3} + (2-\nu)\frac{\partial^3 w}{\partial x \partial y^2}\right]. \\ \frac{\partial w}{\partial x} = \theta_x = 0 \quad & \text{or} \quad M_x = D\left[\frac{\partial^2 w}{\partial x^2} + \nu\frac{\partial^2 w}{\partial y^2}\right] \end{aligned} \tag{7.79}$$

Similarly, we can write the stress resultants V_{yz} and M_{zz} for the edge y *is constant.* Note that the shear resultant V_{xz} specified in Eqn. (7.79) is called *Kirchoff Shear*, which is obtained using the expression

$$V_{xz} = Q_{xz} - \frac{\partial M_{xy}}{\partial y}$$

where Q_{xz} is the shear force, which is needed to balance the imbalances caused by the shear moment M_{xy}. For more details on the theory of plates and the Classical Plate Theory, the reader is advised to refer [36]. Thus, in a plate, there are three motions associated with the deformation of the plate, namely the transverse displacement $w(x,y)$, the rotations $\theta_x = \partial w/\partial x$ and $\theta_y = \partial w/\partial y$.

7.6.1 Spectral Analysis

As was done for 1-D and 2-D waveguides, we first transform the field variable $w(x,y,t)$ in the governing equation Eqn. (7.78) to frequency domain using Discrete Fourier Transform (assuming no loading) given by

$$w(x,y,t) = \sum \hat{w}(x,y,\omega)e^{i\omega t}$$

Substituting this in Eqn. (7.78), we get

$$\nabla^4\hat{w} + \beta^4\hat{w} = 0, \qquad \beta = \sqrt{\omega}\left[\frac{\rho h}{D}\right]^{0.25} \tag{7.80}$$

The above equation can be factorized and split into two second-order equations as

$$\begin{aligned}(\nabla^2\hat{w} + \beta^2\hat{w})(\nabla^2\hat{w} - \beta^2\hat{w}) = 0. \\ \nabla^2\hat{w} + \beta^2\hat{w} = 0, \qquad \nabla^2\hat{w} - \beta^2\hat{w} = 0\end{aligned} \tag{7.81}$$

Plate is a 2-D waveguide, wherein the spatial wave modes are as important as the temporal wave modes and hence it is a very important component in the wave response. This requires introduction of an additional spatial Fourier Transform in the y direction as well as an additional wavenumber as was done for the other in-plane 2-D waveguides studied previously. We call this case as *Wavenumber Transform Solution* . However, if the wave is incident either normally or obliquely along any of the edges where y is constant, the need for an additional transform is not necessary and the wave analysis procedure can be similar as that of the elementary beam. We call this case as the *Plate Edge Wave propagation.* Both these methods have solution of the form

$$\hat{w}(x, y, \omega) = \bar{\hat{w}}(x, \omega)e^{i\eta y}$$

Substituting the above equation in Eqn. (7.81), we get two constant coefficient ordinary differential equation given as

$$\frac{d^2\bar{\hat{w}}}{dx^2} + [-\eta^2 - \beta^2]\bar{\hat{w}} = 0, \qquad \frac{d^2\bar{\hat{w}}}{dx^2} + [-\eta^2 + \beta^2]\bar{\hat{w}} = 0 \tag{7.82}$$

From the above equations, we can write the wavenumbers as

$$k_{1,2} = \pm\sqrt{-\eta^2 - \beta^2}, \qquad k_{3,4} = \pm\sqrt{\eta^2 + \beta^2}$$

7.6.2 Plate Edge Wave Propagation

This case does not require a second transform in the y direction. Consider a edge of a plate shown in Fig. 7.16, where the edge is subjected to an oblique incident wave that is incident at an angle ϕ. We assume that the wave is propagating in the y-direction such that $\eta = \beta \sin\phi = \beta S$, where we designate $S = \sin\phi$ and $C = \cos\phi$. Using these expressions, the wavenumbers become

$$k_{3,4} = \pm\beta C, \qquad k_{1,2} = \sqrt{1 + S^2} \tag{7.83}$$

The complete solution of the wave equation can be written as

$$\bar{\hat{w}}(x, y, \omega) = (\mathbf{A}e^{ik_1x} + \mathbf{B}e^{ik_2x} + \mathbf{C}e^{ik3x} + \mathbf{D}e^{ik_4x})e^{i\beta Sy} \tag{7.84}$$

The above equation can be used to perform wave reconstructions for different boundary conditions as done for the case of 1-D beam in Chapter 4.

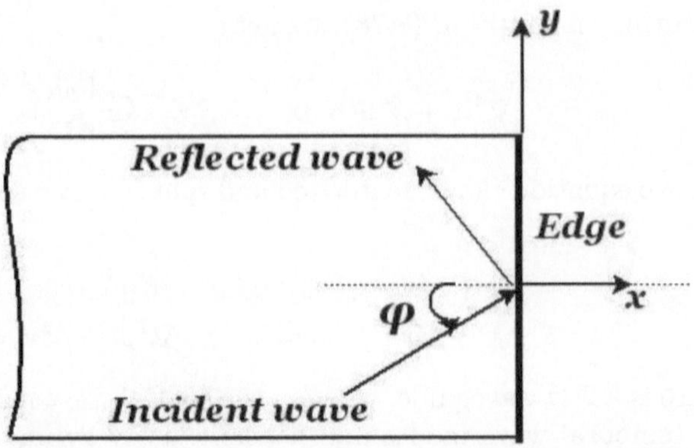

FIGURE 7.16: A plate edge with obliquely incident wave

Wavenumber Transform Solution

In this case, we assume that the wave is incident normally at any location on the surface of the plate. In this case, the spatial modes actively participate in the overall response and hence to account for this, we need to introduce an additional wavenumber in the y direction and an additional transform. The solution of the two uncoupled wave equation (Eqn. (7.81)) is of the form

$$w(x, y, t) = \sum_n \sum_m \bar{\hat{w}}_n(x, y, \omega_n) e^{i\eta_m y} e^{i\omega t}$$

Substituting this in Eqn. (7.81), it will lead to two ordinary differential equations as given in Eqn. (7.82) with wavenumbers given by

$$k_{1,2} = \pm\sqrt{\beta^2 - \eta_m^2}, \qquad k_{3,4} = \pm\sqrt{\beta^2 - \eta_m^2}, \qquad \eta_m = \frac{2m\pi}{L} \tag{7.85}$$

where L is the length of the spatial window for the transform in the y direction. Since the transform is valid for periodic functions, we need L to be very large for the solution to be valid. Hence, we always consider in this analysis that the y direction always extends to infinity. The wavenumber given by Eqn. (7.85) is plotted in Fig. 7.17.

The figure clearly shows that the mode 1 is always propagating, and it is highly dispersive in nature. Mode 2 on the other hand undergoes change, that is, it starts as evanescent wave and starts propagating at higher frequencies. That is, mode 2 exhibits a cut-off frequency, which increases with m and hence η. The plot shown in Fig. 7.17 is generated assuming the plate to be made of aluminum with the plate thickness of 10 mm. After the wavenumbers are computed, we can write the total solution to the governing plate equation as

$$\bar{\hat{w}}(x, y, \omega) = \sum_n \sum_m (\mathbf{A}e^{ik_1 x} + \mathbf{B}e^{ik_2 x} + \mathbf{C}e^{ik_3 x} + \mathbf{D}e^{ik_4 x}) e^{i\eta y} \tag{7.86}$$

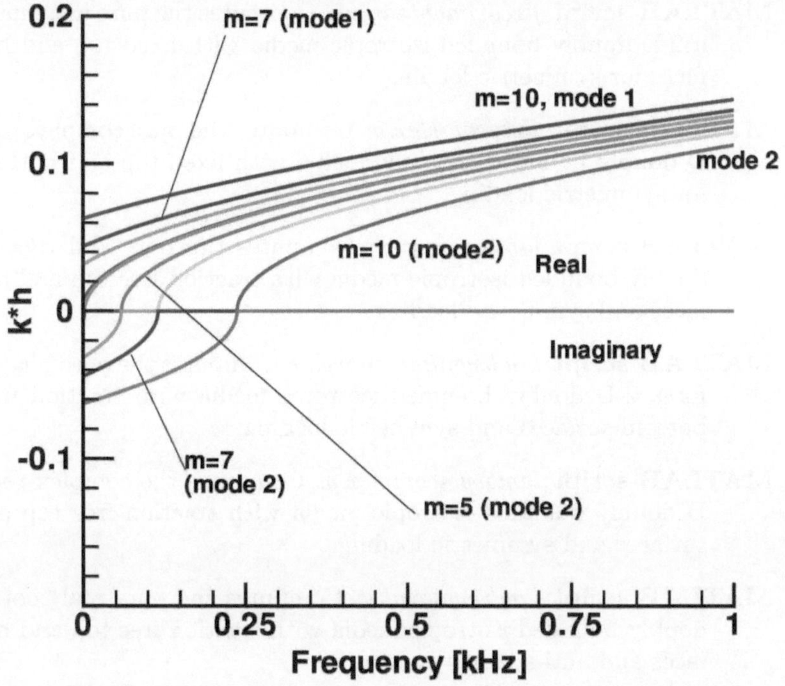

FIGURE 7.17: Spectrum relation for doubly bounded plate

Using the above equation, the reflected responses can be obtained to a variety of different boundary conditions.

Note on MATLAB® scripts provided in this chapter

In this chapter, following two MATLAB scripts are provided for performing wave propagation analysis in 2-D isotropic

MATLAB script *eqn.m* Solves and plots the wavenumbers for Rayleigh waves in a 2D semi-infinite isotropic media. The script uses *roots* function to obtain the wavenumbers.

MATLAB script it seminf2diso.m Computes the wave propagation responses in a 2-D isotropic semi-infinite media subjected to triangular gaussian signal. The signal file *pul.txt* is also provided along with other files.

MATLAB script *fixedreal.m* Computes the pure real roots for a 2-D doubly bounded isotropic media with fixed top and bottom surfaces for symmetric loading.

MATLAB script *fixedrimaginary.m* Computes the pure imaginary roots for a 2-D doubly bounded isotropic media with fixed top and bottom surfaces for symmetric loading.

MATLAB script *fixedcomplex.m* Computes the pure complex roots for a 2-D doubly bounded isotropic media with fixed top and bottom surfaces for symmetric loading

MATLAB script *lambsymreal.m* Computes the pure real roots for a 2-D doubly bounded isotropic media with traction free top and bottom surfaces and symmetric loading.

MATLAB script *lambsymimaginary.m* Computes the pure imaginary roots for a 2-D doubly bounded isotropic media with traction free top and bottom surfaces and symmetric loading.

MATLAB script *lambsymcomplex.m* Computes the complex roots for a 2-D doubly bounded isotropic media with traction free top and bottom surfaces and symmetric loading.

MATLAB script *lambasymreal.m* Computes the pure real roots for a 2-D doubly bounded isotropic media with traction free top and bottom surfaces and anti-symmetric loading.

MATLAB script *lambasymimaginary.m* Computes the pure imaginary roots for a 2-D doubly bounded isotropic media with traction free top and bottom surfaces and anti-symmetric loading.

MATLAB script *lambasymcomplex.m* Computes the complex roots for a 2-D doubly bounded isotropic media with traction free top and bottom surfaces and anti-symmetric loading.

Summary

In this chapter, the methodology to perform wave propagation analysis in an isotropic 2-D waveguides is presented. The chapter begins with the derivation of the *Navier's equation* of motion. For the solution of this equation, the concept of *Helmholtz decomposition* is introduced, which expresses the deformations in terms of a scalar potential and three vector potentials corresponding to the three coordinate directions. The different type of waves, namely the P-wave, S_V wave and S_H wave and their generation in an infinite media, is discussed in detail.

Next, the wave propagation in 2-D semi- infinite media is presented. Here, propagation characteristics are presented for three different surface boundary conditions. One particular boundary condition, that is of great interest to those working in seismic engineering, is the generation of *Rayleigh waves*, which occur when the surface is traction free. The Rayleigh wave solution and its characteristics are discussed in detail. Next, the wave propagation in doubly bounded media for different boundary conditions is presented. The characteristic equation for computation of wavenumbers for certain boundary conditions results in transcendental equations that are difficult to solve analytically. A numerical method based on the bracketing method is used to solve the transcendental equations and this method is explained in detail. Using this method of solution, the wave propagation in a doubly bounded media that has fixed and traction free surfaces is discussed in detail. The stress free boundary conditions generate a wave called the *Lamb wave* and its characteristics for symmetric and anti-symmetric loadings are discussed. Finally, the wave propagation in thin plates is discussed. A number of important MATLAB scripts of the topics discussed in this chapter are provided and the details of the scripts are provided previously.

Exercises

7.1 The S_H wave propagation in a semi infinite isotropic media is governed by Eqn. (7.16). Show that the solution for the vector potentials H_x and H_z is given by

$$\hat{H}_x = \mathbf{A}e^{(kx-\eta z)} + \mathbf{B}e^{(ikx+\eta z)}, \;\; \hat{H}_z = \frac{k}{\eta}\mathbf{A}e^{(ikx-\eta z)} - \frac{k}{\eta}\mathbf{B}e^{(ikx+\eta z)}$$

where $k^2 = k_s^2 - \eta^2$. Applying boundary conditions $\tau_{zy} = 0$ at $z = 0$, the number of wave coefficients can be reduced to one. Assuming this wave coefficient to be a gaussian signal shown in Fig. 4.3, determine the out-of plane responses along $y = 0$ for different x.

7.2 Two semi-infinite media intersect at boundary $z = 0$. The material properties of the lower media are E_1, ν_1 and ρ_1 and the upper media are E_2, ν_2 and ρ_2, respectively. Setup the equation to determine the equation to compute wavenumber for this system and plot the spectrum relation for this system.

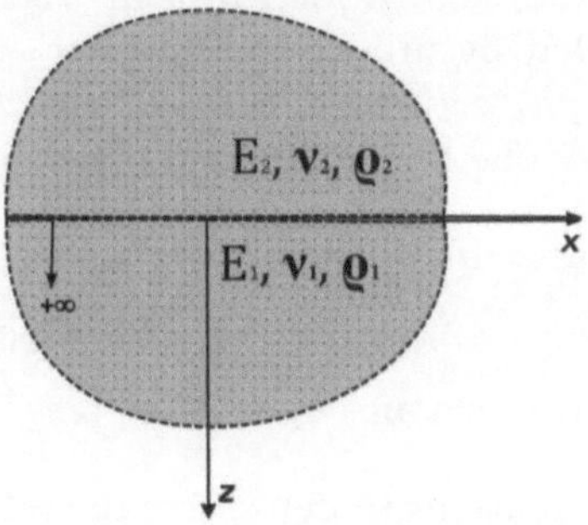

Figure for Problem 7.2

7.3 A semi-infinite medium is overlaid by another strip of thickness h as shown in the figure below The material properties of the semi-infinite

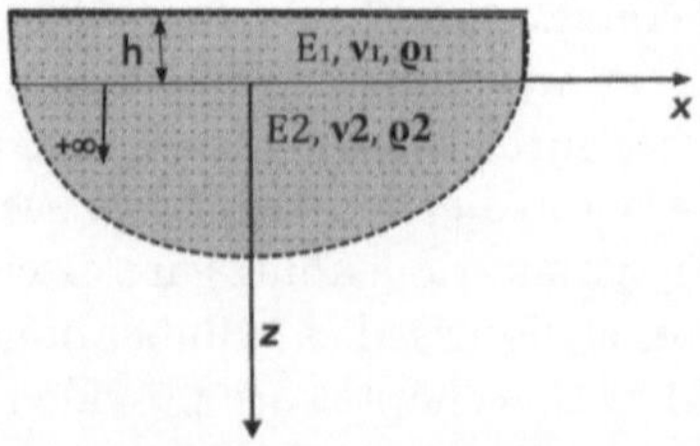

Figure for Problems 7.3 and 7.4

media are E_2, ρ_2 and ν_2, while the property of the finite strip are E_1, ρ_1 and ν_1. For this medium, the S_H wave propagation governing equation given by

$$\mu\left[\frac{\partial^2 w}{\partial x^2}+\frac{\partial^2 w}{\partial z^2}\right]=\rho\frac{\partial^2 w}{\partial t^2}$$

where $\mu = E/(2(1+\nu))$ is the shear modulus. The above equation is used to solve the S_H wave propagation in the above figure. To solve this problem, we enforce the following boundary.

$$\tau_{zx}=0 \quad \text{at} \quad z=-h, \qquad (\tau_{zx})_1=(\tau zx)_2 \quad \text{at} \quad z=0$$
$$z_1=z_2 \quad \text{at} \quad z=0$$

Show that the frequency equation for this problem is given by

$$\mu_1\eta_1-\mu_2\eta_2\tan\beta h=0 \tag{7.87}$$

where η_1 and η_2 are the vertical wavenumbers in the z-direction.

7.4 The same figure shown for Problem 7.3 is again considered here as a finite strip having properties E_2, ν_2, ρ_2 lying over a semi-infinite media having properties E_1, ν_1, ρ_1, respectively. The scalar and vector potentials for the two domains is given by

$$\text{for finite strip} \quad \hat{\Phi}_1=e^{i\eta_1 z}\left[\mathbf{A_1}e^{-ik_1x}+\mathbf{B_1}e^{ik_1x}\right]$$
$$\hat{H}_{y1}=e^{i\eta_1 z}\left[\mathbf{C_1}e^{-ik_1x}+\mathbf{D_1}e^{ik_1x}\right]$$
$$\text{for semi-infinite region} \quad \hat{\Phi}_2=\mathbf{A_2}e^{i\eta_2}ze^{-ik_2x} \quad \hat{H}_{y2}=\mathbf{C_2}e^{i\eta_2}ze^{-ik_2x}$$

Set up the equations to determine the reflected and transmitted coefficients clearly the bringing out the mode coupling of the P and S waves when (a) P wave is incident (b) S wave is incident.

7.5 The solution of the wave equation for a plate having normal incidence is given by

$$\hat{w}(x,\omega) = \left[\mathbf{A}e^{-ikx} + \mathbf{B}e^{-\bar{k}x} + \mathbf{C}e^{ikx} + \mathbf{D}e^{\bar{k}x}\right]\cos\eta z$$

$$\text{where} \qquad k = \sqrt{\beta^2 - \eta^2},\ \bar{k} = \sqrt{\beta^2 + \eta^2},\ \beta = \left[\frac{\rho A}{D}\right]^{0.25}$$

Set up the equation to determine the reflected and transmitted coefficients for the following cases

1. When one end is fixed
2. When one end is simply supported
3. When one end is free
4. Across the stepped plate where the properties of the two plate segments are E_1, A_1 and I_1 and E_2, A_2 and I_2, respectively.

CHAPTER 8

Wave Propagation in Laminated Composite Waveguides

Composites are the heterogeneous materials that exhibit anisotropic character, which means that the deformations or motions in all coordinate directions will be coupled. That is, an axial impact will create a bending motion and or a torsional motion depending of the level and nature of anisotropy. From the wave propagation point view, the coupled motion will make the wave propagation very complex introducing multiple cut-off frequencies and different propagation characteristics. In this chapter, a very brief introduction to the composite materials, laminated composites, their evaluation of material properties and their homogenization is presented. The knowledge and understanding of the basic theory of composites is fundamental to the understanding of wave propagation in this material. This chapter outlines the theory only very briefly and the reader is advised to refer to many advanced text books in this subject such as [69, 138].

The next aspect we need to address is the need for wave propagation analysis in composites. Composite structures as such exhibit poor resistance to impact. Starting from manufacturing, and throughout their design life, these structures are vulnerable to highly transient loading such as tool drop and other kinds of impact. These loading have very small duration (μ-sec range). Hence, the energy of the system is confined over a large frequency band, exciting all the higher-order modes, thus setting up stress waves to propagate over the entire structure. The effect of impact on laminated composite structures is a crucial issue, which researchers have tried to address with an increasing emphasis. The main reason is that the way these laminated fiber reinforced structures are constructed, it contributes to high ratios of the longitudinal to the lateral elastic moduli, and in addition, it has significant layer-wise anisotropy

DOI: 10.1201/9781003120568-8

due to ply orientations. One of the critical aspects is that the steep and discontinuous bending stress gradient at the ply interfaces may cause eventual delamination or debonding of the layers, thus putting the structural integrity in question.

One important property present in laminated composite structure that is missing in isotropic structures is the presence of the stiffness and the inertial coupling . These are normally caused due to unsymmetrical stacking of plies with respect to middle plane. The layer-wise construction of fiber reinforced composite beam has a great advantage for embedding different class of functional materials such as piezoelectric ceramic, relaxor and anti-ferroelectric thin films, magnetostrictive plate strips and particle layers mixed with matrix [16], [50], [79]. Such configurations necessarily lead to cross-sectional unsymmetry. The axial-flexural coupling due to unsymmetric construction give rise to additional progressive waves which are unlikely in a structure made up of isotropic material.

This chapter begins with a brief introduction to composite materials, followed by the procedure to evaluate its material properties. Since these materials are manufactured layer-wise, first the materials properties of the lamina is established using *Micro-Mechanical Analysis.* The mechanical properties of a lamina is established using the properties of the fiber and its binder, namely the matrix. Many laminae are stacked together to form a laminate. To determine the mechanical properties of the full laminate, we will use *Classical Plate Theory*, often known as *Macro-Mechanical Analysis.*

Next, the wave propagation in elementary composite waveguides is presented, where the effects of axial-bending coupling on wave propagation is studied. For laminated composite beams, it is well established that shear deformation and rotary inertia play a key role in its behavior under applied loads. In this regard, [94] have shown the effects of shear deformation introduced by higher-order refined theory on the transient response of laminated composite beams. However, these effects are dependent on length-to-depth ratio (L/h) of the beam. As seen in [27], first order shear deformation theory (FSDT) including rotary inertia, and elementary beam theory produce identical results for a slender beam ($L/h > 100$). These concepts were discussed in Chapter 4 in the context of isotropic materials. In addition, when a 1-D waveguide is subjected to axial loading, some lateral deformation takes place due to Poisson's ratio effect. This lateral contraction causes an additional wave called the *contraction wave.* This aspect again was discussed in Chapter 4 in the context of isotropic materials through Mindlin-Herrmann rod formulation. Hence, this chapter, we will also present wave propagation characteristics in higher-order 1-D structure that includes the effects of both shear deformation and Poisson's contraction, and compare its behavior with the responses obtained using elementary theory.

In the last part of this chapter, the wave propagation in 2-D composite waveguides is presented. Unlike the use of Helmholtz decomposition used to uncouple the 2-D wave equation in isotropic waveguides given in Chapter 7, here the governing equations cannot be uncoupled as the entire wave field cannot be separated as P-wave or S-wave components due to material anisotropy of the system. Here, we use a new method called the *Partial Wave Technique* to get the wave modes in a 2-D composite waveguide.

Several MATLAB scripts are provided, which will enable the reader to perform wave propagation analysis in these waveguides.

8.1 INTRODUCTION TO COMPOSITE MATERIALS

As the name suggests, composite materials are made by combining two or more materials at the macro scale to obtain a useful structural material. Although these materials at the microscopic scale can be inhomogeneous in nature, they can be thought as homogenous material at the macroscopic level. These materials possess best of all the qualities of each of the constituents and the choice of the constituents depends on the application for which these materials are required. These materials are highly preferred in many structural applications due to their light weight and high strength characteristics, and in addition, they possess high corrosion resistance properties. The two main constituents of a composite material are the *Fiber* and the *Matrix.* Depending upon the type of fiber and matrix and the way they are bound together, composites of different properties can be constructed. Due to the difference in the constitutive behavior of these two constituent materials, the constitutive model of the resulting compound material is normally anisotropic. Composites can be classified into three different categories, namely *Fibrous Composites*, *Particulate Composites* and *Laminated Composites*.

For the structural applications, fibrous or particulate composites are seldom used and the most extensively used type is the laminated composites. hence, in this chapter only theory of laminated composites is discussed.

8.2 THEORY OF LAMINATED COMPOSITES

Laminated composites in recent times, have found extensive use as aircraft structural materials due to its high strength to weight and stiffness to weight ratios. The reason for their popularity is due to the fact that they are extremely light-weight and in addition, the laminate construction enables the designer to tailor the strength of the structure in any desired directions depending upon the loading directions in which the structure is subjected to. Besides aircraft structures, they have found their way in automobile, pipe line and civil structures. An additional advantage of the structures made from these materials is that, they have better thermal and acoustic insulation properties over the metallic structures.

The laminated composite structure consists of many laminae (plies) stacked together to form the laminate structure. The number of plies or laminas depends on the loads that the structure is required to sustain. Each lamina contains fibers oriented in the direction where the maximum strength is required. These fibers are bonded together by a matrix material. The laminated composite structure derives its strength by its fibers. The commonly used fibers are made from the following: Carbon, glass, Kevlar and Boron fibers, respectively. The most commonly used matrix material is the Epoxy resin, although other matrix materials such as PEEK (Polyether ether ketone) or Vinyl easter resins have found extensive use these days for a variety of structural applications. Laminated composite materials are assumed to have orthotropic properties at the lamina level while at the laminate level, they exhibit high level of anisotropic behavior. As mentioned earlier, the anisotropic behavior results in stiffness coupling, such as bending-axial-shear coupling in beams and plates, bending-axial-torsion coupling in aircraft thin-walled structures, etc. These coupling effects make the analysis of laminated composite structures more difficult compared to isotropic structures.

Next, the theory of laminated composites is presented, and the analysis consists of three parts consisting of (a) Lamina property determination (called the *Micro-mechanics analysis*), (b) Laminate property determination (called the *Macro-mechanics analysis*) and (c) the Classical Plate Theory to obtain the governing equations of motion.

8.2.1 Micro-mechanical Analysis of Composites

A lamina is a basic element of a laminated composite structure, constructed with the help of fibers that are bonded together with the help of a suitable matrix resin. The lamina and hence the laminate strength depends on the type of the fiber, their orientation and also the volume fraction of the fiber in relation to the overall volume of lamina. This makes the lamina heterogeneous with mixture of fibers dispersed in the matrix. Hence, the determination of material properties of a lamina, which is assumed to be orthotropic in character, is a very involved process. The method of determination of lamina material properties constitute the micro-mechanical analysis. According to Jones [69], micro-mechanics is the study of composite material behavior, wherein the interaction of the constituent materials is examined in detail as a part of the definition of the behavior of the heterogeneous composite material.

Hence, the goal of the micro-mechanics analysis is to determine the elastic moduli of a composite material in terms of elastic moduli of the constituent materials, namely, the fibers and matrix. Hence, the property of a lamina can be expressed as

$$Q_{ij} = Q_{ij}(E_f, E_m, \nu_f, \nu_m, V_f, V_m), \tag{8.1}$$

where E, ν and V are the elastic moduli, Poisson's ratio and the volume fraction, respectively, and f and m in the subscript denote the fiber and matrix, respectively. The volume fraction of the fiber is determined from the expression: V_f = (volume of the fiber)/(total volume of the lamina) and $V_m = 1 - V_f$.

There are two different methods to determine the material properties of a lamina. They can be grouped as: (1) Strength of materials approach and (2) Theory of elasticity approach. The strength of materials method uses the experimental measurement to determine the elastic moduli. The second method actually gives the upper and lower bounds of the elastic moduli and not their actual values. In fact, there are many papers available in the literature that report the theory of elasticity approach to determine the elastic moduli of the composite. In this section, only the strength of materials approach of determining the lamina mechanical properties is presented.

Material properties of a lamina are determined by making certain assumptions as regards its behavior. The fundamental assumption is that the fiber is strong and hence it is the main load bearing member and the matrix is weak and its main function is to protect the fibers from severe environmental effects. Also, the strains in the matrix as well as fiber are assumed the same. Hence, the plane sections before being stressed remain plane after the stress is applied. In the present analysis, we consider a unidirectional, orthotropic composite lamina for deriving the expressions for the elastic moduli. In doing so, we limit our analysis to a small volume element, which is quite small to show the microscopic structural details, yet large enough to represent the overall behavior of the composite lamina. Such a volume is called the Representative Volume (RV). A simple RV is a fiber surrounded by matrix as shown in Fig. 8.1(a).

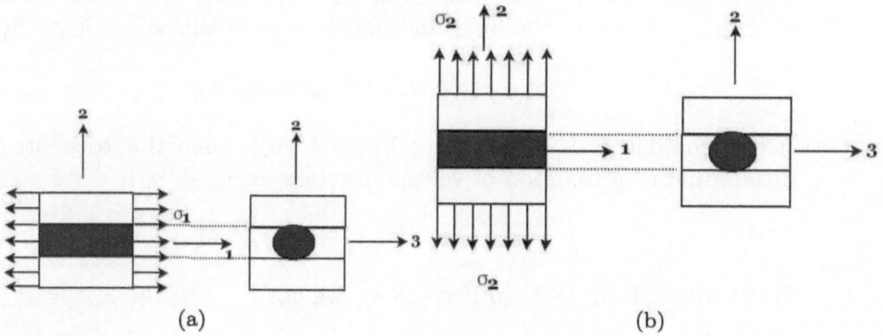

FIGURE 8.1: (a) RV for determination of longitudinal material property. (b) RV for determination of transverse material property

First, the procedure for determining the elastic modulus E_1 is outlined. In Fig. 8.1(a), the strain in 1-direction is given by $\epsilon_1 = \Delta L/L$, where this strain is felt both by the matrix and the fiber, according to our basic assumption. The corresponding stresses in the fiber and the matrix is given by

$$\sigma_f = E_f \epsilon_1\,, \quad \sigma_m = E_m \epsilon_1\,. \tag{8.2}$$

Here E_f and E_m are the elastic modulus of the fiber and matrix, respectively.

The cross-sectional area of the RV is denoted as A and is made of the area of the fiber A_f and the area of the matrix A_m. If the total stress acting on the cross-section of the RV is σ_1, then the total load acting on the cross-section is

$$P = \sigma_1 A = E_1 \epsilon_1 A = \sigma_f A_f + \sigma_m A_m \,. \tag{8.3}$$

From the above expression, we can write the elastic moduli in 1-direction as

$$E_1 = E_f \frac{A_f}{A} + E_m \frac{A_m}{A} \,. \tag{8.4}$$

The volume fraction of the fiber and the matrix can be expressed in terms of areas of fiber and matrix as

$$V_f = A_f/A \,, \quad V_m = A_m/A \,. \tag{8.5}$$

Using Eqn. (8.5) in Eqn. (8.4), we can write the modulus in 1-direction as

$$E_1 = E_f V_f + E_m V_m \,. \tag{8.6}$$

Eqn. (8.6) is the well-known *Rule of mixtures* for obtaining the equivalent modulus of the lamina in the direction of the fibers.

The equivalent modulus E_2 of the lamina is determined by subjecting the RV to a stress σ_2 perpendicular to the direction of the fiber as shown in Fig. 8.1(b). This stress is assumed to be same in both the matrix as well as the fiber. The strains in the fiber and matrix due to this stress is given by

$$\epsilon_f = \sigma_2/E_f \,, \quad \epsilon_m = \sigma_2/E_m \,. \tag{8.7}$$

If h is the depth of the RV (see Fig. 8.1 (b)), then this total strain ϵ_2 gets distributed as a function of volume fraction as

$$\epsilon_2 h = (V_f \epsilon_f + V_m \epsilon_m) h \,. \tag{8.8}$$

Substituting Eqn. (8.7) in Eqn. (8.8), we get

$$\epsilon_2 = V_f \frac{\sigma_2}{E_f} + V_m \frac{\sigma_2}{E_m} \,. \tag{8.9}$$

However, we have

$$\sigma_2 = E_2 \epsilon_2 = E_2 \left(V_f \frac{\sigma_2}{E_f} + V_m \frac{\sigma_2}{E_m} \right) . \tag{8.10}$$

From the above relation, the equivalent modulus in the transverse direction is given by

$$E_2 = \frac{E_f E_m}{V_f E_m + V_m E_f} \,. \tag{8.11}$$

The major Poisson's ratio ν_{12} is determined as follows. If the RV of width W and depth h is loaded in the direction of the fiber, then both strains ϵ_1 and ϵ_2 will be induced in the 1 and 2 directions. The total transverse deformation δ_h is the sum of the transverse deformation in the matrix and the fiber and is given by

$$\delta_h = \delta_{hf} + \delta_{hm}\,. \tag{8.12}$$

The major Poisson's ratio is also defined as the ratio of transverse strain to the longitudinal strain and mathematically expressed as

$$\nu_{12} = -\epsilon_2/\epsilon_1\,. \tag{8.13}$$

The total transverse deformation can also be expressed in terms of depth h as

$$\delta_h = -h\epsilon_2 = h\nu_{12}\epsilon_1\,. \tag{8.14}$$

Following the procedure adopted for the determination of transverse modulus, the transverse displacement in the matrix and fiber can be expressed in terms of its respective volume fraction and Poisson's ratio as

$$\delta_{hf} = hV_f\nu_f\epsilon_1, \quad \delta_{hm} = hV_m\nu_m\epsilon_1\,. \tag{8.15}$$

Using Eqns. (8.14 & 8.15) in Eqn. (8.12), we can write the expression for major Poisson's ratio as

$$\nu_{12} = \nu_f V_f + \nu_m V_m\,. \tag{8.16}$$

By adopting similar procedure as used in the determination of the transverse modulus, we can write the shear modulus in terms of constituent properties as

$$G_{12} = \frac{G_f G_m}{V_f G_m + V_m G_f}\,. \tag{8.17}$$

The next important property of the composite that requires determination is the density. For this we begin with the total mass of the lamina, which is the sum of the masses of the fiber and the matrix. That is, the total mass M can be expressed in terms of the densities (ρ_f and ρ_m) and volumes (V_f and V_m) as

$$M = M_f + M_m = \rho_f V_f + \rho_m V_m\,. \tag{8.18}$$

Density of the composite can then be expressed as

$$\rho = \frac{M}{V} = \frac{\rho_f V_f + \rho_m V_m}{V}\,. \tag{8.19}$$

Once the properties of the lamina are determined, then one can proceed to perform a macro-mechanical analysis of the lamina to characterize the constitutive model of a laminate, which is described in the next sub-section.

8.2.2 Macro-mechanical Analysis of Composites

Determination of the overall constitutive model of a lamina in a laminated composite constitutes the macro-mechanical study of composites. Unlike the micro-mechanical study, where the composite is treated as a heterogeneous mixture, here the composite is presumed to be homogenous and the effects of constituent materials are accounted only as an averaged apparent property of the composite. Following are the basic assumptions used in deriving the constitutive relations.

1. The composite material is assumed to behave in a linear (elastic) manner. That is, the Hooke's law as well as the principle of superposition is valid.
2. At the lamina level, the composite material is assumed to be homogenous and orthotropic. Hence, the material has two planes of symmetry, one coinciding with fiber direction and the other perpendicular to the fiber direction.
3. The state of stress in a lamina is predominantly plane stress.

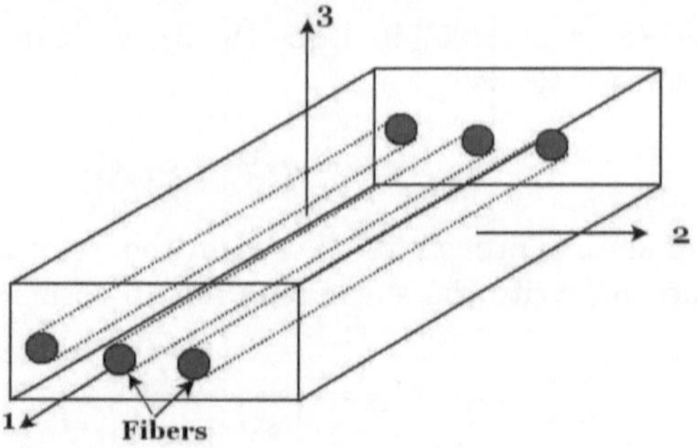

FIGURE 8.2: Principal axes of a lamina

Consider a lamina shown in Fig. 8.2 with its principal axes, which we denote it as 1-2-3 axes. That is, axis 1 corresponds to the direction of the fiber and axis 2 is the axis transverse to the fiber. The lamina is assumed to be in 3-D state of stress with 6 stress components given by $\{\sigma_{11}, \sigma_{22}, \sigma_{33}, \tau_{23}, \tau_{13}, \tau_{12}\}$. For an orthotropic material under 3-D state of stress, 9 engineering constants require to be determined. The macro-mechanical analysis will begin from here. The stress-strain relationship for an orthotropic material is given by (see Chapter 2 of the unified book [54] for more details on the constitutive model for an orthotropic material)

$$\left\{\begin{array}{c}\epsilon_{11}\\ \epsilon_{22}\\ \epsilon_{33}\\ \gamma_{23}\\ \gamma_{13}\\ \gamma_{12}\end{array}\right\}=\left[\begin{array}{cccccc}S_{11} & S_{12} & S_{13} & 0 & 0 & 0\\ S_{12} & S_{22} & S_{23} & 0 & 0 & 0\\ S_{13} & S_{23} & S_{33} & 0 & 0 & 0\\ 0 & 0 & 0 & S_{44} & 0 & 0\\ 0 & 0 & 0 & 0 & S_{55} & 0\\ 0 & 0 & 0 & 0 & 0 & S_{66}\end{array}\right]\left\{\begin{array}{c}\sigma_{11}\\ \sigma_{22}\\ \sigma_{33}\\ \tau_{23}\\ \tau_{13}\\ \tau_{12}\end{array}\right\}. \tag{8.20}$$

Here, S_{ij} are the material compliances. Its relationship with engineering constants is given in [69]. The ν_{ij} is the Poisson's ratio for transverse strain in the jth direction when the stress is applied in the ith direction, and is given by

$$\nu_{ij} = -\epsilon_{jj}/\epsilon_{ii}\,. \tag{8.21}$$

The above condition is for $\sigma_{jj} = \sigma$ and all other stresses being equal to zero. Since the stiffness coefficients $C_{ij} = C_{ji}$, from which it follows that the compliance matrix is also symmetric, that is, $S_{ij} = S_{ji}$. This condition enforces the relation among the Poisson's ratio as

$$\frac{\nu_{ij}}{E_i} = \frac{\nu_{ji}}{E_j} \tag{8.22}$$

Hence, for a lamina under 3-D state of stress, 3 Poisson's ratio namely ν_{12}, ν_{23} and ν_{31} are required to be determined. Other Poisson's ratio can be obtained from Eqn. (8.22).

For most of our analysis, we assume the condition of plane stress. Here, we derive the equations assuming that the condition of plane stress exist in 1-2 plane (see Fig. 8.2). However, if one has to do an analysis of laminated composite beam, which is essentially a 1-D member, the condition of plane stress will exist in the 1-3 plane and the similar procedure could be followed. For the plane stress condition to exist in 1-2 plane, we set the following stresses equal to zero in Eqn. (8.20), that is $\sigma_{33} = \tau_{23} = \tau_{13} = 0$. The resulting constitutive model under plane stress condition can be written as

$$\left\{\begin{array}{c}\epsilon_{11}\\ \epsilon_{22}\\ \gamma_{12}\end{array}\right\}=\left[\begin{array}{ccc}1/E_1 & -\nu_{12}/E_1 & 0\\ -\nu_{21}/E_2 & 1/E_2 & 0\\ 0 & 0 & 1/G_{12}\end{array}\right]\left\{\begin{array}{c}\sigma_{11}\\ \sigma_{22}\\ \tau_{12}\end{array}\right\}. \tag{8.23}$$

Note that the strain ϵ_{33} also exists, which can be obtained from the third constitutive equation

$$\epsilon_{33} = S_{13}\sigma_{11} + S_{23}\sigma_{22}\,. \tag{8.24}$$

From this equation, it also means that Poisson's ratios ν_{13} and ν_{23} should also exist. Inverting Eqn. (8.23), we can expresses stresses in terms of strains, which is given by

$$\left\{\begin{array}{c}\sigma_{11}\\ \sigma_{22}\\ \tau_{12}\end{array}\right\}=\left[\begin{array}{ccc}Q_{11} & Q_{12} & 0\\ Q_{12} & Q_{22} & 0\\ 0 & 0 & Q_{66}\end{array}\right]\left\{\begin{array}{c}\epsilon_{11}\\ \epsilon_{22}\\ \gamma_{12}\end{array}\right\}, \tag{8.25}$$

where Q_{ij} are the reduced stiffness coefficients, which can be expressed in terms of elastic constants as

$$Q_{11} = \frac{E_1}{1-\nu_{12}\nu_{21}},\ Q_{12} = \nu_{21}Q_{11},\ Q_{22} = \frac{E_2}{1-\nu_{12}\nu_{21}},\ Q_{66} = G_{12} \tag{8.26}$$

Next, we will outline the procedure to obtain the constitutive relation for a lamina that have fibers oriented arbitrarily. In most cases, the orientation of the global axes, which we call as x-y axes that are geometrically natural for the solution of the problem, do not coincide with the lamina principal axes, which we have already designated as 1-2 axes. The lamina principal axes and the global axes are shown in Fig. 8.3(a). A small element in the lamina of area dA is taken and the free body diagram is written as shown in Fig. 8.3(b).

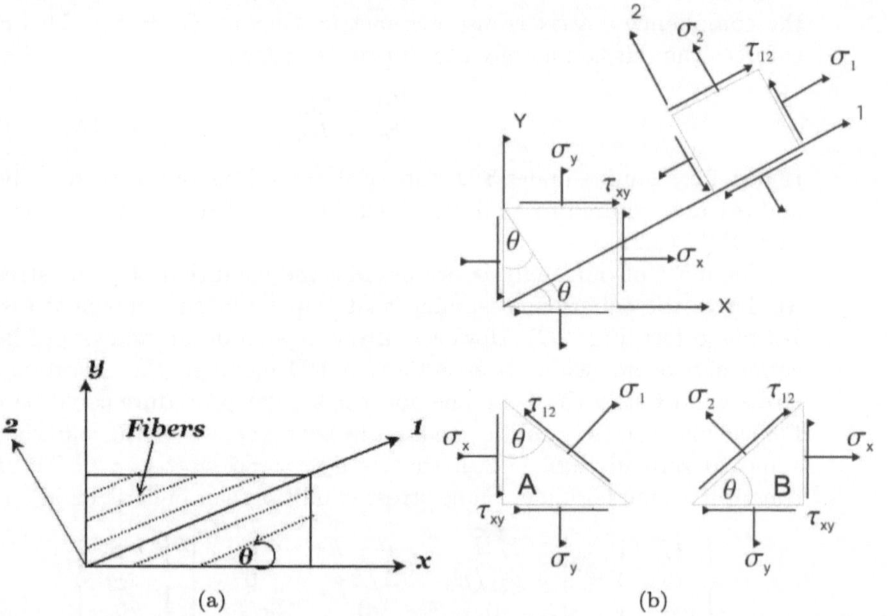

FIGURE 8.3: (a) Principal material axes of a lamina and the global x-y axes. (b) Lamina and laminate coordinate system and FBD of a stressed element

Consider the free body A. Summing of all the forces in the direction of 1-axis, we get

$$\begin{aligned}\sigma_{11}dA \quad &- \quad \sigma_{xx}(\cos\theta dA)(\cos\theta) - \sigma_{yy}(\sin\theta dA)(\sin\theta)\\ &- \quad \tau_{xy}(\sin\theta dA)(\cos\theta) - \tau_{xy}(\cos\theta dA)(\sin\theta) = 0\,.\end{aligned} \tag{8.27}$$

On simplification, the above equation can be written as

$$\sigma_{11} = \sigma_{xx} \cos^2\theta + \sigma_{yy} \sin^2\theta + 2\tau_{xy} \sin\theta \cos\theta\,. \tag{8.28}$$

Similarly, by summing up all the forces along 2-axis (free body A), we get

$$\begin{aligned} \tau_{12} dA \quad &- \quad \sigma_{xx}(\cos\theta dA)(\sin\theta) - \sigma_{yy}(\sin\theta dA)(\cos\theta) \\ &- \quad \tau_{xy}(\sin\theta dA)(\sin\theta) - \tau_{xy}(\cos\theta dA)(\cos\theta) = 0\,. \end{aligned} \tag{8.29}$$

Simplifying the above equation, we get

$$\tau_{12} = -\sigma_{xx} \sin\theta\cos\theta + \sigma_{yy} \sin\theta\cos\theta + \tau_{xy}(\cos^2\theta - \sin^2\theta)\,. \tag{8.30}$$

Following the same procedure and summing up all the forces in the 2-direction in the free body B, we can write

$$\sigma_{22} = \sigma_{xx} \sin^2\theta + \sigma_{yy} \cos^2\theta - 2\tau_{xy} \sin\theta\cos\theta\,. \tag{8.31}$$

Eqns. (8.28, 8.31, & 8.30) can be written in the matrix form as

$$\begin{aligned} &\begin{Bmatrix} \sigma_{11} \\ \sigma_{22} \\ \tau_{12} \end{Bmatrix} = \begin{bmatrix} C^2 & S^2 & 2CS \\ S^2 & C^2 & -2CS \\ -CS & CS & (C^2 - S^2) \end{bmatrix} \begin{Bmatrix} \sigma_{xx} \\ \sigma_{yy} \\ \tau_{xy} \end{Bmatrix} \\ &C = \cos\theta\,, S = \sin\theta\,. \end{aligned} \tag{8.32}$$

or,

$$\boldsymbol{\sigma}_{12} = \mathbf{T}\,\boldsymbol{\sigma}_{xy}$$

In a similar manner, the strains from $1-2$ axis, can be transformed to $x-y$ axis by a similar transformation. Note that for having the same transformation, shear strains are to be divided by two. Without going into much detail, they can be written as

$$\begin{aligned} &\begin{Bmatrix} \epsilon_{11} \\ \epsilon_{22} \\ \frac{\gamma_{12}}{2} \end{Bmatrix} = \begin{bmatrix} C^2 & S^2 & 2CS \\ S^2 & C^2 & -2CS \\ -CS & CS & (C^2 - S^2) \end{bmatrix} \begin{Bmatrix} \epsilon_{xx} \\ \epsilon_{yy} \\ \frac{\gamma_{xy}}{2} \end{Bmatrix} \\ &or,\ \bar{\boldsymbol{\epsilon}}_{12} = \mathbf{T}\,\bar{\boldsymbol{\epsilon}}_{xy}\,. \end{aligned} \tag{8.33}$$

Inverting Eqns. (8.32 & 8.33), we can express the stresses and strains in terms of global coordinates as

$$\begin{aligned} &\begin{Bmatrix} \sigma_{xx} \\ \sigma_{yy} \\ \tau_{xy} \end{Bmatrix} = \begin{bmatrix} C^2 & S^2 & -2CS \\ S^2 & C^2 & 2CS \\ CS & -CS & (C^2 - S^2) \end{bmatrix} \begin{Bmatrix} \sigma_{11} \\ \sigma_{22} \\ \tau_{12} \end{Bmatrix} \\ &\boldsymbol{\sigma}_{xy} = \boldsymbol{T}^{-1}\,\boldsymbol{\sigma}_{12}\,. \end{aligned} \tag{8.34}$$

$$\begin{Bmatrix} \epsilon_{xx} \\ \epsilon_{yy} \\ \frac{\gamma_{xy}}{2} \end{Bmatrix} = \begin{bmatrix} C^2 & S^2 & -2CS \\ S^2 & C^2 & 2CS \\ CS & -CS & (C^2 - S^2) \end{bmatrix} \begin{Bmatrix} \epsilon_{11} \\ \epsilon_{22} \\ \frac{\gamma_{12}}{2} \end{Bmatrix}$$

$$or,\ \bar{\boldsymbol{\epsilon}}_{xy} = \boldsymbol{T}^{-1}\,\bar{\boldsymbol{\epsilon}}_{12}\,. \tag{8.35}$$

Actual strain vectors in both $1-2$ and $x-y$ axes $\{\epsilon\}_{12}$ and $\{\epsilon\}_{xy}$ can be related to $\{\bar{\epsilon}\}_{12}$ and $\{\bar{\epsilon}\}_{xy}$ and through a transformation matrix as

$$\begin{Bmatrix} \epsilon_{11} \\ \epsilon_{22} \\ \gamma_{12} \end{Bmatrix} = \begin{bmatrix} 1 & 0 & 0 \\ 0 & 1 & 0 \\ 0 & 0 & 2 \end{bmatrix} \begin{Bmatrix} \epsilon_{11} \\ \epsilon_{22} \\ \frac{\gamma_{12}}{2} \end{Bmatrix}$$

$$\begin{Bmatrix} \epsilon_{xx} \\ \epsilon_{yy} \\ \gamma_{xy} \end{Bmatrix} = \begin{bmatrix} 1 & 0 & 0 \\ 0 & 1 & 0 \\ 0 & 0 & 2 \end{bmatrix} \begin{Bmatrix} \epsilon_{xx} \\ \epsilon_{yy} \\ \frac{\gamma_{xy}}{2} \end{Bmatrix}. \tag{8.36}$$

$$\boldsymbol{\epsilon}_{12} = \mathbf{R}\,\bar{\boldsymbol{\epsilon}}_{12}, \qquad \boldsymbol{\epsilon}_{xy} = \mathbf{R}\,\bar{\boldsymbol{\epsilon}}_{xy}\,,$$

Now the constitutive equation of a lamina in its principal directions (Eqn. (8.25)) can be written as

$$\boldsymbol{\sigma}_{12} = \mathbf{Q}\,\boldsymbol{\epsilon}_{12}\,. \tag{8.37}$$

Substituting Eqn. (8.32), (8.33) & (8.36) in Eqn. (8.37), we get

$$\mathbf{T}\,\boldsymbol{\sigma}_{xy} = \mathbf{Q}\,\mathbf{R}\,\bar{\boldsymbol{\epsilon}}_{12} = \mathbf{Q}\,\mathbf{R}\,\mathbf{T}\,\bar{\boldsymbol{\epsilon}}_{xy} = \mathbf{Q}\,\mathbf{R}\,\mathbf{T}\,\mathbf{R}^{-1}\boldsymbol{\epsilon}_{xy}\,. \tag{8.38}$$

Hence the constitutive relation in the global $x-y$ axes can now be written as

$$\boldsymbol{\sigma}_{xy} = \bar{\mathbf{Q}}\,\boldsymbol{\epsilon}_{xy} = \mathbf{T}^{-1}\,\mathbf{Q}\,\mathbf{R}\,\mathbf{T}\,\mathbf{R}^{-1}\boldsymbol{\epsilon}_{xy}\,. \tag{8.39}$$

Here the matrix $[\bar{Q}]$ is fully populated. Hence, although the lamina in its own principal direction is orthotropic, in the transformed coordinate direction, it represents complete anisotropic behavior, that is the normal stresses are coupled to the shear stresses and vice versa. The elements of $[\bar{Q}]$ is given by

$$\begin{aligned}
\bar{Q}_{11} &= Q_{11}C^4 + 2(Q_{12} + 2Q_{66})S^2C^2 + Q_{22}S^4\,, \\
\bar{Q}_{12} &= (Q_{11} + Q_{22} - 4Q_{66})S^2C^2 + Q_{12}(S^4 + C^4)\,, \\
\bar{Q}_{16} &= (Q_{11} - Q_{12} - 2Q_{66})SC^3 + (Q_{12} - Q_{22} + 2Q_{66})S^3C\,, \\
\bar{Q}_{22} &= Q_{11}S^4 + 2(Q_{12} + 2Q_{66})S^2C^2 + Q_{22}C^4\,, \\
\bar{Q}_{26} &= (Q_{11} - Q_{12} - 2Q_{66})S^3C + (Q_{12} - Q_{22} + 2Q_{66})SC^3\,, \\
\bar{Q}_{66} &= (Q_{11} + Q_{22} - 2Q_{12} - 2Q_{66})S^2C^2 + Q_{66}(S^4 + C^4)\,,
\end{aligned} \tag{8.40}$$

which gives the constitutive equation of a lamina under plane-stress in the $1-2$ plane.

8.3 CLASSICAL LAMINATION PLATE THEORY

In the previous two subsections, the method of evaluating material properties of the composite at lamina level was outlined. Laminated composites consists of a number of lamina having different fiber orientations that are stacked together to form a laminate. In order to obtain the material property of the complete laminate, one needs a methodology to synthesis each lamina properties and this is done using Classical Lamination Plate Theory (CLPT).

CLPT is a direct extension of Classical Plate Theory (CPT) for isotropic and homogeneous materials proposed by Kirchoff and Love [74, 87]. However, the extension of this theory to laminates requires some modifications to take into account the inhomogeneity in thickness direction due to different ply stacking. As in the case of CPT, CLPT is valid only for thin laminates undergoing small deformations. This theory is based on certain assumption, which are summarized below:

1. The laminate is made of perfectly bonded layers. There is no slip between the adjacent layers. In other words, it amounts to saying that the displacement components are continuous through the thickness.
2. The bonding itself is infinitesimally small (there is no flaw or gap between layers).
3. The bonding is non-shear-deformable (no lamina can slip relative to another).
4. The strength of bonding is as strong as it needs to be (the laminate acts as a single lamina with special integrated properties).
5. Each lamina is considered to be a homogeneous layer such that its effective properties are known.
6. Each lamina is in a state of plane stress.
7. The individual lamina can be isotropic, orthotropic or transversely isotropic.
8. The laminate deforms according to the Kirchhoff-Love assumptions for bending and stretching of thin plates, which can be stated as (a) The normals to the mid-plane remain straight even after deformation, and (b) The normals to the mid-plane do not change their lengths.

Now consider a generic point P on the mid-plane of a laminate, which is located at distance z from the mid-plane. After deformation, the axial displacement of this point along x-direction can be obtained from Fig. 8.4 as

$$u(x, y, z) = u_0(x, y) - z\tan(\alpha) = u_0(x, y) - z\alpha = u_0(x, y) - z\frac{\partial w}{\partial x} \tag{8.41}$$

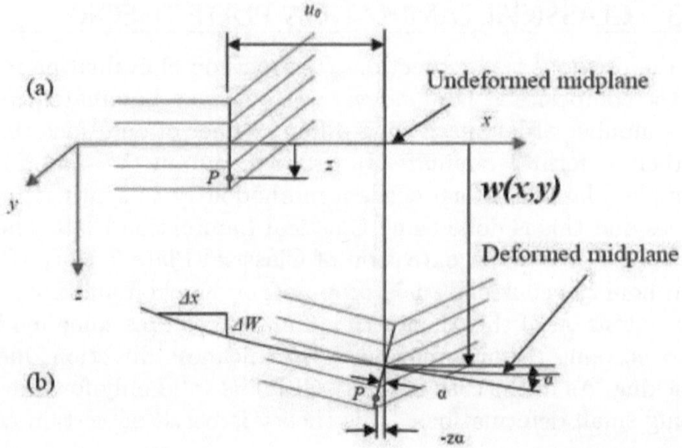

FIGURE 8.4: Laminate geometry: (a) undeformed and (b) deformed

where u_0 is the mid-plane axial deformation, α is the angle between the deformed mid-plane and horizontal and $w(x,y)$ is the transverse deformation in the z direction. Similarly, for the deformation in yz plane, we can express the slope of the deformed mid-plane as $\partial w/\partial y$. Thus, the transverse displacement of a generic point along y axis can be given as

$$v(x,y,z) = v_0(x,y) - z\frac{\partial w}{\partial y} \tag{8.42}$$

where $v_0(x,y)$ is the mid-plane transverse deformation. Thus, the complete deformation field of the laminate is given by

$$u(x,y,z) = u_0(x,y) - z\frac{\partial w}{\partial x}, \qquad v(x,y,z) = v_0(x,y) - z\frac{\partial w}{\partial y}$$
$$w(x,y,z) = w_0(x,y) \tag{8.43}$$

Using strain-displacement relations, strains are obtained using Eqn. (8.43) as

$$\epsilon_{xx} = \frac{\partial u}{\partial x} = \frac{\partial u_0}{\partial x} - z\frac{\partial^2 w}{\partial x^2}, \qquad \epsilon_{yy} = \frac{\partial v}{\partial y} = \frac{\partial v_0}{\partial y} - z\frac{\partial^2 w}{\partial y^2}$$
$$\gamma_{xy} = \frac{\partial u}{\partial y} + \frac{\partial v}{\partial x} = \frac{\partial u_0}{\partial y} + \frac{\partial v_0}{\partial x} - 2z\frac{\partial^2 w}{\partial x \partial y} \tag{8.44}$$

The above equation can be written in matrix form as

$$\boldsymbol{\epsilon}_{xy} = \boldsymbol{\epsilon^{(0)}}_{xy} + z\,\boldsymbol{\kappa}_{xy} \tag{8.45}$$

where

$$\epsilon^{(0)}{}_{xy} = \{\epsilon^{(0)}_{xx} \quad \epsilon^{(0)}_{yy} \quad \gamma^{(0)}_{xy}\}^T = \{\frac{\partial u_0}{\partial x} \quad \frac{\partial v_0}{\partial y} \quad \frac{\partial u_0}{\partial y} + \frac{\partial v_0}{\partial x}\}^T$$

are the mid-plane strains and

$$\boldsymbol{\kappa}_{xy} = \{\kappa_{xx} \quad \kappa_{yy} \quad \kappa_{xy}\}^T = \left\{-\frac{\partial^2 w}{\partial x^2} \quad -\frac{\partial^2 w}{\partial y^2} \quad -2\frac{\partial^2 w}{\partial x \partial y}\right\}^T$$

represents the curvatures. The terms κ_{xx} and κ_{yy} represent bending curvatures, while the term κ_{xy} represents the twisting curvature. From Eqn. (8.45), we see that strains vary linearly across the thickness direction z.

The stresses at any location can be calculated from the strains and lamina constitutive relations (Eqn. (8.39). It is assumed that the lamina properties are known. Hence, the constitutive equation for a k_{th} lamina is known, that is, the reduced stiffness matrices (in principal material directions and global directions) are known. Thus, the stresses in k_{th} lamina can written using Eqn. (8.39) as

$$\boldsymbol{\sigma}^k_{xy} = \bar{\mathbf{Q}}^k \, \boldsymbol{\epsilon}^k_{xy} \tag{8.46}$$

where the elements of the $\bar{\mathbf{Q}}$ is given by Eqn(8.40). Substituting for $\boldsymbol{\epsilon}^k_{xy}$ from Eqn. (8.45) and expanding, we get

$$\boldsymbol{\sigma}^k_{xy} = \bar{\mathbf{Q}}^k \, \boldsymbol{\epsilon}^{(0)}{}_{xy} + \bar{\mathbf{Q}}^k + z\,\boldsymbol{\kappa}_{xy} \tag{8.47}$$

In these equations, the strains are given at a location z where the stresses are required. It should be noted that the strains are continuous and vary linearly through the thickness. If we look at the stress distribution through the thickness, it is clear that the stresses are not continuous through the thickness. This is because the stiffness is different for different laminae in thickness direction. In a lamina, the stress varies linearly. The slope of this variation in a lamina depends upon its moduli. However, at the interface of two adjacent laminae there is a discontinuity in the stresses. The same thing is depicted in Fig. 8.5 for a laminated composite having three layers.

Next, we will consider the stress resultants acting on the laminates, which consists of in-plane forces and the bending moments. These are shown in Fig. 8.6(a) and 8.6(b), respectively.

In-plane forces per unit length is defined in terms of stress components as

$$N_{xx} = \int_h^h \sigma_{xx}\, dz, \qquad N_{yy} = \int_h^h \sigma_{yy}\, dz, \qquad N_{xy} = \int_h^h \tau_{xy}\, dz \tag{8.48}$$

The above equation can be written in matrix form as

$$\boldsymbol{N}_{xy} = \int_h^h \boldsymbol{\sigma}_{xy}\, dz \tag{8.49}$$

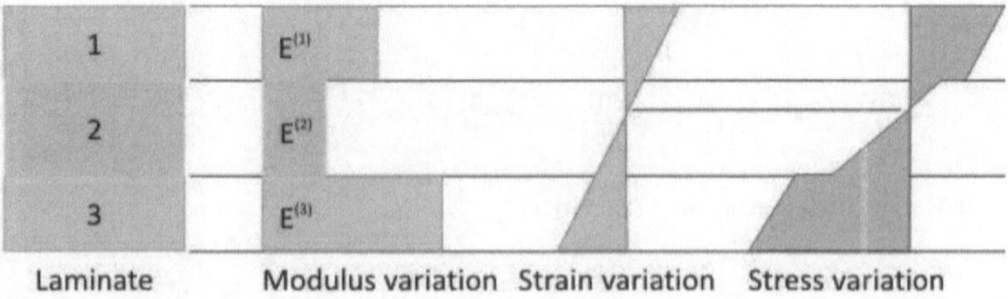

FIGURE 8.5: Strain, Stress and Young's moduli distribution across the depth of a laminate

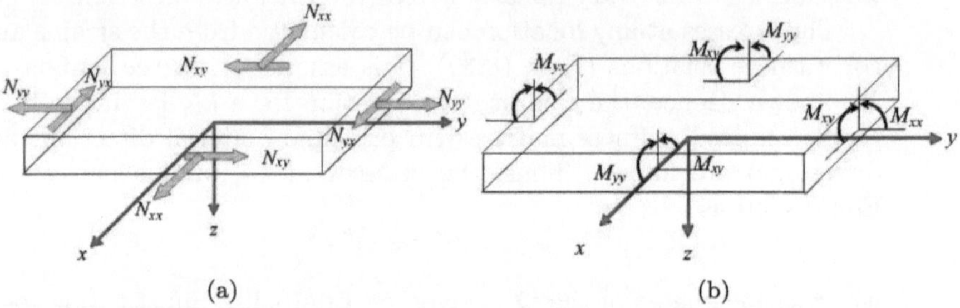

FIGURE 8.6: (a) In-plane stress resultants and (b) Moment resultants

where h is the thickness of the laminate. Using Eqn. (8.47) in Eqn. (8.49), we get

$$\boldsymbol{N}_{xy} = \sum_{k=1}^{N_{lay}} \int_{-z_{k-1}}^{z_k} \bar{\mathbf{Q}}^k \, \boldsymbol{\epsilon}^{(0)}{}_{xy} \, dz + \sum_{k=1}^{N_{lay}} \int_{-z_{k-1}}^{z_k} \bar{\mathbf{Q}}^k \, \boldsymbol{\kappa}_{xy} \, z \, dz \tag{8.50}$$

where N_{lay} is the number of layers in a laminate. Now recall that the mid-plane strains $\boldsymbol{\epsilon}^{(0)}{}_{xy}$ and the curvatures $\boldsymbol{\kappa}_{xy}$ are independent of z coordinate. The reduced transformed stiffness matrix $\bar{\mathbf{Q}}$ is function of thickness and constant over a given lamina thickness. Now, we can replace the integration over the laminate thickness as sum of the integrations over individual lamina thicknesses. Thus, Eqn. (8.50) can be written as

$$\boldsymbol{N}_{xy} = \mathbf{A} \, \boldsymbol{\epsilon}^{(0)}{}_{xy} + \mathbf{B} \, \boldsymbol{\kappa}_{xy} \tag{8.51}$$

where

$$\mathbf{A} = \sum_{k=1}^{N_{lay}} \bar{\mathbf{Q}}^k \, (z_k - z_{k-1}) \,, \qquad \mathbf{B} = \frac{1}{2} \sum_{k=1}^{N_{lay}} \bar{\mathbf{Q}}^k \, (z_k^2 - z_{k-1}^2) \tag{8.52}$$

The matrix **A** represents the in-plane stiffness, that is, it relates the in-plane forces with mid-plane strains and the matrix **B** represents the coupling stiffness, that is, it relates the in-plane forces with mid-plane curvatures. It should be noted that the matrices **A** and **B** are symmetric as the matrix$[\bar{Q}]$.

The moment resultants per unit length is defined in terms of stress components as

$$M_{xx} = \int_h^h \sigma_{xx}\, z\, dz, \qquad M_{yy} = \int_h^h \sigma_{yy}\, z\, dz, \qquad M_{xy} = \int_h^h \tau_{xy}\, z\, dz \tag{8.53}$$

The above equation can be written in matrix form as

$$\boldsymbol{M}_{xy} = \int_h^h \boldsymbol{\sigma}_{xy}\, z\, dz \tag{8.54}$$

Using Eqn. (8.47) in Eqn. (8.54), we get

$$\boldsymbol{M}_{xy} = \sum_{k=1}^{N_{lay}} \int_{-z_{k-1}}^{z_k} \bar{\mathbf{Q}}^k \boldsymbol{\epsilon}^{(0)}{}_{xy}\, z\, dz + \sum_{k=1}^{N_{lay}} \int_{-z_{k-1}}^{z_k} \bar{\mathbf{Q}}^k \boldsymbol{\kappa}_{xy} z^2 dz \tag{8.55}$$

As before, replacing the integration by summation, we can write Eqn. (8.55) as

$$\boldsymbol{M}_{xy} = \mathbf{B}\,\boldsymbol{\epsilon}^{(0)}{}_{xy} + \mathbf{D}\,\boldsymbol{\kappa}_{xy} \tag{8.56}$$

where

$$\mathbf{D} = \frac{1}{3}\sum_{k=1}^{N_{lay}} \bar{\mathbf{Q}}^k \left(z_k^3 - z_{k-1}^3\right) \tag{8.57}$$

The matrix **D** represents the bending stiffness, that is, it relates resultant moments with mid-plane curvatures. Again, the matrix **D** is also symmetric. Further, it is important to note that the matrix **B** relates the resultant moments with mid-plane curvatures as well. Thus, the constitutive matrix for the laminate can be written in matrix form as

$$\begin{Bmatrix} \boldsymbol{N}_{xy} \\ \boldsymbol{M}_{xy} \end{Bmatrix} = \begin{bmatrix} \mathbf{A} & \mathbf{B} \\ \mathbf{B} & \mathbf{D} \end{bmatrix} \begin{Bmatrix} \boldsymbol{\epsilon}^{(0)}{}_{xy} \\ \boldsymbol{\kappa}_{xy} \end{Bmatrix} \tag{8.58}$$

8.4 WAVE PROPAGATION IN 1-D THIN LAMINATED COMPOSITE WAVEGUIDE

Here, we assume that Euler-Bernoulli beam theory is applicable. We begin the derivation by first writing the displacement field, which are given by

$$u(x,y,z,t) = u^o(x,t) - zw(x,t)_x\,, \qquad w(x,y,z,t) = w(x,t)\,, \tag{8.59}$$

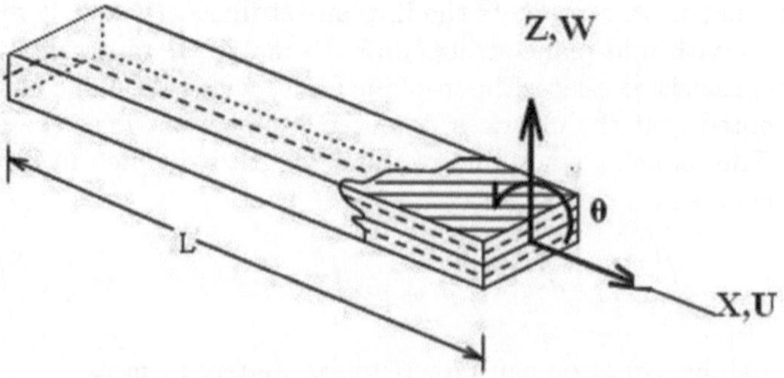

FIGURE 8.7: Laminated composite elementary waveguide with the coordinate system and degrees of freedom

where u^o is the axial displacement along the middle plane and w is the transverse displacement as shown in Fig. 8.7. Here, z is measured from the middle plane and subscript x indicate derivative with respect to x.

The 3-D constitutive model for composites was derived earlier (see Eqn. (8.40)). In the present case, the laminated composite waveguide is in 1-D state of stress and hence the layer-wise constitutive law is defined as

$$\sigma_{xx} = \bar{Q}_{11}\epsilon_{xx} \,, \tag{8.60}$$

where σ_{xx} and ϵ_{xx} are stress and strain in x-direction. The expression for $\bar{Q}_{11}$ as a function of ply fiber angle θ is

$$\bar{Q}_{11} = Q_{11}cos^4\theta + Q_{22}sin^4\theta + 2\left(Q_{12} + 2Q_{66}sin^2\theta cos^2\theta\right) \,, \tag{8.61}$$

where Q_{ij} are the orthotropic elastic coefficients for the individual composite ply and was derived in the previous section.

The strain energy and the kinetic energy are defined as

$$U = \frac{1}{2}\int \sigma_{xx}\epsilon_{xx}dv \,, \qquad t = \frac{1}{2}\int \rho(\dot{u^o}^2 + \dot{w}^2)dv \,, \tag{8.62}$$

where $\dot{u^o}$ and $\dot{w}$ are the axial and transverse velocities, and ρ is the layer-wise density. Applying Hamilton's principle, the governing differential equations are obtained, and they can be expressed as

$$\rho A\ddot{u^o} - A_{11}u^o_{xx} + B_{11}w_{xxx} = 0 \,, \tag{8.63}$$

$$\rho A\ddot{w} - B_{11}u^o_{xxx} + D_{11}w_{xxxx} = 0 \tag{8.64}$$

and the corresponding force boundary conditions are obtained as

$$A_{11}u^o_x - B_{11}w_{xx} = N_x \,, \tag{8.65}$$

$$B_{11}u^o_{xx} - D_{11}w_{xxx} = V_x \,, \tag{8.66}$$

$$-B_{11}u^o_x + D_{11}w_{xx} = M_x \,, \tag{8.67}$$

where

$$[A_{11}, B_{11}, D_{11}] = \int_{-h/2}^{+h/2} \bar{Q}_{11}\left[1, z, z^2\right] b\, dz \,, \tag{8.68}$$

Here, h is the depth of the beam, b is the layer width and A is the cross-sectional area of the beam. $\ddot{u}^o$ and $\ddot{w}$ are the mid-plane longitudinal and transverse accelerations. $\langle . \rangle_x, \langle . \rangle_{xx} \ldots$ represent partial derivatives with respect to x. N_x is the axial force, V_x is the shear force and M_x is the bending moment.

8.4.1 Computation of Wavenumbers

The governing differential Eqns. (8.63 & 8.64) represent a system of coupled linear PDEs, which is first transformed to frequency domain using DFT using the following transformation

$$u^o(x,t) = \sum_{n=1}^{N} \hat{u}(x, \omega_n) e^{i\omega_n t} = \sum_{n=1}^{N} \left(\tilde{u}_p e^{-ik_p x}\right) e^{i\omega_n t} \tag{8.69}$$

$$w(x,t) = \sum_{n=1}^{N} \hat{w}(x, \omega_n) e^{i\omega_n t} = \sum_{n=1}^{N} \left(\tilde{w}_p e^{-ik_p x}\right) e^{i\omega_n t} \tag{8.70}$$

where $i = \sqrt{-1}$. As was done in the previous chapters, all variables with $<\hat{}>$ states that these quantities are frequency dependent. In the above equations $\tilde{u}_p$ and $\tilde{w}_p$ represent the wave coefficients, which are to be evaluated from three displacement and three force boundary conditions. Also, k_p is the wavenumber associated with p^{th} mode of propagation.

Substituting Eqns. (8.69 & 8.70) in Eqns. (8.63 & 8.64), we get

$$\begin{bmatrix} (c_L^2 k_p^2 - \omega_n^2) & i c_c^3 k_p^3 \\ -i \frac{c_c^3}{\omega_n} k_p^3 & \left(\frac{c_b^4}{\omega_n^2} k_p^4 - \omega_n^2\right) \end{bmatrix} \begin{Bmatrix} \tilde{u}_p \\ \tilde{w}_p \end{Bmatrix} = \begin{Bmatrix} 0 \\ 0 \end{Bmatrix} \tag{8.71}$$

In the above equation c_L and c_b represent the axial and flexural speeds, which are given by

$$c_L = \sqrt{\frac{A_{11}}{\rho A}}, \qquad c_b = \sqrt{\omega_n}\left[\frac{D_{11}}{\rho A}\right]^{0.25}$$

The phase speed of the induced dispersive wave due to unsymmetric ply orientation causing axial-flexural coupling is defined as

$$c_c = \left[\omega_n \frac{B_{11}}{\rho A}\right]^{1/3}$$

From Eqn. (8.71), by setting the determinant of the matrix to zero, we get a sixth order characteristic equation for the solution of wavenumber k_p, which is given by

$$(1-r)k_p^6 - k_L^2 k_p^4 - k_b^4 k_p^2 + k_L^2 k_b^4 = 0 \qquad (8.72)$$

where a non-dimensional axial-flexural coupling parameter and the fundamental wavenumbers corresponding to uncoupled axial and flexural modes is introduced as

$$r = \frac{B_{11}^2}{A_{11}D_{11}}, \qquad k_L = \frac{\omega_n}{c_L}, \qquad k_b = \frac{\omega_n}{c_b}$$

Eqn. (8.72) can be written as a cubic polynomial equation and can be solved using *roots* function in MATLAB. From the physics of wave propagation, it can be inferred that one of the roots will be real, while the other two will be complex conjugates. The spectral amplitudes of the displacement field is then explicitly defined as

$$\begin{Bmatrix} \hat{u}(x,\omega_n) \\ \hat{w}(x,\omega_n) \end{Bmatrix} = \begin{bmatrix} R_{11} & R_{12} & R_{13} & R_{14} & R_{15} & R_{16} \\ R_{21} & R_{22} & R_{23} & R_{24} & R_{25} & R_{26} \end{bmatrix} \begin{Bmatrix} \tilde{u}_1 e^{-ik_1 x} \\ \tilde{u}_2 e^{-ik_1(L-x)} \\ \tilde{w}_3 e^{-ik_2 x} \\ \tilde{w}_4 e^{-ik_2(L-x)} \\ \tilde{w}_5 e^{-ik_3 x} \\ \tilde{w}_6 e^{-ik_3(L-x)} \end{Bmatrix} \qquad (8.73)$$

where R_{1j} and R_{2j} are the amplitude ratios for the jth mode of wave propagation. These are derived from Eqn. (8.71) and are given by

$$r_{11} = R_{12} = 1, \qquad R_{13} = -i\sqrt{r}\frac{k_L k_2^3}{k_b^2(k_2^2 - k_L^2)} = -R_{14}.$$

$$R_{15} = -i\sqrt{r}\frac{k_L k_3^3}{k_b^2(k_3^2 - k_L^2)} = -R_{16}, \quad R_{21} = -i\sqrt{r}\frac{k_b^2 k_1^3}{k_L(k_1^3 - k_b^4)} = -R_{22}$$

$$R_{23} = R_{24} = R_{25} = R_{26} = 1 \qquad (8.74)$$

It is interesting to note that for a symmetric ply orientation, the axial-flexural coupling parameter r becomes zero because $B_{11} = 0$. As a consequence, axial and flexural modes become uncoupled and Eqn. (8.72) gives two equations. One is a second-order characteristic equation, similar to that in case of elementary rod with roots $\pm k_L$ and the other is a fourth order characteristic equation, similar to that in case of a homogeneous Euler- Bernoulli beam with roots $\pm k_b$ and $\pm ik_b$ as was discussed in Chapter 4. For this case, the amplitude ratios R_{13}, R_{14}, R_{15}, R_{16}, R_{21} and R_{22} will become zero.

Next, we will now plot the wavenumber and group speeds as a function of frequency. From design consideration, one big advantage of using laminated composite is that it can be tailored to get the required strength and stiffness by different ply orientations. The main objective here is to bring out the effect of coupling on the wave behavior. The maximum axial-flexural coupling that one gets from such natural ply-stacking is when the cross-plies and 0° plies are stacked in separate groups. A generalization of the effect of axial-flexural coupling gives some valuable insights, when the spectrum relation (Fig. 8.8(a)) and the dispersion relation (Fig. 8.8(b)) are studied. AS/3501-6 graphite-epoxy plies (thickness of each layer is 1.0 mm) with three stacking sequences $[0_{10}]$ ($r = 0.0$), $[0_5/30_2/60_3]$ ($r = 0.312$) and $[0_5/90_5]$ ($r = 0.574$) are considered. In Fig. 8.8(a), it can be observed that corresponding to axial mode (Mode 1) and flexural modes (Modes 2 and 3), the wavenumbers increase in magnitude for increasing coupling. However, this increase in Mode 2 (propagating component) is more than that in the Mode 3 (evanescent component). Fig. 8.8(b) also shows the variation of group speed $C_g = d\omega/dk_j$ normalized with the parameter $C_o = \sqrt{E/\rho}$. From these plots, it is clear that the axial speed is reduced by more than 26% due to presence of unsymmetry arising from cross-ply stacking in groups. Also, in the range of 50 kHz, the flexural speed of propagation is reduced by 42% for maximum coupling.

8.5 WAVE PROPAGATION IN THICK 1-D LAMINATED COMPOSITE WAVEGUIDES

In the previous section, we studied the wave propagation in thin or elementary composite beam and showed that the effect of stiffness and inertial coupling is that it reduces the wave speeds by almost 42%. In this section, we will study the wave propagation behavior in thick composite beam based on first order shear deformation theory (FSDT) (or Timoshenko beam theory). The axial motion is modeled using Mindlin-Herrmann rod theory, since due to material anisotropy, both these motions are coupled. Wave propagation in thick isotropic rod and beam were studied in Chapter 4 (see Sections 4.4.1 & 4.4.2). It was shown that the higher-order isotropic waveguide introduces a pair of additional propagating modes (called propagating shear or lateral contraction wave modes) after the cut-off frequency. It will be interesting to know how the higher-order effects influences the wave propagation behavior in a thick composite beam with unsymmetric ply stacking sequence considering that the material anisotropy introduces stiffness and inertial coupling over a wide frequency range of excitation. The present section will present some examples that will help in capturing this special feature in the spectrum and dispersion relations.

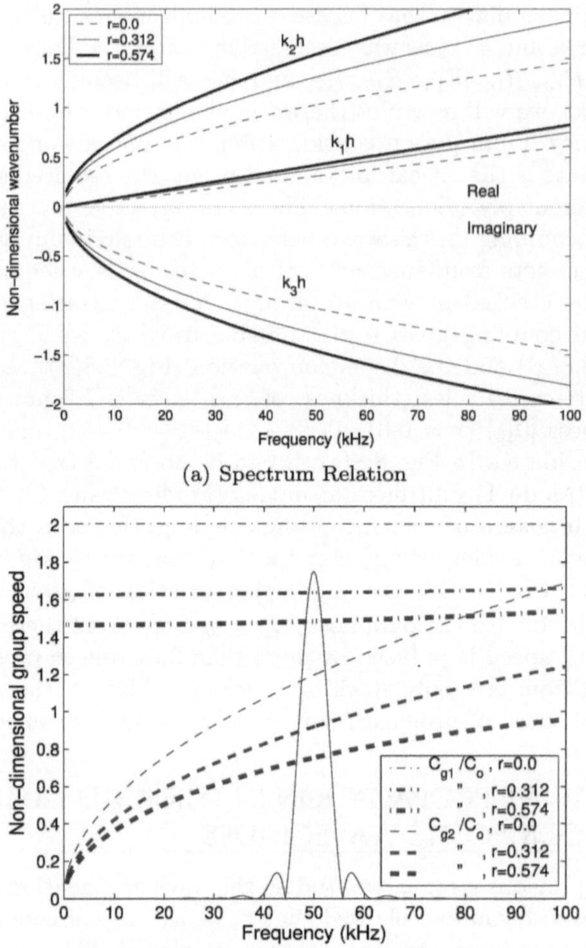

(a) Spectrum Relation

(b) Dispersion relations

FIGURE 8.8: (a) Spectrum relation for various axial-flexural coupling. (b) Dispersion relation for various axial-flexural coupling

One of the characteristic features of higher-order waveguide is the presence of cut-off frequencies due to inherent shear constraints. That is, the shear constraints converts the evanescent flexural mode into propagating shear mode beyond the cut-off frequency. Appearance of higher-order wave modes above certain cut-off frequencies have been studied for metallic waveguides in Chapter 4, and for laminated composite plates, this aspect was studied in reference [71]. In this chapter, contribution of a contractional mode along with shear mode is studied for different types of structural composites. Expression of cut-off frequencies in shear mode and contractional mode in presence of asymmetric ply stacking sequence are also derived.

8.5.1 Governing Equation for a Thick Composite Beam

The displacement field for axial and transverse motion based on FSDT and thickness contraction [93], [46] is given by

$$u(x,y,z,t) = u^o(x,t) - z\theta(x,t)\ , \quad w(x,y,z,t) = w(x,t) + z\psi(x,t) \quad (8.75)$$

where u and w are respectively the axial and transverse displacements at a material point. Here, u^o is the beam axial displacement along the reference plane, w^o is the transverse displacement on the reference plane, θ is the curvature-independent rotation of the beam cross-section about Y-axis and $\psi = \epsilon_{zz}$ is the contraction/elongation parallel to Z-axis (shown in Fig. 8.9).

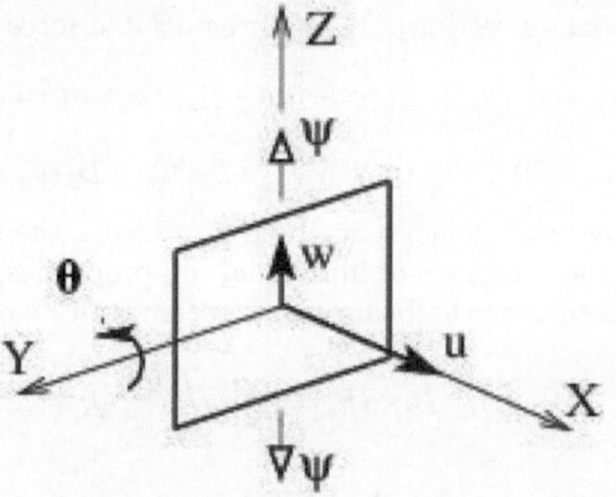

FIGURE 8.9: Beam cross-section in YZ plane and the degrees of freedom.

The thick beam is subjected to the normal stress σ_{xx} and the shear stress τ_{xz}. The corresponding strains are ϵ_{xx} and γ_{xz}. These strains can be expressed in terms of displacements using strain displacement relations. The constitutive model for laminated composites was discussed earlier in this chapter (see Eqn. (8.40)). In the present case, it is given by

$$\begin{Bmatrix} \sigma_{xx} \\ \sigma_{zz} \\ \tau_{xz} \end{Bmatrix} = \begin{bmatrix} \bar{Q}_{11} & \bar{Q}_{13} & 0 \\ \bar{Q}_{13} & \bar{Q}_{33} & 0 \\ 0 & 0 & \bar{Q}_{55} \end{bmatrix} \begin{Bmatrix} \epsilon_{xx} \\ \epsilon_{zz} \\ \gamma_{xz} \end{Bmatrix} . \quad (8.76)$$

The strain energy and kinetic energy expressions are given

$$U = \frac{1}{2}\int_V (\sigma_{xx}\epsilon_{xx} + \tau_{xz}\gamma_{xz})dV, \quad T = \frac{1}{2}\int_V \rho(\dot{u_0}^2 + \dot{w}^2)dV \tag{8.77}$$

Expressing stresses and strains in terms of displacements and substituting these in Hamilton's principle and performing integration by parts, the governing differential equations for thick composite beam is obtained, which are given by

$$\begin{aligned}
\delta u &: \quad I_0\ddot{u}^o - I_1\ddot{\theta} - A_{11}u^o_{xx} + B_{11}\theta_{xx} - A_{13}\psi_x = 0\,, && (8.78)\\
\delta\psi &: \quad I_2\ddot{\psi} + I_1\ddot{w} + A_{13}u^o_x - B_{13}\theta_x + A_{33}\psi \\
&\qquad -B_{55}\left(w_{xx} - \theta_{,x}\right) - D_{55}\psi_{xx} = 0\,, && (8.79)\\
\delta w &: \quad I_0\ddot{w} + I_1\ddot{\psi} - A_{55}\left(w_{xx} - \theta_x\right) - B_{55}\psi_{xx} = 0\,, && (8.80)\\
\delta\phi &: \quad I_2\ddot{\theta} - I_1\ddot{u}^o - A_{55}\left(w_x - \theta\right) - B_{55}\psi_x \\
&\qquad +B_{11}u^o_{xx} - D_{11}\theta_{xx} + B_{13}\psi_x = 0 && (8.81)
\end{aligned}$$

where $<.>_x$ and $<.>_{xx}$ are derivatives of the field variables with respect to x (the axial direction). The four associated force boundary conditions are

$$A_{11}u^o_x - B_{11}\theta_x + A_{13}\psi = N_x\,, \;\; B_{55}\left(w_x - \theta\right) + D_{55}\psi_x = Q_x\,, \tag{8.82}$$

$$A_{55}\left(w_x - \theta\right) + B_{55}\psi_x = V_x\,, \;\; -B_{11}u^o_x + D_{11}\theta_x - B_{13}\psi = M_x\,. \tag{8.83}$$

As in the case of elementary composite beam, the stiffness coefficients here, which are the functions of individual ply properties, ply orientation, etc. and are integrated over the beam cross-section, can be expressed as

$$\left[A_{ij}, B_{ij}, D_{ij}\right] = \sum_i \int_{z_i}^{z_{i+1}} \bar{Q}_{ij}\left[1, z, z^2\right] b dz\,. \tag{8.84}$$

The coefficients associated with the inertial terms can be expressed as

$$\left[I_0, I_1, I_2\right] = \sum_i \int_{z_i}^{z_{i+1}} \rho\left[1, z, z^2\right] b dz\,. \tag{8.85}$$

In Eqns. (8.84 & 8.85), z_i and z_{i+1} are the Z-coordinate of bottom and top surfaces of the ith layer and b is the corresponding width. It can be noticed that for asymmetric ply stacking, all four modes: axial, flexural, shear and thickness contraction are coupled with each other. This makes the problem cumbersome to solve accurately using analytic approach for all boundary conditions. However, at this stage, different approximate methods can be referred (e.g. [2]), which are computationally intensive. Here we use Polynomial eigenvalue problem (PEP) technique to solve for the wavenumbers. Two cases are considered, one with thickness contraction mode and the other without it.

8.5.2 Wave Propagation in a Thick Beam Model with Both Shear Deformation and Lateral Contraction Included

This model due to inherent coupling of transverse displacement, rotation and lateral contraction will exhibit four-way axial-flexural-shear-lateral contraction coupling. For this study, a plane wave type solution is sought, where the displacement field, $\mathbf{u} = \{u^o\ ,\ \psi\ ,\ w\ ,\ \phi\}(x,t)$, can be written as

$$\mathbf{u} = \sum_{n=1}^{N} \{\tilde{u}\ ,\ \tilde{\psi}\ ,\ \tilde{w}\ ,\ \tilde{\phi}\}(x) e^{i\omega_n t} = \sum_{n=1}^{N} \tilde{\mathbf{u}}(x) e^{i\omega_n t}\,, \tag{8.86}$$

where ω_n is the circular frequency at the nth sampling point and N is the frequency index corresponding to the Nyquist frequency in FFT.

Substituting the assumed solution of the field variables in Eqs. (8.78)-(8.81), a set of ordinary differential equations (ODEs) is obtained for $\tilde{\mathbf{u}}(\mathbf{x})$. Since, the ODEs are of constant coefficient type, the solution is of the form $\tilde{\mathbf{u}}_\circ e^{-ikx}$, where k is the wavenumber and $\tilde{\mathbf{u}}_\circ$ is a vector of unknown constants, that is, $\tilde{\mathbf{u}}_\circ = \{u_\circ\ ,\ \psi_\circ\ ,\ w_\circ\ ,\ \phi_\circ\}$. Substituting the assumed form in the set of ODEs, a matrix vector relation is obtained, which gives the characteristic equation

$$\mathbf{W}\,\tilde{\mathbf{u}}_\circ = \mathbf{0}\,, \tag{8.87}$$

where $\mathbf{W}$ is

$$\begin{bmatrix} A_{11}k^2 - I_\circ\omega_n^2 & p & 0 & q \\ -p & t_2 & r & -s \\ 0 & r & A_{55}k^2 - I_\circ\omega_n^2 & -iA_{55}k \\ q & s & iA_{55}k & t_1 \end{bmatrix}. \tag{8.88}$$

where $p = A_{13}k$, $q =_1 \omega_n^2 - B_{11}k^2$, $r = -I_1\omega_n^2 + B_{55}k^2$, $s = iB_{55}k - iB_{13}k$, $t_1 = -I_2\omega_n^2 + D_{11}k^2 + A_{55}$, and $t_2 = -k_I I_2\omega_n^2 + A_{33} + D_{55}k^2$. According to previous discussion, in this case, the order of the PEP, $p = 2$ and N_v (size of the $\mathbf{W}$) is 4. Thus, there are 8 eigenvalues altogether, which are the roots of the polynomial obtained from the singularity condition of $\mathbf{W}$ as

$$Q_1k^8 + Q_2k^6 + Q_3k^4 + Q_4k^2 + Q_5 = 0\,. \tag{8.89}$$

The spectrum relation suggests that the roots can be written as $\pm k_1$, $\pm k_2$, $\pm k_3$ and $\pm k_4$.

Before solving this 8th order characteristic equation (obtained by setting the determinant of the PEP equal to zero), one can obtain an overview of the number of propagating and evanescent modes as follows. By substituting $\omega_n = 0$ in the characteristic equation and solving for k_j, it can be shown that for uncoupled case ($B_{ij} = 0$)

$$k(0)_{1,\cdots,6} = 0\,, \qquad k(0)_{7,8} = \pm\sqrt{\frac{A_{13}^2 - A_{11}A_{33}}{A_{55}D_{55}}}\,. \tag{8.90}$$

This implies that six zero roots starting at $\omega_n = 0$ correspond to the axial, flexural and shear modes, whereas the two non-zero roots must be the wavenumbers associated with contractional mode. Here, it is to be noted that $\sigma_{zz} = 0$ for elementary theory and FSDT and the orthotropic constitutive model with respect to XY plane, whereas in presence of the thickness contraction, $\sigma_{zz} \neq 0$, which requires a plane-stress model in XZ plane reduced from 3D constitutive model. This produces slight difference in the values of A_{55} compared to that in FSDT. However, almost all the conventional fiber reinforced composites used as structural material have $Q_{11} > Q_{13}$, $Q_{33} > Q_{13}$, which imply that the nonzero roots in Eqn. (8.90) must be imaginary at and near $\omega_n = 0$. Therefore, we have two evanescent (one forward and one backward) components in the contractional mode at low frequency regime. Next, by substituting $k_j = 0$ in Eqn. (8.89) and solving for ω_n, we get the cut-off frequencies as

$$\omega_{\text{cut-off}} = \sqrt{\frac{A_{55}}{I_2(1 - s_2^2)}}, \sqrt{\frac{A_{33}}{I_2(1 - s_2^2)}}. \tag{8.91}$$

This shows that initially there are two forward propagating modes (one axial, one flexural), two backward propagating modes (one axial and one flexural), two evanescent flexural mode (forward and backward) and two additional evanescent contractional mode (forward and backward) after $\omega_n = 0$. The shear mode starts propagating after the cut-off frequency corresponding to A_{55} in Eqn. (8.91). The contractional mode starts propagating later, since $A_{33} \geq A_{55}$.

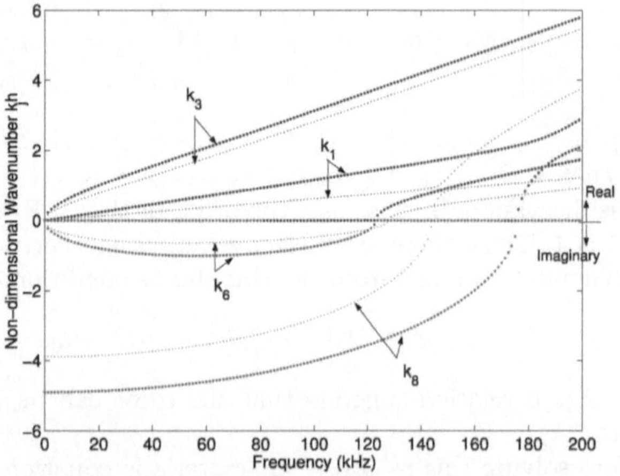

FIGURE 8.10: Nature of wavenumber dispersion in axial (k_1), flexural (k_3), and shear (k_6) with cut-off and contraction (k_8) with cut-off; ....., graphite-epoxy AS/3501 $[0]_{10}$ composite; ....., glass-epoxy $[0]_{10}$ composite; total thickness $h = 0.01m$.

In Fig. 8.10, the wavenumber dispersion is plotted for AS/3501 graphite-epoxy and glass-epoxy $[0]_{10}$ composite beam with total thickness of $h =$

0.01 m. Material properties are taken from [117]. Note that the graphite-epoxy has very high ratio of E_{11}/G_{13} (≈ 20) and moderate stiffness $E_{11} \approx 144$ GPa. On the other hand, the glass-epoxy has very low ratio of E_{11}/G_{13} (≈ 6) and very low stiffness $E_{11} \approx 54$ GPa. For both of these materials, the plot in Fig. 8.10 shows that the propagating components before the cut-off frequency in contraction are similar to that in Fig. 8.11(a) in the absence of the variable ψ. The latter is studied in more detail in Section 8.5.3. Also, the wavenumber associated with the evanescent components in contractional mode before the cut-off frequency is much higher than that due to shear, and therefore decay rapidly. Hence, below the cut-off frequency in contraction (which is always much higher than shear cut-off since $A_{33} > A_{55}$ in case of composite), change in the response due to the addition of contractional mode is negligible. Such behavior is different than that in isotropic waveguides, as the cut-off frequencies for shear and contractional modes in isotropic material are very close to each other. The behavior and restriction of the isotropic beam waveguide model has been discussed in the context of a three mode beam theory and Lamb wave modes in 2D cross-section in [34]. To accommodate such higher-order effects in the present asymmetric composite beam waveguide model, the following corrections can be imposed for high-frequency applications.

Correction factors at high-frequency limit:

For applications where the beam cross-sectional configuration and the range of excitation frequency ω_{max} are such that

$$\omega_{max} > min \left(\sqrt{\frac{A_{33}}{I_2(1-s_2^2)}}, \ \sqrt{\frac{A_{55}}{I_2(1-s_2^2)}} \right), \tag{8.92}$$

four correction factors K_1, K_2, K_3 and K_4 can be introduced as $A_{55} \leftarrow K_1 A_{55}$, $D_{55} \leftarrow K_2 D_{55}$, $I_2 \leftarrow K_3 I_2$ and $A_{33} \leftarrow K_4 A_{33}$. The idea is to estimate these factors by putting upper bound on the cut-off frequencies from appropriate Lamb wave modes (Note that the Lamb wave theory was explained in Chapter 7 in the context of isotropic materials) and by adjusting the propagating wavenumbers from those of the Lamb wave modes at high frequencies. Now, what appears as a contrast between the pairs of longitudinal, flexural, shear and contractional modes as shown in Fig. 8.10 with the propagation of the first three pairs of symmetric Lamb wave modes and the first three pairs of antisymmetric Lamb wave modes in actual 2D cross-section (the YZ plane in Fig. 8.9) is as follows. At low frequencies, the pair of propagating longitudinal modes are identical to the first pair of propagating symmetric Lamb wave modes. Also, at low frequencies, the pair of propagating flexural modes and the pair of evanescent shear modes are identical to the first pair of propagating antisymmetric Lamb wave modes and second pair of evanescent antisymmetric Lamb wave modes, respectively. The rest of the symmetric and antisymmetric Lamb wave modes are evanescent, therefore the effect is highly localized and can be neglected. At high frequencies, the pair of propagating flexural modes and the shear modes are respectively given by

$$k_{3,4} \approx \pm\omega_n\sqrt{\frac{I_0}{K_1 A_{55}}}\,, \quad k_{5,6} \approx \pm\omega_n\sqrt{\frac{K_3 I_2}{D_{11}}}\,. \tag{8.93}$$

These forms differ slightly from the corresponding first and second pair of antisymmetric Lamb wave modes (introduced in Chapter 7 in regard to waves in 2-D isotropic waveguides). The adjustable cut-off frequencies in shear and contraction are now given by

$$\omega_{cs} = \sqrt{\frac{K_1 A_{55}}{K_2 I_2(1-s_2^2)}}\,. \quad \omega_{cc} = \sqrt{\frac{A_{33}}{K_2 I_2(1-s_2^2)}}\,. \tag{8.94}$$

Note from Fig. 8.10, the sudden diversion of the propagating longitudinal modes as they interact with the propagating contractional modes. After such interaction, at higher frequencies, the longitudinal modes and the contractional modes are respectively given by

$$k_{1,2} \approx \pm\omega_n\sqrt{\frac{K_3 I_2}{K_2 D_{55}}}\,, \quad k_{7,8} \approx \pm\omega_n\sqrt{\frac{I_0}{A_{33}}}\,, \tag{8.95}$$

which are not the case due interaction of second and third pairs of propagating symmetric Lamb wave modes, because they first become complex, just above cut-off frequency and break down followed by addition of the fourth pair at higher frequencies. This fourth pair of symmetric Lamb wave modes propagates with similar form as in Eqn. (8.95) for $k_{7,8}$. Now, we denote the Rayleigh wave speeds c_R in the cross-sectional planes YZ. c_R is associated with the non-dispersive wave propagation along x due to impact at the top or bottom surface of the beam. However, for general ply stacking, one needs to compute c_R using a plane-stress model and appropriate averaging in the YZ plane. As a frequency approximation, now we can write

$$|k_{1,2}| = \frac{\omega_n}{c_R}\,, \quad |k_{3,4}| = \frac{\omega_n}{c_R}\,, \quad \omega_{cs} \approx \frac{1}{2}\frac{2\pi}{h}c_s\,, \quad \omega_{cc} \approx \frac{2\pi}{h}c_s\,, \tag{8.96}$$

where $\frac{1}{2}\frac{2\pi}{h}c_s$ is the first non-zero cut-off frequency of the antisymmetric Lamb wave mode pair and $\frac{2\pi}{h}c_s$ is the first non-zero cut-off frequency of the symmetric Lamb wave mode pair. c_s is the shear wave speed, h is the depth of the beam cross-section. Substituting Eqns. (8.93)–(8.95) in Eqn. (8.96), we get

$$K_1 = \frac{c_R^2}{c_s^2}\,, \quad K_3 = \frac{h^2 I_0}{\pi^2 I_2(1-s_2^2)}\frac{c_R^2}{c_s^2}\,, \tag{8.97}$$

$$K_2 = \frac{h^2 I_0}{\pi^2 I_2(1-s_2^2)K_5}\frac{c_R^4}{c_s^4}\,, \quad K_4 = 4\frac{A_{55}}{A_{33}}\frac{c_R^2}{c_s^2}\,. \tag{8.98}$$

In case of isotropic materials, the above equations reduces to

$$K_5 = \frac{A_{55} I_2}{D_{55} I_0} = 1, \; s_2 = 0, \; \left.\frac{c_R^2}{c_s^2}\right|_{\nu=0.3} = 0.86, \; \left.\frac{A_{55}}{A_{33}}\right|_{\nu=0.3} = 0.28 \qquad (8.99)$$

and we recover the so called shear correction factors $K_1 = 0.86$, $K_3 = 1.216$ (discussed in Chapter 4 and as well as in [34], and also previously proposed in [32] as $K_1 = 5/6$, $K_3 = 1$) for Timoshenko beam along with $K_2 = 0.89$ and $K_4 = 0.98$, which accommodate the second pair of antisymmetric Lamb wave modes (shear modes) and second pair of symmetric Lamb wave modes (contractional modes) within the high-frequency limit given by the cut-off frequency of next higher-order Lamb wave mode. Hence, for asymmetric laminated composite, as long as the frequency content of excitation is below the cut-off in contraction, the model of coupled axial-flexural-shear wave modes can be adjusted approximately to obtain results with sufficient accuracy.

Note that in the above discussions, we described all the wave modes in terms of Lamb wave modes, which are the 2-D spatial modes generated due to traction free boundary. The point that is made here is that, by using higher-order 1-D waveguide theories, one can obtain the 2-D equivalent Lamb wave modes. That is, higher the 1-D waveguide theories used, many higher Lamb wave modes can be obtained with good accuracy.

8.5.3 Wave Propagation in a Thick Beams Model That Includes Only the Shear Deformation

This model exhibits three way coupling, that is the axial-flexure and shear coupling. When the thickness contraction term ψ is neglected in the displacement field, and subsequently in the wave equations, the characteristic equation is a 6-th order polynomial in k. Again, either *roots* function or PEP framework in MATLAB can be used to obtain the wavenumber and hence, the group speeds .

In Fig. 8.11(a), wavenumbers corresponding to axial, flexural and shear modes are shown. AS/3501-6 graphite-epoxy beam cross-section with depth $h = 0.01m$ is considered. Beside this, to study how the wave packets travel at different frequencies, group speeds $C_g = Re[d\omega_n/dk_j]$ in axial, flexural and shear modes are plotted in Fig. 8.11(b), where $C_0 = \sqrt{A_{11}/I_0}$ is the constant phase speed in axial mode. From these two plots, only one cut-off frequency appears, above which the shear mode starts propagating, which is otherwise evanescent component contributing to the flexural wave. Fig. 8.11(b) shows that higher stiffness coupling (higher value of r) yields higher group speed of shear wave shortly above the cut-off frequency. At the same instance, the group speed of longitudinal wave drastically falls well before cut-off frequency. The flexural mode remains least affected by both stiffness and mass coupling, and remains almost non-dispersive above cut-off frequency.

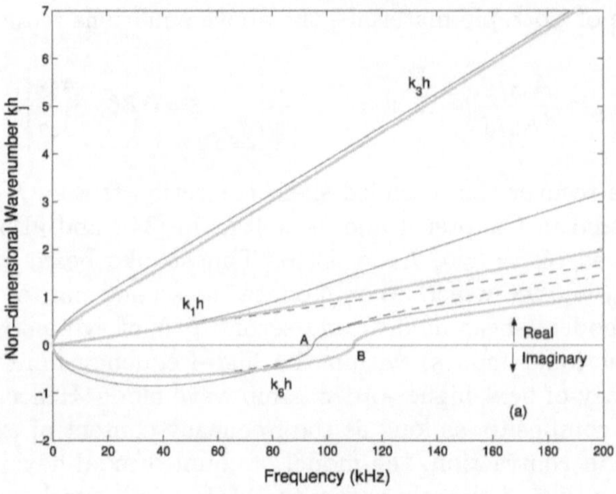

(a) Spectrum Relation

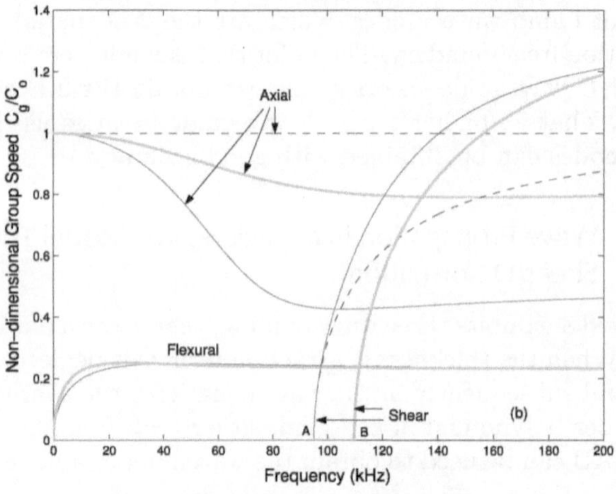

(b) Dispersion relations

FIGURE 8.11: (a) Nature of wavenumber dispersion. (b) Dispersion of group speeds in axial, flexural and shear modes for different stiffness and material asymmetries. - - -, $r = 0.0$, $s2 = 0.0$; —, $r = 0.757$, $s2 = 0.0$; —, $r = 0.0$, $s2 = 0.5$. The locations of cut-off frequency are marked by A and B.

8.6 WAVE PROPAGATION IN TWO-DIMENSIONAL COMPOSITE WAVEGUIDES

In Chapter 7, wave propagation in 2-D Isotropic solids (both semi-infinite and bounded solids) was presented. There, the governing Navier's equation was solved using Helmholtz decomposition, which reduced the governing equations into two independent PDE's in terms of the scalar and vector potentials, one representing the solution to the P-waves and the other representing the solution of S-waves. In the case of laminated composites, the anisotropic character at the laminate level makes the clear distinction between P-waves and S-waves very difficult. The anisotropy makes these two waves coupled. This makes Helmholtz decomposition not applicable to solve the governing equations . Hence we need to choose an alternate methods and one such method by which we can solve the governing equation is the *Partial Wave Technique* (PWT).

In this section, for 2-D composite waveguides, the governing equations are solved using PWT, where the SVD method (described in Chapter 3 see Section 3.5.1) is utilized to obtain the wave amplitudes, which is essential for constructing the partial waves. The general procedure of PWT is as follows. First, a set of partial waves are constructed. Once the partial waves are found, the wave coefficients are made to satisfy the prescribed boundary conditions. For computing the wavenumbers, we can use PEP or *roots* function in MATLAB. The code for the latter is provided with this text.

8.6.1 Formulation of Governing Equations and Computation of Wavenumbers

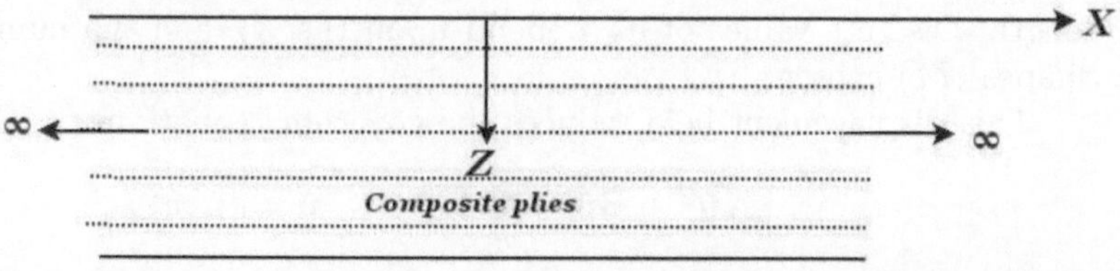

FIGURE 8.12: The coordinate system for 2D laminated composites

In this formulation, it is assumed that there is no heat conduction in and out of the system, displacements are small, material is homogeneous and anisotropic and the domain is in 2-D Euclidean space. The governing equation for a 2-D laminated composites, undergoing wave motions in x and z directions (see Fig. 8.12 for the coordinate system) with deformations $u(x,t)$ and $w(x,t)$ is given by general elasto-dynamic equation of motion in three dimension is given by (The details of this derivation can be found in the author's unified book [54])

$$\begin{aligned} Q_{11}u_{xx} + (Q_{13}+Q_{55})w_{xz} + Q_{55}u_{zz} &= \rho\ddot{u} \\ Q_{55}w_{xx} + (Q_{13}+Q_{55})u_{xz} + Q_{33}w_{zz} &= \rho\ddot{w} \end{aligned} \quad (8.100)$$

where Q_{ij}s are the stiffness coefficients, which depend on the ply layup, its orientation and z coordinate of the layer. The expressions of the Q_{ij}s are given in the earlier subsection of this chapter. The displacement field is assumed to be a synthesis of frequency and wavenumbers, both horizontal and vertical, as

$$u(x,z,t) = \sum_{n=1}^{N-1}\sum_{m=1}^{M-1} \hat{u}(z,\eta_m,\omega_n)\left\{\begin{array}{c}\sin(\eta_m x)\\ \cos(\eta_m x)\end{array}\right\} e^{i\omega_n t}, \tag{8.101}$$

$$w(x,z,t) = \sum_{n=1}^{N-1}\sum_{m=1}^{M-1} \hat{w}(z,\eta_m,\omega_n)\left\{\begin{array}{c}\cos(\eta_m x)\\ \sin(\eta_m x)\end{array}\right\} e^{i\omega_n t}, \tag{8.102}$$

where ω_n is the discrete angular frequency, η_m is the discrete horizontal wavenumber and $i^2 = -1$. Note that we have used Fourier series as the second transform instead of a second Fourier Transform. As the assumed field suggests, for $M \to \infty$, the model will stretch to infinite extent in positive and negative x-direction, although it is of finite extent in the z direction, that is, it will be a layered structure. In particular, the domain can be written as $\Omega = [-\infty, +\infty] \times [0, L]$, where L is the thickness of the layer. The boundaries of any layer will be specified by a fixed value of z. The x dependency of the displacement field (sine or cosine) will be determined based upon the loading pattern. In this study, symmetric load pattern is considered. The real computational domain is $\Omega_c = [-X_L/2, +X_L/2] \times [0, L]$, where X_L is the x window length. Discrete values of η_m depend upon the X_L and the number of mode shapes (M) chosen.

This displacement field reduces the governing equations to

$$\mathbf{A}\hat{\mathbf{u}}'' + \mathbf{B}\hat{\mathbf{u}}' + \mathbf{C}\hat{\mathbf{u}} = \mathbf{0}, \quad \hat{\mathbf{u}} = \{\hat{u}\ \hat{w}\}, \tag{8.103}$$

where prime denotes differentiation with respect to z. The matrices $\mathbf{A}$, $\mathbf{B}$ and $\mathbf{C}$ are

$$\mathbf{A} = \left[\begin{array}{cc} Q_{55} & 0 \\ 0 & Q_{33}\end{array}\right], \mathbf{B} = \left[\begin{array}{cc} 0 & -(Q_{13}+Q_{55})\eta_m \\ (Q_{13}+Q_{55})\eta_m & 0\end{array}\right],$$

$$\mathbf{C} = \left[\begin{array}{cc} -\eta_m^2 Q_{11} + \rho\omega_n^2 & 0 \\ 0 & -\eta_m^2 Q_{55} + \rho\omega_n^2\end{array}\right].$$

The associated boundary conditions are the specifications of stresses σ_{zz} and σ_{xz} at the layer interfaces. From the displacement field, the strains are computed using strain-displacement relation. Using the computed strains, stresses are computed and these stresses are related to the unknowns as

$$\hat{\mathbf{s}} = \mathbf{D}\hat{\mathbf{u}}' + \mathbf{E}\hat{\mathbf{u}}, \hat{\mathbf{s}} = \{\sigma_{zz}\ \sigma_{xz}\}.$$

$$\mathbf{D} = \left[\begin{array}{cc} 0 & Q_{33} \\ Q_{55} & 0\end{array}\right], \quad \mathbf{E} = \left[\begin{array}{cc} \eta_m Q_{13} & 0 \\ 0 & -\eta_m Q_{55}\end{array}\right]. \tag{8.104}$$

The boundary value problem (BVP) reduces to finding $\hat{\mathbf{u}}$, which satisfies Eqn. (8.103) for all $z \in \Omega_c$ and specification of $\hat{\mathbf{u}}$ or $\hat{\mathbf{s}}$ at $z = 0$ or $z = L$. The solution is then obtained for different values of z in the frequency-wavenumber domain, ($z - \eta - \omega$ domain), for given values of ω_n and η_m. Summation over η_m will bring the solution back to $z - x - \omega$ space and inverse FFT will bring the solution back to time domain, that is, $z - x - t$ space.

The solutions to these ODEs are of the form $u_\circ e^{-ikz}$ and $w_\circ e^{-ikz}$, which yields the PEP

$$\mathbf{W}\,\mathbf{u}_\circ = \mathbf{0}, \mathbf{W} = -k^2\mathbf{A} - ik\mathbf{B} + \mathbf{C}\,, \mathbf{u}_\circ = \{u_\circ \; w_\circ\}\,, \tag{8.105}$$

where $\mathbf{W}$, the wave matrix is given by

$$\mathbf{W} = \begin{bmatrix} -k^2Q_{55} - \eta_m^2 Q_{11} + \rho\omega_n^2 & ik\eta_m(Q_{13} + Q_{55}) \\ -ik\eta_m(Q_{13} + Q_{55}) & -k^2Q_{33} - \eta_m^2 Q_{55} + \rho\omega_n^2 \end{bmatrix}. \tag{8.106}$$

The singularity condition of $\mathbf{W}$ yields

$$\begin{aligned} Q_{33}Q_{55}k^4 \; &+ \; \{(Q_{11}Q_{33} - 2Q_{13}Q_{55} - Q_{13}^2)\eta_m^2 - \rho\omega_n^2(Q_{33} + Q_{55})\}k^2 \\ &+ \; \{Q_{11}Q_{55}\eta_m^4 - \rho\omega_n^2\eta_m^2(Q_{11} + Q_{55}) + \rho^2\omega_n^4\} = 0 \end{aligned} \tag{8.107}$$

The above equation which relates vertical wavenumber k to the horizontal wavenumber η and frequency ω gives the spectrum relation. It is to be noted that for each value of η_m and ω_n, there are four values of k, denoted by k_{lmn}, $l = 1..4$, which will be obtained by solving the spectrum relation. Explicit solution of the wavenumber k is $k_{lnm} = \pm\sqrt{-b \pm \sqrt{b^2 - 4ac}}$, where a, b and c are the coefficients of k^4, k^2 and k^0, respectively, in Eqn. (8.107).

There are certain properties of the wavenumbers which will be explored now. As can be seen from Eqn. (8.107), for $\eta_m = 0$, the equation is readily solvable to give the roots $\pm\omega\sqrt{\rho/Q_{33}}$ and $\pm\omega\sqrt{\rho/Q_{55}}$. Since, none of the ρ, Q_{33} or Q_{55} can be negative or zero, these roots are always real and linear with ω. When η_m is not zero, k becomes zero for ω satisfying

$$\begin{aligned} &Q_{11}Q_{55}\eta_m^4 - \rho\omega_n^2\eta_m^2(Q_{11} + Q_{55}) + \rho^2\omega_n^4 = 0 \\ &\text{or} \quad (Q_{11}\eta_m^2 - \rho\omega^2)(Q_{55}\eta_m^2 - \rho\omega^2) = 0 \\ &\text{where,}\; \omega = \eta_m\sqrt{Q_{11}/\rho}, \quad \eta_m\sqrt{Q_{55}/\rho} \end{aligned}$$

Before these frequencies, the roots are imaginary and non-propagating and after these frequencies, the roots are real and propagating. These frequencies are the cut-off frequencies. For isotropic materials they are given by $c_p\eta$ and $c_s\eta$ as reported in [119] and also shown in Chapter 7. The current expressions for the cut-off frequencies are also reducible to that of isotropic materials if we identify Q_{11} and Q_{55} with $\lambda + 2\mu$ and μ, respectively, where λ and μ are the Lame's parameters. These were discussed in Chapter 7. As mentioned earlier,

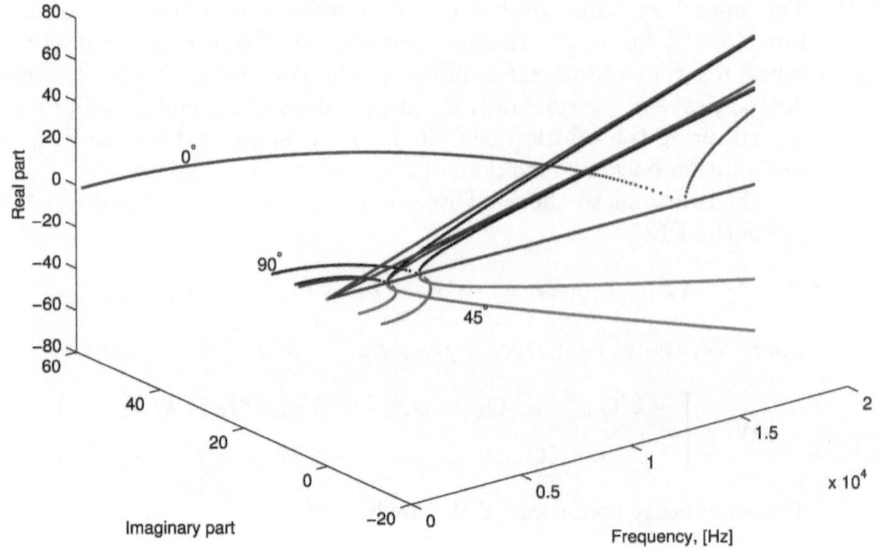

FIGURE 8.13: Variation of wavenumber with ω_n ($\eta_m = 10$)

we do not have explicit P- or S-waves. Instead, we have QP-wave (Quazi P-wave) and QS-wave (Quazi S-wave) If we identify QP-wave with Q_{33} (or Q_{11}) and QSV-wave with Q_{55}, then as the cut-off frequencies suggest, for the same value of η_m, it is the QSV wave that becomes propagating first, since $Q_{11} > Q_{55}$. The wavenumbers of positive roots denote forward propagating modes and the negative roots denote backward propagating modes. In Fig. 8.13, the wavenumbers are plotted for three different ply-angles, 0°, 45° and 90°.

For all the ply-angles, Q_{33} and Q_{55} are assumed 9.69 GPa and 4.13 GPa, respectively. For Q_{11} and Q_{13}, following values are assumed. For 0°, $Q_{11} =$ 146.3 GPa and $Q_{13} =$ 2.98 GPa, for 45°, $Q_{11} =$ 44.62 GPa and $Q_{13} =$ 1.62 GPa and for 90°, $Q_{11} =$ 9.69 GPa and $Q_{13} =$ 2.54 GPa. In Fig. 8.13, imaginary part of the wavenumbers is plotted in horizontal plane and real part in the vertical plane. Further, the imaginary part of the wavenumbers for 0° and 90° are plotted in the positive side, whereas for 45° it is plotted in the negative side, for distinction. Two different η_m values are chosen. The linear variation of the real part of the wavenumbers are seen for $\eta_m = 0$ and rest of the plots are for $\eta_m = 10$. As discussed previously, slope of the linear portion depends upon Q_{33} and Q_{55} and as they are equal for all the ply-angles, this part is common for all the ply-angles. The difference comes in the imaginary part and the cut-off frequencies. Two different cut-off frequencies are seen in the figure for each ply-angle, where the largest value is for 0° ply-angle because it has largest value of Q_{11}. Further, the shear cut-off frequency is same for all the ply-angles as Q_{55} is equal in all the cases.

Once, the required wavenumbers k are obtained, for which the wave matrix $\mathbf{W}$ is singular, the solution $\mathbf{u}_\circ$ as a function of frequency ω_n and wavenumber η_m is given by

$$\begin{aligned} u_{nm} &= R_{11}C_1e^{-jk_1x} + R_{12}C_2e^{-jk_2x} \\ &+ R_{13}C_3e^{-jk_3x} + R_{14}C_4e^{-jk_4x} \end{aligned} \quad (8.108)$$

$$\begin{aligned} w_{nm} &= R_{21}C_1e^{-jk_1x} + R_{22}C_2e^{-jk_2x} \\ &+ R_{23}C_3e^{-jk_3x} + R_{24}C_4e^{-jk_4x} \end{aligned} \quad (8.109)$$

where R_{ij} are the amplitude coefficients to be determined, and they are called wave amplitudes. As outlined before, following the method of SVD, R_{ij} are obtained from the wave matrix $\mathbf{W}$ evaluated at wavenumber k_i.

8.7 WAVE PROPAGATION IN 2-D LAMINATED COMPOSITE PLATES

Laminated composite 2D plate is the 2D counterpart of the 1D composite beam. The previous section explained the procedure to solve the governing wave equation in 2-D laminated composites subjected to in-plane excitations using Partial Wave Technique where companion matrix method using SVD Technique to compute wave amplitudes and PEP to obtain wavenumbers were used. In case of plates, we can use alternative methods of computing wavenumbers and wave amplitudes. Here we explore the use of latent roots and right latent roots and right latent eigenvector of the system matrix (the wave matrix) for computing the wavenumber and the amplitude ratio matrix and the whole problem is posed as a PEP. Eigenvectors are needed only when response needs to be estimated. In this regard, the MATLAB Code *complatewavenum.m*, provided with this text, can be used to generate wavenumbers. Response estimation code can be written using this wavenumber code and the equations provided in the text.

8.7.1 Governing Equations and Wavenumber Computations

According to the CLPT, the displacement field is given by

$$U(x,y,z,t) = u(x,y,t) - z\partial w/\partial x, \quad V(x,y,z,t) = v(x,y,t) - z\partial w/\partial y,$$
$$W(x,y,z,t) = w(x,y,t)$$

where, u, v and w are the displacement components of the reference plane in the x, y and z directions, respectively and z is measured downward positive (see Fig. 8.14). Using these displacement field and Hamilton's Principle, we can derive the governing differential equations for a laminated composite plate. The details of deriving the governing equations is not presented here.

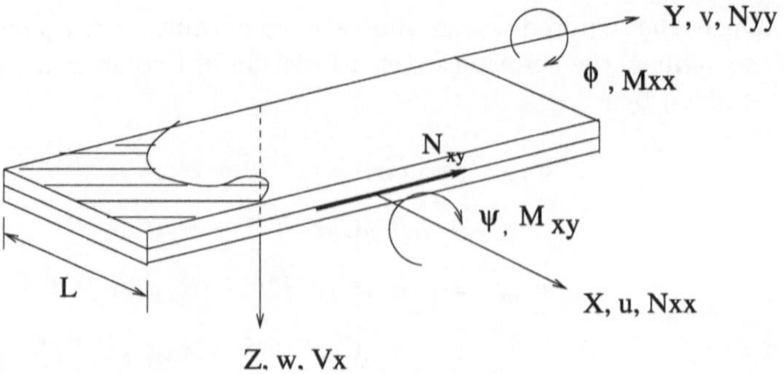

FIGURE 8.14: 2D composite plate with displacements and stress resultants

The reader may refer to the author's unified book [54] for the details of the derivation.

The governing equations written in terms of these force resultants is given by

$$\partial N_{xx}/\partial x + \partial N_{xy}/\partial y = I_\circ \ddot{u} - I_1 \partial \ddot{w}/\partial x\,, \tag{8.110}$$
$$\partial N_{xy}/\partial x + \partial N_{yy}/\partial y = I_\circ \ddot{v} - I_1 \partial \ddot{w}/\partial y\,, \tag{8.111}$$
$$\begin{aligned}&\partial^2 M_{xx}/\partial xx + 2\partial^2 M_{xy}/\partial xy + \partial^2 M_{yy}/\partial yy = I_\circ \ddot{w} \\ &-I_2(\partial^2 \ddot{w}/\partial xx + \partial^2 \ddot{w}/\partial yy) + I_1(\partial \dot{u}/\partial x + \partial \dot{v}/\partial y)\,,\end{aligned} \tag{8.112}$$

where the mass moments are defined as

$$[I_\circ\,, I_1\,, I_2] = \int_A \rho[1\,, z\,, z^2]\, dA\,, \tag{8.113}$$

The governing equations can be further expanded in terms of the displacement components. However, because of their complexity, they are not given here and can be found in [117]. The associated boundary conditions are

$$\bar{N}_{xx} = N_{xx}n_x + N_{xy}n_y\,, \quad \bar{N}_{yy} = N_{xy}n_x + N_{yy}n_y\,, \tag{8.114}$$
$$\bar{M}_{xx} = -M_{xx}n_x - M_{xy}n_y\,, \tag{8.115}$$
$$\begin{aligned}\bar{V}_x &= (\partial M_{xx}/\partial x + 2\partial M_{xy}/\partial y - I_1\ddot{u} + I_2 \partial \ddot{w}/\partial x) n_x \\ &+ (\partial M_{xy}/\partial x + 2\partial M_{yy}/\partial y - I_1\ddot{v} + I_2 \partial \ddot{w}/\partial y) n_y\,,\end{aligned} \tag{8.116}$$

where $\bar{N}_{xx}$ and $\bar{N}_{yy}$ are the applied normal forces in the x and y direction, $\bar{M}_{xx}$ and $\bar{M}_{yy}$ are the applied moments about Y and X axes and $\bar{V}_x$ is the applied shear force in Z direction.

The wavenumber computation begins by assuming the same kind of solution of the displacement field as was considered in the earlier formulation explained in the previous section, that is, the time harmonic waves are sought and Fourier series is employed in y direction (in which the model is unbounded). Thus,

$$u(x,y,t) = \sum_{n=1}^{N}\sum_{m=1}^{M} \hat{u}(x) \left\{ \begin{array}{c} \cos(\eta_m y) \\ \sin(\eta_m y) \end{array} \right\} e^{i\omega_n t}, \qquad (8.117)$$

$$v(x,y,t) = \sum_{n=1}^{N}\sum_{m=1}^{M} \hat{v}(x) \left\{ \begin{array}{c} \sin(\eta_m y) \\ \cos(\eta_m y) \end{array} \right\} e^{i\omega_n t}, \qquad (8.118)$$

$$w(x,y,t) = \sum_{n=1}^{N}\sum_{m=1}^{M} \hat{w}(x) \left\{ \begin{array}{c} \cos(\eta_m y) \\ \sin(\eta_m y) \end{array} \right\} e^{i\omega_n t}, \qquad (8.119)$$

where again the cosine or sine dependency is chosen based upon the symmetry or anti-symmetry of the load about x axis. The ω_n and the η_m are the circular frequency at n^{th} sampling point and the wavenumber in the y direction at the mth sampling point, respectively. The N is the index corresponding to the Nyquist frequency in FFT.

Substituting Eqns. (8.117)–(8.119) in Eqns. (8.110)–(8.112), a set of ODEs is obtained for the unknowns $\hat{u}(x)$, $\hat{v}(x)$ and $\hat{w}(x)$. Since these ODEs are having constant coefficients, their solutions can be written as $\tilde{u}e^{-ikx}$, $\tilde{v}e^{-ikx}$ and $\tilde{w}e^{-ikx}$, where k is the wavenumber in the X direction, yet to be determined and $\tilde{u}$, $\tilde{v}$ and $\tilde{w}$ are the unknown constants. Substituting these assumed forms in the set of ODEs, a PEP is posed as to find $(\mathbf{v}, k)$, such that,

$$\mathbf{\Psi}(k)\mathbf{v} = (k^4\mathbf{A}_4 + k^3\mathbf{A}_3 + k^2\mathbf{A}_2 + k\mathbf{A}_1 + \mathbf{A}_0)\mathbf{v} = 0, \quad \mathbf{v} \neq 0, \qquad (8.120)$$

where $\mathbf{A}_i \in \mathbf{C}^{3\times 3}$, k is an eigenvalue and $\mathbf{v}$ is the corresponding right eigenvector. The matrices $\mathbf{A}_i$ are given by

$$\mathbf{A}_0 = \begin{bmatrix} -A_{66}\eta_m^2 + I_o\omega_n^2 & 0 & 0 \\ 0 & -A_{22}\eta_m^2 + I_o\omega_n^2 & -B_{22}\eta_m^3 + I_1\omega_n^2\eta_m \\ 0 & -B_{22}\eta_m^3 + I_1\omega_n^2\eta_m & -D_{22}\eta_m^4 + I_o\omega_n^2 + I_2\omega_n^2\eta_m^2 \end{bmatrix}$$

$$\mathbf{A}_1 = \begin{bmatrix} 0 & -i\eta_m(A_{12}+A_{66}) & -i\eta_m^2(B_{12}+2B_{66}) + iI_1\omega_n^2 \\ i\eta_m(A_{12}+A_{66}) & 0 & 0 \\ i\eta_m^2(B_{12}+2B_{66}) - iI_1\omega_n^2 & 0 & 0 \end{bmatrix}$$

$$\mathbf{A}_2 = \begin{bmatrix} -A_{11} & 0 & 0 \\ 0 & -A_{66} & -\eta_m(B_{12}+2B_{66}) \\ 0 & -\eta_m(B_{12}+2B_{66}) & -\eta_m^2(2D_{12}+4D_{66}) + I_2\omega_n^2 \end{bmatrix}$$

$$\mathbf{A}_3 = \begin{bmatrix} 0 & 0 & -iB_{11} \\ 0 & 0 & 0 \\ iB_{11} & 0 & 0 \end{bmatrix}, \qquad \mathbf{A}_4 = \begin{bmatrix} 0 & 0 & 0 \\ 0 & 0 & 0 \\ 0 & 0 & -D_{11} \end{bmatrix}.$$

It can be noticed that $\mathbf{A}_4$ is singular, thus the lambda matrix $\mathbf{\Psi}(k)$ is not regular [82] and admits infinite eigenvalues [136].

The PEP is solved by the methods described before. The MATLAB script *complatewavenum.m* is provided to solve these system of equations. In this case, the spectrum relation is a quartic polynomial of $m = k^2$,

$$p(m) = m^4 + C_1 m^3 + C_2 m^2 + C_3 m + C_4 \,, \quad C_i \in \mathbf{C}\,, \tag{8.121}$$

which generates a companion matrix of order 4.

The polynomial governing the wavenumbers (Eqn. (8.121)) is solved by considering a Graphite-Epoxy (Gr.-Ep.) (AS/3501) plate of following material properties:

$$\begin{aligned}
&E_1 = 144.48\,GPa\,, E_2 = 9.63\,GPa\,, E_3 = 9.63\,GPa\,,\\
&G_{23} = 4.128\,GPa\,, G_{13} = 4.128\,GPa\,, G_{12} = 4.128\,GPa\,,\\
&\nu_{23} = 0.3\,, \nu_{13} = 0.02\,, \nu_{12} = 0.02\,, \rho = 1389\,kg/m^3,\ h_\circ = 1.0\,mm,\\
&N_\ell = 10
\end{aligned}$$

where $h_\circ$ is the thickness of each layer and N_ℓ is the total number of layers. Two different ply-stacking sequences are considered, one symmetric $[0_{10}]$ and the other asymmetric $[0_5/90_5]$. The y wavenumber, η_m is fixed at 50 for all the wavenumber computation. The real and imaginary part of the wavenumbers are shown in Figs. 8.15 and 8.16, respectively. The points in the abscissa marked by 1, 2 and 3 denote the cut-off frequencies, and they are at 3, 13.7 and 21 kHz. Two roots are equal before point 1, and they are denoted by $k_{1,2}$. Thus, before point 1, there are only four nonzero real roots ($\pm k_{1,2}$) and eight nonzero imaginary roots ($\pm k_{1,2}$, $\pm k_3$ and $\pm k_4$). After point 1, one of the $k_{1,2}$ becomes pure real and another one becomes pure imaginary, and it is the only imaginary root at high-frequency. These roots correspond to the bending mode, w. It can be further noticed that before point 1, these wavenumbers ($k_{1,2}$) are simultaneously possessing both real and imaginary parts, which implies these modes are attenuated while propagating. Thus, there exists inhomogeneous wave in anisotropic composite plate, [24]. The points marked by 2 and 3 are the two cut-off frequencies, since the roots k_3 and k_4 become real at this point from their imaginary values. These roots correspond to the in-plane motion, that is, u and v displacements.

Next, the asymmetric ply-sequence is considered (Figs. 8.17 and 8.18), for which the wavenumber pattern remains qualitatively same. The cut-off frequencies are coming at 5.3, 13.8 and 60 kHz, where the first one corresponds to the bending mode and the last two correspond to the in-plane motion. In comparison to the symmetric ply-stacking, it can be said that the first and the third cut-off frequencies are of higher magnitude than their symmetric counterparts and the rate of increment is higher in the third cut-off frequency. Further, the magnitudes of all the wavenumbers are increased. Significantly, at higher frequency, the third wavenumber k_3 has lower magnitude than the

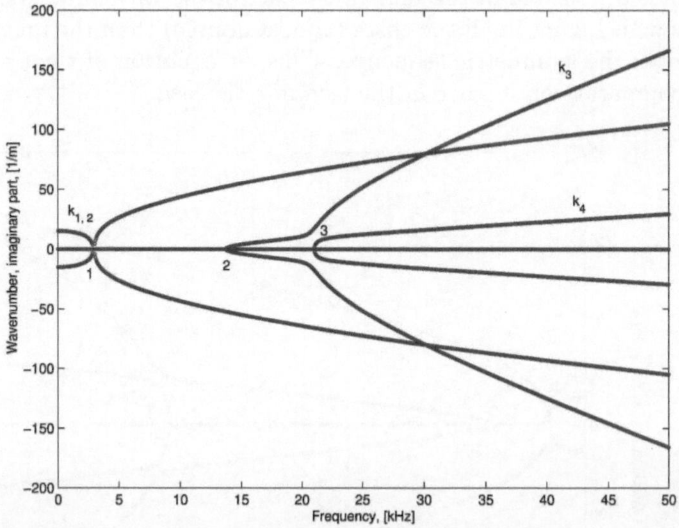

FIGURE 8.15: Real part of wavenumbers, symmetric sequence

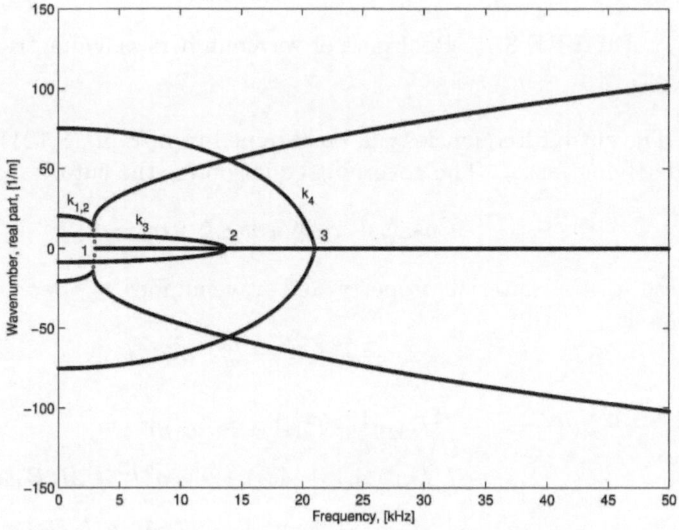

FIGURE 8.16: Imaginary part of wavenumbers, symmetric sequence

bending wavenumbers (one of $k_{1,2}$) as opposed to the symmetric case. Similar trends are visible in the imaginary part of the wavenumbers, where the magnitude is higher in all the cases (almost double) than the imaginary wavenumbers of the symmetric sequence. Thus, attenuation of the propagating modes is comparatively higher in the asymmetric case.

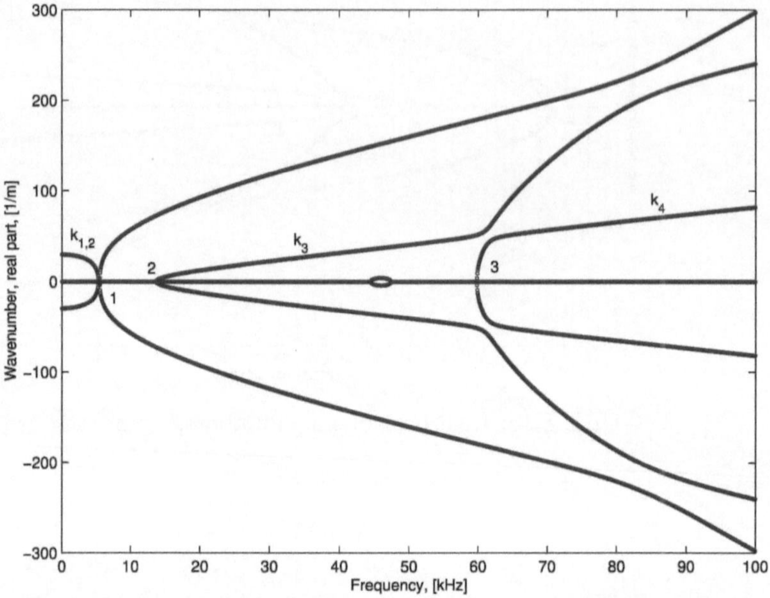

FIGURE 8.17: Real part of wavenumbers, asymmetric sequence

The cut-off frequencies can be obtained from Eqn. (8.121) by letting $k = 0$ and solving for ω_n. The governing equation for the cut-off frequency becomes

$$a_0\omega_n^6 + a_1\omega_n^4 + a_2\omega_n^2 + a_3 = 0\,, \tag{8.122}$$

where a_i are material property and wavenumber η_m dependent coefficients given as

$$a_0 = I_\circ^2 I_2 \eta^2 + I_\circ^3 - I_\circ I_1^2 I_2\,, \tag{8.123}$$

$$\begin{aligned}
a_1 &= -I_\circ^2 D_{22}\eta^4 - I_\circ^2(A_{22} + A_{66})\eta^2 \\
&\quad -I_\circ I_2 \eta^4 (A_{66} + A_{22}) + A_{66}\eta^4 I_1^2 + 2I_\circ B_{22}\eta_4 I_1\,, \\
a_2 &= -I_\circ B_{22}^2 \eta^6 + A_{66}\eta^4 A_{22} I_\circ + A_{66}\eta^6 I_\circ D_{22} \\
&\quad -2A_{66}\eta^6 B_{22} I_1 + I_\circ A_{22}\eta^6 D_{22} + A_{66}\eta^6 A_{22} I_2\,, \\
a_3 &= A_{66}\eta^8(-A_{22}D_{22} + B_{22}^2)\,.
\end{aligned}$$

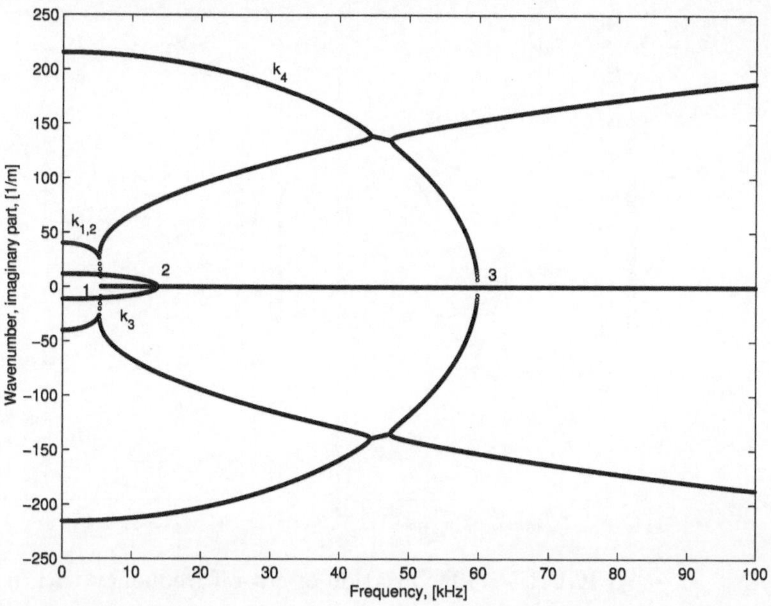

FIGURE 8.18: Imaginary part of wavenumbers, asymmetric sequence

When Eqn. (8.122) is solved for different η_m, the variation of the cut-off frequencies with η_m can be obtained. This variation is given in Fig. 8.19. As shown in the figure, variation of the cut-off frequency for the bending mode, $\omega_{1,2}$, follows a nonlinear pattern, whereas, the other two are increasing linearly. Although apparently not evident from this figure, a close inspection will reveal that the pattern for ω_3 is same for both symmetric and unsymmetric cases. Since, it is the magnitude of A_{12} that has not changed with ply angle, it can be concluded that ω_3 is proportional to the ratio of $\sqrt{A_{12}/\rho}$. Further, there is no variation in $\omega_{1,2}$ for changing ply-stacking, whereas, for ω_4, the effect is maximum. Thus, with the help of this figure, the location of the points 1, 2 and 3 in Figs. 8.15–8.18 can be explained.

Note on MATLAB® scripts provided in this chapter

In this chapter, following MATLAB scripts are provided:

MATLAB script *compwvnum.m* Computes the wavenumbers and group speeds in thin composites waveguide. The *prop* array contains the four material properties and ply angle for n_l number of layers. Using these properties, the program computes the material property Q_{11}, and the stiffness properties A_{11}, B_{11} and D_{11} . These computed properties are then used to determine and plot the wavenumbers and group speeds.

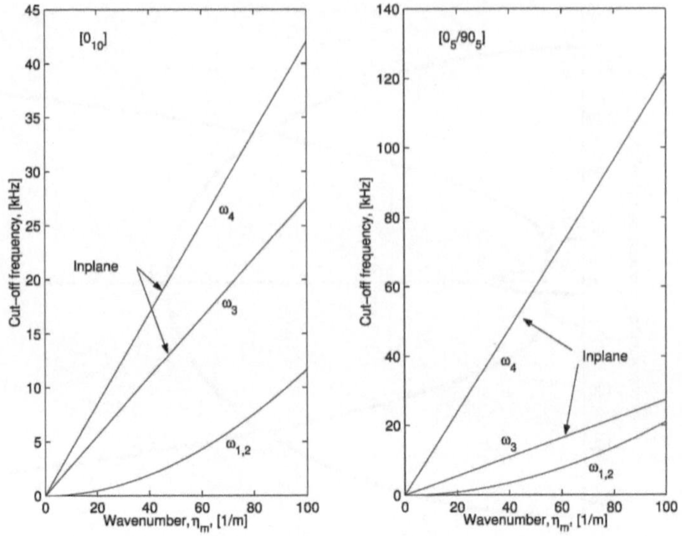

FIGURE 8.19: Variation of cut-off frequency with η_m

MATLAB script *compresponse.m* Computes the wave propagation response in thin laminated composite waveguide subjected to tone burst signal. This program uses the MATLAB function *campmat.m* to compute the material and stiffness properties of the laminate. The program can also be used to determine responses to Gaussian signal. For this program needs to read the file *pul.txt*, which is also provided.

MATLAB script *comphodisp.m* Computes the wavenumber and group speeds for a thick composite unsymmetric waveguide, wherein both higher-order rod and beam approximations are included. It uses the function *test3.m* to compute the coefficients of wave matrix **W**. This function uses MATLAB function*compmat.m* to compute the stiffness properties of the laminate. It also uses the function *detm.m* to compute the coefficients of different orders of wavenumber k in the characteristic polynomial obtained by computing the determinant of the wave matrix **W** symbolically. We have used this method instead of PEP since this program is only plotting wavenumbers and group speeds. If one also requires to determine and plot the responses, then the eigenvectors of the PEP is required to obtain the amplitude ratios.

MATLAB script *comp2dwninplane.m* Computes the wavenumber for a 2-D composite plate subjected to in-plane loading. It uses the properties given in Section 8.6.1 and can plot wavenumber for any given η

MATLAB script *complatewavenum.m* Computes the wavenumber for a 2-D composite plate. This program uses PEP function *polyeig* to compute the wavenumbers

Summary

In this chapter, the first part discusses the constitutive models for laminated composites. In the case of composites, first the micro-mechanics theory is discussed to obtain the material properties at the lamina level. This will lead to derivation of *Rule of Mixtures*. Here, simple strength of materials approach is used to derive the lamina stress-strain relations. Next, the transformation of the lamina constitutive model to the global coordinates is discussed and finally the laminate constitutive model is obtained using Classical Laminate Plate Theory (CLPT).

Following the derivation of the constitutive model, a detailed study on wave propagation in 1-D and 2-D composite waveguides is presented and important physics behind the wave propagation in this media is understood. We saw that the anisotropic nature of construction of laminated composites introduces certain characteristics that are not present in their isotropic counterpart. First, the wave propagation in the Euler-Bernoulli beam was presented, wherein we saw that the stiffness and inertial coupling causing the group speeds of axial and bending modes reducing by nearly 40 percent. Next, wave propagation in thick beam was presented that included the effects of shear deformation, rotary inertia and lateral contraction effects due to Poisson's ratio. Here, it is found the existence of two cut-off frequency, one due to shear deformation and second due to lateral contraction and the cut-frequency shifts if the coupling is changed. Coupling also reduces the group speeds of both axial as well as the bending wave modes. In the last part of the chapter, the wave propagation in 2-D laminated composites is studied, wherein the wavenumbers are obtained for structures with both in-plane loading (membrane type behavior) and out-of-plane loadings (plates). For the composite structures with in-plane loadings, it is shown that the explicit existence of P-wave and S-wave does not occur, and they are usually coupled. Hence, solution using Helmholtz decomposition is not applicable. Here partial wave technique is used to obtain the wave modes as well as the solutions.

The chapter did not provide any plots as regards the response estimation for all the cases. The MATLAB script for determining the responses for the case of thin composite waveguide is provided although these were not presented in this chapter. This code can be used by the reader to understand the additional modes generated by stiffness and inertial coupling in laminated composites. For all the cases of response estimation, it can be easily developed by the reader using the MATLAB scripts provided in the earlier chapters.

Exercises

8.1 An unsymmetric AS/3501-6 thin graphite-epoxy composite waveguide is attached to a spring and as shown in figure below. This is a case where

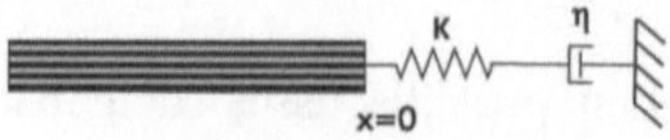

Figure for Problem 8.1

the bending and axial motion will be coupled. Set up the equations to determine the reflected coefficients (a) when the loading is axial (b) when the loading is transverse. Plot the reflected coefficients for both the loading cases as a function of different values of spring constant K, damping coefficient η, ply angles and coupling coefficient r. In the case of transverse loading, assume the wave coefficient associated with second bending wavenumber to be zero. Use the values of the properties for AS/3501-6 graphite -epoxy used in the example problems.

8.2 For the case in previous problem, the orientation of the spring-damper system is rotated by 90° as shown below. Plot the reflected coefficients

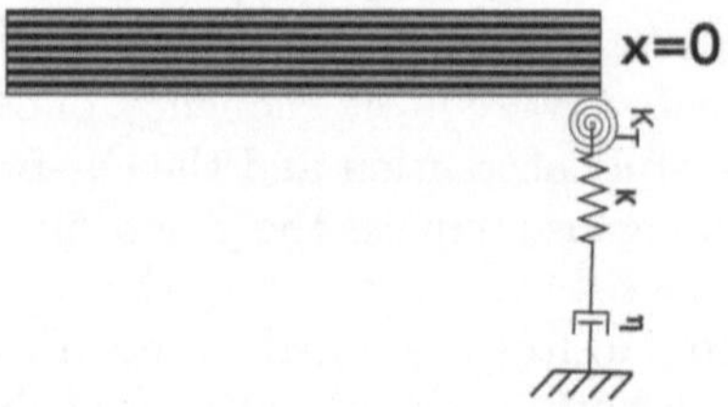

Figure for Problem 8.2

for the cases as indicated in Problem 1 for (a) $K_T = 0$ and K, η not equal to zero (b)$K = 0$ and K_T, η not equal to zero for both loading cases

8.3 A stepped thin composite beam is shown below. Such a structure is called a ply drop structure, wherein the total thickness of composite is dropped from higher value of 10 mm to of lower value 1.0 mm. For this stepped waveguide, set up the equations to determine the reflected and transmission coefficients for (a) axial loading (b) transverse loading. In the case of transverse loading, assume the wave coefficient associated with second bending wavenumber to be zero. Plot these as a function ply angles

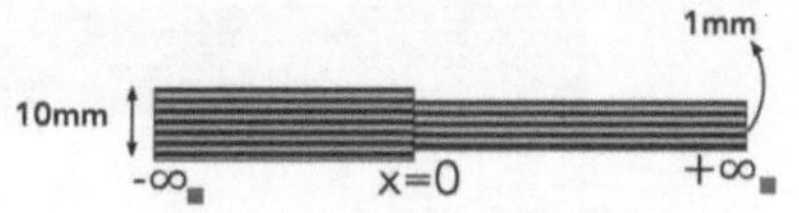

Figure for Problem 8.3

assuming (a) both segments are unsymmetric (b) when both segments are symmetric

8.4 A thin composite waveguide is connected to a mass M as shown in the figure below. Set up the equations to determine the reflected and trans-

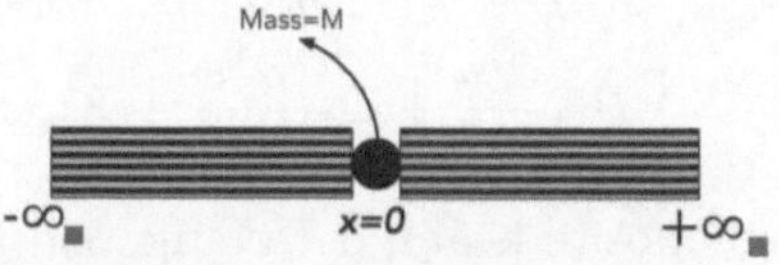

Figure for Problem 8.4

mission coefficients for (a) axial loading (b) transverse loading. In the case of transverse loading, assume the wave coefficient associated with second bending wavenumber to be zero. Plot these as a function ply angles assuming (a) both segments are unsymmetric (b) when both segments are symmetric

8.5 The governing equations for an axially loaded thin laminated composite beam is given by

$$A_{11}\frac{\partial^2 u}{\partial x^2} + B_{11}\frac{\partial^3 w}{\partial x^3} = \rho A\frac{\partial^2 u}{\partial t^2}$$
$$-B_{11}\frac{\partial^3 u}{\partial x^3} + D_{11}\frac{\partial^4 w}{\partial x^4} \pm P\frac{\partial^2 w}{\partial x^2} = \rho A\frac{\partial^2 w}{\partial x^2}$$

where $u(x,t)$ and $w(x,t)$ are the axial and transverse displacement, P is the axial load, which can be either compressive or tensile and the sign of the term associated with term containing P is accordingly changed. A_{11}, B_{11}, D_{11} and ρ are the material properties of the AS/3501-6 composite, which is available in numerical section of this chapter. Determine the spectrum and dispersion relations and plot the same as function of ply stacking sequence for both cases of loading, namely *Pretension* and *Pre-compression*

8.6 A laminated composite thin beam is resting on two-parameter elastic foundations, namely Winkler Foundation with stiffness coefficient K_w

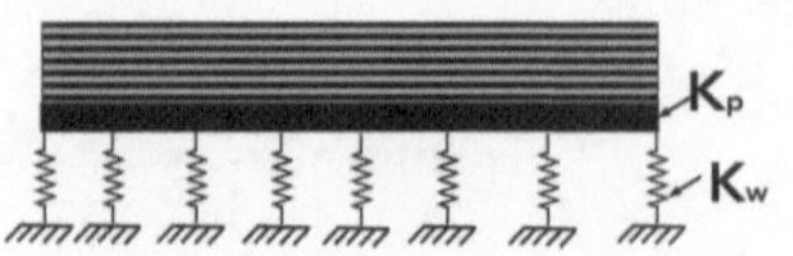

Figure for Problem 8.6

and Pasternak Foundation with stiffness coefficient K_p as shown below The governing differential equation for this 2 parameter foundation is given by

$$A_{11}\frac{\partial^2 u}{\partial x^2} + B_{11}\frac{\partial^3 w}{\partial x^3} = \rho A\frac{\partial^2 u}{\partial t^2}$$

$$-B_{11}\frac{\partial^3 u}{\partial x^3} + D_{11}\frac{\partial^4 w}{\partial x^4} + K_w w - K_p\frac{\partial^2 w}{\partial x^2} = \rho A\frac{\partial^2 w}{\partial x^2}$$

For this composite beam, determine and plot the spectrum and dispersion relation as a function of ply-stacking sequence under the following conditions:

1. When $K_p = 0$ and vary K_w
2. When $K_w = 0$ and vary K_p
3. Vary both K_w and K_p

CHAPTER 9

Wave Propagation in Graded Material Waveguides

This chapter outlines the study of elastic wave propagation in functionally graded materials (FGM) structures. FGM structures are essentially used when two different material structures, whose Young's modulus differ by a large margin, are joined together. When such materials are joined together, then the huge difference in the Young's modulus causes steep stress gradients, which eventually peels one material layer from other. Providing an FGM layer between these two material layer will alleviate the stress gradients and thus will avoid any peel stress developing in the material interface. The properties of FGM layer will vary between the properties of the material layer between which it is introduced, which means that the FGM layer is inhomogeneous in character. That is, the elastic properties and densities will be the function of spatial coordinates. These properties, from the point of view of wave propagation will make the waves inhomogeneous, that is the waves, as it propagates, it will also attenuate (see [21], [144]). That is, one can expect new wave propagation characteristics that is not seen in the isotropic or composite waveguides. In this chapter, the spectral analysis is again used to study the wave behavior in 1-D and 2-D FGM waveguides.

9.1 INTRODUCTION TO FUNCTIONALLY GRADED MATERIALS (FGM)

The idea of FGM was first conceived in Japan in the early eighties, where the material concept was proposed to increase adhesion and minimize the thermal stresses in metallic-ceramic composites. In this material, the percentage content of metal or ceramic is varied in a controlled way to achieve desired

DOI: 10.1201/9781003120568-9

property gradation in spatial direction. FGM is used extensively in thermal insulation and thermal shock resistance system wherein graded Thermal Barrier Coating (TBC) is normally used to prevent failure. This coating is used to shield the metallic structure (copper alloy) from temperature as high as 3500 K (in jet engines in rocket) or 1500 K (in internal combustion engines or gas turbine engines) [125]. Other than TBC systems, FGM is used for wear resistance of cutting tools, tribological applications in automobile industries and oxidation protection systems. They are also finding applications in graded thermoelectric and dielectrics, piezoelectrically graded materials, applied for broad-band ultrasonic transducers (UT) and graded composite electrode for solid oxide fuel cells (SOFC) [101].

FGMs are innovative composite materials whose composition and microstructure vary in space following a predetermined material law. The gradual change in composition and microstructure leads to gradient change of properties and performances [76]. They belong to a class of advanced material characterized by variation in properties as the dimension varies. The overall properties of FGM are unique and different from any of the individual material that forms it. FGM structures have a dual-phase graded layer in which two different constituents are mixed continuously and functionally according to a given volume fraction. In many applications, it is necessary to join metal and ceramic, metal with metal having higher Young' s modulus. The concept of FGM is shown in Fig. 9.1. We have a ceramic material say Zirconium oxide (ZrO2) on the top to be connected to a metal at the bottom as shown in Fig. 9.1(a). This is done by combining these two materials using various fabrication techniques to produce a graded structure shown in Fig 9.1(b).

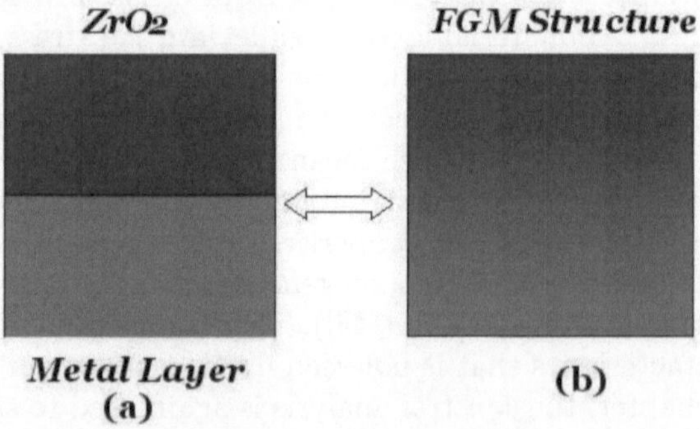

FIGURE 9.1: (a) Joining a metal to ceramic and (b) A FGM structure

Many structural components are subjected to harsh loading environment such as mechanical, thermal or chemical loads that are unevenly distributed across their section. Gradient materials or FGM offer the possibility to combine two materials properties avoiding most of the disadvantages of a bi-material. In contrast, traditional materials such as composites are homogeneous mixtures, and they therefore involve a compromise between the desirable properties of the component materials. Since significant proportions of an FGM contain the pure form of each component, the need for compromise is eliminated. The properties of both components can be fully utilized. For example, the toughness of a metal can be mated with the refractoriness of a ceramic, without any compromise in the toughness of the metal side or the refractoriness of the ceramic side. Consider for example a turbine blade, which must withstand high non-stationary heat fluxes and centrifugal accelerations. An ideal structure for this application would consist of a tough metal core and a heat and corrosion resistant ceramic at the hot surface of the blade. If the ceramic is directly bonded to the metal, spilling may occur during thermal cycling as very high thermal stresses occur at the interface. A gradient material that has a smooth transition from the ceramic surface to the metal core can avoid the thermo mechanical stress concentration at the interface. The gradient material is what is shown in Fig. 9.1(b). The main feature of a gradient material is that its properties changes gradually with position. When used as coatings and interfacial zones, they help to reduce mechanically and thermally induced stresses caused by the material property mismatch and improve the bonding strength.

Fabrication and processing of FGM's are quite involved. The brief outline of some of the fabrication methods is given in the unified book [54] . They are either made as surface coatings or in bulk form. A good overview of fabrication techniques is given in [88]. There are other fabrication methods for functionally graded materials; readers can refer to the review studies by Kieback and Neubrand [72].

9.2 MODELING OF FGM STRUCTURES

Modeling of FGM structures, due to their inherent inhomogeneity, is quite complex and challenging. Several analytical and computational models are available in the literature (see [132], [92]) that discussed the issue of finding suitable functions for modulus variation in inhomogeneous materials. The criteria for selecting them are several. They are desired to be continuous, simple and should have the ability to exhibit curvature, both *concave upward* and *concave downward* ([92]). Here, two types of variations are considered which generally cover all the existing analytical models. The exponential law, which is more common in fracture studies in FGM structures (see [73], [124]), but does not show curvature in both directions. The exponential law can be written as

$$\mathcal{P}(z) = \mathcal{P}_t\, exp(-\delta(1 - 2z/h)), \qquad \delta = \frac{1}{2} log\left(\frac{\mathcal{P}_t}{\mathcal{P}_b}\right). \tag{9.1}$$

The power law, for commonly adopted Voigt type estimate ([92]), having all the desired properties and introduced by Wakashima *et. al.* [146], is given by

$$\mathcal{P}(z) = (\mathcal{P}_t - \mathcal{P}_b)\left(\frac{z}{h} + \frac{1}{2}\right)^n + \mathcal{P}_b\,, \tag{9.2}$$

where $\mathcal{P}(z)$ denotes a typical material property (E, G, α, ρ). $\mathcal{P}_t$ and $\mathcal{P}_b$ denote values of the variables at the topmost and the bottom most layer of the structure, respectively and n is a variable parameter, the magnitude of which determines the curvature. The working range of n is taken as $1/3$ to 3, as any value outside this range will produce an inhomogeneous material having too much of one phase (see [104]).

Another way of estimating material properties is the rule-of-mixture which was derived for composite materials in Section 8.2.1. The concept of equivalent homogeneity yields a three phase composite sphere model, and a composite cylinder model (see [29]). Composite sphere and cylinder models can be improved further by the method as given in [84], which we call it as SBS method. Among all the methods, the method adopted for particle reinforced composite material is best suited for use in most of the cases. In summary, the inhomogeneous materials, such as, FGM can be treated as a matrix-particle mixture of different particle volume fraction, which is smoothly varied along the beam depth. The two different materials at the top and bottom of the beam play the role of the matrix and the particle.

These different models for material property variation are compared in Fig. 9.2, where the variation of Young's modulus along the depth is plotted. Top and bottom materials (particle and matrix, respectively for SBS method) are taken as steel and ceramic with Young's modulus ratio 1.857. The figure clearly shows the different trends of distribution for different models. In the SBS method, *constant area composition* is used and the particle volume fraction V_{p1}, is taken as 0.001. Since, the SBS method predicts only the elastic and thermal properties, in the subsequent calculations where SBS method is used, the inertial properties are evaluated by the power law model with suitable value of the exponent n. We adopt the constitutive model given by Eqn. (9.1) for most of the examples reported in this chapter.

Structural applications of FGM, such as TBC or in UT, etc., demand accurate prediction of their response to high-frequency (thermal/mechanical) loading, that is, the wave propagation analysis. As in the case of isotropic structures, two different cases are to be considered, one, wave propagating in the direction of gradation and second, propagation in the direction normal to the gradation. The analysis procedure is same as that is normally done for isotropic or anisotropic materials case, except for one important difference. In FGM structures, asymmetric (about reference plane) gradation in density may result in first mass moment, which will be absent in composite materials case.

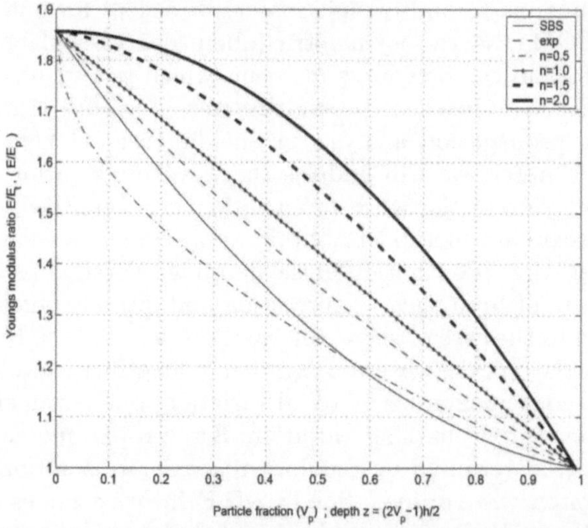

FIGURE 9.2: Variation of Young's modulus for different material models for FGM structure

We begin the FGM wave analysis by writing the form of homogeneous plane wave as

$$f = \Phi(\omega)\exp[i(k_x x + k_y y + k_z z - \omega t)] = \Phi(\omega)\exp[i(\mathbf{k}\cdot\mathbf{r} - \omega t)]\,,$$

where $\Phi(\omega)$ is the wave amplitude (real) and $\mathbf{k}$ is the wave vector (real). This form describes the wave propagation phenomena in homogeneous, linear and non-dissipative solids. Inhomogeneous wave is described by the above form with complex Φ and $\mathbf{k}$ (see [24]).

Although it is known that the plane wave exact solution does not exist for heterogeneous media, the present chapter discusses the possibility of an approximate plane wave solution with complex wavenumber (wave vector in one-dimension) for wave propagating in the direction of inhomogeneity. This wave is called herein as inhomogeneous wave, whose behavior is similar to that of the wave propagating through viscoelastic material, although the present inhomogeneous material (FGM) is linear and elastic.

With the advent of FGM in several structural applications, it is necessary to know their behavior for high-frequency mechanical and thermal loading to which they are frequently subjected. However, as mentioned earlier, two different cases will be considered here, one, when the wave propagating in the direction of gradation and the second, when the propagation is in the direction normal to the gradation. The first case gives rise to an inhomogeneous wave, where the wave amplitude decreases while propagating. This characteristic is not seen in the second case. The analysis procedure is the same as that is normally done for anisotropic materials, except for one important difference. In the FGM case, the asymmetric (about reference plane) gradation in density may result in the first mass moment, which will be absent in the anisotropic materials. It is needless to say that the literature is minimal on the aspect of wave propagation analysis in one-dimensional heterogeneous waveguide. In this chapter, we will address the wave propagation for both these cases. We first address the wave propagation in a lengthwise graded rod and use an approximate method to obtain approximate wavenumbers. Next, we address the wave propagation in the depth wise graded beam, where we consider the effects of both shear deformation and lateral contraction. We reduce the problem to the case of shear deformation case and study the wave characteristics. Following this, wave propagation in lengthwise graded beam is presented, wherein for exponential material variation, one parameter variation, one can obtain exact wavenumber variation. For two-parameter variation, wave characteristics determination requires approximate solutions. In the last part of the chapter, wave propagation in 2-D FGM waveguides is presented, wherein, again, the method of *Partial wave Technique* will be used to obtain the wave parameters, namely the wavenumbers and the group speeds.

9.3 WAVE PROPAGATION IN LENGTHWISE GRADED RODS

The governing differential equation for an inhomogeneous 1-D rod waveguide is derived as follows. A rod with different materials along the length and thickness, shown in Fig. 9.3 is composed of essentially three types of materials. Material 1 and 2 are homogeneous (material properties do not vary with spatial coordinates), whereas the FGM has varying material properties in the x-direction. Moreover, material properties of the FGM is such that it has material 1 at the left edge and material 2 at the right edge. As for the spatial variation, following expressions are assumed for Young's modulus E and density ρ,

$$E(x,z) = E_\circ f(x) g(z), \qquad \rho(x,z) = \rho_\circ s(x) t(z)\ , \tag{9.3}$$

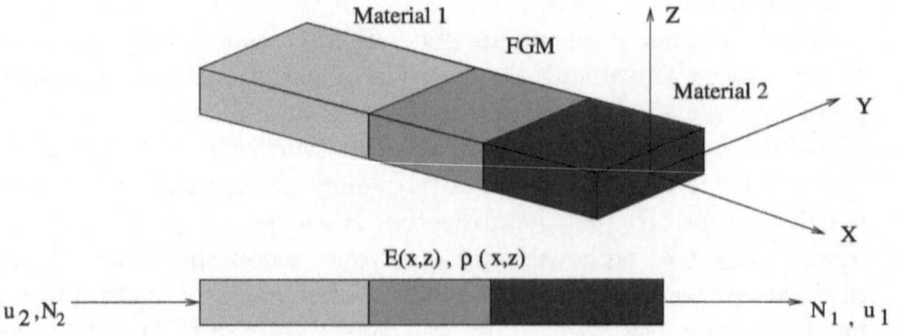

FIGURE 9.3: Lengthwise graded rod, coordinate axes and the stress resultants

where $E_\circ$ and $\rho_\circ$ are constant over length and thickness of the rod. Since only a longitudinal motion is considered, the relevant stress is σ_{xx} and the corresponding longitudinal displacement is u.

The stress is related to the displacement gradient by

$$\epsilon_{xx} = u_x = \frac{\partial u}{\partial x}, \qquad \sigma_{xx} = E(x,z)u_x \;, \tag{9.4}$$

where x as subscript denotes first derivative with respect to the spatial variable x. Applying Hamilton's principle, the governing partial differential equation equation for the FGM rod, in the absence of body force, is given by

$$(A_{11}f(x)u_x)_x = I_\circ s(x)\ddot{u} \;, \tag{9.5}$$

where, for simplicity, the cross-sectional area A is assumed constant over the length and dot over a variable denotes the differentiation with respect to time. The A_{11} and $I_\circ$ are the depthwise (z) integrated properties and are defined as

$$A_{11} = E_\circ \int_A g(z)\,dA, \qquad I_\circ = \rho_\circ \int_A t(z)\,dA \,. \tag{9.6}$$

The natural boundary condition, obtained from the variational principle, is given by

$$A_{11}f(x)u_x = F \;, \tag{9.7}$$

where F is the applied concentrated axial load at the boundary. Although, $g(z)$ and $t(z)$ introduced in Eqn. (9.3) can have arbitrary variations, in the subsequent studies they are kept constant, that is, there is no depthwise variation for E and ρ. This does not simplify the problem to a great extent as any variation is ultimately integrated to obtain the constants A_{11} and $I_\circ$. The lengthwise variation is described by $f(x)$ and $s(x)$, which for polynomial variation are described as:

$$f(x) = (1+\alpha x)^n, \qquad s(x) = (1+\beta x)^n \;, \tag{9.8}$$

and for exponential variation:

$$f(x) = e^{\alpha x}, \qquad s(x) = e^{\beta x} \;. \tag{9.9}$$

Frequency domain analysis starts by transforming the governing equation to the frequency domain using DFT. That is, the field variable can be written as $\tilde{u}(x,\omega_n)e^{i\omega_n t}$. Hence, Eqn. (9.5) becomes

$$(A_{11}f(x)\tilde{u}_x)_x + I_\circ s(x)\omega_n^2\tilde{u} = 0 \;, \tag{9.10}$$

for each value of ω_n. The equation is a classical example of the Sturm-Liouville boundary value problem or SL system whose generic form is given by

$$-d\big[p(x)dy/dx\big]/dx + q(x)y = \lambda r(x)y \;,\; p(x) > 0 \,, r(x) > 0 \;,\; x \in [a,b] \;. \tag{9.11}$$

The non-negative requirements are automatically satisfied for exponential variations laws. For a linear variation, these requirements are satisfied by suitably choosing the values of α and β. The associated boundary conditions are

$$a_1 y(a) + a_2 p(x) y'(a) = 0\,, \qquad b_1 y(b) + b_2 p(x) y'(b) = 0\,. \tag{9.12}$$

For a cantilever rod problem, for example, $a_1\,, b_2 = 1$, $a_2 = 0\,, b_1 = 0$, the system is only satisfied for a discrete set of eigenvalues λ_j with $j = 0, 1, \ldots$, and the corresponding eigenfunctions $y_j(x)$. The SL problem can be solved efficiently by using the Prufer transformation (see [14]), whose details can be found in [25].

Introducing the homogeneous wavenumber $k_\circ$, Eqn. (9.10) becomes

$$f(x)\tilde{u}_{xx} + f'(x)\tilde{u}_x + k_\circ^2 s(x)\tilde{u} = 0, \qquad k_\circ^2 = I_\circ \omega^2 / A_{11}\,. \tag{9.13}$$

The above equation is very similar to the tapered rod wave equation discussed in Section 4.6 and hence one can expect the exact solution to be in terms of Bessel's functions. This equation in terms of linear and exponential material variation laws takes the form

$$\text{Linear } (n=1) : (1+\alpha x)\tilde{u}_{xx} + \alpha \tilde{u}_x + k_\circ^2 (1+\beta x)\tilde{u} = 0\,, \tag{9.14}$$

$$\text{Exponential} : \tilde{u}_{xx} + \alpha \tilde{u}_x + k_\circ^2 e^{\gamma x}\tilde{u} = 0, \qquad \gamma = \beta - \alpha\,. \tag{9.15}$$

There is no closed form solution for the polynomial variation, even in the particular case of $n = 1$ given in Eqn. (9.14). Reference [111] gives this solution in terms of the solution of the degenerate hypergeometric equation, which is again another complicated series solution. For the exponentially varying material property (Eqn. (9.15), the exact solution is also given in [111] and is expressed in terms of the Bessel function of fractional order. This solution gives no notion of wavenumber, which is essential for wave propagation studies and no conclusion can be drawn regarding the effect of inhomogeneity on the dispersion relation. Hence, we introduce a concept of *Naive wavenumber* to understand the physics behind the wave propagation in inhomogeneous waveguides.

The solution of a differential equation with constant coefficients can be written in the form of e^{-ikx}. In the present case, the equation is ODE with variable coefficient. We naively assume that the solution of Eqn. (9.15) still can be written as e^{-ikx} and try to find k. Substituting this assumed solution into Eqn. (9.15), the dispersion relation becomes

$$k^2 + jk\alpha = k_\circ^2 e^{\gamma x}\,. \tag{9.16}$$

The wavenumber k is a function of both x and frequency ω, and it is implicitly assumed that $\partial k / \partial x$ is negligible. We assume that the solution k has a real part a and an imaginary part b. Then the governing equations for a and b are

$$a^2 - b^2 - \alpha b = k_\circ^2 e^{\gamma x}\,, \qquad \alpha a + 2ab = 0\,, \qquad a, b \in \mathbf{R}\,. \tag{9.17}$$

The solution can then be written as

$$k_{1,2} = \pm\Upsilon - i\alpha/2\,, \qquad \Upsilon = \sqrt{|k_\circ^2 e^{\gamma x} - \alpha^2/4|}\,, \tag{9.18}$$

so that the solution of Eqn. (9.15) is

$$\tilde{u}(x) = e^{\alpha x/2}(Ae^{-i\Upsilon x} + Be^{+i\Upsilon x})\,. \tag{9.19}$$

From Eqn. (9.18), we see that the wavenumber is a function of both the space and the frequency. In addition, we see that when $\alpha^2/4 > k_\circ^2 e^{\gamma x}$, the wavenumber will become purely imaginary and one can expect no propagation. This will happen at a value of x given by

$$x < \frac{2\log\alpha - 2\log 2k_\circ}{\beta - \alpha} \tag{9.20}$$

In the above equation, α will always be less than 1, which means Eqn. (9.20) will never be satisfied. This implies that the waves in this waveguide will always be inhomogeneous in nature. Also, in Eqn. (9.19), the first term $e^{\alpha x/2}$ is a hyperbolic term that tends to a large value even for small distances x when α is positive. That is, the solution escapes to infinity even for small distances. This requires that the values of α be really small for meaningful propagation of waves to take place. However, the given solution is not exact and it was introduced to purely bring out the nature of wave propagation one can expect in this type of waveguide.

9.4 WAVE PROPAGATION IN DEPTHWISE GRADED FGM BEAM

Here, we first develop the governing equations considering both the shear deformation and lateral contraction in the model. The effect of temperature is also considered in deriving the equation of motion. The depth wise graded beam along with different motions and stress resultants are shown in Fig. 9.4.

The displacement field for this case is similar to that of the higher-order composite beam, which is given by

$$u(x,y,z,t) = u^o(x,t) - z\phi(x,t)\,, \quad w(x,y,z,t) = w(x,t) + z\psi(x,t) \tag{9.21}$$

The linear strains in the presence of temperature field, obtained from Eqn. (9.21) are

$$\begin{aligned} \epsilon_{xx} &= u_{,x} - z\phi_{,x} - \alpha(z)T(z)\,, \epsilon_{zz} = \psi - \alpha(z)T(z) \\ \gamma_{xz} &= -\phi + w_{,x} + z\psi_{,x}\,, \end{aligned} \tag{9.22}$$

where α and T are the depth dependent coefficient of thermal expansion and temperature field, respectively and $(.)_{,x}$ represents $(\partial/\partial x)$. The nonzero

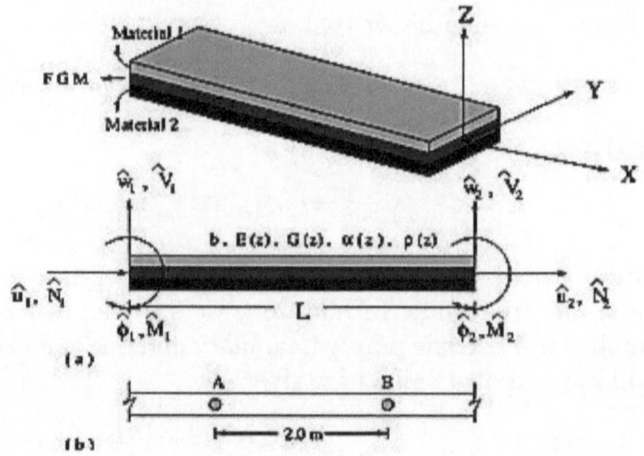

FIGURE 9.4: Lengthwise graded beam, coordinate axes and the stress resultants

stresses are related to these strains by

$$\boldsymbol{\sigma} = \left\{ \begin{array}{c} \sigma_{xx} \\ \sigma_{zz} \\ \tau_{xz} \end{array} \right\} = \left[\begin{array}{ccc} \bar{Q}_{11}(z) & \bar{Q}_{13}(z) & 0 \\ \bar{Q}_{13}(z) & \bar{Q}_{33}(z) & 0 \\ 0 & 0 & \bar{Q}_{55}(z) \end{array} \right] \left\{ \begin{array}{c} \epsilon_{xx} \\ \epsilon_{zz} \\ \gamma_{xz} \end{array} \right\} = \bar{\mathbf{Q}}(\mathbf{z})\boldsymbol{\epsilon} \tag{9.23}$$

where the $\bar{Q}_{ij}(z)$s are the depth dependent elements of the constitutive matrix. For inhomogeneous (but isotropic) materials, the $\bar{\mathbf{Q}}$ becomes

$$\frac{E(z)}{1-\nu^2} \left[\begin{array}{ccc} 1 & \nu & 0 \\ \nu & 1 & 0 \\ 0 & 0 & (1-\nu)/2 \end{array} \right] \tag{9.24}$$

where $E(z)$ and ν are the Young's modulus and the Poisson's ratio. Here, the Poisson's ratio is assumed constant. However, the Young's modulus and the density ρ, vary over the depth of the beam.

Following the regular procedure of using Hamilton's principle, the four governing equations corresponding to four dof, u, ψ, w and ϕ are

$$I_o\ddot{u} - I_1\ddot{\phi} - A_{11}u_{,xx} + B_{11}\phi_{,xx} - A_{13}\psi_{,x} = 0\,, \tag{9.25}$$

$$k_I I_2\ddot{\psi} + I_1\ddot{w} + A_{31}u_{,x} - B_{31}\phi_{,x} + A_{33}\psi - B_{55}\left(w_{,xx} - \phi_{,x}\right) - k_d D_{55}\psi_{,xx} - L_T = 0\,, \tag{9.26}$$

$$I_o\ddot{w} + I_1\ddot{\psi} - A_{55}\left(w_{,xx} - \phi_x\right) - B_{55}\psi_{,xx} = 0\,, \tag{9.27}$$

$$I_2\ddot{\phi} - I_1\ddot{u} - A_{55}\left(w_{,x} - \phi\right) - B_{55}\psi_{,x} + B_{11}u_{,xx} - D_{11}\phi_{,xx} + B_{13}\psi_{,x} = 0\,, \tag{9.28}$$

and the four associated force boundary conditions are

$$
\begin{aligned}
&A_{11}u,_x - B_{11}\phi,_x + A_{13}\psi - N_T = N_x \\
&B_{55}\,(w,_x - \phi) + k_d D_{55}\psi,_x = Q_x \qquad (9.29)\\
&A_{55}\,(w,_x - \phi) + B_{55}\psi,_x = V_x. \\
&B_{11}u,_x + D_{11}\phi,_x - B_{13}\psi + M_T = M_x \,. \qquad (9.30)
\end{aligned}
$$

In the above equations, the stiffness coefficients and the mass moments integrated over the cross-section are defined in the same way as before (see Eqns (8.84 & 8.85)).

In the above equations, k_I and k_d are the correction factors introduced to compensate for the approximations introduced in the analysis [46]. Compared to the equation governing thick laminated composite beams, the governing equations and force boundary conditions of a depth wise graded beams also have terms relating to thermal properties. Otherwise, the form of equations looks similar.

The contributions from the temperature field come from the thermal forces N_T, M_T and L_T defined as

$$
\begin{aligned}
&[N_T\,,\ M_T] = \int_{z_1}^{z_2} \alpha(z)\{Q_{11}(z) + Q_{13}(z)\}[T(z)\,,\ zT(z)]\,bdz \\
&L_T = \int_{z_1}^{z_2} \alpha(z)\{Q_{13}(z) + Q_{33}(z)\}T(z)\,bdz \qquad (9.31)
\end{aligned}
$$

It is to be noted that the effect of the temperature field is limited to the force boundary conditions in the FSDT theory. However, in the present formulation, the Poisson's contraction results in a term in the governing equation (L_T), which can be treated as a body force.

9.4.1 Reduction to FSDT

The displacement field of FSDT is obtained by omitting the contractional degrees of freedom ψ from the description of W as

$$
U(x,y,z,t) = u^\circ(x,t) - z\phi(x,t)\,,\ W(x,y,z,t) = w^\circ(x,t)\,. \qquad (9.32)
$$

The resulting governing PDEs in terms of the three degrees of freedom (u°, w° and ϕ) are:

$$
\begin{aligned}
&\delta u: && I_\circ \ddot{u}^\circ - I_1\ddot{\phi} - A_{11}u^\circ_{,xx} + B_{11}\phi,_{xx} = 0\,, && (9.33)\\
&\delta w^\circ: && I_\circ \ddot{w}^\circ - A_{55}\left(w^\circ_{,xx} - \phi_x\right) = 0\,, && (9.34)\\
&\delta\phi: && I_2\ddot{\phi} - I_1\ddot{u}^\circ + B_{11}u^\circ{,_{xx}} - D_{11}\phi,_{xx} && \\
& && -A_{55}\left(w^\circ_{,x} - \phi\right) = 0\,. && (9.35)
\end{aligned}
$$

The associated force boundary conditions are

$$A_{11}u^{\circ}{}_{,x} - B_{11}\phi_{,x} - N_T = N_x \tag{9.36}$$

$$A_{55}\left(w^{\circ}{}_{,x} - \phi\right) = V_x \tag{9.37}$$

$$-B_{11}u^{\circ}{}_{,x} + D_{11}\phi_{,x} + M_T = M_x \tag{9.38}$$

where all the coefficients and mass moments are as defined before.

Next, we will determine the wave parameter for this depth-wise graded higher-order beam. For wavenumber computation, the governing equations will be considered in absence of any body force and temperature rise. We seek plane wave type solution, where the displacement field, $\{\mathbf{u}\} = \{u\ ,\ \psi\ ,\ w\ ,\ \phi\}(x,t)$, can be written as

$$\mathbf{u} = \sum_{n=1}^{N}\{\tilde{u}\ ,\ \tilde{\psi}\ ,\ \tilde{w}\ ,\ \tilde{\phi}\}(x)e^{i\omega_n t} = \sum_{n=1}^{N}\tilde{\mathbf{u}}(x)e^{i\omega_n t}, \tag{9.39}$$

where ω_n is the circular frequency at the nth sampling point and $i^2 = -1$. The N is the frequency index corresponding to the Nyquist frequency. When Eqn. (9.39) is substituted in the governing equations, a set of ordinary differential equations is obtained for $\tilde{\mathbf{u}}(\mathbf{x})$. Since, the ODEs are of constant coefficient type, the solution is of the form $\tilde{\mathbf{u}}_{\circ}e^{-ikx}$, where k is the wavenumber and $\tilde{\mathbf{u}}_{\circ}$ is a vector of unknown constants, that is, $\tilde{\mathbf{u}}_{\circ} = \{u_{\circ}\ ,\ \psi_{\circ}\ ,\ w_{\circ}\ ,\ \phi_{\circ}\}$. Substituting the assumed form in the set of ODEs, a matrix vector relation is obtained

$$\mathbf{W}\tilde{\mathbf{u}}_{\circ} = \mathbf{0}, \tag{9.40}$$

where

$$\mathbf{W} = \begin{bmatrix} W_{11} & W_{12} & W_{13} & W_{14} \\ W_{21} & W_{22} & W_{23} & W_{24} \\ W_{31} & W_{32} & W_{33} & W_{34} \\ W_{41} & W_{42} & W_{43} & W_{44} \end{bmatrix} \tag{9.41}$$

And

$$\begin{aligned}
&W_{11} = A_{11}k^2 - I_{\circ}\omega_n^2, \quad W_{12} = iA_{13}k, \quad W_{13} = 0,\\
&W_{14} = I_1\omega_n^2 - B_{11}k^2, \quad W_{21} - iA_{13}k, \quad W_{22} = -k_I I_2\omega_n^2 + A_{33} + k_d D_{55}k^2\\
&W_{23} = -I_1\omega_n^2 + B_{55}k^2, \quad W_{24} = -iB_{55}k + iB_{13}k, \quad W_{31} = 0\\
&W_{32} = -I_1\omega_n^2 + B_{55}k^2, \quad W_{33} = A_{55}k^2 - I_{\circ}\omega_n^2, \quad W_{34} = -iA_{55}k\\
&W_{41} = I_1\omega_n^2 - B_{11}k^2, \quad W_{42} = iB_{55}k - iB_{13}k, \quad W_{43} = iA_{55}k\\
&W_{44} = -I_2\omega_n^2 + D_{11}k^2 + A_{55}
\end{aligned}$$

According to the discussion in Section 3.5.2, in this case, the order of the PEP $p = 2$ and N_v (size of the $\mathbf{W}$) is 4. Thus, there are 8 eigenvalues altogether, which are the roots of the polynomial obtained from the singularity condition of $\mathbf{W}$ as

$$Q_1k^8 + Q_2k^6 + Q_3k^4 + Q_4k^2 + Q_5 = 0\,.$$

The spectrum relation suggests that the roots can be written as $\pm k_1$, $\pm k_2$, $\pm k_3$ and $\pm k_4$. Variation of these roots with ω_n is discussed next for a particular material distribution.

For the FSDT case, the wave matrix $\mathbf{W}$ is given by

$$\mathbf{W} = \begin{bmatrix} A_{11}k^2 - I_o\omega_n^2 & 0 & -B_{11}k^2 + I_1\omega_n^2 \\ -B_{11}k^2 + I_1\omega_n^2 & iA_{55}k & D_{11}k^2 + A_{55} - I_2\omega_n^2 \\ 0 & A_{55}k^2 - I_o\omega_n^2 & -iA_{55}k \end{bmatrix},$$

where all the integrated parameters are as defined before. Thus, in this case, $N_v = 3$ and $p = 2$, i.e., there are 6 wavenumbers (eigenvalues).

The coefficients of the spectrum relation in both the cases are real valued and frequency dependent although the roots can be complex for some values of ω_n. There are no closed form solutions of these roots and they must be found numerically, and they are solved by the method of companion matrix explained in Section 3.5.1. Once the variation of the wavenumbers with frequency is known, the variation of group velocities C_g^i, defined as $\Re(d\omega/d\ k_i)$ can be computed numerically, where $\Re$ denotes the real part of a complex number. In the following example, we take a beam structure and analyze it for its spectrum and dispersion relation and draw several important conclusions.

The beam considered here is assumed to have 0.001 m width and 0.05 m depth. There are three layers in the beam. The top layer is made up of steel and 0.01 m thick. The bottom layer is made up of ceramic and 0.031 m thick. In between, there is an FGM layer of 0.009 m thickness, in which properties vary smoothly from that of steel to ceramic according to the power law, where the exponent n takes different values. Material properties of steel is taken as $E = 210$ GPa, $\rho = 7800$ kg/m^3, while that of ceramic is $E = 390$ GPa and $\rho = 3950$ kg/m^3. It well known that FGM behavior is also dictated by the exponent n in its constitutive model given in Eqn. (9.2). Here, we plot the wavenumbers and group speeds first for $n = 1$ and subsequently $n = 1/3$ and $n = 3$. The plots of wavenumber and group speeds for $n = 1$ shown in Fig. 9.5. In the above figure, k_a, k_b, k_c and k_s denote axial, bending, contraction and shear wavenumber, respectively, while C_a, C_b, C_c and C_s denote their corresponding group speeds

In the case of material properties defined by the power law, the power exponent n in Eqn. (9.2) can influence the wave behavior. First, we plot these relations for $n = 1$ and these plots are shown in Fig. 9.5. Fig. 9.5 (a) shows the spectrum relation for this beam with $n = 1$ as stated before. To elicit the difference between the higher-order beam with lateral contraction and designated as HMT and the FSDT, the spectrum relation for the same beam in terms of FSDT is plotted (in dashed lines) in the same figure. In the figure, k_a, k_b, k_s and k_c denote the axial, bending, shear and the contraction wavenumbers. The wavenumbers plotted in the negative side of the ordinate denote the imaginary part of the wavenumbers. The frequencies, where the imaginary wavenumbers become real are the cut-off frequencies. More discussions

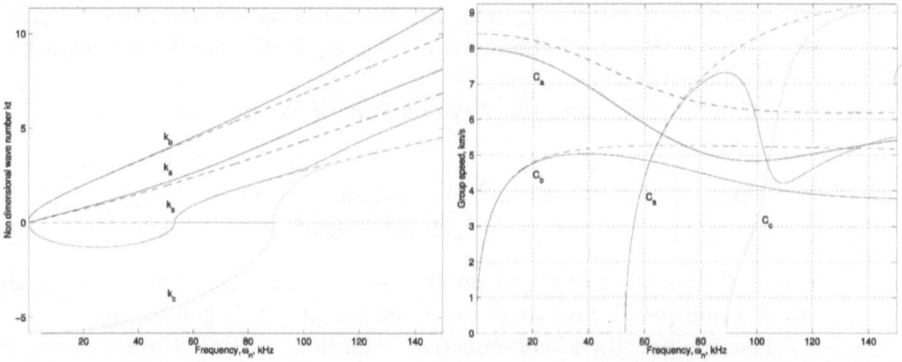

(a) Wavenumber $n = 1$ (b) Group Speed $n = 1$

FIGURE 9.5: Spectrum and Dispersion relation for higher-order and FSDT FGM beam for exponent $n = 1$

on these frequencies are given next. As is seen from the figure, the bending mode of FSDT match with the bending mode of HMT up to 40 kHz. There is always a difference in the axial modes of HMT and FSDT, which is more pronounced above 20 kHz. Although this difference is not clear in this figure, for the frequencies below 20 kHz, the dispersion relation (Fig. 9.5 (b)) will reveal this feature with more clarity. In both the axial and bending modes, the HMT predicts higher gradient of the wavenumbers. The shear modes of the HMT and the FSDT match well up to 90 kHz. After this frequency, the shear mode of FSDT overlap with the contraction mode of the HMT. However, the dispersion of these modes are quite different.

The dispersion relation of this beam model for both the HMT and the FSDT is plotted in Fig. 9.5 (b). The figure suggests that there is always a difference between the axial speeds predicted by the HMT and the FSDT. This difference is maximum near the second cut-off frequency and later decreases. The bending speed predicted by the HMT and the FSDT is equal up to 20 kHz and beyond this frequency, the difference between the two increases. The shear speeds are equal up to about 70 kHz but later they behave quite differently. The shear speed predicted by the HMT dips and slowly regains a lower value than that of FSDT. This behavior can be attributed to the presence of the contraction mode. The contraction speed, at high frequencies match with the shear speed of the FSDT. It is important to note that, there is propagating axial and bending modes for all frequencies, whereas, shear and contraction mode appear only when the frequency exceeds the respective cut-off frequencies.

Next, we plot the spectrum and dispersion relations for $n = 1/3$ and $n = 3$. It would be interesting see how the wavenumber and group speed vary for fractional values of exponent and for very high values of exponent. These plots are shown in Fig. 9.6.

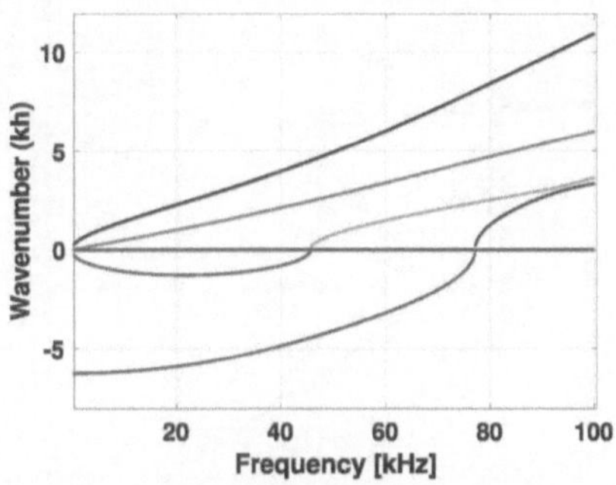

(a) Wavenumber $n = 1/3$

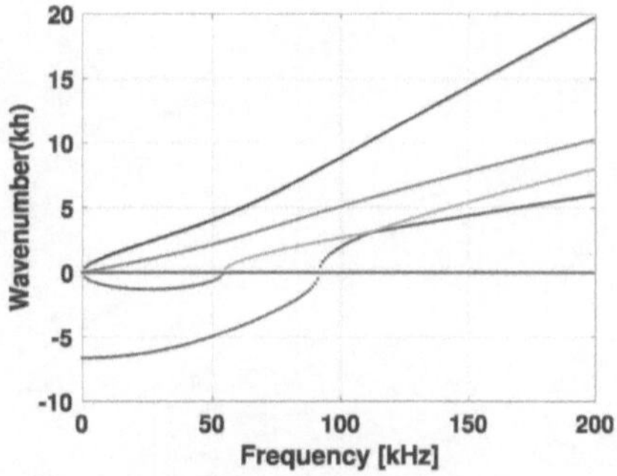

(b) Wavenumber $n = 3$

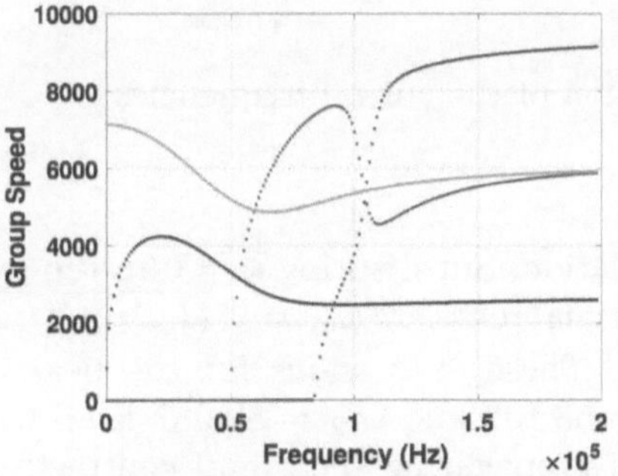

(c) Group Speed $n = 1/3$

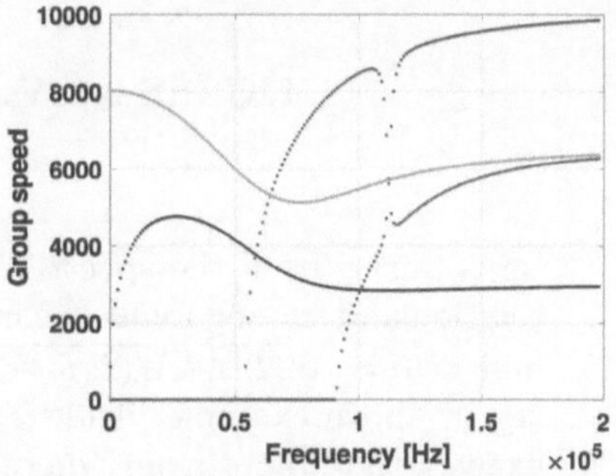

(d) Group Speed $n = 3$

FIGURE 9.6: Spectrum and Dispersion relation for higher-order and FSDT FGM beam for exponent $n = 1/3$ and $n = 3$

For, $n = 1/3$ and $n = 3$, higher the value of n, does not shift the shear cut-off frequency, however, we can see a small change of contraction mode cut-off frequency. Fractional value of exponent reduces the axial mode speed by around 10% while the bending mode speed does not change much. Substantial difference in the shear mode can be visible. That is for $n = 3$, the shear mode increases up to 8300 m/sec and drops sharply to 5000 m/sec around 110 kHz. In the case of $n = 1/3$, the shear mode rises only up to 7500 m/sec and the speed drop at around 105 kHz, which is much smaller than for the case of $n = 3$ and is equal to 2200 m/sec. Hence, change of value of exponent changes mostly the shear mode.

Since the propagating modes appear only when the loading frequency exceeds the cut-off frequency, variation of the later with FGM parameters is important for response prediction. Explicit forms of the cut-off frequencies can

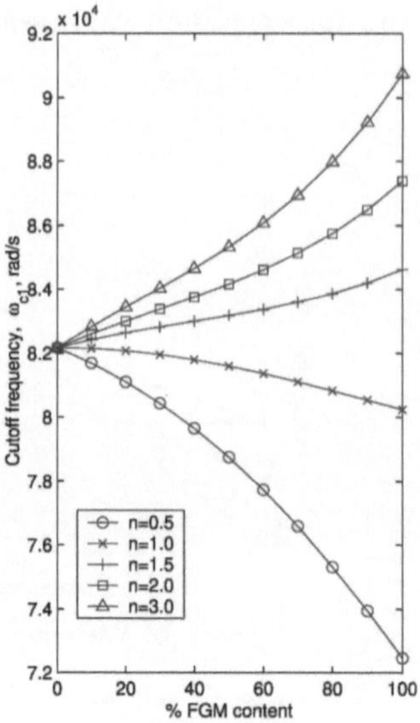

FIGURE 9.7: Variation of the cut-off frequencies

be obtained from the spectrum relation. Substituting $k = 0$ in the spectrum relation and solving for ω, the nontrivial roots are $\omega_{c1} = \sqrt{(I_\circ A_{55})/(I_\circ I_2 - I_1^2)}$ and $\omega_{c2} = \sqrt{(I_\circ A_{33})/(I_\circ I_2 - I_1^2)}$. These expressions for the beam geometry in above example yield, $\omega_{c1} = 53.132$ and $\omega_{c2} = 89.087$ kHz, which are exactly the points from where the propagating shear and contraction mode generate.

As the expressions for the cut-off frequencies suggest, they depend upon the FGM content of the beam (quantified by h_{fgm}/h, where h_{fgm} is the thickness of the FGM layer and h is the total beam thickness) and the gradation in FGM layer (that is, the exponent n for power law variation in Eqn. (9.2)). For the same beam, the thickness of the FGM layer is varied from 0% to 100% for a range of values of the parameter n. The variation is plotted in Fig. 9.7. From the figure, we see that ω_{c1} shows a nonlinear variation with FGM content, whereas, ω_{c2} shows the linear variation. Both the frequencies decrease with the FGM content when n is less than 1. When $n = 1$, ω_{c2} becomes independent of the FGM content, whereas, ω_{c1} decreases slowly. Although it is not shown here, ω_{c1} becomes independent of the FGM content for $n = 1.2$. For all values of n above 1, the cutoff frequencies increase monotonically with the FGM content. Hence, it is an added advantage of using FGM, where propagation of higher-order modes can be suppressed by increasing the FGM content, thereby increasing the range of validity of the FSDT.

9.5 WAVE PROPAGATION ON LENGTHWISE GRADED BEAM

In this case, we consider only the shear deformation and hence adopt FSDT based formulation. In the previous formulation based on FSDT theory, the material properties were assumed constant over the length of the beam, which is not the case in the lengthwise graded beam. In this section, we extend the previous formulation of depth-wave graded beam to the present case, which will bring out several new features of the inhomogeneous wave. In this formulation, the nonzero stresses are related to the nonzero strains by

$$\boldsymbol{\sigma} = \begin{Bmatrix} \sigma_{xx} \\ \tau_{xz} \end{Bmatrix} = f(x) \begin{bmatrix} \bar{Q}_{11}(z) & 0 \\ 0 & \bar{Q}_{55}(z) \end{bmatrix} \begin{Bmatrix} \epsilon_{xx} \\ \gamma_{xz} \end{Bmatrix} = f(x)\mathbf{Q}\boldsymbol{\epsilon} \tag{9.42}$$

where $f(x)$ denotes the x dependency of the inhomogeneity, which in general can be any function. Similarly, density of the material is also assumed to vary in both x and z direction as

$$\rho(x,z) = \rho_o(z)s(x) \, . \tag{9.43}$$

The governing equations in terms of the unknown displacement field are obtained after using the usual Hamilton's procedure, which are given by

$$(f(x)A_{11}u_x - f(x)B_{11}\phi_x)_x - I_o s(x)\ddot{u} + I_1 s(x)\ddot{\phi} = 0 \, , \tag{9.44}$$

$$-(f(x)B_{11}u_x - f(x)D_{11}\phi_x)_x + f(x)A_{55}\,(w_x - \phi) \, .$$
$$-I_2 s(x)\ddot{\phi} + I_1 s(x)\ddot{u} = 0 \, , \tag{9.45}$$

$$(f(x)A_{55}\,(w_x - \phi))_x - I_o s(x)\ddot{w} = 0 \, , \tag{9.46}$$

and the three associated force boundary conditions are

$$f(x)(A_{11}u_x - B_{11}\phi_x) = N_x \tag{9.47}$$

$$f(x)A_{55}\,(w_x - \phi) = V_x \tag{9.48}$$

$$-f(x)B_{11}u_x + f(x)D_{11}\phi_x = M_x \tag{9.49}$$

where N_x, V_x and M_x are the applied nodal axial force, shear force and bending moments, respectively and *subscript* x indicates the derivative with respect to x. The stiffness coefficients and the mass moments are as defined before.

The governing PDEs, Eqns. (9.44)–(9.46) are not solvable readily for arbitrary $f(x)$ and $s(x)$ and some assumption on their forms is necessary before proceeding further. In this study, exponential variation is assumed for both the functions, that is,

$$f(x) = e^{\alpha x} \, , \quad s(x) = e^{\beta x} \, , \tag{9.50}$$

where α and β are the inhomogeneous parameters, which control the gradation. They may or may not be equal to each other. However, when they are equal, the governing PDEs are exactly solvable as they become equations with constant coefficients. When $\beta \neq \alpha$, some approximation is necessary to keep the equations in the same constant coefficient form.

Substituting Eqn. (9.50) in Eqns. (9.44)–(9.46), the new set of governing PDEs are obtained, which are given by

$$A_{11}(\alpha u_x + u_{xx}) - B_{11}(\alpha\phi_x + \phi_{xx}) - I_\circ\gamma\ddot{u} + I_1\gamma\ddot{\phi} = 0 \quad (9.51)$$

$$-B_{11}(\alpha u_x + u_{xx}) + D_{11}(\alpha\phi_x + \phi_{,xx}) + A_{55}(w_x - \phi).$$

$$-I_2\gamma\ddot{\phi} + I_1\gamma\ddot{u} = 0 \quad (9.52)$$

$$A_{55}(\alpha(w_x - \phi) + w_{xx} - \phi_{,x}) - I_\circ\gamma\ddot{w} = 0 \quad (9.53)$$

where $\gamma = e^{(\beta-\alpha)x}$ and double subscript xx indicates the double derivative. When $\alpha = \beta$, $\gamma = 1$, the equations are exactly solvable in frequency domain. When $\alpha \neq \beta$, γ can be evaluated approximately at some representative point in the element, x_c, as $\gamma = e^{(\beta-\alpha)x_c}$, thus rendering the equations once again solvable exactly.

For the wavenumber computation, the new set of governing equations, Eqns. (9.51)–(9.53) will be considered. The wave matrix ($N_v = 3, p = 2$) with $\ell^2 = k^2 + ik\alpha$ becomes,

$$\mathbf{W} = \begin{bmatrix} A_{11}\ell^2 - I_\circ\gamma\omega_n^2 & 0 & -B_{11}\ell^2 + I_1\gamma\omega_n^2 \\ -B_{11}\ell^2 + I_1\gamma\omega_n^2 & jA_{55}k & D_{11}\ell^2 + A_{55} - I_2\gamma\omega_n^2 \\ 0 & A_{55}\ell^2 - I_\circ\gamma\omega_n^2 & -A_{55}(ik - \alpha) \end{bmatrix} \quad (9.54)$$

which yields the spectrum relation as

$$Q_1k^6 + Q_2k^5 + Q_3k^4 + Q_4k^3 + Q_5k^2 + Q_6k + Q_7 = 0\,. \quad (9.55)$$

Variation of these roots with ω_n is discussed in the next for a particular material. The coefficients $Q_1 - Q_7$ are

$$\begin{aligned}
Q_1 &= A_{11}D_{11}A_{55} - B_{11}^2A_{55}, \qquad Q_2 = 3iA_{11}\alpha D_{11}A_{55} - 3iB_{11}^2\alpha A_{55} \\
Q_3 &= -3A_{11}D_{11}\alpha^2A_{55} - A_{11}D_{11}I_\circ\gamma\omega^2 + 2B_{11}I_1\gamma\omega^2A_{55} \\
&\quad -I_\circ\gamma\omega^2D_{11}A_{55} + 3B_{11}^2\alpha^2A_{55} - A_{11}\gamma I_2\omega^2A_{55} + \\
&\quad +B_{11}^2I_\circ\gamma\omega^2 \\
Q_4 &= 2iB_{11}^2\alpha I_\circ\gamma\omega^2 - 2iI_\circ\gamma\omega^2D_{11}\alpha A_{55} + iB_{11}^2\alpha^3A_{55} \\
&\quad +4jB_{11}\alpha I_1\gamma\omega^2A_{55} - jA_{11}\alpha^3D_{11}A_{55} \\
&\quad -2iA_{11}\alpha D_{11}I_\circ\gamma\omega^2 - 2jA_{11}\alpha\gamma I_2\omega^2A_{55}
\end{aligned}$$

$$
\begin{aligned}
Q_5 &= -I_1^2\gamma^2\omega^4 A_{55} - A_{11}A_{55}I_\circ\gamma\omega^2 + I_\circ^2\gamma^2\omega^4 D_{11} \\
&\quad + A_{11}\alpha^2 D_{11} I_\circ\gamma\omega^2 - 2B_{11}\alpha^2 I_1\gamma\omega^2 A_{55} + I_\circ\gamma^2\omega^4 I_2 A_{55} \\
&\quad + A_{11}\gamma^2 I_2\omega^4 I_\circ - 2B_{11}I_1\gamma^2\omega^4 I_\circ + I_\circ\gamma\omega^2 D_{11}\alpha^2 A_{55} \\
&\quad - B_{11}^2\alpha^2 I_\circ\gamma\omega^2 + A_{11}\alpha^2\gamma I_2\omega^2 A_{55} \\
Q_6 &= -2iB_{11}\alpha I_1\gamma^2\omega^4 I_\circ - iI_1^2\gamma^2\omega^4 A_{55}\alpha + iA_{11}\alpha\gamma^2 I_2\omega^4 I_\circ \\
&\quad + iI_\circ\gamma^2\omega^4 I_2 A_{55}\alpha + iI_\circ^2\gamma^2\omega^4 D_{11}\alpha - iA_{11}\alpha A_{55} I_\circ\gamma\omega^2 \\
Q_7 &= -I_\circ^2\gamma^3\omega^6 I_2 + I_1^2\gamma^3\omega^6 I_\circ + I_\circ^2\gamma^2\omega^4 A_{55}
\end{aligned}
$$

The spectrum relation is solved using the companion matrix method described in Chapter 3.. A steel beam with following material properties is considered: Young's modulus $E = 210$ GPa, shear modulus $G = 80.76$ GPa, Poisson's ratio $\nu = 0.3$ and density $\rho = 7800$ kg/m^3. The beam has 0.1 m width and 0.1 m thickness and is considered homogeneous in the thickness direction and inhomogeneous in the longitudinal direction. The inhomogeneous parameters α and β (with $\beta = \alpha$) are varied from 0 to 30 in four steps.

Fig. 9.8 shows the spectrum relation of the beam for different values of the α and β. In the figure, k_a, k_b and k_s denote the axial, bending and shear wavenumbers. The wavenumbers plotted in the negative side of the ordinate denote the imaginary part of the wavenumbers. The frequency at which the imaginary component of the wavenumber changes to real is called the cut-off frequency. For $\alpha = 0 = \beta$, the beam is homogeneous and the wavenumbers are exactly the solutions of the spectrum equation known previously and discussed in Chapter 4. There is one property of the wavenumbers, which requires attention. That is, the wavenumbers do not posses nonzero real and imaginary parts, simultaneously. However, the situation changes dramatically, when the nonzero values are assigned to α (and β). As the figure suggests, with increase in the magnitude of α (as indicated by the arrows), cut-off frequencies appear for all the modes. The wavenumbers simultaneously posses both nonzero real and imaginary part, which implies attenuation of the wave magnitude while propagating. At high frequencies, real part of the wavenumbers converge to their homogeneous counterpart, whereas, the imaginary parts take the value of $\alpha/2$. Also, it is evident that the effect of α is more pronounced in axial mode as shown by comparatively large shifting of the axial cut-off frequency.

The dispersion relation is plotted in Fig. 9.9, where axial, bending and shear modes are denoted by C_a, C_b and C_s, respectively. For a given frequency, the presence of positive group speed indicates propagation of that particular mode. As the figure suggests, for $\alpha = 0$, the axial and the bending modes are propagating, whereas, the shear mode is propagating only when the frequency exceeds the cut-off frequency, before which it does not exist. However, nonzero α (and β) introduces cut-off frequencies in axial and bending modes, which means, if the highest frequency content of the loading is less than the lowest

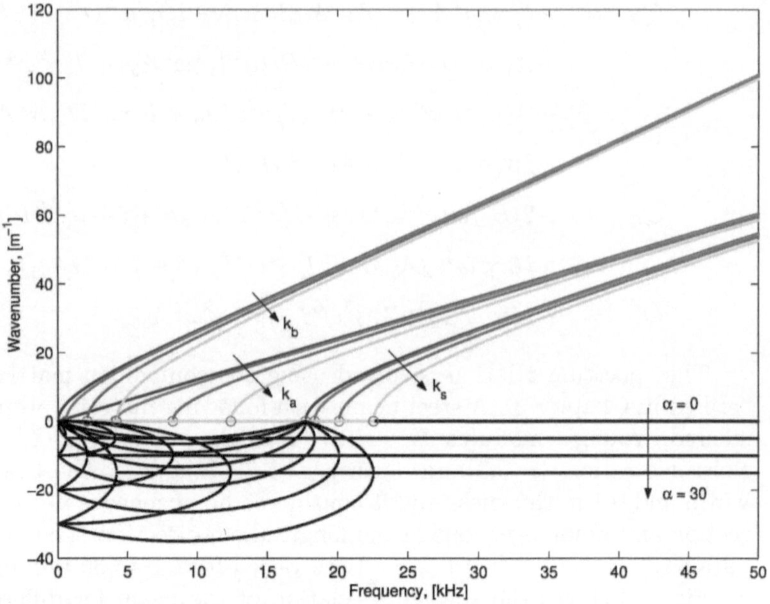

FIGURE 9.8: Spectrum relation: k_a, k_b and k_s denote axial, bending and shear wavenumber, respectively (α increases in the direction of arrows and $\beta = \alpha$).

cut-off frequency (given by the bending mode), there will be no response in the structure, that is, gradation of materials will act as a high-pass filter.

Next, we will examine the effect of gradation on the cut-off frequencies. Since the propagating modes appear only when the loading frequency exceeds the lowest cut-off frequency, the variation of the later with FGM parameters is important for response prediction. As opposed to the earlier relatively simple cases (FSDT and HMT), the explicit forms of the cut-off frequencies cannot be obtained from the spectrum relation. This is because, in this case the cut-off frequency cannot be obtained by substituting $k = 0$ in the spectrum relation and solving for ω. To solve for the cut-off frequencies, we note that they are the frequencies where the imaginary wavenumbers take the value of $-\alpha/2$. Hence, following the two steps described below, the equation governing the cut-off frequencies can be obtained. The steps are: first, we need to substitute $k = k_r + ik_i$ in the spectrum relation and then substitute zero for k_r, thus obtaining the governing equation for the imaginary part of the wavenumbers. Next, we need to substitute $k_i = -\alpha/2$ and arrange the equation in descending powers of ω_n. The equation can be written as

$$A_1\omega_n^6 + A_2\omega_n^4 + A_3\omega_n^2 + A_4 = 0\,, \tag{9.56}$$

where A_is are complex valued and is a function of material properties and the α. Thus, for a fixed base material (steel in this case), the cut-off frequencies

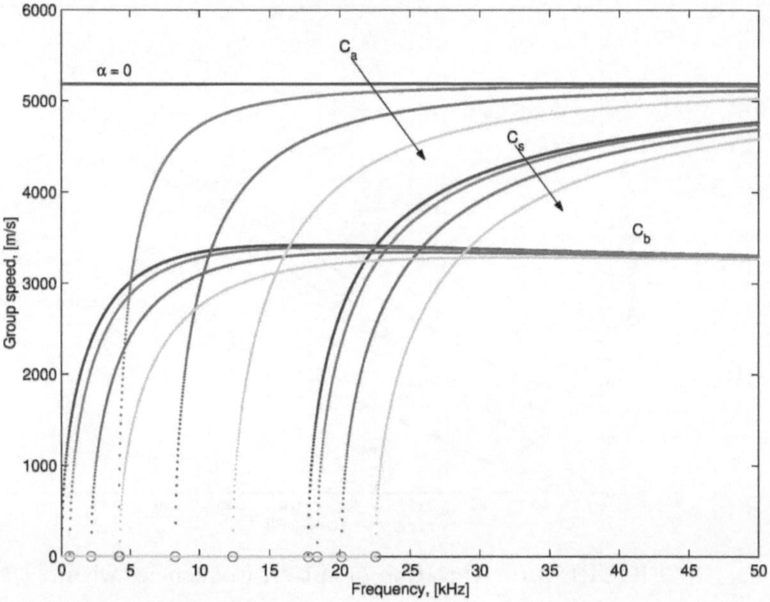

FIGURE 9.9: Dispersion relation: C_a, C_b and C_s denote axial, bending and shear group speeds, respectively (α increases in the direction of arrows and $\beta = \alpha$)

can be varied by varying α, and the required response can be obtained. In this way, the mode selection for a particular loading frequency as well as complete blocking is possible by tuning the inhomogeneous parameter.

To see the variation of the cut-off frequencies with α, we solve Eqn. (9.56), for a range of values of α and plotted in Fig. 9.10. As is shown in the figure, for $\alpha = 0$, there is only one cut-off frequency (which belongs to the shear mode). For nonzero α, both axial and bending modes do have their cut-off frequencies. All these frequencies are varying in a nonlinear fashion for low α value and at higher values, they are fairly linear with α. The gradient $\partial\omega_c/\partial\alpha$ is maximum for axial mode and gradually the axial cut-off frequency reaches the shear cut-off frequency. The bending mode cut-off frequency has the lowest value for a given α and is least affected by gradation (in terms of the gradient).

For a given tone-burst signal (single frequency excitation), the gradation can be used to obtain the desired modal response from the structure. Suppose, the loading has a centre frequency of 30 kHz, then, for $\alpha < \alpha_1$, all the modes will participate in the response, as is shown in the figure. For $\alpha_1 < \alpha < \alpha_2$, only axial and bending modes will participate and for $\alpha_2 < \alpha < \alpha_3$, the response will be due to bending mode only. If the α is increased beyond α_3, there will be no response in the structure as all the modes are effectively

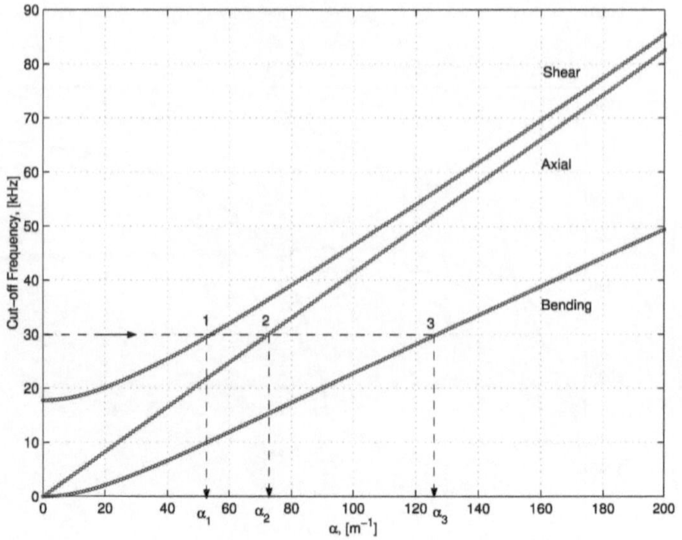

FIGURE 9.10: Variation of cut-off frequencies with α ($\beta = \alpha$)

blocked (damped out). Thus, the gradation can be used effectively for selecting the wave modes, which one wants to propagate or otherwise

9.6 WAVE PROPAGATION IN 2-D FUNCTIONALLY GRADED STRUCTURES

In the previous sections, we have shown that the FGM structures generates spatially dependent coefficients. We will now look into the wave propagation in these structures when the structure themselves are two-dimensional. As in the case of composites (discussed in Chapter 8), we will first derive the governing equation taking into account the spatial variation of the properties and then use spectral analysis to obtain the wave parameters. The governing equation for 2-D laminated composites was solved using *Partial wave Technique* as the wave motions are coupled. As mentioned in Chapter 8, PWT uses SVD method to obtain the wave amplitudes, which is essential for constructing the partial waves. Once the partial waves are found, the wave coefficients are made to satisfy the prescribed boundary conditions, that is, two non-zero tractions specified at the top and bottom of the layer. Here, it differs from other formulations based on the PWT, as no specific problem oriented boundary conditions are imposed. In this section, we will again consider PWT to solve the governing equation. Here, the material property variation is assumed as exponential in the direction of propagation and derive the governing equation of motion. If the gradation parameter for density and elastic moduli are same, then the wave parameters obtained from spectral analysis is exact as in the case of beam.

The governing equation for 2-D FGM waveguides can be derived as was done for 2-D isotropic and composite waveguides. However, here, one has to take care of spatial property distribution. The complete derivation is given in the unified book [54]. Here only the final equations are provided, which is given by

$$
\begin{aligned}
Q_{11}u_{xx} + (Q_{13} + Q_{55})w_{xz} + Q_{55}u_{zz} + Q'_{55}(u_z + w_x) &= \rho(z)\ddot{u}\,, \\
Q_{55}w_{xx} + (Q_{13} + Q_{55})u_{xz} + Q_{33}w_{zz} + Q'_{13}u_x + Q'_{33}w_z &= \rho(z)\ddot{w}
\end{aligned}
\tag{9.57}
$$

where Q_{ij}s are the stiffness coefficients, which depend on the layer layup, its orientation and z coordinate, and the prime in the above equations denotes differentiation with respect to z. The displacement field is assumed to be a synthesis of frequency and wavenumbers, both horizontal and vertical, as

$$
u(x,z,t) = \sum_{n=1}^{N-1}\sum_{m=1}^{M-1} \hat{u}(z,\eta_m,\omega_n) \left\{ \begin{array}{c} \sin(\eta_m x) \\ \cos(\eta_m x) \end{array} \right\} e^{I\omega_n t}\,, \tag{9.58}
$$

$$
w(x,z,t) = \sum_{n=1}^{N-1}\sum_{m=1}^{M-1} \hat{w}(z,\eta_m,\omega_n) \left\{ \begin{array}{c} \cos(\eta_m x) \\ \sin(\eta_m x) \end{array} \right\} e^{i\omega_n t}\,, \tag{9.59}
$$

where ω_n is the discrete angular frequency, η_m is the discrete horizontal wavenumber and $i^2 = -1$. As in the composites case, the assumed field suggests, for $M \to \infty$, the model will have infinite extent in positive and negative x-direction, although of finite extent in the z direction, that is, it will be a layered structure. In particular, the domain can be written as $\Omega = [-\infty, +\infty] \times [0, L]$, where L is the thickness of the layer. The boundaries of any layer will be specified by a fixed value of z. The x dependency of the displacement field (sine or cosine) will be determined based upon the loading pattern. In all the subsequent formulation and computation, the symmetric load pattern is considered. The real computational domain is $\Omega_c = [-X_L/2, +X_L/2] \times [0, L]$, where X_L is the X window length. Discrete values of η_m depend upon the X_L and the number of mode shapes (M) chosen.

To get the expression for $\hat{u}(z)$ and $\hat{w}(z)$, Eqns. (9.58 & 9.59) need to be substituted in Eqn. (9.57), which results in two ordinary differential equations with $\hat{u}(z)$ and $\hat{w}(z)$ as dependent variables, where ω_n and η_m will be appear as parameters. The equation in matrix vector notation is

$$
\mathbf{A}\hat{\mathbf{u}}'' + \mathbf{B}\hat{\mathbf{u}}' + \mathbf{C}\hat{\mathbf{u}} = \mathbf{0} \qquad \hat{\mathbf{u}} = \{\hat{u}\;\hat{w}\} \tag{9.60}
$$

The matrices **A**, **B** and **C** (all functions of z) are

$$\mathbf{A} = \begin{bmatrix} Q_{55} & 0 \\ 0 & Q_{33} \end{bmatrix},$$

$$\mathbf{B} = \begin{bmatrix} Q'_{55} & -(Q_{13}+Q_{55})\eta_m \\ (Q_{13}+Q_{55})\eta_m & Q'_{33} \end{bmatrix} \qquad (9.61)$$

$$\mathbf{C} = \begin{bmatrix} -\eta_m^2 Q_{11} + \rho\omega_n^2 & -Q'_{55}\eta_m \\ Q'_{13}\eta_m & -\eta_m^2 Q_{55} + \rho\omega_n^2 \end{bmatrix} \qquad (9.62)$$

Thus, the effect of inhomogeneity manifests in terms of diagonal terms in **B** and off-diagonal terms in **C**, which are zero for homogeneous material. The associated boundary conditions are the specifications of stresses σ_{zz} and σ_{xz} at the layer interfaces. From the constitutive model and the computed strains, the stresses are related to the unknowns as

$$\hat{\mathbf{s}} = \mathbf{D}\hat{\mathbf{u}}' + \mathbf{E}\hat{\mathbf{u}}\,\hat{\mathbf{s}} = \{\sigma_{zz}\ \sigma_{xz}\}\,.$$

$$\mathbf{D} = \begin{bmatrix} 0 & Q_{33} \\ Q_{55} & 0 \end{bmatrix}, \quad \mathbf{E} = \begin{bmatrix} \eta_m Q_{13} & 0 \\ 0 & -\eta_m Q_{55} \end{bmatrix} \qquad (9.63)$$

where **D** and **E** are also functions of z. The boundary value problem (BVP) reduces to finding $\hat{\mathbf{u}}$, which satisfies Eqn. (9.60) for all $z \in \Omega_c$ and specification of $\hat{\mathbf{u}}$ or $\hat{\mathbf{s}}$ at $z = 0$ or $z = L$. Once the solution is obtained for different values of z in the frequency-wavenumber domain, ($z - \eta - \omega$ domain), for given values of ω_n and η_m), the summation over η_m will bring the solution back to $z - x - \omega$ space and inverse FFT will bring the solution back to time domain that is, $z - x - t$ space. Thus, any kind of inhomogeneity can be tackled in this formulation, provided the BVP is solved numerically. However, there is a special case for which the BVP is exactly solvable, and this is taken up next.

Let us now assume that the material property variation is exponential, that is,

$$Q_{ij}(z) = Q_{ij\circ}e^{\alpha z}\,, \quad \rho(z) = \rho_\circ e^{\beta z}\,, \qquad (9.64)$$

where $Q_{ij\circ}$ and $\rho_\circ$ are the constant properties of the background homogeneous material. Substituting Eqn. (9.64) in Eqn. (9.62) we get

$$\mathbf{A} = \begin{bmatrix} Q_{55\circ} & 0 \\ 0 & Q_{33\circ} \end{bmatrix} e^{\alpha z}, \quad \gamma = e^{(\beta-\alpha)z}.$$

$$\mathbf{B} = \begin{bmatrix} \alpha Q_{55\circ} & -(Q_{13\circ}+Q_{55\circ})\eta_m \\ (Q_{13\circ}+Q_{55\circ})\eta_m & \alpha Q_{33\circ} \end{bmatrix} e^{\alpha z},$$

$$\mathbf{C} = \begin{bmatrix} -\eta_m^2 Q_{11\circ} + \rho_\circ\omega_n^2\gamma & -\alpha Q_{55\circ}\eta_m \\ \alpha Q_{13\circ}\eta_m & -\eta_m^2 Q_{55\circ} + \rho_\circ\omega_n^2\gamma \end{bmatrix} e^{\alpha z} \qquad (9.65)$$

Substituting Eqn. (9.65) in Eqn. (9.60) and canceling the $e^{\alpha z}$ term, another equation is obtained in which the elements of the matrices **A** and **B** are constant, but the elements of **C** has z dependency in terms of γ. If β is equal to α, then $\gamma = 1$ and all the matrices become constant. Then the solutions are in the form of $u_\circ e^{-ikz}$ and $w_\circ e^{-ikz}$, where $u_\circ$, $w_\circ$ and k, the vertical (Z direction) wavenumbers, are unknowns.

Substituting these solutions in Eqn. (9.60) for the matrices given by Eqn. (9.65), the problem reduces to one of finding nontrivial $u_\circ$, $w_\circ$ from the equation

$$\mathbf{W}\{\mathbf{u}_\circ\} = \mathbf{0}\ , \mathbf{W} = -k^2\mathbf{A} - ik\mathbf{B} + \mathbf{C}\, , \{\mathbf{u}_\circ\} = \{u_\circ\ w_\circ\}\, , \tag{9.66}$$

where $\mathbf{W}$ is the wave matrix. Thus, in the 2D layered media, $N_v = 2$ and order of the PEP (see Section 3.5.2) $p = 2$, which yields four eigenvalues (wavenumbers). The wave matrix in explicit form is

$$\mathbf{W} = \begin{bmatrix} W_{11} & W_{12} \\ W_{21} & W_{22} \end{bmatrix} \tag{9.67}$$

where

$$\begin{aligned}
W_{11} &= -k^2 Q_{55\circ} - \eta_m^2 Q_{11\circ} + \rho_\circ \omega_n^2 \gamma - ikQ_{55\circ}\alpha \\
W_{12} &= ik\eta_m(Q_{13\circ} + Q_{55\circ}) - Q_{55\circ}\alpha\eta_m \\
W_{21} &= -ik\eta_m(Q_{13\circ} + Q_{55\circ}) + Q_{13\circ}\alpha\eta_m \\
W_{22} &= -k^2 Q_{33\circ} - \eta_m^2 Q_{55\circ} + \rho_\circ \omega_n^2 \gamma - ikQ_{33\circ}\alpha
\end{aligned}$$

The singularity condition of $\mathbf{W}$ yields

$$\begin{aligned}
& Q_{33\circ}Q_{55\circ}k^4 + 2iQ_{55\circ}Q_{33\circ}k^3 \\
+ \quad & \{(Q_{11\circ}Q_{33\circ} - 2Q_{13\circ}Q_{55\circ} - Q_{13\circ}^2)\eta_m^2 - \rho_\circ\omega_n^2(Q_{33} + Q_{55})\gamma - Q_{55\circ}Q_{33\circ}\alpha^2\}k^2 \\
+ \quad & ij\alpha(Q_{11\circ}Q_{33\circ} - 2Q_{13\circ}Q_{55\circ} - Q_{13\circ}^2)\eta_m^2 - i\alpha\rho_\circ\omega_n^2(Q_{33\circ} + Q_{55\circ})\gamma\}k \\
+ \quad & \{Q_{55\circ}Q_{13\circ}\alpha^2\eta_m^2 + Q_{11\circ}Q_{55\circ}\eta_m^4 \\
- \quad & \rho_\circ\omega_n^2\eta_m^2(Q_{11\circ} + Q_{55\circ})\gamma + \rho_\circ^2\omega_n^4\gamma^2\} = 0\, ,
\end{aligned} \tag{9.68}$$

which is the required characteristic equation for determination of spectrum relation. It is to be noted that for each value of η_m and ω_n, there are four values of k, denoted by $k_{lmn}, l = 1..4$, which will be obtained by solving the spectrum relation.

There are several extra features of this spectrum relation compared to its homogeneous material counterpart. First, the coefficient of k^3 and k are nonzero and complex, which means that the roots are not complex conjugate to each other, as opposed to the homogeneous case. This implies that the notion of forward and backward moving wave is somewhat blurred in this case. However, this notion, which is effective for understanding the physics can be retained by looking at the signs of the real part only. Thus, the wavenumber with positive real part denotes forward propagating wave and the negative real part denotes the backward propagating wave, although their imaginary parts may have any sign. Further, no closed form solution is available in this case.

Secondly, in the case of homogeneous material, the roots are either fully real or totally complex that is, at no value of ω_n or η_m the wavenumbers posses both real and imaginary (nontrivial) part. However, in the present situation, this kind of solution is quite natural because of the existence of nonzero coefficients of k^3 and k. For the same reason, the notion of cut-off frequency is also absent. However, one can find the cut-off frequencies by setting the constant part in Eqn. (9.68) equal to zero, that is,

$$\rho_\circ^2\omega_n^4\gamma^2 - \rho_\circ\omega_n^2\eta_m^2(Q_{11\circ} + Q_{55\circ})\gamma + Q_{55\circ}Q_{13\circ}\alpha^2\eta_m^2 + Q_{11\circ}Q_{55\circ}\eta_m^4 = 0 \tag{9.69}$$

Here, the ω that satisfies this relation will be called as the cut-off frequency. Thus, both the ideas of forward (or backward) propagating wave and cut-off frequency need to be modified in the inhomogeneous case.

The behavior of the roots will be more clearly visible if a particular material is considered and the spectrum relation is solved for its material properties. For this purpose, the layer is assumed to be of 0.1 m thickness, and material properties are varied from steel to ceramic. The material properties of steel is taken as follows: Young's modulus E = 210 GPa, Poisson's ratio $\nu = 0.3$ and density ρ = 7800 kg/m^3. Similarly, for ceramic, E = 390 GPa, $\nu = 0.3$ and ρ = 3950 kg/m^3. For these material properties and exponential variation, inhomogeneous parameters, α and β become 6.19 m^{-1} and 6.80 m^{-1}, respectively.

Fig. 9.11 shows the variation of the wavenumbers with ω, for $\eta_m = 0$. Both homogeneous (marked by $\pm H_1$ and $\pm H_2$) and inhomogeneous cases are plotted. For the background homogeneous (steel) material, the wavenumbers are linearly varying with ω (they are actually given by $\pm\omega\sqrt{\rho/Q_{33}}$ and $\pm\omega_n\sqrt{\rho/Q_{55}}$) as shown in the figure. Also, the roots are symmetric about $k = 0$ and real. Compared to them, the wavenumbers of inhomogeneous materials are simultaneously possessing both real and imaginary parts. The real parts are symmetric about $k = 0$ and imaginary parts are symmetric about $k = -3.158$. At higher frequencies, along with the propagating real parts, there is a constant imaginary part, which will be responsible for attenuation of wave amplitudes.

For nonzero η_m, the wavenumbers are plotted in Fig. 9.12. Also, the wavenumbers corresponding to homogeneous material are plotted in the same figure and their real and imaginary parts are marked by RH_i and IH_i, respectively. The symmetric behavior of the roots (about $k = 0$) is evident here. Further, the cut-off frequencies can be identified clearly, before which the roots are imaginary. However, the behavior is considerably different in the inhomogeneous case. Here again, the real and imaginary parts coexist. Further, the real and imaginary parts are symmetric about the previous values, that is, $k = 0$ and $k = -3.158$, respectively. Both the figures suggest that, at higher frequencies, the roots become closer to their homogeneous counterpart, along with an imaginary part of constant magnitude. In the homogeneous case, the

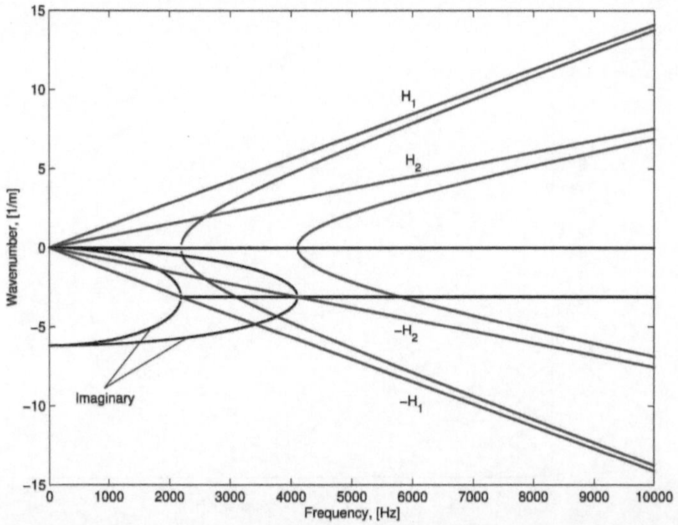

FIGURE 9.11: Variation of wavenumber with ω_n and $\eta_m = 0$, (RH_j – real homogeneous)

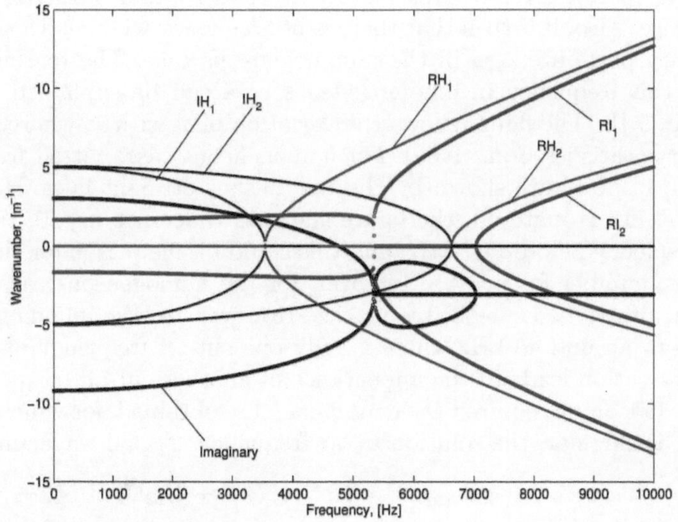

FIGURE 9.12: Variation of wavenumber with ω_n and $\eta_m = 5$ (RH_j – real homogeneous, IH_j – imaginary homogeneous, RI_j – real inhomogeneous)

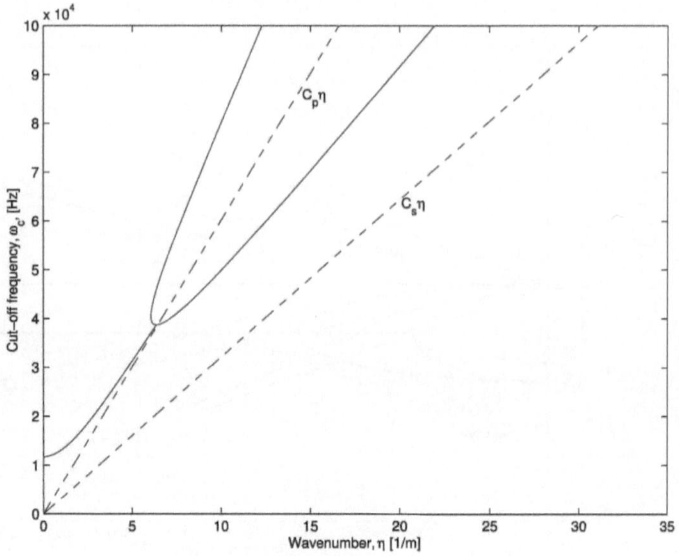

FIGURE 9.13: Variation of cut-off frequency, solid line – homogeneous material, dashed line – inhomogeneous material

cut-off frequencies denote the point of zero wavenumber (both real and imaginary part), given approximately at 3548 Hz and 6613 Hz, as shown in the figure. Also, it is seen that there is no frequency where both the real and imaginary parts are zero in the inhomogeneous case. The reason for the absence of this frequency in inhomogeneous case can be explained with the help of Fig. 9.13. The figure shows the variation of ω with η, where any point on the curve satisfies Eqn. (9.69). For homogeneous case, cut-off frequencies are $c_p\eta$ and $c_s\eta$ and are shown by the two dashed straight lines. However, for inhomogeneous material, the figure suggests that once $\eta \neq 0$, there is no cut-off frequency before 11.7 kHz, and this could be the reason for the absence of zero wavenumber for $\eta = 5$. Moreover, for the homogeneous case, at any η, there are always two cut-off frequencies. However, for the inhomogeneous material, up to around 40 kHz, there is only one cut-off frequency and afterwards the bifurcation leads to the appearance of another cut-off frequency.

Once, the required wavenumbers k are obtained, for which the wave matrix $\mathbf{W}$ is singular, the solution $\mathbf{u}_\circ$ ar frequency ω_n and wavenumber η_m is

$$\begin{aligned} u_{nm} = \; & R_{11}C_1e^{-ik_1x} + R_{12}C_2e^{-ik_2x} + \\ & R_{13}C_3e^{-ik_3x} + R_{14}C_4e^{-ik_4x}. \end{aligned} \tag{9.70}$$

$$\begin{aligned} w_{nm} = \; & R_{21}C_1e^{-ik_1x} + R_{22}C_2e^{-ik_2x} + \\ & R_{23}C_3e^{-ik_3x} + R_{24}C_4e^{-ik_4x} \end{aligned} \tag{9.71}$$

where R_{ij} are the amplitude coefficients to be determined, and they are called wave amplitudes. As outlined before, following the method of SVD, R_{ij} are obtained from the wave matrix $\mathbf{W}$ evaluated at wavenumber k_i.

Once the four wavenumbers and wave amplitudes are known, the four partial waves can be constructed and the displacement field can be written as a linear combination of the partial waves. Each partial wave is given by

$$\mathbf{a}_i = \left\{ \begin{array}{c} u_i \\ w_i \end{array} \right\} = \left\{ \begin{array}{c} R_{1i} \\ R_{2i} \end{array} \right\} e^{-ik_i z} \left\{ \begin{array}{c} \sin(\eta_m x) \\ \cos(\eta_m x) \end{array} \right\} e^{-i\omega_n t}, \quad i = 1 \ldots 4,$$

and the total solution is

$$\mathbf{u} = \sum_{i=1}^{4} C_i \mathbf{a}_i \tag{9.72}$$

Note on MATLAB® scripts provided in this chapter

In this chapter, following two MATLAB scripts are provided:

MATLAB script *fgmhodisp.m*: This script computes the spectrum and dispersion relation for a depth wise graded beam. This program uses the functions *fgmdetr.m*, which uses the function *detm.m* that computes the determinant of the wave matrix W given by Eqn. (9.41) to obtain the wavenumbers and hence the group speeds. This program also uses function *fgmmat.m*, which computes the material properties for a depth-wise graded beam. This program does not use PEP with Companion Matrix method to solve for wavenumbers since, the computer code for the response estimation is not provided in this chapter. In case the determination of responses are necessary, then PEP will be needed to solve for the wavenumbers and the eigenvectors will give the amplitude ratios, which will be necessary.

MATLAB script *fgmlgwavnum.m* : This script computes the spectrum and dispersion relation as well as axial and bending responses for a lengthwise graded beam with exponential variation of properties. Note that, this chapter had not provided any examples on wave propagation responses. However, the reader can run this code and can plot responses for lengthwise graded beam and see the effect of various parameters on the wave propagation responses. Using similar approaches, the reader should attempt to write MATLAB code for wave propagation in depth wise graded beam or for the 2-D wave propagation problem reported in this chapter

Summary

In this chapter, the wave propagation in functionally graded material structure is presented. The first part presents the basic introduction to functionally graded materials and some details on it manufacturing. Next, some basic concepts of its modeling is presented. Two different material variation, namely the power variation and the exponential variation is considered for deriving the governing equation of motion in such structures. For most of the cases, the governing equations are so complicated that the exact solution will be very difficult to obtain. However, for certain class of problems, for example, FGM structure with the exponential variation of the material property can give exact solutions for the wavenumber computations.

The concept of inhomogeneous wave, that is a wave that has complex wavenumber, is introduced here. That is, waves in such structures always attenuates as it propagates. Here first the wave propagation in 1-D waveguide is presented, where, first, the wave propagation in lengthwise graded rod is presented. Here, the concept of naive wavenumber is introduced to understand the physics of wave propagation. Next, the wave propagation in depth wise graded beam is presented followed by lengthwise graded beam. In the latter case, the exponential variation of the material is considered to obtain the exact wavenumber variation. Following this, waves in 2-D FGM waveguides is presented, wherein the partial wave technique is extended to handle this inhomogeneous case. Again, exponential variation is assumed to obtain exact wavenumber variation.

Exercises

9.1 The governing equation for a lengthwise graded rod undergoing axial motion in the frequency domain is given by Eqn. (9.14) for a linear variation of Young's modulus and density, that is $E(x) = E_0(1+\alpha x)$ and $\rho = \rho_0(1+\beta x)$, where E_0 and ρ are the Young's modulus and density of background material, which in this case is Steel, which is transitioning to Ceramic in a linear fashion. Here, α and β are the gradation factors for Young's modulus and density. Eqn. (9.14) is not exactly solvable and hence, by using series solution of the differential equation, which is given by (see [48] for more details)

$$\hat{u}(x,\omega) = \mathbf{A} - \frac{P_2}{P_1}x + \frac{1}{2}\left(\alpha\frac{P_2}{P_1} - k_0^2\mathbf{A}\right)x^2 + \left(\frac{1}{3}\alpha^2\frac{P_2}{P_1} + \frac{1}{3}\alpha k_0^2\mathbf{A} + \frac{1}{6}k_0^2\frac{P_2}{P_1} - \frac{1}{6}\beta k_0^2\mathbf{A}\right)x^3$$

where $$P_1 = L - \frac{1}{2}L^2\alpha + \frac{1}{3}L^3\alpha^2 - \frac{1}{6}L^3k_0^2$$

$$P_2 = \mathbf{A}\left(1 - \frac{1}{2}L^2k_0^2 + \frac{1}{3}L^3\alpha k_0^2 - \frac{1}{6}L^3k_0^2\beta\right) - \mathbf{B}$$

In the above equation, L is the length of the FGM rod over which the Young's modulus and density of the rod varies from steel properties in the left end (E_s and ρ_s) to the ceramic properties (E_c and ρ_c) at the right end of the rod. **A** and **B** are the wave coefficients and $k_0 = \omega\sqrt{\rho_s/E_s}$. This FGM rod segment is attached to two isotropic infinite rod segment as shown in the figure below. The left infinite segment is made of steel while the right side infinite segment is made of ceramic. This rod system is loaded at $x = 0$ axially using a gaussian pulse

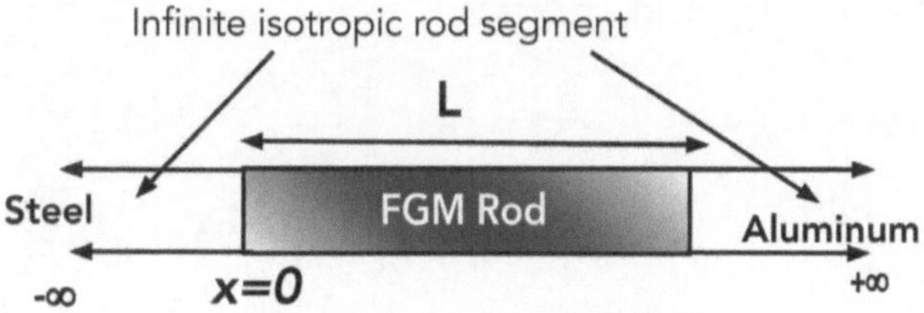

Figure for Problem 9.1

1. Set up the matrix equation to solve for the wave coefficients and hence the responses
2. By modifying the MATLAB script *fgmlgwavnum.m*, determine the time response in the FGM rod segment at different x and for different values of α and β

9.2 The governing equation for a lengthwise graded rod undergoing axial motion in the frequency domain is given by Eqn. (9.15) for an exponential variation of Young's modulus and density, that is $E(x) = E_0 e^{\alpha x}$ and $\rho = \rho_0 e^{\beta x}$, where E_0 and ρ are the Young's modulus and density of background material, which is steel in this case. When, $\alpha = \beta$, the governing equation Eqn. (9.15) is exactly solvable since the equation reduces to constant coefficient equation. For this case,

1. Discuss the spectrum and dispersion relation as a function of gradation parameter α
2. Obtain time responses for an infinite FGM beam shown in Problem 1 for different x and α by modifying the MATLAB script *fgmlgwavnum.m.*

9.3 The governing equation for exponential variation of the material properties (Eqn. (9.15)) is exactly solvable and the exact solution is given in [110], which is given by

$$\hat{u}(x.\omega) = e^{-\alpha x/2}\left((\mathbf{A}J_\nu(2\gamma^{-1}k_0 e^{\gamma x/2}) + \mathbf{B}Y_\nu(2\gamma^{-1}k_0 e^{\gamma x/2})\right)$$

$$\text{where} \qquad \nu = \frac{\alpha}{\beta - \alpha}$$

Here, J_ν and Y_ν are the Bessel functions of first and send kind. For the above solution, obtain responses at different x for different α and β for an infinite FGM beam shown in Problem 9.1

9.4 Again, let us consider the governing equation with exponential variation (Eqn. (9.15)). The approximate series solution for this equation is given by (see [48]for more details)

$$\begin{aligned}\hat{u}(x,\omega) &= \mathbf{A} - \frac{P_2}{P_1}x + \frac{1}{2}\left(\alpha\frac{P_2}{P_1} - k_0^2\mathbf{A}\right)x^2 \\ &+ \left(-\frac{1}{6}\alpha^2\frac{P_2}{P_1} + \frac{1}{3}\alpha k_0^2\mathbf{A} + \frac{1}{6}k_0^2\frac{P_2}{P_1} - \frac{1}{6}\beta k_0^2\mathbf{A}\right)x^3 \\ \text{where} \quad & P_1 = L - \frac{1}{2}L^2\alpha + \frac{1}{6}L^3\alpha^2 - \frac{1}{6}L^3k_0^2 \\ & P_2 = \mathbf{A}\left(1 - \frac{1}{2}L^2k_0^2 - \frac{1}{3}L^3\alpha k_0^2 - \frac{1}{6}L^3k_0^2\beta\right) - \mathbf{B}\end{aligned}$$

where L is again the length of the FGM rod (see figure shown in problem 1) over which the Young's modulus and density of the rod varies from steel properties in the left end (E_s and ρ_s) to the ceramic properties (E_c and ρ_c) at the right end of the rod. **A** and **B** are the wave coefficients and $k_0 = \omega\sqrt{\rho_s/E_s}$. Modify the MATLAB code *fgmlgwavnum.m* to determine the responses at different x for different gradation parameter. How does the response compare with the exact solution (Problem 9.3)

9.5 Consider the bi-material beam made of two segments, one of which is steel and the other one is made of aluminum as shown in the first of figure shown below

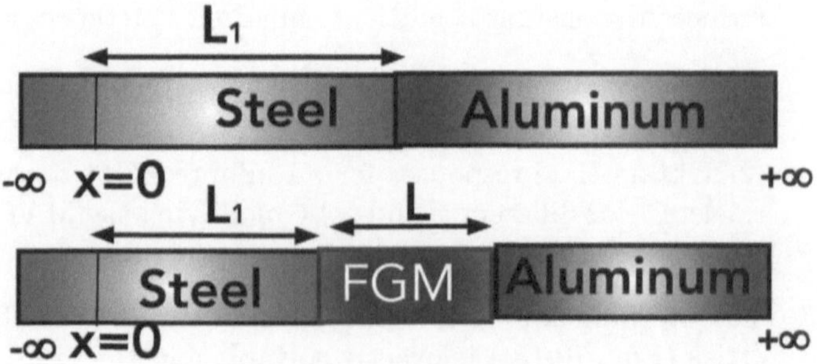

Figure for Problem 9.5

1. The beam is loaded transversely at $x = 0$ and the bi-material interface is located at a distance of L_1 from the loading site. A gaussian

pulse impacted at (either axially or transversely) will induce a reflection at the bi-material interface due to impedance mis-match caused by steep difference in the material properties. Set up the necessary equations for the solutions of the wave coefficients in this bi-material system and determine both axial and flexural response clearly showing the interface reflections

2. The second of the figure shows a small FGM layer of length L, over which the material properties transit from steel to Aluminum for certain gradation properties α and β, respectively. Firstly, FGM layer is assumed to be lengthwise graded, determine the value of L for a given value of α and β (modeled with exponential variation) that makes the interface axial and flexural reflections minimum. Secondly, the same problem is modeled as depth wise graded FGM segment of length L with gradation parameters α and β (modeled as linear polynomial variation). Again for this case, determine the value of L that makes the interface axial or flexural reflections, minimum. In both the cases assume $\alpha = \beta$.

CHAPTER 10

Wave Propagation in Granular Medium

According to Wikipedia [61], a granular material is a conglomeration of discrete solid, macroscopic particles characterized by a loss of energy whenever the particles interact (the most common effect of interaction is the friction caused by colliding grains). The size of the grains vary across different materials with the lowest grain size limit being 1 μm. Some examples of granular materials are the snow, nuts, coal, sand, rice, coffee, corn flakes, fertiliser, and bearing balls. Understanding of elastic wave propagation in granular medium is very important in many disciplines such as geotechnical, geophysical, oil exploration and also in army applications such as sand bunker designs. In this chapter, among all the granular material, we will study the propagation of elastic waves only in dry sands.

Performing wave propagation analysis requires material characterization of the media. The granular media such as dry sands is characterized by the presence of pores, the sectional properties of the containment in which the dry sand is housed. In addition, the material properties are dependent on the level at which the sand is compacted or in other words, the confinement pressure $\bar{\sigma}$, which normally varies from $1\,MPa$ to $10\,MPa$. The pores in dry sand are characterized by void ratio e, which is the ratio of the volume of the pores V_p to the total volume of the containment V. Higher the confined pressure $\bar{\sigma}$, higher will be values of the material properties. These properties will vary for sands from different regions as the level of porosities are different in different regions. Researchers are dependent on extensive experimentation to arrive at the properties. Several researchers have used different modeling techniques that involve inter grain interactions to obtain the material properties. In this chapter, we will use the spectral analysis techniques developed in Chapter 4 along with the experimentally characterized dry sands available in the literature to study the wave motion in containment having compacted dry sands. One of the key applications of the dry sands is in the sand bunker design,

DOI: 10.1201/9781003120568-10

wherein the key motivation is to reduce or attenuate the wave amplitudes in the downstream of the sand bunker. This requires that the transmitted wave amplitude on structures attached to this sand bunker to be minimum. In this chapter, we will study the transmitted wave in such connected structures and come up with some design guidelines on the selection of the type of dry sands that gives minimum transmitted attenuation coefficient in the down stream of the sand-bunker attached structure.

In the next section, we will provide the details on the material property characterization of the dry sand, following which, we study the wave dispersion in dry sands as a function of void ratio and the confined pressure. Following this some wave propagation responses is different dry sands is presented. The last part of the chapter presents some fundamentals of the sand bunker design.This part of the chapter will also deal with the estimation of blast load and its response in the sand-bunker reinforced structure and show how these bunkers are useful in mitigating the effect of blast.

10.1 MECHANICAL PROPERTIES EVALUATION FOR DRY SANDS

Soils in general have a highly non-linear constitutive model and the degree of non-linearity is characterized by Plasticity Index (PI) [145]. The soils in general will have linear stress-strain relations if the $PI < 5$. For dry stands, which is the focus of this chapter, $PI = 0$. An alternate way of classifying the type of soils is to look at the strain levels at which the soils are subjected based on shear stress-shear strain plots. At low strain levels (typically $< 0.01\%$), the stress-strain plots are found to be linear as reported by several researchers [13, 75, 91]. Since, this book deals only with elastic wave propagation, we will limit our discussion only to small strain problems. Unlike metallic structures, shear modulus is the fundamental material property that is determined experimentally for different soils. Using the shear modulus, other material properties of the soil is determined using the following relations

$$E = 2G(1+\nu), \qquad K = \frac{2G(1+\nu)}{3(1-2\nu)}, \qquad M = E\frac{(1-\nu)}{(1+\nu)(1-2\nu)} \tag{10.1}$$

where G is the shear modulus of the soil, E is the Young's modulus, K is the bulk modulus, M is the storage modulus and ν is the Poisson's ratio.

Several methods are adopted to measure the material properties of the soils in general and dry sands in particular. Some of these methods can be categorised into following three methods

- Resonant Column Method
- Quasi-static loading method with high strain resolution method, and
- Wave propagation based method

In the Resonant Column method, the soil container column is fixed in the base and for this configuration, the fundamental longitudinal and torsional mode is determined. The shear modulus G is then found using torsional vibrations, while the Young"s modulus is found using longitudinal vibrations. After measuring G and E, other properties are determined using the relations given in Eqn. (10.1). In the quasi-static loading method, the dry sand soil container is subjected to slowly increasing loading increment and the corresponding strains are measured. Using the stress-strain curve, the shear modulus G and the Young's modulus E are determined. In the wave propagation method, which is most common method of determination of material properties of the dry sands, bender elements are used to measure the longitudinal wave velocity (C_p) and the shear wave velocity (C_s) to a pre-determined time input, from which the shear and the constraint modulus can be determined using the following relation

$$G = \rho C_s^2, \qquad M = \rho C_P^2 \tag{10.2}$$

where ρ is the density of the sand. The details of the wave propagation method of determining the material properties in sands and different soils is given in references [9, 56]. These references also provide the details of other references (not covered here) on the experimental methods used in the determination of material properties. Many of the these references have provided empirical relationship in terms of different sand properties such as void ratio, confined pressure, etc. We will use such empirical relations in this chapter to determine material properties, which are required to perform wave propagation analysis in granular medium.

Most of the references reports that the shear modulus of the sand is estimated using the empirical relation as given below

$$G = A_0 F(e) \left(\frac{\bar{\sigma}}{\sigma_0}\right)^n \tag{10.3}$$

where A_0 is the sand dependent material constant having unit *Pascal* (N/m^2), $F(e)$ is the void ratio function, which is different for different soil types, $\bar{\sigma}$ is the confining pressure, σ_0 is the standard pressure, which is about 1 kPa, and n is the exponent again depends on the type of the sand. From Eqn. (10.3), we see that the sand properties are dependent on the void ratio e and the confined pressure . The value of exponent n, according to [9], depends on the type of sands or soil. For dense sands, the value of n is lower compared to loose sands

Several researchers have experimentally obtained the value of A_0, F_e and n for different types of sands and soils, which are summarized in Table 10.1. The references of the papers where the properties of these soils are reported, is also given in Table 10.1.

Table 10.1 gives the values of A_0, exponent n and the void function $F(e)$ for different sands and for some clayey sands. We see that the values and hence the material properties is vastly different. This table shows the values of these

TABLE 10.1: Void ratios and constants for material property estimation using Eqn. (10.3)

Sl no:	Type of Sand	A_0 (MPa)	$F(e)$	n	Reference
1	Round Grained	6900	$\frac{(2.17-e)^2}{1+e}$	0.5	[58]
2	Angular Grained Sand	3270	$\frac{(2.97-e)^2}{1+e}$	0.5	[58]
3	Clean Sand	41,600	$0.67-\frac{e}{1+e}$	0.5	[127]
4	Clean Sand-1 ($C_u < 1.8$)	14,100	$\frac{(2.17-e)^2}{1+e}$	0.4	[66]
5	Clean Sand-2	9000	$\frac{(2.17-e)^2}{1+e}$	0.4	[67]
6	Toyoura Sand	8400	$\frac{(2.17-e)^2}{1+e}$	0.5	[77]
7	Clean Sand-3	7000	$\frac{(2.17-e)^2}{1+e}$	0.5	[149]
8	Clean Sand-4	9300	$\frac{1}{e^{1.3}}$	0.45	[86]
9	Undisturbed Clay	90	$\frac{(7.32-e)^2}{1+e}$	0.6	[78]
10	Sand and Clay	6250	$\frac{1}{(0.3+0.7e^2)}$	0.5	[57]
11	Several Soils	5700	$\frac{1}{e}$	0.5	[18]

Note: e represents the void ratio of the sand, C_u represents the Coefficient of Uniformity and PI represents Plasticity Index

parameters for example in the case *clean sands* from different regions and these values significantly changes from region to region. Note that the void function for some of the *clean sands* are also significantly different. This is the biggest difference between sands and metals or composites, which were studied in the earlier chapters. The table also shows that for some types of clays or the soil with mixture of dry sand and clay, the empirical relation given in Eqn. (10.3) can be used to estimate the material properties. Also, *item 11* in the table (named as *Several soils*), is a soil type obtained by mixing different types dry soils. In this case, we see that the void function is very different compared to other types of sand. In addition, we see from the table that the value of A_0, which represents the stiffness, is significantly different among different sands. Among the *clean stands*, some are very stiff with the value of A_0 as high as

41600 MPa, while some other *clean sand* from a different region, the stiffness is as low as 7000 MPa.

To determine the properties of the soil, determination of shear modulus alone is not sufficient. Assuming that the soil is isotropic and Eqn. (10.1) is valid, one more property needs to be determined to obtain all other properties, which is normally the Poisson's ratio. However, for dry sands in general, many researchers have found that the Poisson's ratio is also a function of the void ratio e [77]. An empirical relation relating Poisson's ratio to shear modulus is derived in [56], which is adopted in this chapter for performing wave propagation analysis in sands. Such relations will do away of experimentally determining one more material constant for estimating all the material properties of the soil, as is done in structural materials.

Next, we will plot the variation of the material properties (shear modulus, and Young's modulus) for some of the dry sand types given in Table 10.1 using the formula given by Eqns. (10.3 & 10.1). These are shown in Figs. 10.1 & 10.2. These figure are 3-D plots, which are plotted as a function of confined pressure $\bar{\sigma}$ and the void ratio e. From the table, we see there are different types of clean sand from different parts of the world, soil with mixture of sand and clay as well as undisturbed clay. Figure 10.2 shows the Young's modulus variation for different types of clean sands given in Table 10.1. The aim of this plot is to show how the material properties vary significantly even among different clean sands and the effect of void ratio and confined pressures on their material properties.

From Figs. 10.1 and 10.2, it is clear that the material properties for different sands are significantly different and their variations are dependent on the void function $F(e)$ and the confined pressures. Some of materials exhibit linear stress behavior with void ratio e especially at low confined pressure $\bar{\sigma}$. Some soils, namely the *Several soils* and the *Clean sand-4* exhibit significant non-linear behavior with the void ratio e. The value of material property increases with increase in confined stress $\bar{\sigma}$ for low values of e. As the value of e increases, we see that the value of material property decreases, which is expected.

In this chapter, we consider the behavior of sands as isotropic, which means that there are only two independent material property that requires to be determined; one is the shear modulus and the other is either the Poisson's ratio or the Young's modulus or the Bulk modulus. However, for certain class of sands, reference [56] have shown through experimentation that there is a clear non-linear relationship between the Poisson's ratio ν and the shear modulus G, which is given by

$$\nu = 0.620G^{-0.200} \tag{10.4}$$

This equation means, for the soils, a single property is sufficient to characterize all the material properties of the soil. Also, the Poisson's ratio value will

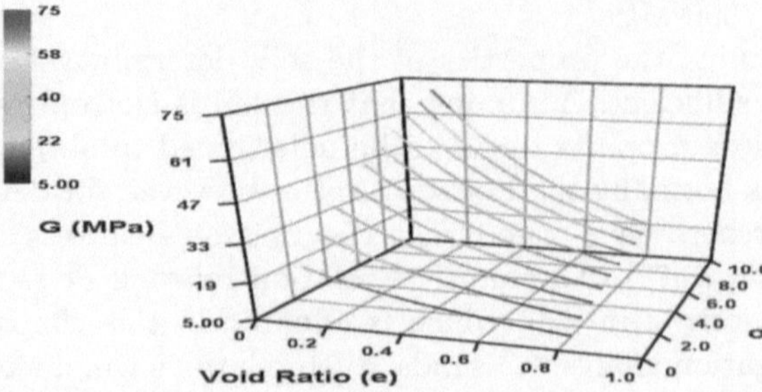

(a) Round Grained Soil

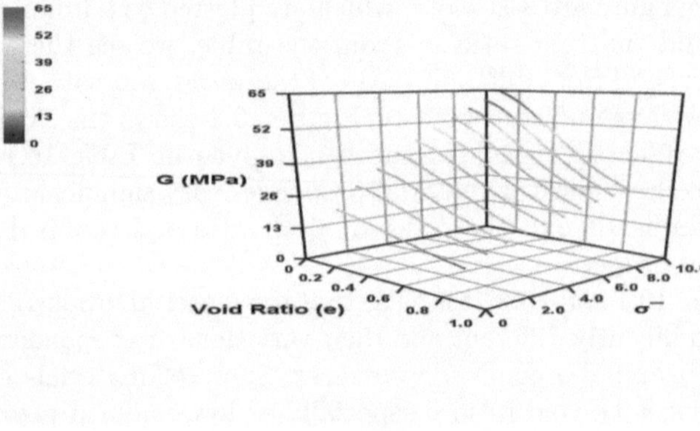

(b) Sand & Clay

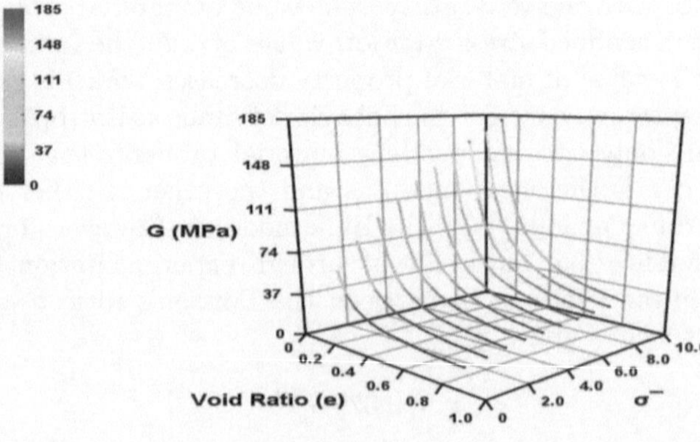

(c) Several Soils

FIGURE 10.1: Shear modulus of some of the soils listed in Table 10.1

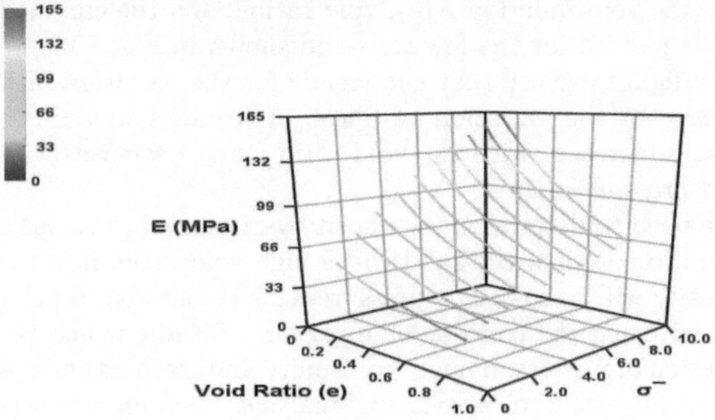

(a) Clean Sand

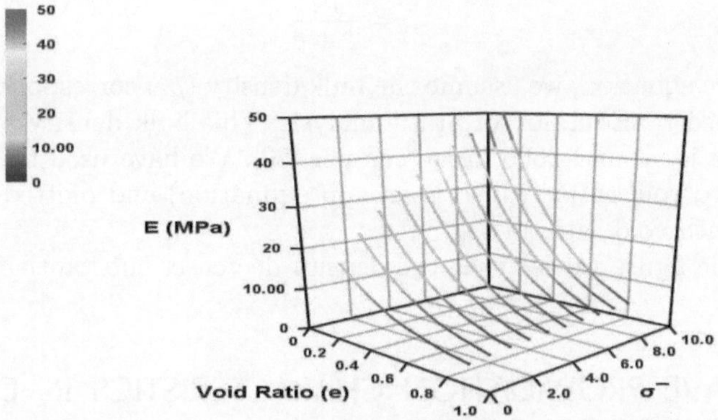

(b) Clean Sand -3

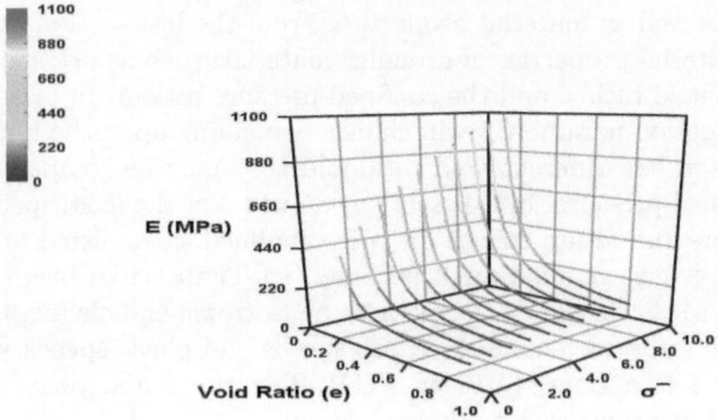

(c) Clean Sand -4

FIGURE 10.2: Young's Modulus of Clean sands listed in Table 10.1

change with the void function $F(e)$, void ratio e and the confined pressure $\bar{\sigma}$. Eqn. (10.4) is plotted for the few soils and shown in Fig. 10.3.

From the figure, we see that the trends for the variation of the Poisson's ratio is reverse of the variation of Shear or Young's modulus. That is the Poisson's ratio increases with the void ratio e and decreases with increase in the confined pressure ratio $\bar{\sigma}$.

Next, we need to determine the density variation of the sand as a function of the void ratio e. When the sands have high void ratio, it is to be expected that the density will be low. The pressure-density relation in principle have to be determined using the concept of *Equation of State*, which is a method of solving equation of physics, namely continuity and momentum equations. This process essentially leads to non-linear analysis. Such an analysis is reported in [80]. Here, we are not performing this analysis to determine the density. If we know the bulk density of the soil ρ_s, and the void ratio e, the density of the soil is then given by

$$\rho = \frac{\rho_s}{1+e} \tag{10.5}$$

In the above equation, we assume the bulk density (ρ_s) corresponding to pressure obtained by Equation of State analysis. This bulk density at a pressure of 600 MPa is around 2600 kg/m^3 as per [80]. We have used this as density value at zero void ratio (which is an approximation) and plotted the density variation with void ratio in Fig. 10.4.

From the figure we see that the density decreases substantially with void ratio.

10.2 WAVE PROPAGATION CHARACTERISTICS IN DIFFERENT TYPES OF DRY SANDS

From the previous chapters, we know that the wave propagation parameters, namely wavenumbers, phase speeds and group speeds are functions of the frequency as well as material properties. From the last section, we have seen that the material properties of granular materials such as dry sands are functions of the void ratio e and the confined pressure ratio $\bar{\sigma}$. In other words, the wave propagation parameters will change depending upon the type of the soil since each soil has different void ratio and they may be compacted with different confined pressure. In this section, we will plot the axial speeds, bending wavenumbers and group speeds for some confined sands listed in Table 10.1. The bending wave numbers will be based on Timoshenko beam theory. Assuming dry sand waveguide behavior to be isotropic and elastic, the equation for computing the wavenumber, group speeds and phase speeds were derived in Chapter 4 (see Eqns. (4.10 & 4.11)). The group and phase speed for a longitudinal dry sand waveguide is given by

$$C_p = C_g = C_0 = \sqrt{E/\rho}, \qquad E = 2(1+\nu)G \tag{10.6}$$

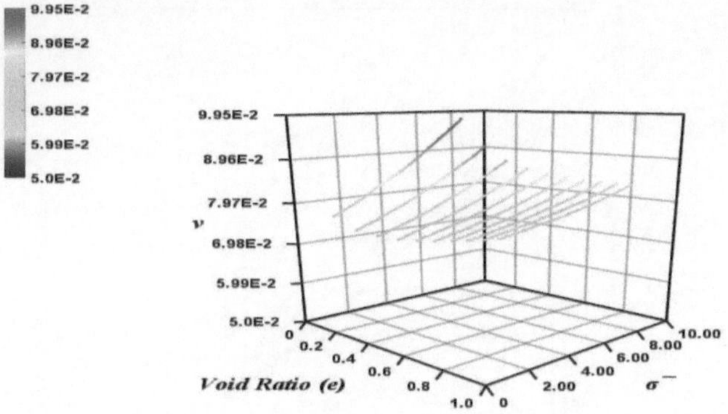

(a) Clean Sand -2

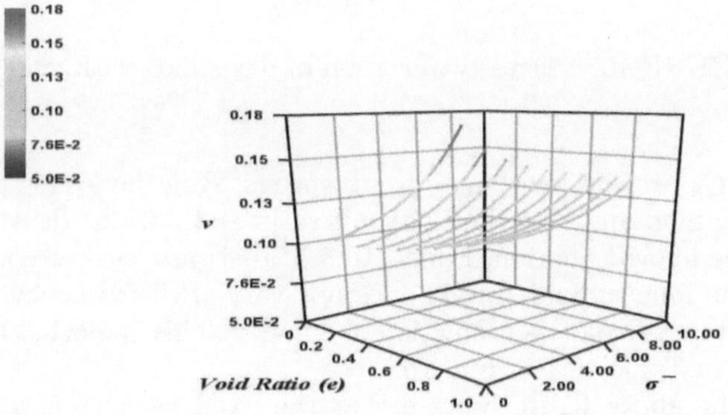

(b) Clean Sand -3

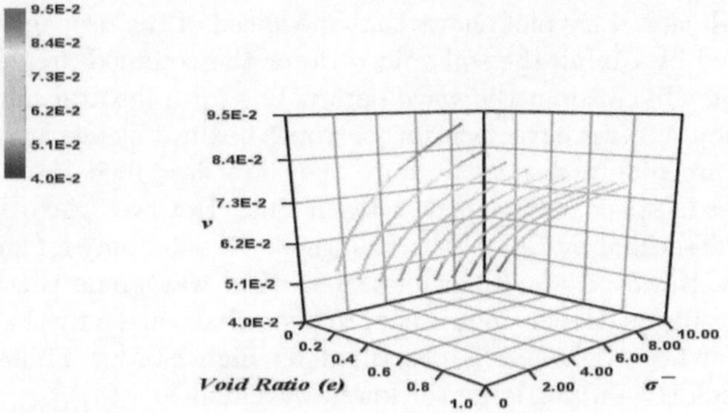

(c) Clean Sand -4

FIGURE 10.3: Poisson's ratio for some Clean sands listed in Table 10.1

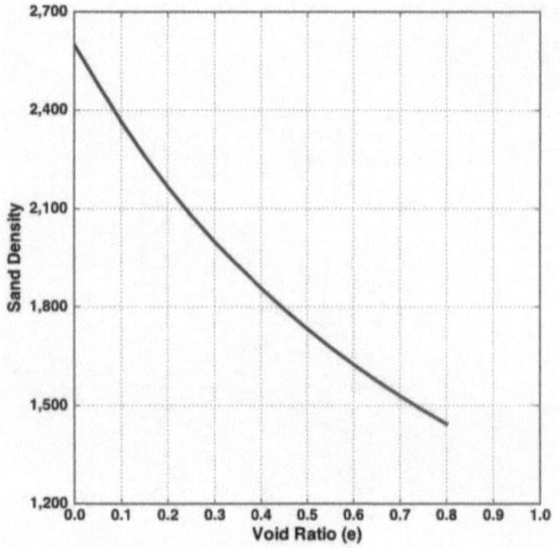

FIGURE 10.4: Density variation of dry sands with void ratio

where C_p, C_g are the phase and group speeds, E is the Young's modulus, G is the shear modulus, ν is the Poisson's ratio and ρ is the density of the dry sand. In the above equation, Eqns. (10.3 & 10.4) are used to compute G and ν. Since the longitudinal speeds does not vary with frequency, they can be represented in 3-D plot as a function of e and $\bar{\sigma}$. This is plotted for 3 different soils listed in Table 10.1 in Fig. 10.5.

From the above figure, we see that the axial velocity is low in the soil mixture of sand and clay, while for the other mixture of several soils, the peak velocity for high confined pressure ratio is nearly 3 time higher. In addition, the variation of axial velocities in several soil mixture or in Clean Sand-4 is highly non-linear. This plot shows that the speed of the propagating wave can be controlled by tuning the soil void ratio or the confined pressure.

Next, we will consider the wave parameters for a flexural sand waveguide. From Section 4.3, we have seen for isotropic flexural elementary waveguides, the waves are highly dispersive since they are described by a PDE that is fourth order in space and second-order in time. However, flexural elementary waveguide described by Euler-Bernoulli theory is valid only for small thickness waveguides. However, small thickness 1-D sand waveguide is seldom used in practical problems. Hence, in this chapter we will discuss only the flexural sand waveguide, whose behavior is described by higher-order Timoshenko beam theory, which is valid for large thickness waveguides.

Timoshenko beam theory was presented in Chapter 4 in Section 4.4.1. All the relevant equation for determination of wavenumbers and group speeds were derived in this section and hence not repeated here. Here, we will adapt these equation to obtain the wavenumber and group speeds for different types sands and soils listed in Table 10.1. As in the case of axial waveguides presented

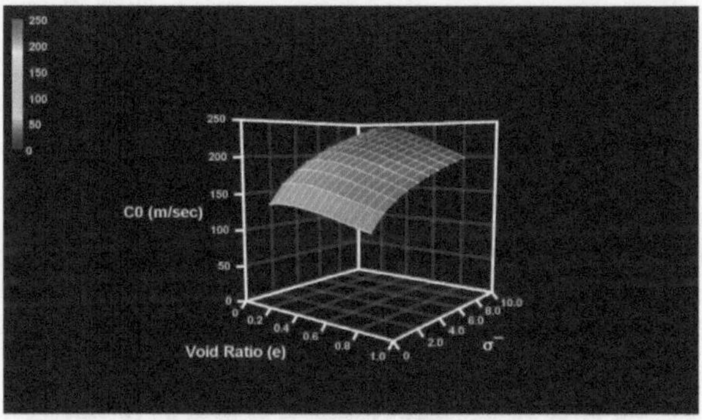

(a) Sand and Clay

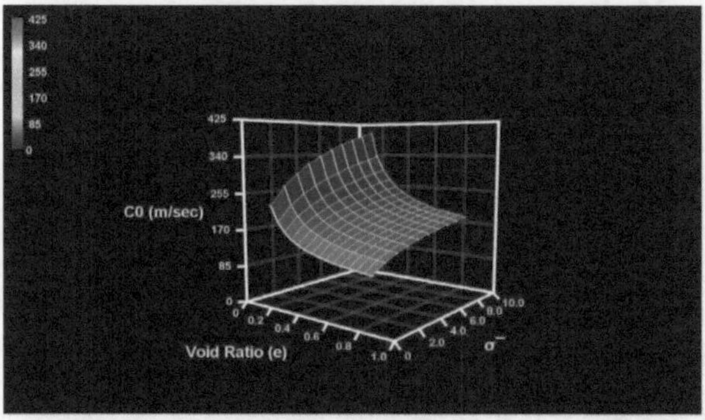

(b) Several Soils

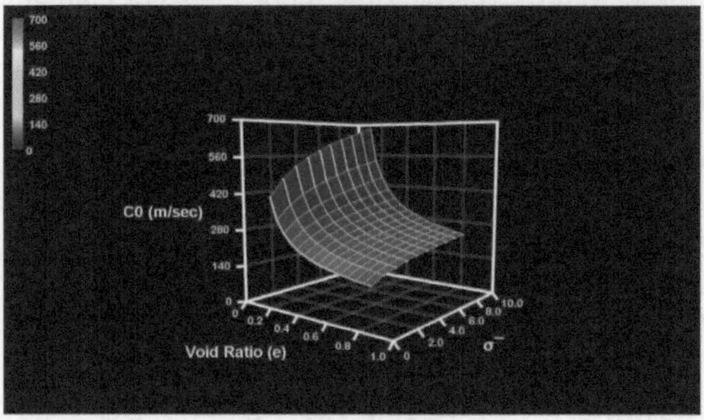

(c) Clean Sand -4

FIGURE 10.5: Axial speed variation for some sands listed in Table 10.1

earlier, the wavenumber and speeds behavior will be function of both void ratio e and confined pressure ratio $\bar{\sigma}$. The important aspect of Timoshenko beam theory is the existence of two propagating wave spectrums. This aspect was detailed in Chapter 4 Section 4.4.1. Here the second frequency starts propagating after cut-off frequency $\omega_c = GAK/\rho A$, where G is the shear modulus, A is area of cross section, ρ is the density of the sand and K is the shear correction factor, whose value for circular cross section is 0.9. We will be using such circular sand waveguide later in this chapter. It will be interesting to see how the cut-off frequency changes with soil types.

The wavenumber k for Timoshenko beam is obtained by solving the Polynomial Eigen Value Problem given in Eqn. (4.77) and group speeds are obtained as $C_g = d\omega/dk$. The wavenumber variation is plotted for two different soils listed in Table 10.1 and the group speeds are plotted for round grained sands. These plots are shown in Fig. 10.6.

Since the speeds are functions of frequency, the variations is represented only as 2-D plots. Hence, the spectrum and dispersion plots were generated for a confined pressure ratio of $\bar{\sigma} = 5$. The cross section of the waveguide is assumed circular with a diameter of 0.4 m. Thick cross section is assumed so that the shear effects are predominant in the waveguide behavior. From the plots, we can infer the following:

1. Different sands yields different wave behavior in terms of speeds and cut-off frequencies.
2. Increase in void ratio, pushes the cut-off frequencies towards the origin.
3. The speeds of both model and mode 2 increases with increase in e for the two different clean sands. This is seen as increase in slope in Figs. 10.6(a) & (b). However, for the round grained sand, whose dispersion plot shown in Fig. 10.6(c), the bending mode does not change with void ratio, while the speed of the shear mode increases with increase in the value of e.
4. The shear effects are negligible for soils with high void ratio.

The wavenumber and dispersion relations for other soils can be similarly obtained. Once the wavenumber is determined, the procedure outlined in Chapter 4 can be followed to obtain responses for any input signal and the propagation of waves and its interaction with boundaries can be studied. This exercise is left to the reader.

10.3 DESIGN CONCEPT FOR SAND BUNKERS FOR EFFICIENT BLAST MITIGATION

One of the motivations to study wave propagation in granular materials such as dry sands is that it has direct application in the design of sand bunkers. Sand bunkers are normally used in a situation that requires protection of humans or structures from high intensity blasts. A bunker is a defensive fortification provided to shield a structures or personal from the ill effects of harsh loadings such as impact or blast. Bunkers can be underground or overground. There are different types of bunkers and one of the common materials used in construction of bunkers is dry sands because of its great ability to damp

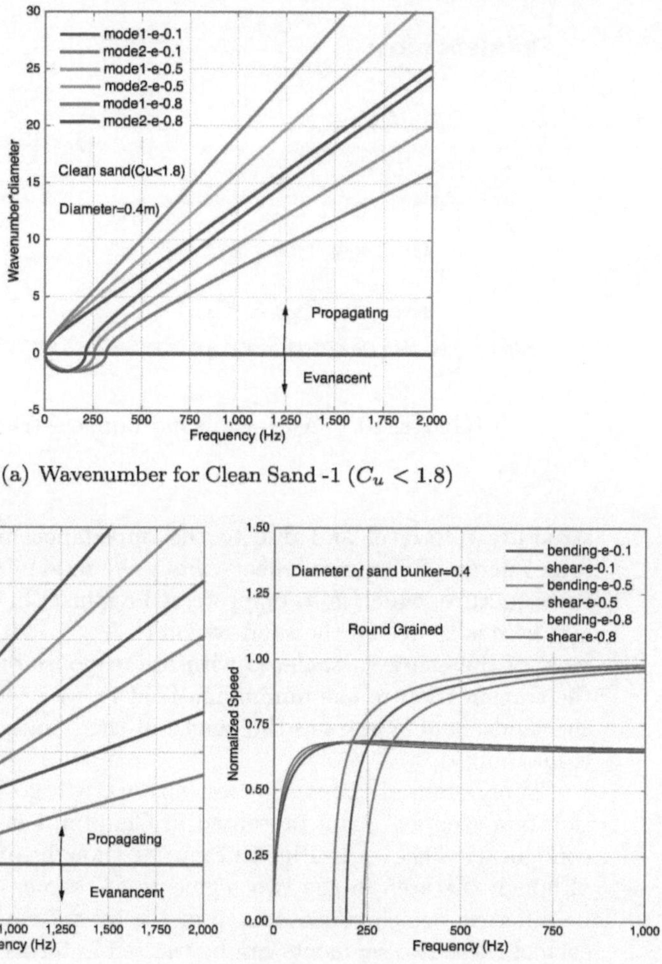

(a) Wavenumber for Clean Sand -1 ($C_u < 1.8$)

(b) Wavenumber for Clean Sand -4 (c) Group speeds for Round grained sand

FIGURE 10.6: Spectrum and Dispersion relations for some sands listed in Table 10.1

out the high energy input coming from either gun fire or blast. A typical sand bunker-structure is shown in Fig. 10.7, which is idealized as connected 1-D waveguides.

In this figure, the region between $-L < x < 0$ represents the sand bunker region (where L is the length of the bunker) and region between $0 < x < \infty$ represents the structure that requires protection from the high or low velocity impact caused by load $P(t)$. Such a load on the structure will cause a stress wave, which propagates across the structure and interacts with the sand-

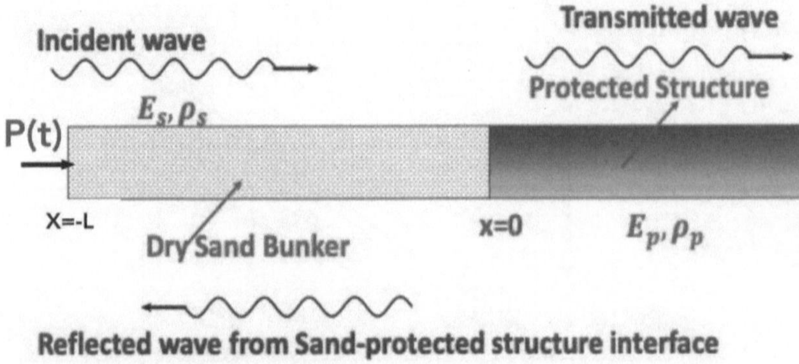

FIGURE 10.7: A typical sand bunker-structure configuration

structure interface, and due to the impedance mis-match at the interface, some energy of response reflects, while the remaining energy is transmitted to the structure, which is to be protected against this failure causing loads. The aim here is to design the sand region ($-L < x < 0$) of length L such that the most of the energy associated with the response due $P(t)$ is reflected keeping the transmission to the minimum. That is, we need to choose the material of the sand (sand type) and the length of the bunker L such that the reflection is maximized.

To understand the design concept, we will go back to the wave propagation in a stepped beam presented in Chapter 4 in Section 4.2.6. The bunker configuration shown in Fig. 10.7 can be thought of as a stepped beam, where although the area in the two segments are same, their Young's modulus can be different, which causes a step at the sand-structure interface. Solution for each of these two segments can be written in terms of the incident wave coefficients $\mathbf{A_{1n}}$, reflected wave coefficient in sand segment $\mathbf{B_{1n}}$ and the transmitted wave coefficient $\mathbf{A_{2n}}$. In order to maximize the reflection and minimize the transmission, we need the reflection coefficient (RC) ($\mathbf{B_{1n}}/\mathbf{A_{1n}}$) should be maximum while the transmitted coefficient (TC) ($\mathbf{A_{2n}}/\mathbf{A_{1n}}$) should be minimum. The reflected and transmitted coefficients for a stepped rod was derived in Chapter 4 Section 4.2.6, and they are given by Eqn. (4.22). This equation is reproduced here for the sake of clarity in the present situation and is given by

$$\mathbf{B_{1n}} = \frac{(1 - S\alpha)}{(1 + S\alpha)}\mathbf{A_{1n}}, \qquad \mathbf{A_{2n}} = \frac{2}{(1 + S\alpha)}\mathbf{A_{1n}} \tag{10.7}$$

where in the present case

$$S = \frac{E_p A_p}{E_s A_s}, \qquad \alpha = \frac{k_{2n}}{k_{1n}}, \qquad k_{1n} = \omega\sqrt{\frac{\rho_s A_s}{E_s A_s}}, \qquad k_{2n} = \omega\sqrt{\frac{\rho_p A_s}{E_p A_s}}$$

TABLE 10.2: Values of Young's Modulus E_p and density ρ_p used for generating reflected and transmitted coefficients

Sl no:	**Material**	E_p (GPa)	ρ_p (kg/m^3)
1	Steel	210	7800
2	Aluminum	70	2800
3	Concrete	40	2400

where as per Fig. 10.7, E_s and E_p are the Young's modulus of the sand and the protected structure and their corresponding densities are ρ_s and ρ_p, respectively, and A_s and A_p are their respective area of cross section.

From Eqn. (10.7), we see that for maximizing the reflection (or reflection coefficient $\mathbf{B_{1n}}$), it is necessary that the value of $(1 + S\alpha)$ be minimum. The maximum value of reflection coefficient is 1 and for this to happen we should have $S\alpha << 1$. This can be achieved only when $E_P >> E_s$. Hence, we need to choose the type of the sand that has very low Young's modulus. This is a fundamental design criteria when designing a sand bunker. The expression of reflection and transmitted coefficient does not involve the parameter L. However, from the wave propagation point of view, the longer the length of the bunker, stress wave takes longer time to reach the bunker-structure interface and in the process the stress wave generated will attenuate significantly due to inherent damping present in the medium.

We also see from Eqn. (10.7) that both reflected and transmitted coefficient for a this case is frequency independent and hence these parameters can be plotted for different sand types listed in Table 10.1 as 3-D plots, function of both void ratio e as well as the confined pressure ratio $\bar{\sigma}$. These plots are shown in Figs. 10.8 and 10.9, respectively, for two different soils, namely Clean Sand-4 and Several soils, the properties of which are listed in Table 10.1. Also, in these plots, three different protected structures are considered, namely the steel, aluminum and concrete, respectively, and the properties of these are shown in Table 10.2.

The reflection and transmitted coefficients for other sands listed in Table 10.1 can be similarly obtained. From Figs. 10.8 and 10.9, we can draw following conclusions:

1. The variation of reflected and transmitted coefficients with void ratio e and the confined pressure ratio $\bar{\sigma}$ is non-linear for both the soils. We can expect similar behavior for other type of sand bunkers connected to different protective structures.

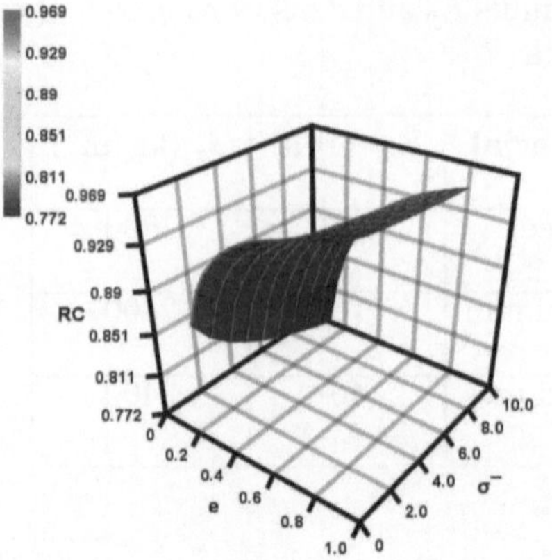

(a) Reflected Coefficient-Steel

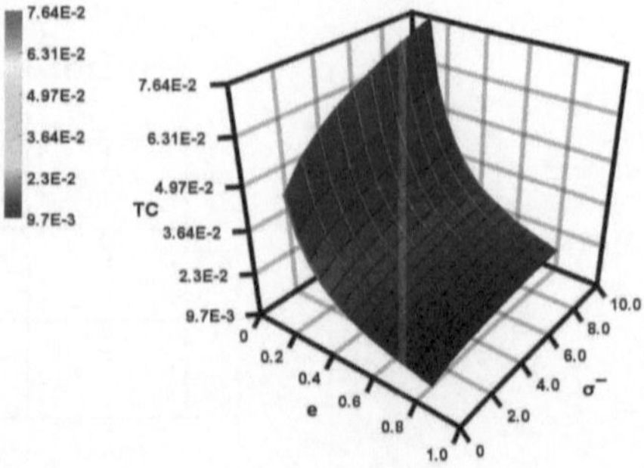

(b) Transmitted Coefficient-Steel

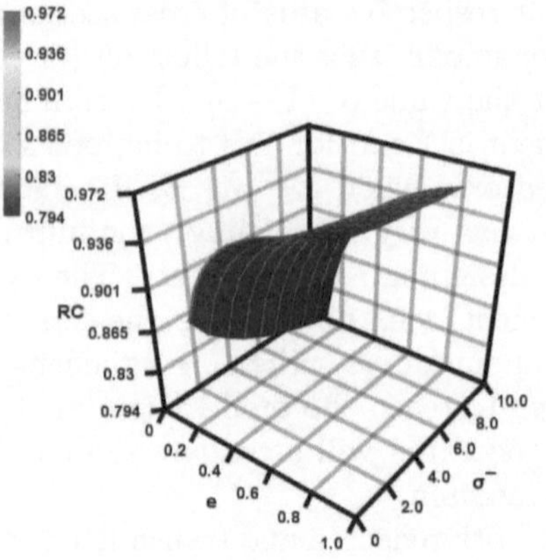

(c) Reflected Coefficient-Aluminum

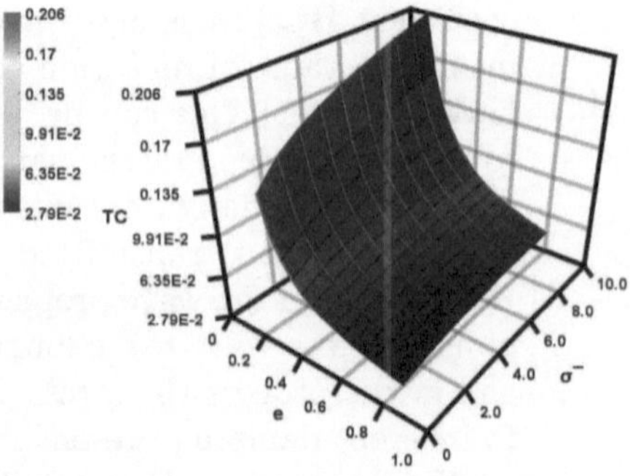

(d) Transmitted Coefficient-Aluminum

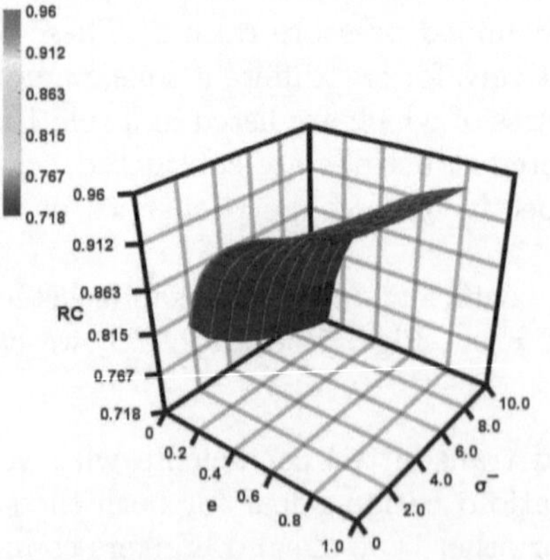

(e) Reflected Coefficient-Concrete

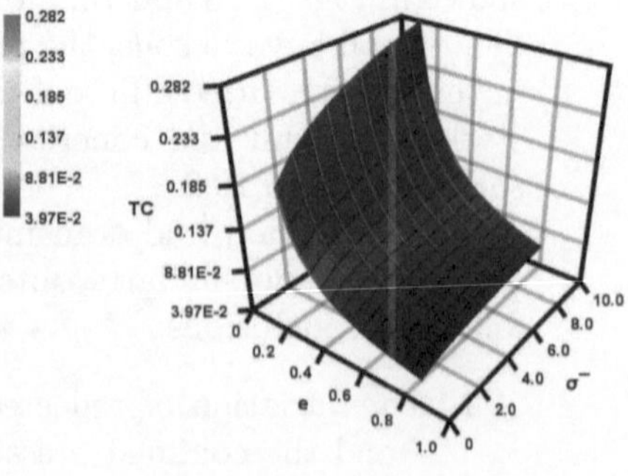

(f) Transmitted Coefficient-Concrete

FIGURE 10.8: Reflected and Transmitted coefficient for sand bunker made from Clean Sand-4

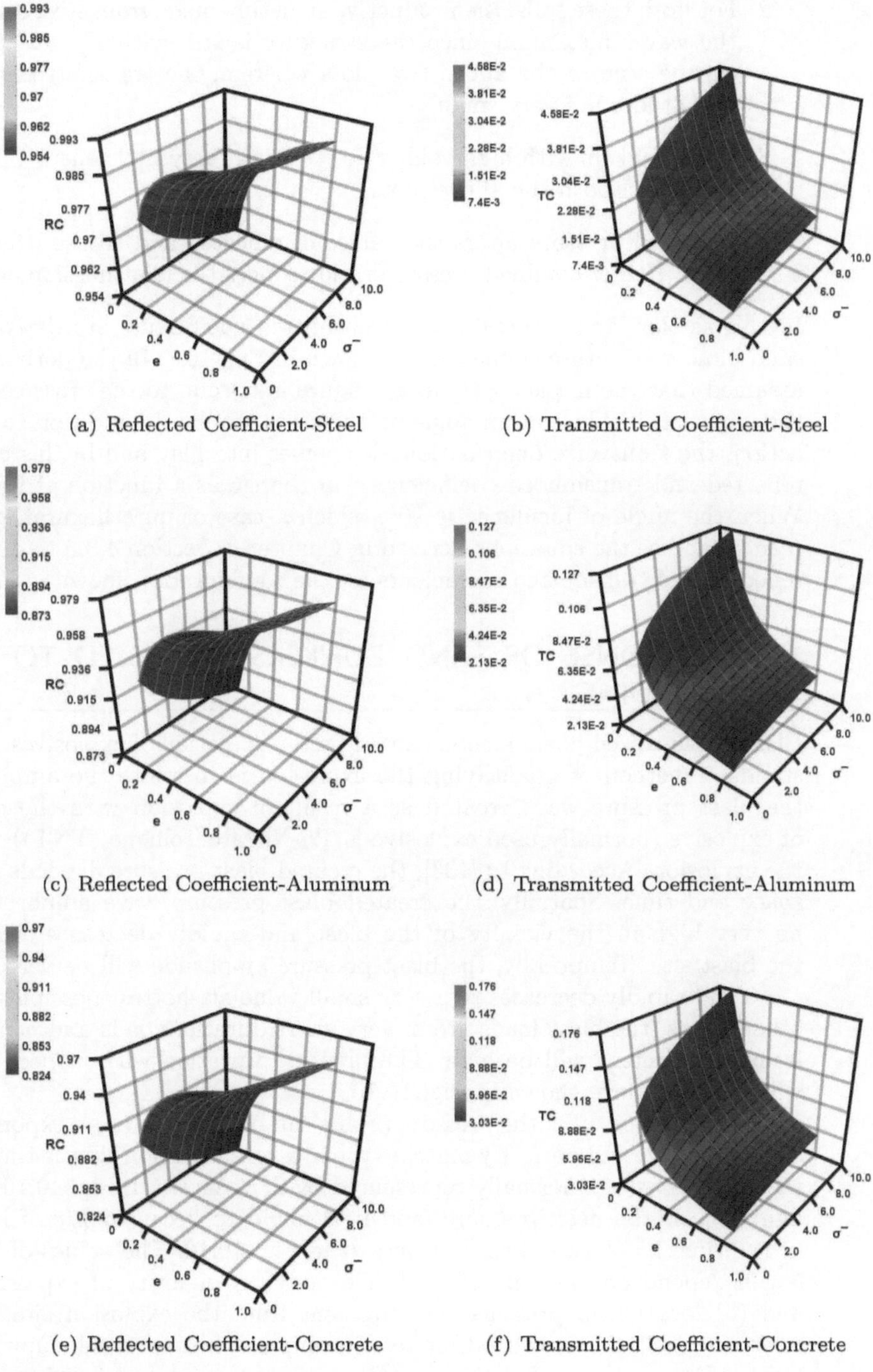

(a) Reflected Coefficient-Steel
(b) Transmitted Coefficient-Steel
(c) Reflected Coefficient-Aluminum
(d) Transmitted Coefficient-Aluminum
(e) Reflected Coefficient-Concrete
(f) Transmitted Coefficient-Concrete

FIGURE 10.9: Reflected and Transmitted coefficient for sand bunker made from Several soils

2. For both these soils, the productive structure made from steel attenuates the waves maximum since the structure is stiff with $E_p >> E_s$. This can be seen in the above two plots wherein, the transmitted response amplitudes are very small.

3. Sand bunkers with high void ratios transmit very little energy. This can be seen in both the above plots.

4. Confined pressure alters the values of reflected and transmitted coefficient. Higher confined pressure ratio reduces the transmission efficiency.

In this section, the reflection and transmitted coefficients are derived for a sand bunker-structure configuration shown in Fig. 10.7. In the derivation, we assumed that the impact $P(t)$ in the figure is normal to the cross section. If the impact is incident at an angle on the cross section, in addition to the rod action, the transverse deformation also comes into play and in this case, the reflected and transmitted coefficients will change as a function of frequency. When the angle of incidence is 90°, which a case of pure flexural load, the reader can use the equations derived in Chapter 4, Section 4.3.5 to derive the reflected and transmitted coefficients for the configuration shown in Fig. 10.7.

10.4 RESPONSE OF SAND BUNKERS SUBJECTED TO BLAST LOADING

Blast is a localized phenomenon caused by the explosion of explosives. Normal standard method of quantifying the explosion is to relate the amplitude of the blast pressure wave created as a result of explosion with the quantity of explosive (normally used explosive is Tri-Nitrate Toluene (TNT)) used for the explosion. According to [133], the created blast pressure depends on both space and time. Spatially, the created blast pressure wave amplitudes will be very high in the vicinity of the blast and rapidly decreases away from the blast site. Temporally, the blast pressure amplitude will be very high at $t = 0$ and rapidly decreases to a very small value at shortest possible time. In other words, the blast loads are of very short duration loads and hence their frequency content will be high. The pressure variation with respect to time will be of the form shown in Fig. 10.10.

Hence, as per [133], the pressure profile can be modeled as an exponentially decaying curve and perfectly matches the pressure variation till the time called the *decay constant*, normally represented by θ, as shown in Fig. 10.10. In this figure P_m is the peak pressure, and it is found to decay to $1/e$ of its value at the time $t = \theta$ as shown as *Curve II* in Fig. 10.10. The values of P_m and θ will depend on (1) type of explosive used (2) quantity of explosive used and (3) location of pressure measurement from the explosion site. As per reference [133], the pressure-time relationship can be expressed empirically in terms of the quantity of explosives (Q) and the distance of explosion from the

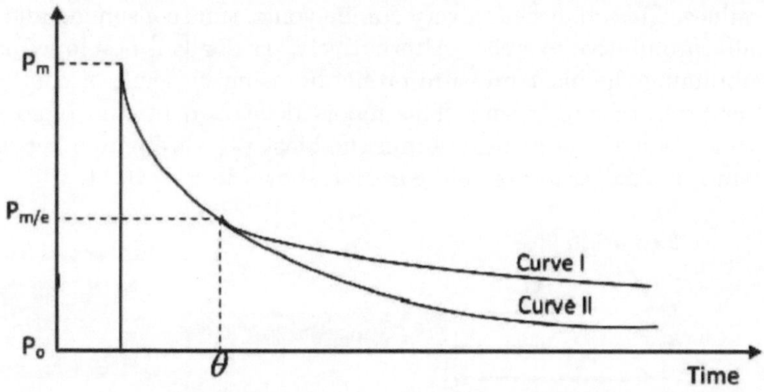

FIGURE 10.10: Temporal variation of blast pressure. Here, *Curve I* is the actual pressure variation and *Curve II* is the exponential function fit of the actual pressure variation

measurement point R as

$$P(t) = P_m e^{\frac{-t}{\theta}}, \tag{10.8}$$

where

$$P_m = 52.4\left(\frac{m^{1/3}}{R}\right)^{1.13} \quad MPa, \qquad \theta = 0.084 m^{1/3}\left(\frac{m^{1/3}}{R}\right)^{-0.23} \quad millisecs$$

In the above equation, m is the mass of the TNT in *kgs* used in the explosion. We will use these expressions to obtain the blast pressure profile, which is derived in the next subsection.

10.4.1 Estimation of Blast Pressure Profile on the Sand Bunker

An explosion will create a pressure or shock wave (given by Eqn. (10.8)) that propagates into the medium (in the present case it is air) and interacts with the sand bunker located at a certain distance and creates a reflected wave. The combination of the incident and the reflected wave forms the total blast pressure profile, which acts on the sand bunker structure. The aim of this section is to obtain the expression for the total pressure profile. Normally this will require two step fluid-structure analysis normally performed iteratively. In the first step, using the far field velocity condition as input and using Computational Fluid Dynamics (CFD) simulation, the pressure profile at the fluid-solid interface is obtained by solving the non-linear Navier-Stokes equation. The obtained pressure profile is then applied to the structure to obtain the velocities at the fluid-solid interface, which will act as the input for the fluid analysis.

This process is repeated till the error between the steps reaches the tolerant values. This analysis is a very cumbersome, time consuming and computationally prohibitive exercise. Alternatively, Taylor [44] developed a novel way of obtaining the blast pressure profile by using principle of continuity and conservation of momentum. This model developed by him is normally referred to as Taylor's model. To obtain the blast pressure profile, let us consider the sand bunker structure configuration shown in Fig. 10.11.

FIGURE 10.11: Estimation of blast pressure profile: Fluid Structure Interaction analysis

In Fig. 10.11, P_I is the incident wave and its expression is given by Eqn. (10.8). This incident wave, will induce a reflected pressure wave P_R, and it is found by considering the momentum conservation equation at the sand bunker-structure interface, which is given by

$$P_I + P_R = m_s \dot{v}_s \tag{10.9}$$

where m_s is the mass of the sand bunker, v_s is the velocity of the sand bunker created as a result of the pressure created by the blast. From the above equation, we have two unknowns, namely the reflected wave P_R and the velocity v_s and we need two equations to determine it. They are obtained as follows. Let $V_I(t)$ be the velocity of the incident pressure wave and $V_R(t)$ is the velocity of the reflected pressure wave. They are evaluated using the impedance equation, which is given by

$$V_I = \left(\frac{P_I}{\rho_a c_a}\right), \qquad V_R = -\left(\frac{P_R}{\rho_a c_a}\right) \tag{10.10}$$

where ρ_a is the density of air and c_a is the velocity of sound in air. At the sand bunker-structure interface, from the equation of continuity, we have

$$V_I(t) + V_R(t) = v_s(t) \tag{10.11}$$

Using Eqns. (10.8, 10.10 & 10.11) in Eqn. (10.9), we can solve for V_R as

$$V_R(t) = \mathbf{A}e^{\frac{-t}{\alpha}} + \frac{P_m(1+a\alpha)}{\rho_a c_a(1-a\alpha)}e^{-at}, \qquad a = \frac{1}{\theta}, \qquad \alpha = \frac{m_s}{\rho_a c_a} \tag{10.12}$$

where **A** is a constant, which is determined by the initial condition. That is

$$v_s(t=0) = V_I(t=0) + V_R(t=0) = 0$$

Applying this condition, the constant **A** and the reflected velocity $V_R(t)$ can be written as

$$\mathbf{A} = \frac{2P_m}{(1-a\alpha)\rho_a c_a}$$
$$V_R(t) = \frac{2P_m}{(1-a\alpha)\rho_a c_a} e^{\frac{-t}{\alpha}} + \frac{P_m(1+a\alpha)}{\rho_a c_a(1-a\alpha)} e^{-at} \tag{10.13}$$

Using $V_I(t)$ and $V_R(t)$ in Eqn. (10.13), we can determine the reflected pressure $P_R(t)$, which is given by

$$P_R(t) = \frac{2P_m}{(1-a\alpha)} e^{\frac{-t}{\alpha}} - \frac{P_m(1+a\alpha)}{(1-a\alpha)} e^{-at} \tag{10.14}$$

the total blast pressure variation, as a function of both space and time can now be written as

$$P(x,t) = P_I\left(t - \frac{x}{c_a}\right) + P_R\left(t + \frac{x}{c_a}\right) \tag{10.15}$$

where substituting $x = 0$ in the above equation will give the blast pressure profile incident on the bunker, which is essentially $P(t) = P_I(t) + P_R(t)$. The blast pressure profile is plotted for fixed quantity of TNT of 1 kg exploding at different distances R from bunker. This is shown in Fig. 10.12 (a) . In the second plot, the distance of explosion is fixed as 10.0m and the blast pressure profile is plotted for different quantity of explosive. This plot is shown in Fig. 10.12(b).

In the above analysis, for determining the sand mass m_s, we have used the expression given by Eqn. (10.5), which is a function of void ratio e. In this simulation, void ratio of $e = 0.5$ is used. Higher void ratio, does not significantly alter the blast pressure profile. The blast pressure time history peaks at $t = 0$ and has a region of negative pressure. This is typical of most blast-pressure profile. This negative pressure region is caused due to the fluid-structure interaction arising due to reflected pressure wave. In this analysis, it is assumed that the pressure wave does not propagate into the soil due to their high inherent damping. However, this assumption can be relaxed by again assuming that the propagation in the soil and further enforcing continuity and conservation of momentum equations in the soil-air interface. Such a model is called the *Haymann Model* and more details of this can be obtained in [15].

The next task here is to determine the response to such blast pressure. Responses due to high blast pressures are not within the scope of this book since such pressures require truly non-linear analysis. Since this book deals only with elastic wave propagation, we perform response analysis for very low

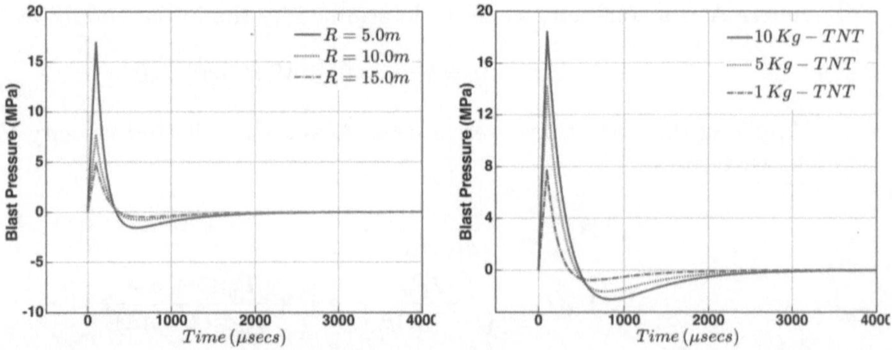

(a) Blast pressure profile for Fixed quantity of $1kg\,TNT$

(b) Blast pressure profile for a fixed distance of $10m$ from explosion

FIGURE 10.12: Blast pressure-time history for different quantity of explosives and for different distances from explosive

intensity blast occurring fairly farther away from the blast site. The obtained blast pressure is converted to force in Newtons by multiplying the pressure with cross-sectional area of the sand bunker for performing response analysis. Here we assume that the cross section is circular with a diameter of the $400mm$. The direction of blast force is shown in Fig. 10.11. In order to simulate low intensity blast, 50 grams of TNT at a distance of 500 m is considered for simulating the blast pressure profile, which is converted to Newtons and plotted in Fig. 10.13. The inset in the figure shows the FFT of the signal, which has frequency content of over 50 kHz.

Here, we perform two different response analysis; in the first, we will see how will the blast pressure will propagate in an an infinite sand medium. For this analysis, we will consider only one type of sand (Clean sand-4 listed in Table 10.1) and for two different confined pressure ratio of $\bar{\sigma} = 2\,\&8$. Here, we will consider two different void ratio $e = 0.3\,\&0.6$ for analysis. The material properties of the Clean-sand-4 obtained through expression and void ratio function for the two different confined pressure ratio is as given in Table 10.3 Axial velocity amplitudes are plotted at three different locations, which are shown in Figs. 10.14 (a) & (b), respectively.

The response are plotted to same scale to see the difference in behavior. Following observations can be made from the plot

- Wave travel non-dispersively although one can see small amount of dispersion. This may be due to FFT caused by abrupt change in the force in the blast force profile.
- Higher confined pressure significantly reduces the amplitude since it increases the stiffness of the sand

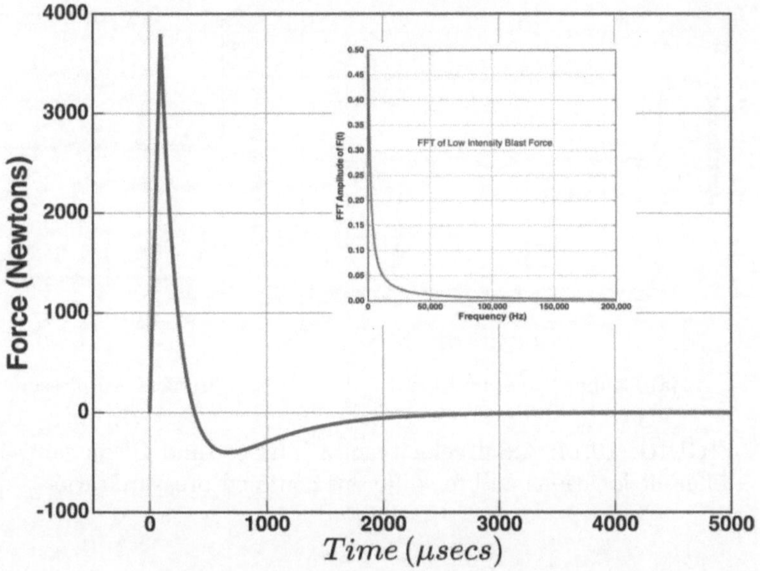

FIGURE 10.13: Low intensity blast force and its FFT for response analysis

TABLE 10.3: Values of material properties for Clean Sand-4 used in simulation studies

Sl no:	Void ratio e	Confined pressure ratio $\bar{\sigma}$	Young"s modulus (MPa)
1	0.3	2	130
2	0.6	2	53.5
3	0.3	8	240
4	0.6	8	99

- Loose sand (sand with higher void ratio) increases the response amplitude
- Void ratio as well as the confined pressure ratio has significant impact on the group velocity of the wave. Wave in loose sand waveguide confined at lower pressure travels slower than the corresponding waveguide at higher confined pressure

Understanding this physics helps us to control the flow of energy in a sand waveguide.

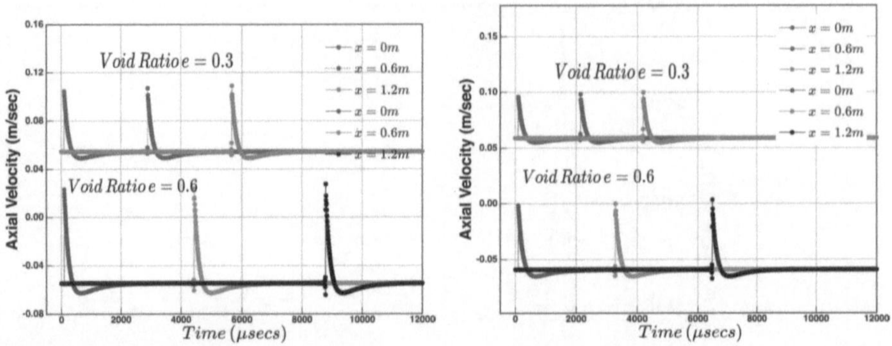

(a) Confined pressure ratio $\bar{\sigma} = 2$ (b) Confined pressure ratio $\bar{\sigma} = 8$

FIGURE 10.14: Axial velocity in a infinite sand Clean sand-4 waveguide at different locations and for different confined pressure ratio

Next, we will perform response analysis in a sand bunker-structure configuration shown in Fig. 10.11. Here we will consider that the sand bunker in the figure is made of *Several Soils* listed in Table 10.1. Again, two different void ratio of $e = 0.3$ and $e = 0.6$ is considered at a confined pressure ratio $\bar{\sigma} = 2$. The Young's modulus of this soil at $\bar{\sigma} = 2$ for these void ratios are 58 MPa and 30 MPa, respectively. Two different protected structure, namely steel (which is very stiff) and concrete are considered and their material properties are given in Table 10.2. As per the design consideration derived earlier, sand waveguides having stiffness many orders smaller than the protected structure is expected to reflect most of the incident energies and allow very little energy to pass through the protected structure.The aim of this study is to see if this phenomenon happens. Figs. 10.15 (a) and (b) show the low intensity blast response for the system having steel and concrete protected structures.

From the figure, we can clearly see that, for both the type of protected structures, most of the incident waves are reflected. A small transmission can be seen in the case of sand bunker with concrete as protected structure. This is because, the stiffness and density of concrete are lower compared to steel. Hence, it can be concluded that dry sand bunkers are highly effective in mitigating the effects of blasts. A more detailed non-linear analysis can shed into more physics associated with this problem.

Summary

In this chapter, propagation of elastic waves in a granular medium such as dry sands was presented. For sands to act as a medium, it requires to be confined in structure by subjecting this structure to some confined pressure. In this chapter, we have assumed that the soil is confined in a cylindrical rubber

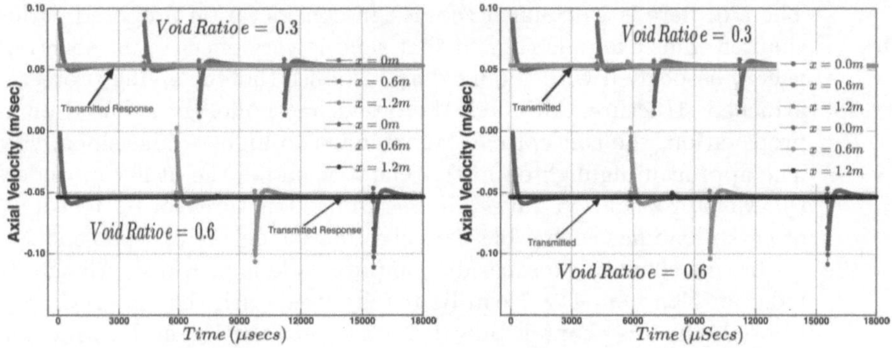

(a) Steel protected structures

(b) Concrete protected structure

FIGURE 10.15: Axial velocity in a sand bunker at different locations for two different protected structures at a confined pressure ratio $\bar{\sigma} = 2$

membrane, whose cross section is a circle. Performing wave propagation analysis requires determination of the material properties of different sands. The chapter begins with material characterization of different types of dry sands, which requires determination of the void ratio function and the confined pressure. The material property was evaluated for 10 different sands reported in the literature using a simple empirical relation that is a function of void ratio and the confined pressure. Following this wave propagation analysis is performed considering both longitudinal and flexural motions for these granular waveguides. The effect of confined pressures and void ratio on the wavenumbers and group speeds were evaluated and some of the results were plotted as 3-D plots. Sand waveguides or bunkers are extensively used to protect major public infrastructures such as buildings and personal in a battle field or in case of harsh environment such as blasts. Blast loads are small duration high-frequency loads, which are caused by explosion of explosives. Hence, a section on preliminary design of sand bunker is presented, where using the concept of stepped rod analysis. Following this, using a simplified Taylor fluid-structure interaction model, the blast pressure time profile is estimated. The amplitude of the pressure wave is shown to be function of distance of the bunker from the explosion site as will as the quantity of explosives used to create this blast pressure wave. It is shown that the blast load, due to fluid-structure interaction effects, always have a negative pressure region. Using a simulated low intensity blast load, response analysis was performed on a typical infinite sand waveguide of different sand parameters (void ratios and confined pressure). It is found that both these parameters have tremendous effect in terms of group speed reduction in these waveguides. Following this, 1-D connected waveguide analysis with one of the soils listed in Table 10.1 (several soils) for two different protected structure (steel and concrete), response analysis is performed.

The aim here is to confirm the design consideration proposed earlier in this chapter. This example showed that sand is very effective in reflecting most of energy associated with the incident response thereby giving protection to the attached structure. Although the design methodology is based on 1-D wave propagation, the concept can be extended to higher dimensional waveguides. The approach highlighted in this chapter, shows the utility of understanding the wave propagation physics in designing structures for mitigating the effect of harsh loadings such a blast on civil infrastructure and personal.

In this chapter, no separate compute code is provided. All the codes provided in Chapters 4 & 7 can be readily used with the material properties of the soils. In this chapter, only 1-D wave propagation in dry sand waveguides was presented. Using the concepts presented in Chapter 7, the analysis can be extended to 2-D waveguides. Some of the problems provided in the exercises below may help the reader in extending the approach presented in this chapter to 2-D waveguides.

Exercises

10.1 The figure below show an infinite dry sand soil system made of two different dry sands, namely *Soil-1* and *Soil-2*, where these two soils can be any of the sands given in Table 10.1. The loading point is at $x = 0$.

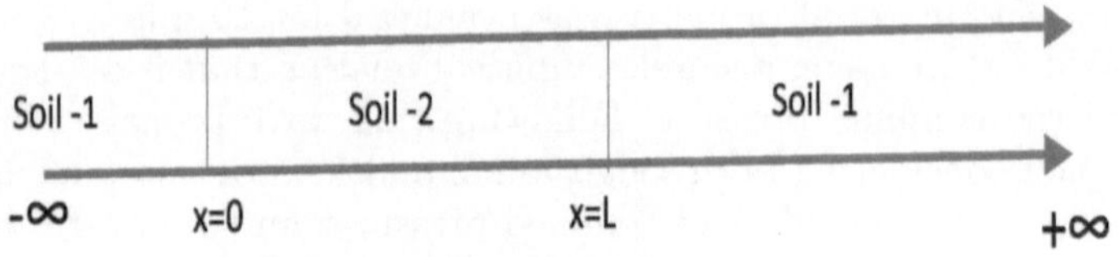

Problem 10.1

The loading can be either axial or transverse. For this dry sand system,

1. By using the equations of equilibrium and compatibility at the interface of the two soils, set up the equations for solving the wave propagation response for both axial as well as flexural responses
2. Modify the MATLAB script given in Chapter 4, to handle wave propagation in this two soil system and determine the axial and flexural response at $x = 0$, $x = 0.5L$ and $x = L$ for any two different soil system given in Table 10.1.

10.2 In Section 10.4, we derived the blast load pressure time profile acting on sand bunker (please refer to Fig. 10.7) where we assumed that the blast pressure wave will get completely reflected from the sand bunker with no propagation in the protected region. This is a reasonable assumption since the stiffness of the sand is very less compared to the protected structure. However, some energy always leaks into the soil. To obtain the blast pressure profile assuming that the propagation exist in the sand region, we can split the problem into three stages as shown below

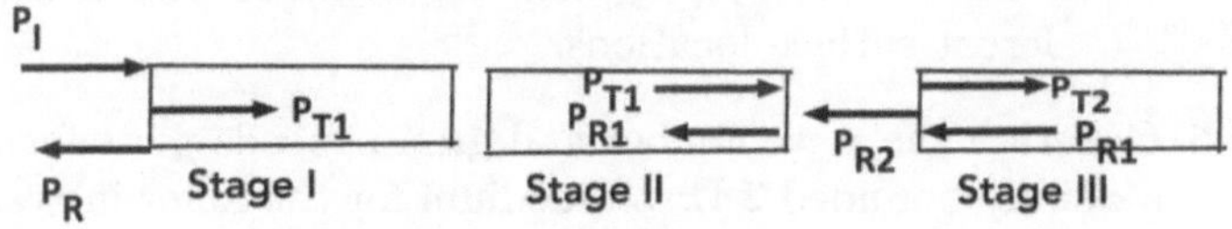

Figure for Problem 10.2

In the figure, P_I is the incident pressure pulse, which is known (see Eqn. (10.8)), P_R is the reflection from the interface. If we allow propagation of waves in the sand, a wave P_{T1} will be generated, which will hit the other sand interface with the protected structure and will generate a reflected wave P_{R1}. This reflected wave will travel back and will be partially reflected as P_{R2} and the rest will be radiated in air as P_{T2}.

1. Using the equations of equilibrium, compatibility and conservation of momentum (impedance relations), determine the unknown pressures P_R, P_{T1}, P_{R1}, P_{T2} and P_{R_2} and hence determine the pressure profile that will act on the sand bunker. Generate this pressure profile for few types of soils given in Table 10.1. Compare the obtained pressure profile with the profile obtained assuming no propagation in the sand region.
2. Use the obtained pressure profile and apply this both longitudinally and transversely on an infinite two soil system described in Problem 10.1 and obtain some responses for different soil types.

10.3 In this chapter, wave propagation in 1-D granular medium was discussed. Using the concepts developed in Chapter 7, the propagation of elastic waves in 2-D granular medium can be studied. The wave propagation in semi-infinite 2-D isotropic waveguides were studied in Section 7.4. The wavenumber for the P-wave and S-wave is given by Eqn. (7.27). Using this equation, plot the wavenumber and group speed variation for a granular 2-D medium made of soils given in Table 10.1 for different horizontal wavenumbers η and $\bar{\eta}$.

10.4 Rayleigh surface wave propagation in soils is a common phenomenon during earthquakes. The Rayleigh wave theory was presented in Section 7.4.3 and the examples of the surface wave propagation in isotropic solids was also presented in this section. Using the same equations presented in this section and modifying the MATLAB program for Rayleigh propagation given in Chapter 7, plot the following:

1. Rayleigh wavenumbers for 2-D granular medium made of different soils given in Table 10.1.
2. Using the blast pressure profile obtained in Problem 10.2, plot the surface wave propagation responses for different soils (sand) at different surface locations.

10.5 Generate the wavenumber and group speeds plots for different soils for a doubly bounded 2-D soil medium for the following cases by modifying the MATLAB program provided in Chapter 7:

1. When the top and the bottom surfaces are fixed.
2. When the top and the bottom surfaces are stress free (Lamb Wave case).

For the above two cases, plot the spectrum and dispersion plots for both symmetric and antisymmetric cases. All the relevant equations are given in Section 7.5.

CHAPTER 11

Wave Propagation in Non-Local One-Dimensional Waveguides

In all the previous chapters, we dealt with continuum theories defined by the theory of elasticity, wherein the stresses and strains are related by a constitutive model defined by Hooke's Law . The theories developed using the principles of theory of elasticity are called *Local Theories.* However, when one deals with structures of very small scales such as *nanostructures*, continuum theories based on theory of elasticity will not be able capture the wave physics at the these scales accurately. Analysis of structures at nano scale requires highly sophisticated mathematical tools such as *Molecular Dynamics* simulations, *Density Functional Theory*, etc. However, wave propagation simulations at nano scale using these sophisticated tools is extremely time consuming and computationally prohibitive. In the analysis of structures at nanoscale, the small scale effects that occurs at these scales, play a very important role in its behavior to applied loads. These small scale effects can be elegantly introduced within the framework of continuum theories using *Non-Local Theories.*

The continuum models, which describe a system in terms of a few variables such as mass, temperature, voltage and stress, are highly suited for direct measurements of these variables. Unlike these continuum models, the physical world is composed of atoms moving under the influence of their mutual interaction forces. Atomistic investigation helps to identify macroscopic quantities and their correlations, and enhance our understanding of various physical theories. In the new modeling philosophy, due to computationally expensive atomistic simulation, many researchers have tried to embed the micro

DOI: 10.1201/9781003120568-11

level properties such as bond-lengths into the continuum theories. A theory, which embeds micro-structural properties into theory of elasticity or local theory is called the non-local theory. There are two different non-local theories reported in the literature, namely, *Integral Theories* and *Gradient Based Theories.* In the integral theories, the governing equations are in the form of integral equations, which requires to be solved for obtaining responses. In the case of gradient theories, the constitutive models of the solid are in the form of integral equation, which are then converted to either stress/strain gradient forms using some lattice dynamics approximations. The examples of integral theory is the *peridynamic theory* and the details of the theory can be obtained from [102, 129]. In this chapter, we will not be reporting any wave propagation studies based on the integral theory based models as this topic is quite advanced and is beyond the scope of this book. The aim here is to present the wave physics in nanoscale structures that can be understood with the wave propagation theories presented until now in the earlier chapters. Hence, we will present the wave propagation studies only on some gradient based 1-D models, as these models can be handled with the knowledge gained from the matter presented in the earlier chapters. Among the many gradient non-local theories reported in the literature, in this chapter, we will present wave propagation studies in only two different gradient models, namely *Stress Gradient model* and *Strain Gradient model*, respectively. More detailed treatment of these theories and their application to nanostructures such as Carbon Nanotubes are given in the author's unified book [54] and [53].

Next, we will to address the need for non-local elasticity theories. In the atomistic regimes, the experimental evidence and observations with developed devices such as atomic force microscopes have suggested that classical theory of elasticity is not sufficient for an accurate and detailed description of the wave propagation phenomena at the nanometer scales. More notably, size effects could not be captured by theory of elasticity. Moreover, in the classical theory of elasticity, due to the assumption of point-to-point correspondence of stresses and strains, the elastic singularities arising due to application of point loads or occurring at dislocation lines and crack tips cannot be removed, and the same is true for discontinuities occurring at interfaces. Another important class of problems that could not be treated with classical theory is when the homogeneous stress strain curve contains regions where the slope is negative. This happens in regions where the material undergoes strain softening or a phase transformation. These are normally known as elastic or plastic instabilities and some examples of these instabilities are the twinning, martensitic transformations (elastic instabilities) and necking and shear banding (plastic instabilities). These instabilities are due to absence internal length parameter, which is a characteristic of microstructure of any material system appearing anywhere in the continuum theories. Non-local elasticity theories can address these situations effectively.

While analysing structures at the nanometer scale using continuum models, one would wonder if such models are valid at atomistic scale, although a significant amount of literature has shown that such models indeed predict the wave parameters to some reasonable accuracy in certain cases. Such continuum models use local theory of elasticity to derive the governing equations. There is a second school of thought that questions very much the use of continuum models for such atomistic simulations arguing that scale effects significantly affect the wave behavior at the nanometer scale and hence advocates molecular dynamics or some *ab initio* modeling principle, which is computationally prohibitive. The via media between these two approaches is the non-local elasticity models, which incorporates the scale effects in the continuum models.

In most of the non-local theories, it is assumed that stresses depend on the strains not only at an individual point under consideration but also at all surrounding points of this individual point. In this theory, the internal size or scale could be represented in the constitutive equations simply as material parameter. Such a non-local continuum mechanics model has been widely accepted and has been applied to many problems including wave propagation, dislocation, crack problems, etc. [68]. Application of non-local continuum theory to nanotechnology problems was initially addressed by Peddison et al. [68], in which the static deformation of a beam structures based on a simplified non-local model was analyzed. In this chapter, we will address wave propagation studies in some gradient based non-local 1-D waveguides. The models are so chosen such that they can be easily addressed within the framework of spectral analysis, which is essentially form the heart of wave propagation analysis in this book.

This chapter is organized as follows. First, the two different gradient elasticity theories and their respective constitutive models are briefly explained. Following this, the governing equation for these non-local models, their respective boundary condition derived. Following this, some response analysis is presented, which essentially brings out some unique wave propagation physics, which hitherto are not present in the local models.

11.1 ERINGEN'S STRESS GRADIENT THEORY

This theory was conceived by Eringen and he reported his investigations in [37, 39]. Eringen's Stress Gradient Theory (ESGT) of elasticity assumes an equivalent effect due to nearest neighbor interaction and beyond the single lattice in the sense of lattice average stress and strain. In this theory, it is assumed that the stress state at any reference point given by coordinates $\mathbf{x} = (x_1, x_2, x_3)$ in the body is assumed to be dependent not only on the strain state at this point but also on the strain states at all other neighboring points given by coordinate vector $\mathbf{x}'$ of the body. This is in accordance with atomic theory of lattice dynamics and experimental observations on phonon dispersion. The most general form of the constitutive relation in the non-local elasticity type representation involves an integral over the entire region of interest. The integral contains a non-local kernel function, which describes the relative influences of the strains at various locations on the stress at a given location. The constitutive equations of linear, homogeneous, isotropic, non-local elastic solid with zero body forces are given by [38].

$$\sigma_{kl},_k + \rho(f_l - \ddot{u}_l) = 0 \tag{11.1}$$

$$\sigma_{kl}(\mathbf{x}) = \int_{\Omega} \alpha(|\mathbf{x} - \mathbf{x}'|, \xi)\sigma^c_{kl}(\mathbf{x}')d\Omega(\mathbf{x}') \tag{11.2}$$

$$\sigma^c_{kl}(\mathbf{x}') = \lambda e_{rr}(\mathbf{x}')\delta_{kl} + 2\mu e_{kl}(\mathbf{x}') \tag{11.3}$$

$$e_{kl}(\mathbf{x}') = \frac{1}{2}\left(\frac{\partial u_k(\mathbf{x}')}{\partial x'_l} + \frac{\partial u_l(\mathbf{x}')}{\partial x'_k}\right) \tag{11.4}$$

Eqn. (11.1) is the equation of equilibrium, where σ_{kl}, ρ, f_l and u_l are the stress tensor, mass density, body force density and displacement vector at a reference point $\mathbf{x}$ in the body, respectively, at time t. Eqn. (11.3) is the classical constitutive relation where $\sigma^c_{kl}(\mathbf{x}')$ is the classical stress tensor at any point $\mathbf{x}'$ in the body, which is related to the linear strain tensor $e_{kl}(\mathbf{x}')$ at the same point through the Lamé constants λ and μ. Eqn. (11.4) is the classical strain-displacement relationship. The only difference between Eqns. (11.1)–(11.4) and the corresponding equations of classical elasticity is the introduction of Eqn. (11.2), which relates the global (or non-local) stress tensor σ_{kl} to the classical stress tensor $\sigma^c_{kl}(\mathbf{x}')$ using the modulus of non-localness. The modulus of non-localness or the non-local modulus $\alpha(|\mathbf{x} - \mathbf{x}'|, \xi)$ is the kernel of the integral Eqn. (11.2) and contains parameters which correspond to the non-localness [112]. A dimensional analysis of Eqn. (11.2) clearly shows that the non-local modulus has dimensions of (length)$^{-3}$ and hence it depends on a characteristic length ratio a/ℓ where a is an internal characteristic length (lattice parameter, size of grain, granular distance) and ℓ is an external characteristic length of the system (wavelength, crack length, size or dimensions of sample) [37], and Ω is the region occupied by the body. Therefore, the non-local modulus can be written in the following form:

$$\alpha = \alpha(|\mathbf{x} - \mathbf{x}'|, \xi), \quad \xi = \frac{e_0 a}{\ell} \tag{11.5}$$

where e_0 is a constant appropriate to the material and has to be determined for each material independently [37].

After making certain assumptions [37], the integro-partial differential equations of the stress gradient theory can be simplified to partial differential equations. For example, Eqn. (11.2) takes the following simple form:

$$(1 - \xi^2\ell^2\nabla^2)\sigma_{kl}(\mathbf{x}) = \sigma^c_{kl}(\mathbf{x}) = C_{klmn}\epsilon_{mn}(\mathbf{x}) \tag{11.6}$$

where C_{ijkl} is the elastic modulus tensor of classical isotropic elasticity and ϵ_{ij} is the strain tensor. where ∇^2 denotes the second-order spatial gradient applied on the stress tensor $\sigma_{kl,k}$ and $\xi = e_0 a/\ell$. The validity of Eqn. (11.6) is established by comparing the expressions for frequency of waves from the above ESGT model with those of the Born-Karman model of lattice dynamics [37]. Eringen reports a maximum difference of 6% and a perfect match for non-local constant value of $e_0 = 0.39$ [37]. A numerical model to predict the value of the non-local parameter for CNT type nanostructure is given in [53]. As $\xi \rightarrow 0$, α must revert to the Dirac delta measure so that classical elasticity limit is included in the limit of vanishing internal characteristic length, that is,

$$\lim_{\xi \rightarrow 0} \alpha(|\mathbf{x} - \mathbf{x}'|, \xi) = \delta(|\mathbf{x} - \mathbf{x}'|) \tag{11.7}$$

In ESGT, the stress at a reference point x is considered to be a function of the strain field at every point x' in the body. For homogeneous, isotropic bodies, the linear theory leads to a set of integro-partial differential equations for the displacement field, which are generally difficult to solve. For a special class of kernels, these equations are reducible to a set of singular partial differential equations. The selection of the appropriate class of kernels is not *ad hoc* but fairly general, based on mathematical conditions of admissibility and physical conditions of verifiability. For example, the dispersion curves available from lattice dynamics and phonon dispersion experiments provide excellent testing on the success of these kernels. Ultimately, these kernels should be expressed in terms of interatomic force potentials or correlation functions. Presently, several solutions obtained for various problems support the theory advanced by Eringen. For example, the dispersion curve, obtained for plane waves are in excellent agreement with those of the Born-Karman theory of lattice dynamics. The dislocation core and cohesive (theoretical) stress predicted by ESGT are close to those known in the physics of solids. Moreover, ESGT reduces to classical (local theory) in the long wavelength limit and to atomic lattice dynamics in the short wavelength limit. These and several other advantages make ESGT, an excellent approximation for a large class of physical phenomena with characteristic lengths ranging from microscopic to macroscopic scales. This situation becomes specially promising in dealing with imperfect solids, dislocations, and fracture, since in these problems, the internal (atomic) state of the body is difficult to characterize.

The constitutive model for ESGT can be obtained by expanding Eqn. (11.6). In the ID case, which is the subject of this chapter, it becomes

$$\sigma_{xx} - (e_0 a)^2 \frac{d^2 \sigma_{xx}}{dx^2} = E\epsilon_{xx} \tag{11.8}$$

In the above equation, σ_{xx} is the axial stress, ϵ_{xx} is the axial strain, e_0 is the material constant that is dependent on a particular material, and a is the characteristic length of the material (or lattice parameter). The parameter $e_0 a$ represents the small scale effect on the material stress-strain behavior. This equation will be used to obtain the governing equation as well as the wave parameters and wave propagation responses.

11.2 STRAIN GRADIENT THEORY

There are many strain gradient theories reported in the literature. The unified book of the author [54] gives many of these references. The most notable of them is the Mindlin theory [97] formulated in 1964. This theory's constitutive model gives a total of 1764 coefficients in the constitutive matrix out of which 903 are independent. Even for isotropic material assumption, the number of independent constants to be determined is equal to 18. It is indeed a daunting task to determine them and hence, this theory, although is rich in completeness and theoretical basis, has very little practical use. Hence, in order to increase its utility and reduce the number of constants to be determined, many researchers including Mindlin have made series of assumptions on deformation field of the microstructure and many such theories have been reported in the literature. For wave propagation analysis, we need a simple theory that not only have less number of constants in the constitutive model but also is stable. That is, one of the problems associated with the strain gradient theories is that, in some gradient theories, the governing differential equations are such that they give unstable or divergent solutions (see [8] for more details on stability issues associated with strain gradient theories). Hence, in this text, we will outline a very simple strain gradient theory that can provide good insight into propagation of waves in these non-local waveguides. One such theory is the *Laplacian based Strain Gradient Theory.*

Laplacian-based strain gradient theories are extensively used in static analysis, especially in those structures involving cracks, primarily to overcome the effects of stress singularities near the crack tips. However, its use in dynamic analysis is quite different, where it is primarily used to describe the dispersive wave propagation in a heterogeneous media. Laplacian based strain gradient theory can be derived using simple lattice model consisting of springs and masses as shown in Fig. 11.1.

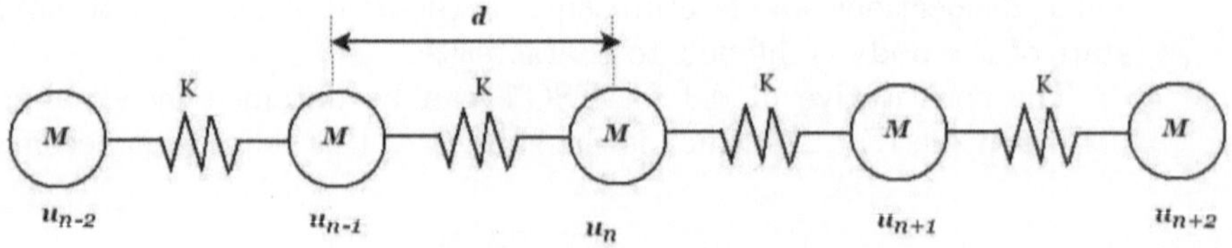

FIGURE 11.1: 1-D discrete lattice model

The figure shows the deformation of the 1-D lattice at discrete points *n+2*, *n+1*, *n*, *n-1* and $n-2$. Let us consider the 3 particles in this lattice at points *n,n-1, and n+1.* If we isolate these points, draw the free body diagram and apply Newton's second law, we get

$$M\ddot{u}_n = K(u_{n+1} - u_n) + K(u_{n-1} - u_n) = K(u_{n+1} - 2u_n + u_{n-1}) \quad (11.9)$$

Next, we will convert this discritized equation motion into a continuum equation. For this, we will expand the deformation of the $n+1$ and $n-1$ particles having a spacing d between them in Taylor's series as

$$u_{n-1} = u_n - d\frac{du_n}{dx} + d^2\frac{1}{2}\frac{d^2u_n}{dx^2} - ....$$

$$u_{n+1} = u_n + d\frac{du_n}{dx} + d^2\frac{1}{2}\frac{d^2u_n}{dx^2} + ... \quad (11.10)$$

We assume that homogenized lattice has a material property (Young's modulus) E and density ρ and they can be expressed in terms of the lattice spacing constant K and lattice mass M as $E = Kd/A$ and $M = \rho Ad$, where A is the area of the equivalent continuum and d is the spacing of particles in the lattice. Using Eqn. (11.10) in Eqn. (11.9) and ignoring the higher-order terms in the Taylor series and assuming $u_n = u(x,t)$, we get

$$E\left(\frac{d^2u}{dx^2} + \frac{1}{12}d^2\frac{d^4u}{dx^4}\right) = \rho\frac{d^2u}{dt^2} \quad (11.11)$$

Eqn. (11.11) can be rewritten using the strain displacement relations as

$$E\left(\frac{d\epsilon}{dx} + \frac{1}{12}d^2\frac{d^3\epsilon}{dx^3}\right) = \rho\frac{d^2u}{dt^2} \quad (11.12)$$

One can easily see that the terms on the left hand side of Eqn. (11.12) is nothing but $d\sigma/dx$, where

$$\sigma = E\left(\epsilon + \frac{1}{12}d^2\frac{d^2\epsilon}{dx^2}\right) \quad (11.13)$$

What we have derived is the constitutive model based on strain gradient elasticity for a simple 1-D lattice. The Eqn. (11.13) can be generalized to a body in a 3-D state of stress as

$$\sigma_{ij} = C_{ijkl}\left(\epsilon_{kl} + g^2\frac{d^2\epsilon_{kl}}{dm^2}\right) \quad (11.14)$$

where g is the length scale parameter, which is normally expressed in terms of lattice parameter d and C_{ijkl} is the fourth order tensor. The constitutive model given in Eqn. (11.14) has been proposed by a number of researchers [11, 28, 131, 141]. The main problem with the above constitutive models is that the final solutions provided by certain order of this constitutive model is neither unique nor stable. For example, the constructive model for the second-order strain gradient mode of 1-D waveguide, which is given by

$$\sigma(x) = E\left(\epsilon(x) + g^2 \frac{d^2\epsilon(x)}{dx^2}\right)$$

is shown to be unstable (see [53]). Similarly, the fourth order strain gradient model of the same nano rod, whose constitutive model is given by

$$\sigma(x) = E\left(\epsilon(x) + g^2 \frac{d^2\epsilon(x)}{dx^2} + g^4 \frac{d^4\epsilon(x)}{dx^4}\right)$$

is also found to be unstable (again see [102]). We will also show this aspect of stability from wave propagation point of view in this chapter. In the above expressions E is the Young's modulus of the non-local waveguide.

However, Aifantis and his coworkers [6, 7, 122] derived the strain gradient elasticity constitutive model as

$$\sigma_{ij} = C_{ijkl}\left(\epsilon_{kl} - g^2 \frac{d^2\epsilon_{kl}}{dm^2}\right) \tag{11.15}$$

Note that the main difference between Eqn. (11.14) and Eqn. (11.15) is that the negative sign before the higher-order strain terms. The above models said to be highly stable. We will also show this aspect from the wave propagation studies in this chapter. The sign of the gradient term and the issues of uniqueness and stability as opposed to their ability to describe dispersive wave propagation have opened up serious debates across the elasticity community; for example, see for instance the early study of Mindlin and Tiersten [96] and Yang and Gao [148], where this dilemma was aptly named as *sign paradox*. Comparative studies between the models with positive and negative sign are presented in [10, 139, 140]. As a result of this sign paradox, we study the wave propagation behavior in these two models and see how the stability and issue of non-uniqueness of solution affect the wave propagation behavior in the next section. We call the first model with positive sign before the higher-order strain terms as *SOSGR-P* model and the model with the negative sign before the higher-order strain terms as *SOSGR-N* model.

11.3 WAVE PROPAGATION IN NON-LOCAL WAVEGUIDES

In this chapter, we only study wave propagation behavior in a few non-local gradient models. Several other models are provided for the reader as exercises at the end of the chapter. The aim is to give a flavour of nature of wave propagation physics in these non-local waveguides. In addition, the nature of governing equations in these waveguides are such that they are easily tractable within the framework of spectral analysis. In this chapter, we will study wave propagation in the following non-local waveguides. The model abbreviation is indicated in brackets:

- Elementary Eringen Stress Gradient Rod (ESGR).
- Second-order Strain Gradient Rod with positive sign in constitutive model (SOSGR-P)
- Second-order Strain Gradient Rod with negative sign in constitutive model (SOSGR-N)
- Fourth order Strain Gradient Rod (FOSGR)
- Euler-Bernoulli Eringen Stress Gradient Beam (ESGB)

For each of these models, the governing partial differential equations is derived followed by their dispersion relations and finally the wave response analysis is studied in some of these waveguides.The complete MATLAB scripts is provided for wave propagation analysis for each of these waveguides. These codes is housed in the publisher website. The reader can run these scripts and can visualize the propagation of waves in these waveguides.

11.3.1 Wave Propagation in Eringen Stress Gradient Rod (ESGR)

The governing partial differential equation for ESG rod is first derived. Before proceeding further, following points have to be borne in mind while performing wave propagation in non-local waveguides:

1. Non-local waveguide analysis are normally performed at atomistic scales. The scale parameter present in constitutive model (e_0a) has a dimension of nanometer. Correspondingly, in order to propagate waves at these spatial scales, the temporal scale should be of the order of *pico seconds*, instead of microseconds that we used in the analysis of waveguides modeled using local theory. Hence, the frequency content of the pico-seconds input will be of the order of Terra Hertz.
2. To propagate waves in the Terra Hertz range, it is required that the waveguide material be stiff as is case in carbon nanotubes. Hence, in all the simulation performed in this chapter, following material and sectional properties are used: Young's modulus $E = 1.06\,TPa$, density $\rho = 2240\,\text{kg/m}^3$, width $b = 5\,\text{nm}$ and depth $d = 5\,\text{nm}$.

The constitutive model for Eringen theory was discussed in the previous section and is given by Eqn. (11.8). If x is the axial coordinate, the displacement and strain fields for this model is given by

$$u = u(x,t), \qquad \epsilon_{xx} = \frac{\partial u}{\partial x} \tag{11.16}$$

For thin rods, Eqn. (11.8) can be written in the following form

$$\sigma_{xx} - (e_0 a)^2 \frac{\partial^2 \sigma_{xx}}{\partial x^2} = E\epsilon_{xx} = E\frac{\partial u}{\partial x} \tag{11.17}$$

where E is the modulus of elasticity, σ_{xx} and ϵ_{xx} are the local stress and strain components in the x-direction, respectively. The equation of motion for an axial rod can be obtained as

$$\frac{\partial N}{\partial x} = \rho A \frac{\partial^2 u}{\partial t^2} \tag{11.18}$$

where N is the axial force per unit length, and they are defined by the standard definition given by

$$N = \int_A \sigma_{xx} dA \tag{11.19}$$

Using Eqns. (11.19 & 11.17), we have

$$N - (e_0 a)^2 \frac{\partial^2 N}{\partial x^2} = EA\frac{\partial u}{\partial x} \tag{11.20}$$

Substitution of the first derivative of N from Eqn. (11.18) into Eqn. (11.20), we obtain

$$N = EA\frac{\partial u}{\partial x} + (e_0 a)^2 \rho A \frac{\partial^3 u}{\partial x \partial t^2} \tag{11.21}$$

where, A is the area of cross section of the rod. Substituting N from Eqn. (11.21) into the equation of motion (11.18), we obtain

$$EA\frac{\partial^2 u}{\partial x^2} + (e_0 a)^2 \rho A \frac{\partial^4 u}{\partial x^2 \partial t^2} = \rho A \frac{\partial^2 u}{\partial t^2} \tag{11.22}$$

Eqn. (11.22) is the governing equation of motion for non-local ESG rod model. When $e_0 a = 0$, it is reduced to the equation of classical rod model.

Next, in order to determine the wave parameters, namely the wavenumber and the group speeds, we will perform spectral analysis on Eqn. (11.22). That is, we will assume the spectral form of solution for the dependent variable $u(x,t)$ as

$$u(x,t) = \sum_{n=0}^{M} \hat{u}(x, \omega_n) e^{i\omega t} \tag{11.23}$$

where $\hat{u}(x,\omega)$ is the DFT of $u(x,t)$. Substituting Eqn. (11.23) in Eqn. (11.22), the governing PDE is converted to n ODE's with frequency ω as a parameter. This is given by

$$EA\frac{d^2 \hat{u}_n}{dx^2} - \omega^2 (e0a)^2 \rho A \frac{d^2 \hat{u}_n}{dx^2} - \rho A \omega^2 \hat{u_n} = 0 \tag{11.24}$$

The above ordinary differential equation is of constant coefficient type and hence we can assume the solution of the type $u(x,t) = \mathbf{A}e^{ikx}$, where k is the wavenumber. This solution is substituted in Eqn. (11.24) to get the characteristic equation for the solution of wavenumber, which is given by:,

$$-k^2 + (e_0a)^2\eta^2\omega^2k^2 + \eta^2\omega^2 = 0 \tag{11.25}$$

where $\eta = \sqrt{\frac{\rho}{E}}$. This is a quadratic equation in k, whose solution is given by

$$k_{1,2} = \pm\sqrt{\frac{\eta^2\omega^2}{1-(e_0a)^2\eta^2\omega^2}} \tag{11.26}$$

We see that the wavenumber is a function of length scale parameter e_0a and when $e_0a \to 0$, the wavenumber $k \to \omega\sqrt{\rho A/EA}$, which is the wavenumber of a uniform local rod. Looking at Eqn. (11.26), we can see that, wavenumber escape to infinity when

$$\omega = \omega_{escape} = \frac{1}{e_0a\eta} = \frac{1}{e_0a}\sqrt{\frac{E}{\rho}} \tag{11.27}$$

This frequency at which the wavenumber escape to infinity is called *Escape Frequency*. We will plot the wavenumber variation with frequency as a function of e_0a little later.

The phase and group speed expression is obtained as follows. As before, the phase speed of the ESG model is given by

$$C_p = \frac{\omega}{k} = \frac{1}{\eta}\left[1-(e_0a)^2\eta^2\omega^2\right]^{1/2} \tag{11.28}$$

The group speed is computed by differentiating Eqn. (11.25) and simplifying, we get

$$C_g = \frac{\partial\omega}{\partial k} = \frac{k(1-(e_0a)^2\eta^2\omega^2)}{\omega\eta^2} \tag{11.29}$$

Looking at Eqns. (11.25, 11.28 & 11.29), we can see that the waves are dispersive in nature and the wave does not propagate beyond escape frequency. That is, the axial response in a ESG model is band limited and this frequency band depends on the value of e_0a. To understand the wave propagation phenomenon better, we plot the wavenumber and group speeds as a function of scale parameter e_0a. This plot is shown in Fig. 11.2 (a) & (b).

From Fig. 11.2 (a), we can clearly see the escape frequency and this value decreases with increase in the value of e_0a. This is also clearly seen in Fig. 11.2 (c). This is also evident from Eqn. (11.27), that the escape frequency value is inversely proportional to the scale parameter.

Fig. 11.2 (b) shows the group speed variation as a function of scale parameter e_0a. The speeds go to zero at escape velocity and one important aspect

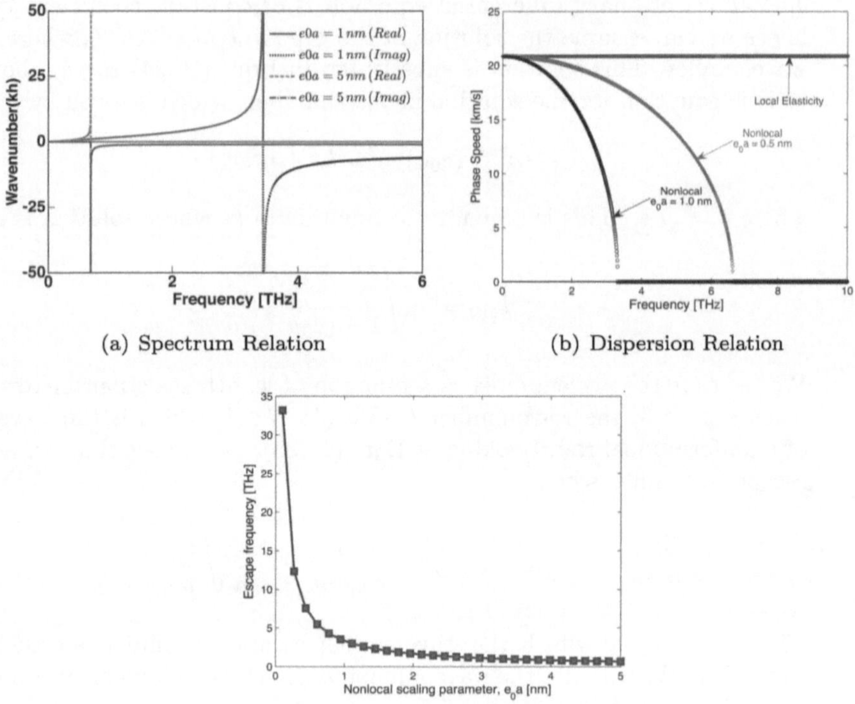

(a) Spectrum Relation (b) Dispersion Relation

(c) Escape Frequency Variation

FIGURE 11.2: Spectrum, Dispersion and Escape frequency variation in ESG rod

is that there is absolutely no propagation beyond escape frequency. Such an interesting behavior is not seen in the local rod.

Next, we will study the wave propagation responses on an infinite ESG rod. To study this, we will again consider a non-local waveguide of material and sectional property as indicated earlier. That is, E is assumed 1.06GPa with density of $2270\,\text{kg/m}^3$. Both width and depth of the non-local waveguide is assumed at 5 nm. To determine the response, we need to solve Eqn. (11.24), which is given by

$$\hat{u}(x, \omega) = \mathbf{A}\, e^{-ikx} \tag{11.30}$$

where k is the wavenumber given by Eqn. (11.26). We can consider any type of signal to study the propagation. However, we will use a tone burst signal. The reason for using this signal will become clear little later. In the earlier chapter, we had used the input signal as the wave coefficient $\mathbf{A}$. However, we can determine the value of $\mathbf{A}$ by considering the force equilibrium at $x = 0$. The internal force expression ($\hat{N}$) for a non-local ESM rod (obtained using

Eqn. (11.21)) is given by

$$\hat{N} = EA(1 - (e_0 a)\rho A\omega^2)\frac{d\hat{u}}{dx} \tag{11.31}$$

To obtain the wave coefficient $\mathbf{A}$ in Eqn. (11.30), we impose the boundary condition at $x = 0$, as $\hat{N} = \hat{P}$, where $P(x,t)$ is the input signal and $\hat{P}$ is its Fourier Transform. Differentiating Eqn. (11.30) and using it in Eqn. (11.32) and imposing the above boundary conditions, will yield the wave coefficient $\mathbf{A}$ as

$$\mathbf{A} = \frac{i}{(EA - (e_0 a)\rho A\omega^2)k} \tag{11.32}$$

Using the computed wave coefficient, we can now evaluate the wave responses. The goal here is to not only see the effect of scale parameter $e_0 a$ on the wave propagation response but also to confirm the group speed plot given in Fig. 11.29(b) that the propagation does not happen beyond the escape frequency. Hence, two tone burst signal is considered, with one having a central frequency of 1 THz and the second having a central frequency of $5\,Tz$. The non-local scale parameter $e_0 a$ is assumed to be 1 nanometer. From the group speed plot shown in Fig. 11.2(b) for $e_0 a = 1\,\text{nm}$, we see full propagation at the frequency of 1 THz, while at the frequency of 5 THz, which is beyond the escape frequency, there is no propagation. The two different tone burst signals is shown in Figs. 11.3 (a) and (b), respectively. The inset in these figures shows the FFT of the signal.

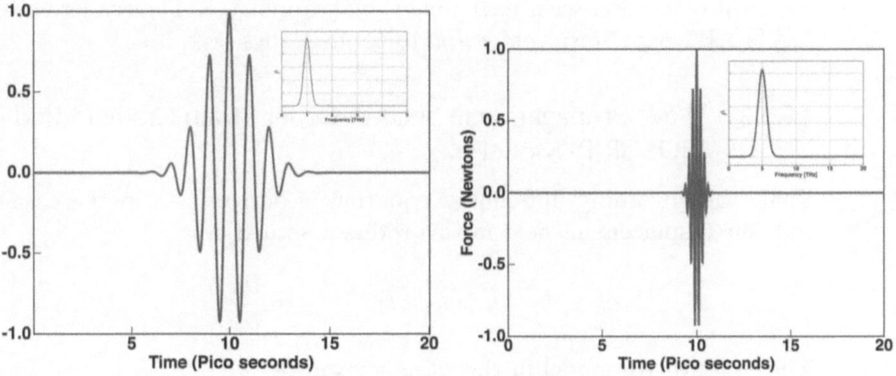

(a) Signal with central frequency of 1 THz (b) Signal with central frequency of 1 THz

FIGURE 11.3: Two different tone burst signals and their FFT

Next, we will input these signals on an infinite non-local ESG rod having $e_0 a = 1\,\text{nm}$ and the wave is allowed to propagate a distance of 5 nm. The responses (axial velocities) are plotted in Fig. 11.4. From the figure we see

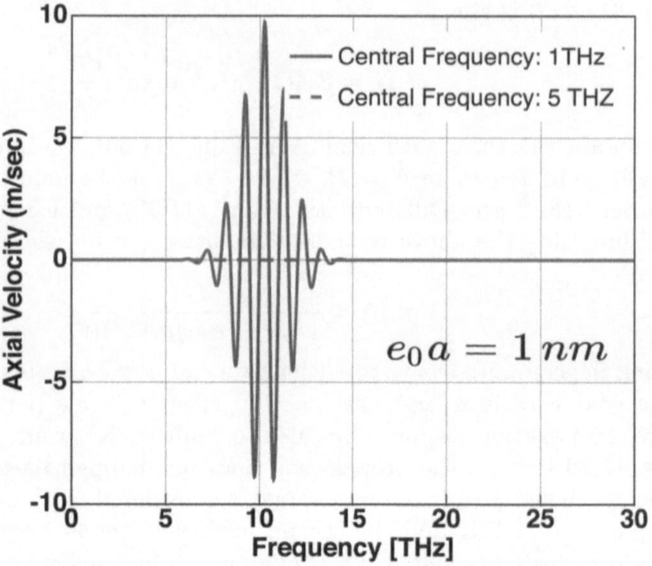

FIGURE 11.4: Axial velocity response at $5\,nm$ distance in an infinite ESM rod having $e_0a = 1\,\text{nm}$.

the propagation of signal having central frequency of 1 THz (Fig. 11.3 (a)), while for the signal having a central frequency of 5 THz (Fig. 11.3 (a)), we see no propagation with negligible signal amplitude. The reader can run the MATLAB script *RespNonlocal.m* to confirm these results.

11.3.2 Wave Propagation in Second-Order Strain Gradient Rod (SOSGR-P) Model

First, the governing differential equation is derived. As in the case of ESG rod, the displacement field in this rod is assumed as

$$u = u(x,t), \qquad \epsilon = \frac{\partial u}{\partial x} \tag{11.33}$$

The constitutive model in this case is given by

$$\sigma(x) = E\left[\epsilon(x) + g^2 \frac{d^2\epsilon(x)}{dx^2}\right] \tag{11.34}$$

As before, the equation of equilibrium is given by

$$\frac{\partial N}{\partial x} = \rho A \frac{\partial^2 u}{\partial t^2} \tag{11.35}$$

where N is the axial force per unit length, and they are defined by the standard definition given by

$$N = \int_A \sigma_{xx} dA \tag{11.36}$$

Substituting for $\sigma(x)$ from Eqn. (11.34) in Eqn. (11.35) and using the expression for ϵ from Eqn. (11.33), the axial force N can be written as

$$N = EA\left[\frac{\partial u}{\partial x} + g^2\frac{\partial^3 u}{\partial x^3}\right] \tag{11.37}$$

Substituting Eqn. (11.37) in Eqn. (11.35), we get the following governing PDE, that governs the motion of SOSGR-P model

$$EAg^2\frac{\partial^4 u}{\partial x^4} + EA\frac{\partial^2 u}{\partial x^2} - \rho A\frac{\partial^2 u}{\partial t^2} = 0 \tag{11.38}$$

where E, A, and ρ are the Young's modulus, area of cross section and density, respectively.

Before studying wave propagation aspects, we will see the uniqueness of the solution given by this model. To do this, we will look at only the static solutions. That is, we solve Eqn. (11.38) without the inertial terms, which is a ordinary differential equation given by

$$EAg^2\frac{d^4 u}{dx^4} + EA\frac{d^2 u}{\partial x^2} = 0 \tag{11.39}$$

The solution of this equation is given by

$$u(x) = C_1 + C_2 x + C_3 \sin\left(\frac{x}{g}\right) + C_4 \cos\left(\frac{x}{g}\right) \tag{11.40}$$

where C_i are constants that have to be determined according to the boundary conditions. The response of the classical continuum local rod will be governed by constants C_1 and C_2 only. Following [8], the uniqueness of the static analytical solution is investigated by imposing the condition $u(x) = 0$ at $x = 0$, L and $\frac{du}{dx} = 0$ at $x = 0$, L. Imposing this condition, we get the following matrix equation:

$$\begin{bmatrix} 1 & 0 & 0 & 1 \\ 1 & L & \sin\left(\frac{L}{g}\right) & \cos\left(\frac{L}{g}\right) \\ 0 & 1 & \frac{1}{L} & 0 \\ 0 & 1 & \frac{1}{L}\cos\left(\frac{L}{g}\right) & -\frac{1}{L}\sin\left(\frac{L}{g}\right) \end{bmatrix} \begin{Bmatrix} C_1 \\ C_2 \\ C_3 \\ C_4 \end{Bmatrix} = \begin{Bmatrix} 0 \\ 0 \\ 0 \\ 0 \end{Bmatrix} \tag{11.41}$$

where L is a representative length of the non-local rod. For the solution to be unique, we need determinant of the matrix in Eqn. (11.41) not to vanish for any L. Expanding and simplifying, the determinant of the matrix is given by

$$\frac{L}{g}\sin\left(\frac{L}{g}\right) + 2\cos\left(\frac{L}{g}\right) - 2 = 0 \tag{11.42}$$

Looking at Eqn. (11.42), we see that determinant vanishes for $L = 2\pi ng$, where n is an arbitrary integer. Thus, we can clearly infer that the constitutive model given by Eqn. (11.34) does not guarantee unique solution.

Next, the wave propagation behavior is studied using spectral analysis. The governing PDE (Eqn. (11.38)) is transformed to frequency domain using DFT given in Eqn. (11.23), The transformed equation becomes an ordinary differential equation, which is given by

$$EAg^2\frac{d^4\hat{u}}{dx^4} + EA\frac{d^2\hat{u}}{dx^2} + \rho A\omega^2\hat{u} = 0 \tag{11.43}$$

This differential equation is of constant coefficient type and its solution is of the type $\mathbf{A}e^{-ikx}$, where k is the wavenumber that needs to be determined. Using this in Eqn. (11.43), we get the characteristic equation for determining the wavenumber, which is given by

$$g^2k^4 - k^2 + \eta^2\omega^2 = 0 \tag{11.44}$$

where, $\eta = \sqrt{\rho/E}$. The above equation is quadratic in k^2 and its solution for wavenumber is given by

$$k_1^2 = \frac{1+\sqrt{1-4g^2\eta^2\omega^2}}{2g^2}, \qquad k_1^2 = \frac{1-\sqrt{1-4g^2\eta^2\omega^2}}{2g^2}$$

$$k_1 = \pm\sqrt{k_1^2}, \qquad k_2 = \pm\sqrt{k_2^2} \tag{11.45}$$

The phase speed as before is obtained by the relation $C_P = Real(\omega/k)$, where k is substituted with k_1 or k_2 (Eqn. (11.45) to obtain the phase speeds of the two modes of propagation. The group speed is obtained using the expression $C_g = Real(d\omega/dk)$. Here, it can be computed by differentiating Eqn. (11.44) and simplifying and the resulting expression is given by

$$C_g = Real\left[\frac{k(1-2g^2k^2)}{\omega\eta^2}\right] \tag{11.46}$$

From the group speed expression given in Eqn. (11.46), the group speed goes to zero when $(1-2g^2k^2) \to 0$ for both the wave modes. Substituting for k_1 and k_2 from Eqn. (11.45), we see that, the frequency at which the group speed goes to zero (we call it as transition frequency ω_{trans} is given by

$$\omega_{trans} = \frac{1}{2\eta g} \tag{11.47}$$

We now plot the wavenumbers and group speeds for a non-local rod modeled using second-order strain gradient theory. Here as before, we use the following properties

$$E = 1.06\,\text{TPa}, \qquad \rho = 2270\,\text{kg}^3, \qquad g = 1.0\,\text{nm}$$

Fig. 11.5 shows the wavenumber and group speeds for the two wave modes.

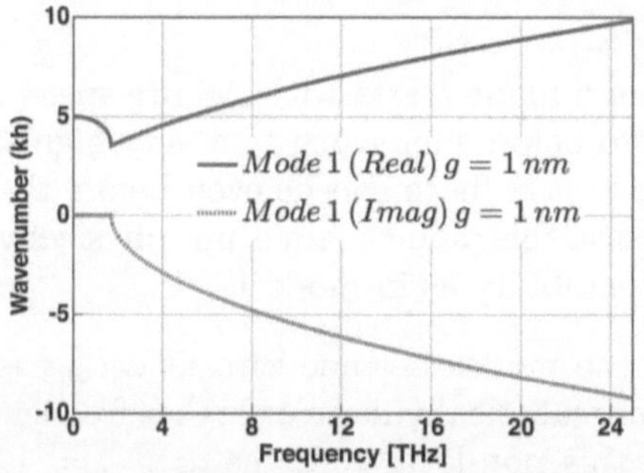

(a) Wavenumber: Mode 1

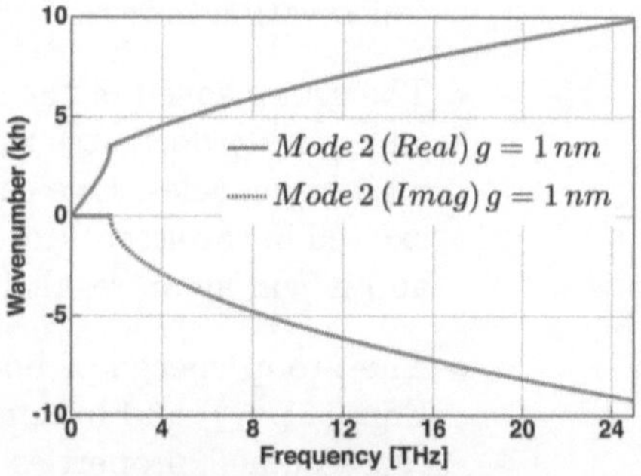

(b) Wavenumber: Mode 2

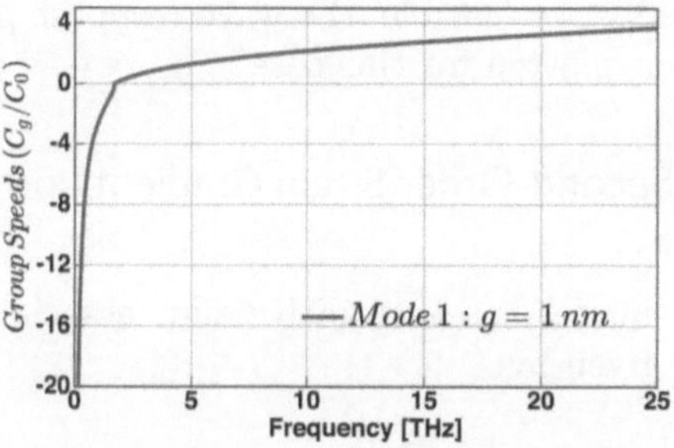

(c) Group speeds: Mode 1

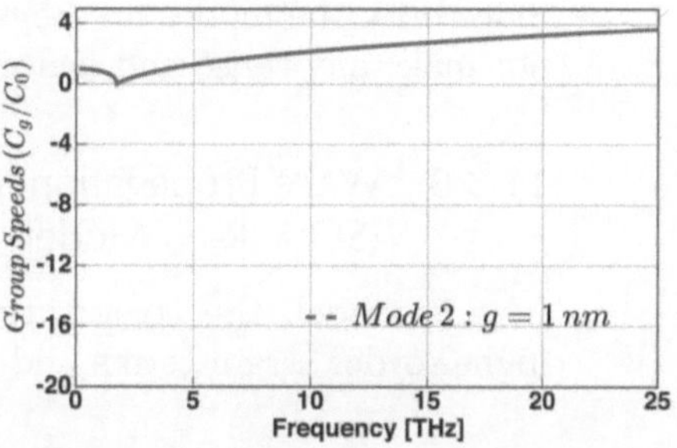

(d) Group speeds: Mode 2

FIGURE 11.5: Dispersion relations for a SOSGR-P model

Figs. 11.5(a) & (b) show the wavenumbers of mode 1 and mode 2. Both the wavenumbers and group speeds (shown in Figs. 11.5(c) & (d)) are plotted to the same scale to know the relative difference between them. From these plots, following observations can be made as regards the wave propagation in SOSGR-P model:

- The wavenumbers of both the modes are complex.
- Both wavenumbers have substantial imaginary part. This imaginary part will make the propagation of wave very difficult.
- Both wavenumbers abruptly changes its slope. This is one of the indications that group speed can become negative at certain frequency.

- Fig. 11.5 (c) & (d) show the group speeds for the two wave modes. The speed of model1 start at $-\infty$ at $\omega = 0$ and increases to unrealistic infinite speed at high-frequency, which is not a feasible mode. This is due to non-unique solution provided by the model (that was also discussed in this section) and in addition, the positive sign in the higher-order strain terms in the constitutive model of SOSGR-P, causes mode instability and such a propagation cannot happen as it violates the energy conservation laws.

- The group speed of the second mode starts with a finite speed at $\omega = 0$ and quickly decreases to zero before increasing to a very high speeds at high frequencies. One cannot have finite speeds even before the wave is excited in the structure. Hence, this mode again is not physically feasible and it will again result in instability as in mode 1.

- The group speeds of both the modes become zero at ω_{trans} (given by Eqn. (11.47)) and become unrealistically large at higher frequencies. For the assumed properties of this non-local rod, the $\omega_{trans} = 1.72THz$, which is clearly seen in Fig. 11.5.

Since both the modes are physically not feasible, the wave response plot does not make any sense and hence not plotted for this case

11.3.3 Wave Propagation in Second-Order Strain Gradient Rod (SOSGR-N) Model

In this model, the constitutive model has a negative sign associated with higher-order strain terms and is given by

$$\sigma(x) = E\left[\epsilon(x) - g^2\frac{d^2\epsilon(x)}{dx^2}\right] \tag{11.48}$$

Following the procedure followed for SOSGR-P model, the internal force N is given by

$$N = EA\left[\frac{\partial u}{dx} - g^2\frac{d^3\partial u}{dx^3}\right] \tag{11.49}$$

and the governing PDE becomes

$$-EAg^2\frac{\partial^4 u}{\partial x^4} + EA\frac{\partial^2 u}{\partial x^2} - \rho A\frac{\partial^2 u}{\partial t^2} = 0 \tag{11.50}$$

We will perform spectral analysis on Eqn. (11.50) by transforming the equation to frequency domain as was done in the case of SOSGR-P model in the previous subsection. The resulting characteristic equation for computing wavenumber is given by

$$g^2k^4 + k^2 - \eta^2\omega^2 = 0 \tag{11.51}$$

The wavenumber expression for the two modes is given by

$$k_1^2 = \frac{-1+\sqrt{1+4g^2\eta^2\omega^2}}{2g^2}, \qquad k_1^2 = \frac{-1-\sqrt{1+4g^2\eta^2\omega^2}}{2g^2}$$
$$k_1 = \pm\sqrt{k_1^2}, \qquad k_2 = \pm\sqrt{k_2^2} \tag{11.52}$$

Group speeds expressions are obtain in a similar manner, and they are given by

$$C_g = Real\left[\frac{k(1+2g^2k^2)}{\omega\eta^2}\right] \tag{11.53}$$

From above expression, we can see that the group speeds does not transition to zero as was seen in SOSGR-P model. The wavenumbers and group speeds for the two modes are shown in Fig. 11.6 for two different values of scale parameter, namely $g = 1\,\text{nm}$ and $g = 5\,\text{nm}$, respectively.

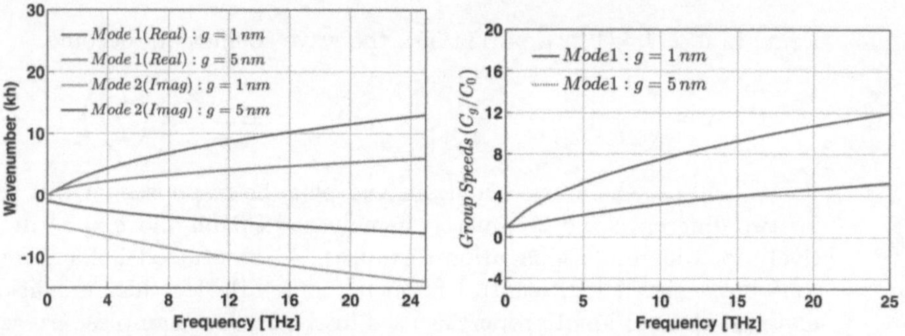

(a) Wavenumber (b) Group speeds

FIGURE 11.6: Spectrum and Dispersion relations for a SOSGR-N model for different scale parameter g

From Figs. 11.6 (a) and (b), we can infer the following:

- The wave propagation behavior, unlike SOSGR-P model, is very stable, although model gives very high group speed at high-frequency.
- The wave propagation behavior is more like that of Euler-Bernoulli beam, where in one mode is propagating, while the second mode is completely evanescent . Hence, there is only on propagating mode.
- The wavenumber of the second mode is completely imaginary and the imaginary component decreases with increase in the scale parameter g. In other words, the wave amplitude will be significantly lower for lower value of g.

- The group speed variation is similar to that of Euler-Bernoulli beam indicating the dispersive nature of the waves in this non-local rod model.

Since the model is a stable model, we will now compute the wave response for the same $5\,pico - secs$ tone burst signal shown in Fig. 11.3 (b) on an infinite SOSGR-N model. The expression for the response is the solution of the governing PDE (Eqn. (11.50)), which is given by

$$\hat{u}(x,\omega) = \mathbf{A}e^{-ik_1x} + \mathbf{B}e^{-ik_2x} \tag{11.54}$$

Note that the solution contains only the forward moving waves as we are considering only an infinite rod. Here wavenumbers k_1 and k_2 is given by Eqn. (11.52). We need to compute the wave coefficients **A** and **B** from the essential and natural boundary conditions, which is given by

$$\text{At } x = 0 \; \frac{d^2\hat{u}}{dx^2} = 0, \qquad \text{and} \qquad EA\frac{d\hat{u}}{dx} - EAg^2\frac{d^3\hat{u}}{dx^3} = \hat{F}/2 \tag{11.55}$$

Using Eqn. (11.54) in Eqn. (11.55), the wave coefficients become

$$\mathbf{A} = \left(\frac{ik_2}{2EAk_1(k_1-k_2)(1-g^2)}\right), \qquad \mathbf{B} = -\mathbf{A}\left(\frac{k_1^2}{k_2^2}\right) \tag{11.56}$$

We will use the above equations and plot the responses for this rod model for two different scale parameter, namely $g = 1.0\,\text{nm}$ and $g = 5.0\,\text{nm}$, respectively for different propagation distances. As mentioned before, we will use tone burst signal with central frequency of 5 THz for this simulation. Same material and sectional properties used in the earlier examples, are again used here. The plot of axial velocity is shown in Fig. 11.7.

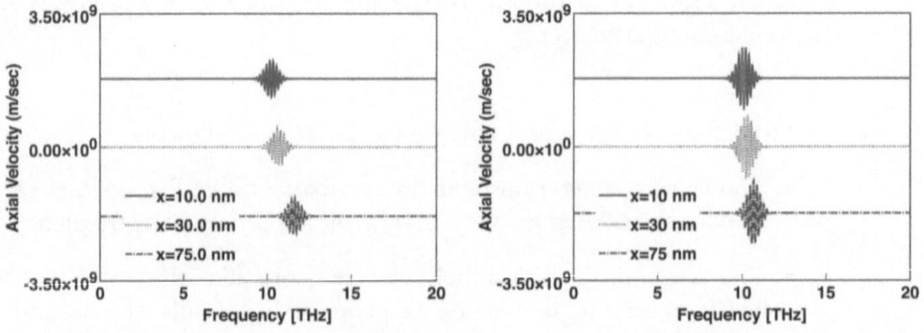

(a) Response for $g = 1.0\,nm$ (b) Response for $g = 5.0\,nm$

FIGURE 11.7: Axial velocity response for an infinite SOSGR-N model for different scale parameter g at different propagation distances

From Fig. 11.7, we see that, the amplitude of velocity response is very high. This is because, the obtained displacement $\hat{u}$ is multiplied by $i\omega$ to obtain velocities. Since, the frequencies are of Terra Hertz range, causing velocity amplitudes to be high. Figure 11.7(a) and (b) are plotted to same scale so that the difference in the responses for the two cases can be easily understood. The amplitude of the response for the case of $g = 1.0\,nm$ (Fig. 11.7(a)) is much higher than the response obtained for the case of $g = 5.0\,nm$ (Fig. 11.7(b)). This is because, the imaginary component of wavenumber decreases with increase in the value of g (please see Fig. 11.6). Also, the group speed magnitude increases with increase in g for a given frequency ω. This manifest into early arriving of the incident wave at a particular location for the case of high values of scale parameter g in the response plots.

11.3.4 Wave Propagation in Fourth-Order Strain Gradient Rod (FOSGR) Model

Next, we will study the propagation of elastic waves in the fourth order strain gradient model rod, where the constitutive equation for this non-local model will have both second-order and fourth order strain terms. That is, the constitutive equation for this model is given by

$$\sigma(x) = E\left(\epsilon(x) + l_1^2 \frac{\partial^2 \epsilon(x)}{\partial x^2} + l_2^4 \frac{\partial^4 \epsilon(x)}{\partial x^4}\right) \tag{11.57}$$

where $l_1 = d^2/12$ and $l_2 = d^4/360$ and d is the atomistic distance expressed normally in the *nanometer* range. The constitutive model was derived in [12], and it is adapted here. As before, we calculate the internal normal force N, which is given by

$$N = \int_A \sigma(x) dA \qquad = EA\left(\frac{\partial u}{\partial x} + l_1 \frac{\partial^3 u}{\partial x^3} + l_2 \frac{\partial^5 u}{\partial x^5}\right) \tag{11.58}$$

The governing equation for this model as before is given by Eqn. (11.18), where the expression for N from Eqn. (11.58) is substituted to get the following governing equation

$$EA\frac{\partial^2 u}{\partial x^2} + EAl_1 \frac{\partial^4 u}{\partial x^4} + EAl_2 \frac{\partial^6 u}{\partial x^6} = \rho A \frac{\partial^2 u}{\partial t^2} \tag{11.59}$$

This is the sixth order equation in space and second-order in time. Hence, there will be three forward moving modes, while rest three are backward moving wave modes.

The next step is to determine the dispersion relations for which we will perform spectral analysis. The first step in the spectral analysis is to transform the governing equation to frequency domain using DFT (Eqn. (11.23)). The transformed equation is now an ordinary differential equation, given by

$$EA\frac{d^2\hat{u}}{dx^2} + EAl_1\frac{d^4\hat{u}}{dx^4} + EAl_2\frac{d^6\hat{u}}{dx^6} + \rho A\omega^2\hat{u} = 0 \tag{11.60}$$

Eqn. (11.60) is a differential equation with constant coefficient and will have solution of the form $\hat{u} = \mathbf{A}e^{-ikx}$, which when substituted in Eqn. (11.60) will yield the characteristic polynomial equation for determining the wavenumber k, which is given by

$$l_2k^6 - l_1k^4 + k^2 - \eta^2\omega^2 = 0 \tag{11.61}$$

where $\eta = \sqrt{\rho/E}$. This equation is solved using *roots* function in MATLAB to find all the 3 forward moving wavenumbers, k_1, k_2, & k_3. MATLAB script *dispNonlocal.m* provided with this book will solve the dispersion relations for this model. The equation is quadratic in k^2, and hence we expect, k_2^2 to be complex conjugate of k_1^2, which means, there will be a real mode k_3 while modes k_1 and k_2 would be complex conjugates

The group speed $C_g = Real(\partial\omega/\partial k)$ is computed by differentiating Eqn. (11.61) and simplifying, and it is given by

$$C_g = \frac{3l_2k^5 - 2l_1k^3 + k}{\eta^2\omega} \tag{11.62}$$

where, the speeds for the three wave modes are obtained from Eqn. (11.62) by substituting the values of k_1, k_2 and k_3 obtained from solving Eqn. (11.61). The wavenumbers of the three modes and their respective group speeds is shown in Fig. 11.8. The same properties as used in the earlier simulation is again used here. The atomistic distance $d = 1.0e{-}6m$ is used in the simulation.

From Fig. 11.8, following can be inferred

- Magnitude of wavenumber is very small and hence group speeds are unreasonable high
- The group speeds of Mode 1 and Mode 2 start at negative infinity at $\omega = 0$ and increase to very unreasonably very high speeds at higher frequencies
- There is only one possible propagating mode for this model

In summary, FOSGR is not a stable model and hence response analysis is not performed on this model.

11.3.5 Wave Propagation in Euler-Bernoulli Eringen Stress Gradient Beam (ESGB)

The four previous non-local model studied in this chapter dealt with waveguides undergoing axial motion. In this section, we will study the wave propagation in non-local waveguides that undergo transverse motion. As before,

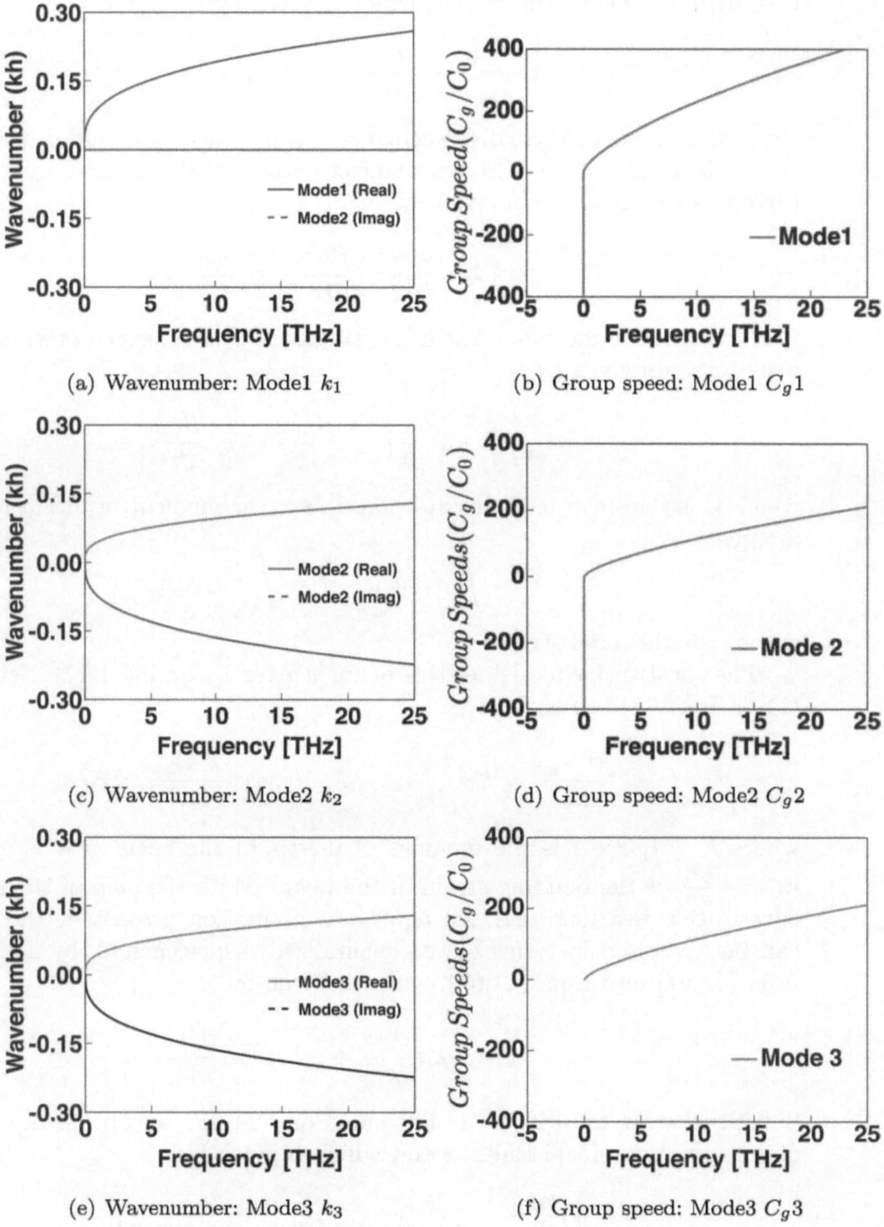

(a) Wavenumber: Mode1 k_1

(b) Group speed: Mode1 C_g1

(c) Wavenumber: Mode2 k_2

(d) Group speed: Mode2 C_g2

(e) Wavenumber: Mode3 k_3

(f) Group speed: Mode3 C_g3

FIGURE 11.8: Spectral and Dispersion relations for fourth order strain gradient rod model we will be first derive the governing differential equation for which we need to first write the beam kinematics, which is given by

$$u(x,z,t) = u^0 - z\frac{\partial w}{\partial x}, \qquad w(x,z,t) = w(x,t) \tag{11.63}$$

where w is the transverse displacements of the point on the middle plane (that is, $z = 0$) of the beam. The only nonzero strain of the Euler-Bernoulli beam theory, is the axial strain given by

$$\epsilon_{xx} = \frac{\partial u}{\partial x} = \frac{\partial u^0}{\partial x} - z\frac{\partial^2 w}{\partial x^2} \tag{11.64}$$

The equations of motion of the Euler-Bernoulli beam theory in terms of stress resultants are given by

$$\frac{\partial V}{\partial x} = \rho A\frac{\partial^2 u^0}{\partial t^2}, \quad \frac{\partial^2 M}{\partial x^2} = \rho A\frac{\partial^2 w}{\partial t^2} \tag{11.65}$$

where V is the shear force resultant and M is the moment resultant and these are given by

$$V = \int_A \sigma_{xx} dA, \;\; M = \int_A z\sigma_{xx} dA \tag{11.66}$$

and σ_{xx} is the axial stress.

The constitutive model for this beam is given by Eqn. (11.17). Using Eqns. (11.66 & 11.17), we have

$$V - (e_0 a)^2\frac{\partial^2 V_{xx}}{\partial x^2} = EA\frac{\partial u^0}{\partial x}, \quad M - (e_0 a)^2\frac{\partial^2 M_{xx}}{\partial x^2} = EI\kappa_e \tag{11.67}$$

where $I = \int_A z^2 dA$ is the moment of inertia of the beam cross section and $\kappa_e = -\frac{\partial^2 w}{\partial x^2}$ is the bending strain of the beam. With the help of the non-local constitutive relations and the equations of motion presented, the moment can be expressed in terms of the generalized displacements, by substituting Eqn. (11.67) into Eqn. (11.65), which is given by

$$M = -EI\frac{\partial^2 w}{\partial x^2} + (e_0 a)^2 \rho A\frac{\partial^2 w}{\partial t^2} \tag{11.68}$$

Substituting M from Eqn. (11.68) into Eqn. (11.65), we obtain the equation of motion of non-local Euler beam, which is given by

$$EI\frac{\partial^4 w}{\partial x^4} + \rho A\frac{\partial^2 w}{\partial t^2} - \rho A(e_0 a)^2\frac{\partial^4 w}{\partial x^2 \partial t^2} = 0 \tag{11.69}$$

where $w = w(x,t)$ is the flexural deflection, ρ is the mass density, A is the cross-sectional area, EI is the bending rigidity of the beam structure and $e_0 a$ is the non-local scaling parameter. It is observed that if the internal length scale a is identically zero, then the local Euler-Bernoulli beam model is recovered.

We begin the spectral analysis for computing the wave parameters, namely the group speeds and wavenumber by transforming the governing equation Eqn. (11.69) to frequency domain. To do this, we first take DFT on the transverse displacement $w(x,t)$ as

$$w(x,t) = \sum_{n=1}^{N} \hat{w}(x,\omega)e^{i\omega t} \tag{11.70}$$

Substitution of Eqn. (11.70) into Eqn. (11.69), we get an ordinary differential equation having constant coefficients

$$EI\frac{d^4\hat{w}}{dx^4} - \rho A\omega^2\left(\hat{w} - (e_0 a)^2\frac{d^2\hat{w}}{dx^2}\right) = 0 \tag{11.71}$$

where $\hat{w}$ is the amplitude of the wave motion, k is the wavenumber, and ω is the frequency. The solution of this constant coefficient equation is of the form [48]:

$$\hat{w}(x) = \tilde{w}e^{-ikx} \tag{11.72}$$

Substitution of Eqn. (11.72) into Eqn. (11.71) yields fourth order characteristic equation for the determination of the wavenumbers, which is given by

$$EIk^4 - \rho A\omega^2\left(e_0 a\right)^2 k^2 - \rho A\omega^2 = 0 \tag{11.73}$$

The fourth order differential equation will give four wavenumbers, two of which is forward moving wavenumber and the rest two are the backward moving wavenumbers. Eqn. (11.73) is a quadratic equation in k^2 and the solution for the four wavenumbers is given by

$$k_{1,2,3,4} = \pm\sqrt{\frac{\rho A\omega^2(e_0 a)^2 \pm \sqrt{\rho A\omega^2\left(4EI + \rho A\omega^2(e_0 a)^4\right)}}{2EI}} \tag{11.74}$$

These wavenumbers are function of the non-local scaling parameter, wave frequency and other material parameters of the beam. Out of these four wavenumbers, two are purely real and the other two are purely imaginary. From Eqn. (11.74) it is obvious that, there is no possibility for a cut-off frequency. This is the main difference in the wave propagation behavior when compared to ESGR model presented in Section 11.3.1.

The phase and group speeds of this beam is calculated before. The phase speed is calculated from the expression $C_p = Real(\omega/k)$, $k = k_1\, or\, k_2$. The expression for group speed $C_g = d\omega/dk$ is given by

$$C_{g_\alpha} = \frac{2EIk_\alpha^3 - \rho A\omega^2\left(e_0 a\right)^2 k_\alpha}{\rho A\omega\left(1 + \left(e_0 a\right)^2 k_\alpha^2\right)}, \qquad \alpha = 1, \quad 2 \tag{11.75}$$

The wavenumbers and group speeds are plotted for two different values of $e_0 a = 1\,nm$ and $e_0 a = 5\,\text{nm}$. The plots are shown in Fig. 11.9.

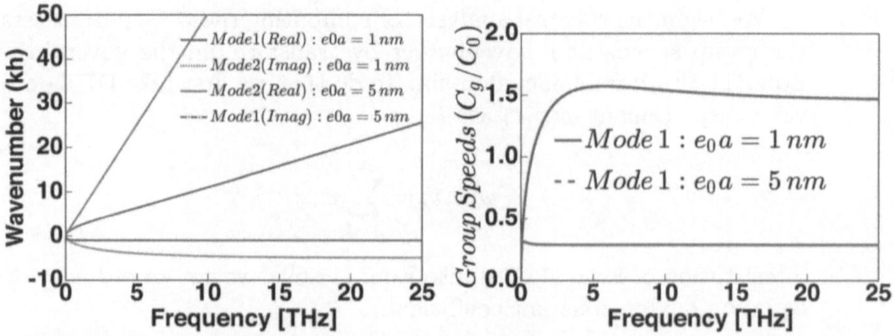

(a) Wavenumber: Mode1 and Mode 2 (b) Group speed: Mode1 and Mode 2

FIGURE 11.9: Spectral and Dispersion relations for Erigen stress gradient Euler Bernoulli beam (ESGB model)

The figure shows some interesting results, which are summarized below:

- The model shows only one propagating mode, while the other is a purely imaginary evanescent mode. This behavior is similar to local Euler-Bernoulli beam.
- When the value of e_0a increases, the wavenumber shown in Fig. 11.9(a) tends to be linear with frequency even at low frequencies. This effect is shown in the group speed plot shown in Fig. 11.9(b), where the speed becomes nearly constant with frequency
- Increase in the value of e_0a causes significant reduction of imaginary component of the wavenumber and beam tends to exhibit non-dispersive character as in the case of local rod waveguide.
- Increase in the value of e_0a causes significant decrease in the value of group speeds. For example, when the e_0a is increased from $1\,nm$ to $5\,nm$, the group speed reduces by almost 75%.

In summary, unlike the local elementary beam, which exhibits highly dispersive behavior, one can make the non-local elementary beam based on Erigen model nearly non-dispersive by increasing the value of e_0a.

Next, we will study the wave propagation response on this non-local infinite beam. From the dispersion analysis, we have only one propagating incident wave mode and the second being evanescent mode. Hence, the frequency domain transverse displacement for an infinite non-local Eringen stress gradient beam is given by

$$\hat{w}(x,\omega) = \mathbf{A}e^{-ikx} \tag{11.76}$$

Here **A** is assumed to represent the gaussian signal same as shown in Fig: 4.3 given in Chapter 4 except that the pulse width is assumed as 50 pico seconds so that we can have a Terra Hertz signal propagating in he beam. The response is plotted for two different values of $e_0a = 1\,\text{nm}$ and $e_0a = 5\,\text{nm}$ and the response is shown in Fig. 11.10. From the figure, we can clearly see, not

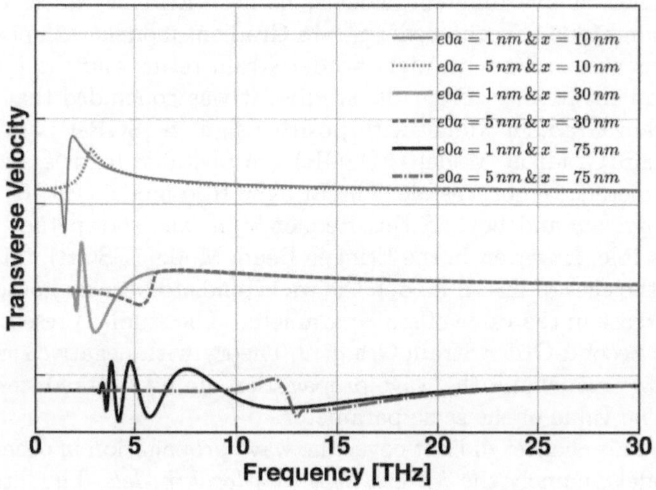

FIGURE 11.10: Transverse wave propagation response in a ESGB model

only significant late arrival of waves as well as significant loss of dispersiveness with the increase in the value of e_0a.

Note on MATLAB® scripts provided in this chapter

Two different MATLAB computer scripts, namely *dispNonlocal.m* and *RespNonlocal.m*, are provided and are housed at the publisher website. These codes can be executed by the reader to not only verify the results provided in this chapter but also modify these scripts to obtain wave propagation behavior of newer non-local waveguides given in the exercises at the end of this chapter.

Summary

This chapter begins with a brief introduction to non-local elasticity and the outline of two different non-local theories, namely the *Eringen Stress Gradient Theory* and the *Strain Gradient Theory.* The aim of this chapter is to see how the non-local scale parameter influences the wave propagation behavior at atomistic scales in one-dimensional rod and beam models. Hence, all the input signals are at the pico-second levels, which makes the frequency range in the Terra-Hertz levels.

There are number of models based on the strain gradient theories and this chapter addressed the second and the fourth order strain gradient models.

In the second-order strain gradient theory, the *sign paradox* associated with the higher-order strain terms in its constitutive models is an issue and this chapter has shown the stability of the model associated with this sign paradox using wave propagation studies. In all, this chapter presents wave propagation studies in five different models, namely Eringen Stress Gradient Rod and Beam Model, Second-order Strain Gradient models with positive and negative signs associated with higher-order strain terms and the Fourth Order Strain Gradient Models. From the studies, it was concluded that the Second-Order Strain Gradient Model with positive sign (SOSGR-P) and the Fourth order Strain Gradient Model (FOSGR) are unstable models and the wave modes predicted are not feasible. The Eringen Rod model (ESGR) introduces *Escape Frequency* and beyond this frequency, it was shown that no propagation is possible. However, in the Eringen Beam Model (ESGB), this feature is absent. In the case of ESGB model, the wave tends to become non-dispersive with the increase in the value of scale parameter. The studies presented also shows that the Second-Order Strain Gradient Theory with negative sign (SOSGR-N) is a stable model and the wave propagation amplitudes reduces with the increase in the value of the scale parameter.

This chapter did not cover the wave propagation in other class of non-local models, namely the *integral type non-local models.* The interested reader can refer to [102, 103] to get more information on the wave propagation in these waveguides

Exercises

11.1 The governing equation for a rod undergoing axial motion $u(x,t)$ with lateral inertia, modeled using Eringen stress gradient theory, is given by

$$\begin{aligned}\rho A\frac{\partial^2 u}{\partial t^2} &= (\rho A\nu^2\psi^2 + (e_0 a)^2\rho A)\frac{\partial^4 u}{\partial x^2\partial t^2} \\ &\quad -(e_0 a)^2\rho A\nu^2\psi^2\frac{\partial^6 u}{\partial x^4\partial t^2} + EA\frac{\partial^2 u}{\partial x^2}\end{aligned}$$

where the cross section is a circle of radius R and the rod is of length L. Here EA is the axial rigidity, ρA is the mass of the rod, ψ is the slenderness ratio (L/R). Determine the wavenumbers and group speeds for this model. The detail of this model can be found in [53].

11.2 The governing equation of Timoshenko beam modeled using Eringen theory is given by

$$GA\kappa\left(\frac{\partial\theta}{\partial x} - \frac{\partial^2 w}{\partial x^2}\right) + \rho A\frac{\partial^2 w}{\partial t^2} - \rho A(e_0 a)^2\frac{\partial^4 w}{\partial x^2\partial t^2} = 0$$

$$EI\frac{\partial^2\theta}{\partial x^2} + GA\kappa\left(\frac{\partial w}{\partial x} - \theta\right) - \rho I\left(\frac{\partial^2\theta}{\partial t^2} - (e_0 a)^2\frac{\partial^4\theta}{\partial x^2\partial t^2}\right) = 0$$

where E, G, I and A are the Young's modulus, Shear modulus, Moment of Inertia and area of cross section of the beam. Here, κ is the shear correction factor and ρ is the density of the beam. The associated stress resultant is given by

$$\begin{aligned} M &= EI\frac{\partial \theta}{\partial x} + \rho A (e_0 a)^2 \frac{\partial^2 w}{\partial t^2} + \rho I (e_0 a)^2 \frac{\partial^3 w}{\partial x \partial t^2} \\ Q &= GA\kappa\left(\frac{\partial w}{\partial x} - \theta\right) + \rho A (e_0 a)^2 \frac{\partial^3 w}{\partial x \partial t^2} \end{aligned}$$

This model is discussed in detail in the unified book [54]. Determine the following:

1. Spectrum and dispersion relations as function of scale parameter $e_0 a$ and show the evolution of escape frequencies in both bending and shear modes.
2. Determine the wave propagation response to a tone burst signal with central frequency placed before and after the cut-off frequencies.

11.3 The governing equation of a rod governed by fourth-order Eringen Stress gradient theory is given

$$\rho A\left(\frac{\partial^2 u}{\partial t^2} - \beta\frac{\partial^4 u}{\partial x^2 \partial t^2} + \gamma\frac{\partial^6 u}{\partial x^4 \partial t^2}\right) = EA\frac{\partial^2 u}{\partial x^2}$$

where β and γ are parameters associated with scale. If $\beta = \gamma = e_0 a$, investigate the stability of this model by plotting the spectrum and dispersion function for different values of $e_0 a$. This model is discussed in [102].

11.4 Reference [1] had came up with the following simplified second-order strain gradient model of a micro-structural solid containing three material parameters A_{23}, A_{22} and A_{21}. By appropriately choosing the material parameters, the propagation characteristics associated with this model can be changed. The governing partial differential equation is given by

$$\rho A\frac{\partial^2 u}{\partial t^2} + \frac{\rho^2 A g^2 A_{23}}{E}\frac{\partial^4 u}{\partial t^4} = EA\frac{\partial^2 u}{\partial x^2} + \rho A g^2 A_{21}\frac{\partial^4 u}{\partial x^2 \partial t^2} - EAg^2 A_{22}\frac{\partial^4 u}{\partial x^4}$$

Determine the spectrum and dispersion relation for the following cases and discuss their stability from the wave propagation point of view (that is existence -negative group speeds or unrealistic group speeds)

1. $A_{23} = 0$, $A_{22} \neq 0$, and $A_{21} \neq 0$.
2. $A_{23} \neq 0$, $A_{22} \neq 0$, $A_{21} \neq 0$.

3. $A_{21} \neq 0$, $A22 = 0$, $A_{23} = 0$.
4. $A_{21} \neq 0$, $A_{23} \neq 0$, $A_{22} = 0$

The macrostructural material properties can be chosen according to the reference [95].

CHAPTER 12

Introduction to Spectral Finite Element Formulation

In Chapters 4–9, we have seen that the solution to the governing wave equations in the frequency domain involves determination of the multiple wave coefficients, which are to be obtained from the boundary conditions of the problem. Each wave coefficient represents a wave mode and if the structure can accommodate more wave modes, the problem will lead to a situation where multiple wave modes need to be determined. Using ad hoc procedures based on equilibrium and compatibility of responses, as done in the previous chapters to determine the wave coefficients, the book keeping of the interaction of different wave modes becomes very cumbersome and unwieldy, especially if the number of wave modes increases. Also, in Chapters 4, 5 and 7, we had used the incident wave coefficient as the input force and used this to determine the reflected and transmitted responses. This process will give only the response trends but not the actual response. That is, the time history of the response obtained by this method will show only the first reflections from the boundary and not the repeated reflections that the incident wave actually undergoes when it interacts with the boundary. This aspect was clearly brought in Chapter 4 and it was shown that the actual response histories were obtained after the incident wave coefficient was determined using the force boundary condition at the response site.

A better approach to handle interaction of multiple wave modes is to adopt matrix methodology, which allows formulation and solution of the wave propagation problem in a compact manner that can enable automation of the entire solution process. Spectral finite element method (SFEM) is one such method, which is ideally suited for solution of wave propagation problem as this method results in problem sizes that are orders of magnitude smaller than

DOI: 10.1201/9781003120568-12

other numerical methods such as finite element method. SFEM is frequency domain-based matrix method. There are three different variants of SFEM, namely, the Fourier Transform-based, Wavelet Transform-based and Laplace Transform-based, respectively, based on the type of integral transform used to transform the problem from the time to frequency domain. In this book, we will address only Fourier Transform-based SFEM, while the details of the other two variants of SFEM is dealt in detail in the unified wave propagation book [54]. In the next section, we will explain the fundamental principles of spectral finite element formulation.

12.1 FUNDAMENTAL PRINCIPLES OF SPECTRAL FINITE ELEMENT FORMULATION

A spectral element method is essentially a finite element method formulated in the frequency domain. However, the method of implementation is quite different. The basic differences between the SFEM and conventional FEM are highlighted in the following paragraphs.

FEM is based on the assumed polynomial for displacement variation. These assumed displacement polynomials are forced to satisfy the weak form of the governing differential equation, which would yield two different matrices, namely the *Stiffness matrix* and the *Mass matrix*, respectively. These elemental matrices are assembled to obtain the global stiffness and mass matrices. The assembly process ensures equilibrium of forces between adjacent elements. This procedure will give discretized form of the governing equation of the form

$$\mathbf{M}\ddot{\mathbf{u}} + \mathbf{C}\dot{\mathbf{u}} + \mathbf{K}\mathbf{u} = \mathbf{F}(\mathbf{t})$$

where $\mathbf{M}$ and $\mathbf{K}$ are the global mass and stiffness matrix and $\ddot{\mathbf{u}}$, $\dot{\mathbf{u}}$ and $\mathbf{u}$ are the acceleration, velocity and displacement vectors, respectively. Matrix $\mathbf{C}$ is the damping matrix, which is normally obtained from the Rayleigh's damping equation, which is formed by combining the stiffness and mass matrix as $\mathbf{C} = \alpha\mathbf{K} + \beta\mathbf{M}$, where α and β are the stiffness and mass proportional damping factors, and the damping scheme is called the *proportional damping* scheme. FEM obtains solutions using either the *model methods* or using suitable time marching schemes. The wave propagation problems being a multimodal phenomenon, a modal method of solution is not convenient for wave propagation analysis. The detail of this method is presented in the unified book [54]. The most preferred method of solution of wave propagation problems under FEM is by using suitable time marching, also called the *direct time integration* scheme of solution. Under the time marching schemes, there are ***explicit methods*** and the ***implicit methods***. For wave propagation and highly transient dynamics problems, explicit methods are normally preferred. In the time marching scheme, the solution process takes place over a small time step ΔT. The solution process is repeated for N time steps till the total time $T = N\Delta T$ is reached. The solution time directly depends on the number of degrees of freedom in the model, which is usually very high for wave propagation problems. All of these aspects are described in detail in the unified book [54]

SFEM requires that the governing equations be transformed to frequency domain. The integral transform we use to transform the governing equation dictates the SFEM formulation procedure. The most common integral transform used is the Fourier Transform. The SFEM uses in most cases the exact solution to the wave equation in the transformed frequency domain as its interpolating function for element formulation. Many of the solutions for different waveguides were derived in Chapters 4–9 for 1- and 2-D waveguides. For example, the solutions to the second and fourth-order 1-D governing equations are given in Chapter 4 (Eqs. 4.12 and 4.30). We see complex exponentials as solutions since the governing equations were constant coefficient equations. Variable coefficients ODE's arise for cases when the material inhomogeneity in the structure is in the direction of wave propagation and also for circular waveguides. The solution of some of these types can be determined and they will be mostly in terms of Bessel's functions. For a constant coefficient ODE, the exact solution can be found for any order of the equation. SFEM employs this exact solution as an interpolating function for element formulation. Unlike polynomials as in the case of FEM, here, we need to deal with complex exponentials as the interpolating functions. The exact solution will have wave coefficients corresponding to initial and reflected wave components. If one wants to model an infinite domain, then the reflected components can be dropped from the interpolating functions. Such solutions having coefficients corresponding to only incident waves will lead to what are called the *throw-off elements*. This is a great advantage that SFEM gives over FEM. Formulating infinite element in conventional FEM is not easy and straight forward. Using the interpolating functions for the displacements, dynamic element stiffness is formulated. Unlike conventional FEM, where two matrices are formulated, namely, the mass and stiffness matrices, in SFEM, only one dynamic stiffness matrix needs to formulated. One can formulate this stiffness matrix as in the case of conventional FEM, using the weak form of governing equation and variational methods. This approach will involve complex integration. Alternatively, one can formulate the dynamic stiffness matrix using the stress or force resultant expressions. This method is normally suitable since it does not involve complex integration.

12.2 GENERAL FORMULATION PROCEDURE OF SFEM

Before formulating some spectral elements, we will explain the steps involved in the spectral element formulation of a waveguide. First, the given forcing function is transformed to frequency domain using forward FFT. In doing so, we need to choose the time sampling rate (Δt) and number of FFT points (N) to decide on the time window of analysis. Care should be take to see that the chosen window is good enough to avoid *Signal wraparound* problems [52] as explained in Chapter 6. The FFT output of the input signal will yield the frequency, the real and imaginary part of the forcing function. Some of the forcing functions used in wave propagation analysis have time profiles that are rectangular, gaussian or modulated sine signal. MATLAB functions are available to generate these signals. Over a big frequency loop, the element dynamic stiffness matrix is generated, assembled and solved as in the case of conventional FEM. However, these operations have to be performed at each sampled frequency. This does not pose a major computational hurdle since the problem sizes are many orders smaller than conventional FEM. The solution process is first done for unit impulse, which would directly yield *frequency response function*(FRF). The FRF is then convolved with the load to get the required output in the frequency domain. This output is then transformed to time domain using inverse FFT.

There are many advantages that SFEM offers over conventional FEM. SFEM can give results both in time and frequency domain in a single analysis. Obtaining FRF is one single big advantage of SFEM. This enables solving inverse problems such as force or system identification problems in a straight forward manner. Since, many damping properties are frequency dependent, damping in structures can be treated more realistically. Linear viscoelastic analysis can be performed with minimum alteration of the spectral element code. Since the approach gives FRF first, responses to different loadings can be obtained using a single analysis. In summary, SFEM is a method in which transform algorithms such as the FFT is an essential part and gives problem sizes of many orders smaller than the conventional FEM. SFEM for some 1-D and 2-D waveguides based on Fourier Transform is given in this chapter. We call the SFEM formulated based on Fourier Transform as FSFEM. Most of the SFEM formulations for different 1-D and 2-D waveguides are given in the textbook [48, 54].

SFEM for 1-D waveguides begins with the solution to the strong form of the governing equation, which is then used to formulate the *dynamic stiffness matrix* and the force-displacement relations in the frequency domain can be written as

$$\hat{\mathbf{K}}\hat{\mathbf{u}} = \hat{\mathbf{F}}$$

where both the matrix $\hat{\mathbf{K}}$ and the vector $\hat{\mathbf{F}}$ are frequency dependent. The dynamic stiffness matrix can also be obtained using regular FEM by taking the Fourier Transform of the governing matrix differential equation without damping given by $\mathbf{M\ddot{u}} + \mathbf{Ku} = \mathbf{F(t)}$, using the stiffness matrix $\mathbf{K}$ and the consistent mass matrix $\mathbf{M}$ as

$$\hat{\mathbf{K}}_n = \mathbf{K} - \omega_n^2\mathbf{M}, \tag{12.1}$$

where n in the suffix indicates the value of dynamic stiffness matrix at any ω_n. In most cases, $\hat{\mathbf{K}}_n$ in SFEM is obtained using the exact solution to the governing in the transformed domain as interpolating functions, whereas $\hat{\mathbf{K}}_n$ from FEM is just an approximation. The $\hat{\mathbf{K}}_n$ in FEM approaches the $\hat{\mathbf{K}}_n$

from SFEM in the limiting process of taking the number of finite elements to infinity. Further, the matrix–vector structure of the SFEM gives the flexibility of FE modeling, where large structures can be assembled in terms of many spectral waveguides. The assemblage and imposition of boundary condition in SFEM is the same as in FEM, which makes the method attractive. With the use of Ritz method and the theorem of Minimum Potential Energy in frequency domain, many approximate spectral elements can be formulated. Examples of these are reported in [26,49]. Furthermore, there is the possibility of coupling SFE and FE in complex structures as reported in [126].

The formulation of SFEs for 2-D structural waveguides poses extra complexity. The reduced equation in the frequency domain is no longer an ODE, but remains a PDE in terms of the two space variables. This PDE is not readily solvable and another transform is necessary to reduce the equation to ODE with dependent variable being function of one spatial dimension. This was explained in Chapter 7 for isotropic waveguides and Chapter 8 & 9 for laminated composite and functionally graded waveguides. Instead of the Fourier Transform in the spatial direction, one can conveniently apply Fourier Series (FS) in the spatial direction for easy mathematical handling. Thus, the unknown variable is further decomposed, using FS representation as

$$\hat{\mathbf{u}}(x,y,\omega_n) = \sum_{m=0}^{M-1} \tilde{\mathbf{u}}(x,\eta_m,\omega_n) \left\{ \begin{array}{c} \sin(\eta_m y) \\ \cos(\eta_m y) \end{array} \right\}, \tag{12.2}$$

where M is the number of FS points, and η_m is the discrete wavenumber related to the spatial window Y by

$$\eta_m = m\Delta\eta = \frac{m\eta_f}{M} = \frac{m}{M\Delta y} = \frac{m}{Y}, \tag{12.3}$$

with Δy denoting the spatial sampling rate and η_f is the highest wavenumber captured by Δy. The spatial variation of the load determines M. Using this representation, the governing equation becomes an ODE in x and again can be solved exactly for some cases. This exact solution is again used as the interpolating function for the spectral element formulation. Thus, for each frequency ω_n and wavenumber η_m, the dynamic stiffness matrix is formed and assembled and the unknown variable is solved for its amplitude $\tilde{\mathbf{u}}_{n,m}$ as

$$\tilde{\mathbf{K}}_{n,m}\tilde{\mathbf{u}}_{n,m} = \tilde{\mathbf{f}}_{n,m}, \tag{12.4}$$

where $\tilde{\mathbf{f}}_{n,m}$ is the Frequency-wavenumber amplitude of applied load. First, $\tilde{\mathbf{u}}_{n,m} = \tilde{\mathbf{u}}(x,\eta_m,\omega_n)$ is recovered by the FS and $\mathbf{u}(x,y,t)$ is computed by the IFFT algorithm.

12.3 SPECTRAL FINITE ELEMENT FORMULATION

In this section, we outline the spectral finite element formulation of both 1-D and 2-D waveguides. We begin with the spectral finite element formulation for an elementary rod and a beam and study the difference between the obtained dynamic stiffness matrix and the conventional finite element dynamic stiffness matrix. This is followed by the formulation of higher-order composite beam element that includes the effect of both shear deformation and lateral contraction. The aim here is to show how mode coupling can be handled under SFEM environment. The last part of the section deals with spectral element formulation for 2-D Isotropic and composite waveguides that can handle in-plane loading. The procedure outlined for spectral element formulation here can be extended to other 1-D & 2-D waveguides not covered in this chapter.

12.3.1 Spectral Rod Element

As in the case of conventional finite element, we represent the spectral element in terms of the degrees of freedom that the rod can support. A longitudinal rod element with its associated degrees of freedom is shown in Fig. 12.1. The rod can support two degree of freedom, namely the $\hat{u}_1$ and $\hat{u}_2$ at its two ends and the length of the rod is assumed as L, with axial rigidity EA and density ρ as shown in Fig. 12.1(a) The spectral element formulation requires

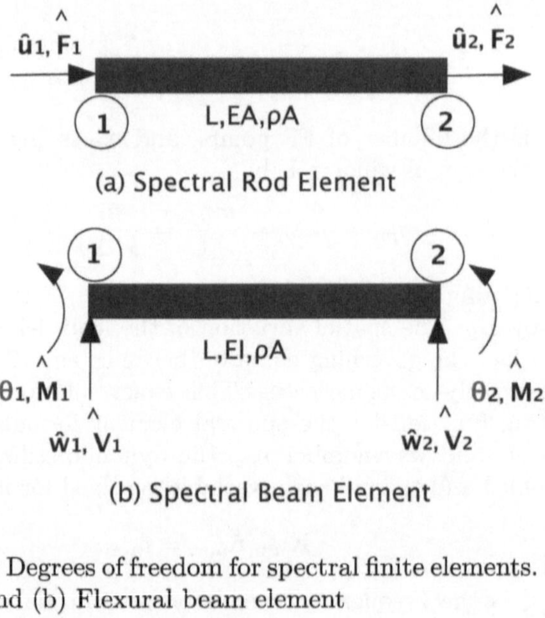

FIGURE 12.1: Degrees of freedom for spectral finite elements. (a) Longitudinal rod element and (b) Flexural beam element

the strong form of the governing differential equation. The homogeneous form of the governing equation for an isotropic homogeneous rod of density ρ and Young's modulus E is

$$\frac{\partial^2 u}{\partial t^2} = c_0^2 \frac{\partial^2 u}{\partial x^2} \tag{12.5}$$

where $u = u(x,t)$ is the axial displacement and $c_0^2 = E/\rho$ is the square of the wave speed in the material. The governing equation is supplemented by the force (natural) boundary condition

$$F(x,t) = AE\frac{\partial u}{\partial x} \tag{12.6}$$

where A is the cross-sectional area of the rod and $F(x,t)$ is the axial force. The displacement (essential) boundary conditions is the specification of the displacement u at the boundaries.

We assume the solution of axial deformation as

$$u(x,t) = \sum_{n=1}^{N} \hat{u}(x,\omega_n)e^{i\omega_n t} \tag{12.7}$$

This transformation allows replacing the time dependency with the parameter ω_n. The summation is carried out up to the Nyquist frequency ω_N. Substituting Eqn. (12.7) in (12.5), the reduced governing ordinary differential equation becomes

$$c_0^2\frac{d^2\hat{u}}{dx^2} + \omega_n^2\hat{u} = 0 \tag{12.8}$$

the solution of the above equation takes the form $u_\circ e^{-ikx}$. Upon substitution in Eqn. (12.8), we get the characteristic equation for computing wavenumber k, which is given by

$$(-c_0^2k^2 + \omega_n^2)u_\circ = 0 \tag{12.9}$$

which gives wavenumber as $k_n = \pm\omega_n/c$. The determination of wavenumber using spectral analysis was discussed in detail in the previous chapters. Thus, the complete solution is given by

$$\hat{u}(x,\omega_n) = \mathbf{c_1}e^{-ik_nx} + \mathbf{c_2}e^{-ik_n(L-x)} \tag{12.10}$$

where $\mathbf{c_1}$ and $\mathbf{c_2}$ are coefficients to be determined from the boundary conditions, with L being the length of the element. Note that in the above equation, is same as what was derived in Chapter 4 (see Eqn. (4.12)) except that the reflected component is replaced with $e^{-ik(L-x)}$. This form also satisfies the governing equation exactly. Specifically, the boundary conditions can be expressed in terms of the nodal displacements $\hat{u}_1 = \hat{u}(x_1,\omega_n)$ and $\hat{u}_2 = \hat{u}(x_2,\omega_n)$ as

$$\left\{\begin{array}{c}\hat{u}_1\\ \hat{u}_2\end{array}\right\} = \left[\begin{array}{cc} e^{-ik_nx_1} & e^{+ik_nx_1}\\ e^{-ik_nx_2} & e^{+ik_nx_2}\end{array}\right]\left\{\begin{array}{c}\mathbf{c_1}\\ \mathbf{c_2}\end{array}\right\} = \mathbf{T_1c} \tag{12.11}$$

Similarly, the force in the frequency domain, $\hat{F}(x,\omega_n)$ obtained from Eqn. (12.6), can be evaluated at x_1 and x_2 to relate the nodal forces to the unknown coefficients

$$\left\{\begin{array}{c}\hat{F}_1\\\hat{F}_2\end{array}\right\} = AE(ik_n)\left[\begin{array}{cc}e^{-ik_n x_1} & -e^{+jk_n x_1}\\-e^{-ik_n x_2} & e^{+ik_n x_2}\end{array}\right]\left\{\begin{array}{c}\mathbf{c_1}\\\mathbf{c_2}\end{array}\right\} = \mathbf{T_2 c} \qquad (12.12)$$

Thus, the nodal forces are related to the nodal displacements as

$$\left\{\begin{array}{c}\hat{F}_1\\\hat{F}_2\end{array}\right\} = \mathbf{T_2 T_1}^{-1}\left\{\begin{array}{c}\hat{u}_1\\\hat{u}_2\end{array}\right\}, \qquad (12.13)$$

Hence, the dynamic stiffness matrix (DSM) for the rod at frequency ω_n is given by

$$\mathbf{D_{SFEM}} = \mathbf{T_2 T_1}^{-1}$$

The explicit form of the dynamic stiffness matrix is given by

$$\mathbf{D_{SFEM}} = \frac{EA}{L}\frac{ikL}{(1-e^{-i2k_nL})}\begin{bmatrix}1+e^{-i2k_nL} & -2e^{-ikL}\\-2e^{-ik_nL} & 1+e^{-i2k_nL}\end{bmatrix} \qquad (12.14)$$

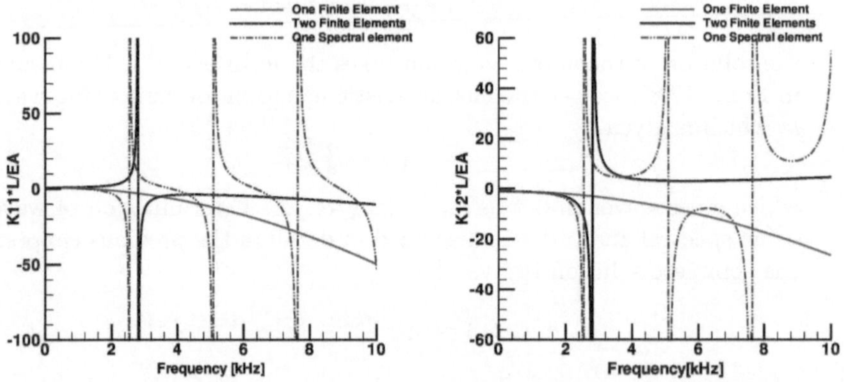

(a) $\bar{\hat{k}}_{11} = \hat{k}_{11}L/EI$ (b) $\bar{\hat{k}}_{12} = \hat{k}_{12}L/EI$

FIGURE 12.2: Comparison of finite element and spectral element dynamic stiffness matrix elements as a function of frequency for an elementary rod (a) Normalized $\hat{k}_{11}$ and (b) Normalized $\hat{k}_{12}$

In comparison, the DSM for conventional FEM will be $\mathbf{D_{FEM}} = \mathbf{K} - \omega_n^2\mathbf{M}$, where $\mathbf{K}$ and $\mathbf{M}$ are the stiffness and mass matrices, respectively obtained from polynomial approximations. That is, the FEM matrix relations can be written as

$$\begin{Bmatrix}\hat{F}_1\\\hat{F}_2\end{Bmatrix} = \left[\frac{EA}{L}\begin{bmatrix}1 & -1\\-1 & 1\end{bmatrix} - \omega^2\frac{\rho AL}{6}\begin{bmatrix}2 & 1\\1 & 2\end{bmatrix}\right]\begin{Bmatrix}\hat{u}_1\\\hat{u}_2\end{Bmatrix} \qquad (12.15)$$

The above equation can be simplified as

$$\begin{Bmatrix} \hat{F}_1 \\ \hat{F}_2 \end{Bmatrix} = \frac{EA}{L} \begin{bmatrix} (1-\frac{\lambda^2}{3}) & -(1+\frac{\lambda^2}{6}) \\ -(1+\frac{\lambda^2}{6}) & (1-\frac{\lambda^2}{3}) \end{bmatrix} \begin{Bmatrix} \hat{u}_1 \\ \hat{u}_2 \end{Bmatrix} \tag{12.16}$$

where k and $\lambda^2 = \omega^2 \frac{\rho A}{EA} L = kL$ is the longitudinal wavenumber

The plot of the elements of $\mathbf{D_{FEM}}$ and $\mathbf{D_{SFEM}}$ is shown Figs. 12.2(a) and (b) respectively. From the figure we see that the stiffness values from FEM and SFEM match only at very low frequencies. At high frequencies, the stiffnesses from $\mathbf{D_{SFEM}}$ goes through multiple zeros crossings and peaks, where the frequency at which peaks occurs signify its *resonant frequency.* The $\mathbf{D_{FEM}}$ obtained from single element of length L does not exhibit any zero crossings and the stiffness of the element keeps decreasing with increasing frequency. We will now see how the accuracy of $\mathbf{D_{FEM}}$ can be increased by modeling the finite element of length L as two elements of length $L/2$. Noting that the force in the middle node (node 3) is zero, the assembled dynamic stiffness equation in frequency domain is given by

$$\begin{Bmatrix} \hat{F}_1 \\ 0 \\ \hat{F}_2 \end{Bmatrix} = \left[\frac{2EA}{L} \begin{bmatrix} 1 & -1 & 0 \\ -1 & 2 & -1 \\ 0 & -1 & 1 \end{bmatrix} - \omega^2 \frac{\rho AL}{12} \begin{bmatrix} 2 & 1 & 0 \\ 1 & 4 & 1 \\ 0 & 1 & 2 \end{bmatrix} \right] \begin{Bmatrix} \hat{u}_1 \\ \hat{u}_3 \\ \hat{u}_2 \end{Bmatrix} \tag{12.17}$$

Now, we need to express the above relation only in terms of the end displacements $\hat{u}_1$ and $\hat{u}_2$ by eliminating the middle displacement $\hat{u}_3$ and this is done by interchanging the rows and column and bringing the middle displacement $\hat{u}_3$ to the bottom of the displacement vector. Now the modified dynamic stiffness relations become

$$\begin{Bmatrix} \hat{F}_1 \\ \hat{F}_2 \\ 0 \end{Bmatrix} = \frac{2EA}{L} \begin{bmatrix} (1-2\beta) & 0 & -(1+\beta) \\ 0 & (1-2\beta) & -(1+\beta) \\ -(1+\beta) & -(1+\beta) & (2-4\beta) \end{bmatrix} \begin{Bmatrix} \hat{u}_1 \\ \hat{u}_2 \\ \hat{u}_3 \end{Bmatrix} \tag{12.18}$$

where

$$k = \omega\sqrt{\frac{\rho A}{EA}}, \qquad \beta = \frac{k^2 L^2}{24}$$

We will now partition the matrix in Eqn. (12.18) as

$$\begin{Bmatrix} \hat{\mathbf{F}} \\ 0 \end{Bmatrix} = \frac{2EA}{L} \begin{bmatrix} \mathbf{K_{11}} & \mathbf{K_{12}} \\ \mathbf{K_{21}} & \mathbf{K_{22}} \end{bmatrix} \begin{Bmatrix} \hat{\mathbf{u}} \\ \hat{u}_3 \end{Bmatrix} \tag{12.19}$$

where

$$\hat{\mathbf{F}} = \begin{Bmatrix} \hat{F}_1 \\ \hat{F}_2 \end{Bmatrix}, \qquad \hat{\mathbf{u}} = \begin{Bmatrix} \hat{u}_1 \\ \hat{u}_2 \end{Bmatrix}, \mathbf{K_{11}} = \begin{bmatrix} (1-2\beta) & 0 \\ 0 & (1-2\beta) \end{bmatrix},$$

$$\mathbf{K_{12}} = \begin{bmatrix} -(1+\beta) \\ -(1+\beta) \end{bmatrix}, \ \mathbf{K_{21}} = \begin{bmatrix} -(1+\beta) & -(1+\beta) \end{bmatrix}, \ \mathbf{K_{22}} = 2-4\beta$$

Expanding Eqn. (12.19), we get the following:

$$0 = \frac{2EA}{L}\mathbf{K_{21}}\hat{\mathbf{u}} + (2-4\beta)\hat{u}_3 \tag{12.20}$$

$$\hat{\mathbf{F}} = \frac{2EA}{L}\mathbf{K_{11}}\hat{\mathbf{u}} + \frac{2EA}{L}\mathbf{K_{12}}\hat{u}_3 \tag{12.21}$$

From the second of the above equation, we can write $\hat{u}_3$, the deformation of the middle node as

$$\hat{u}_3 = \frac{1}{4\beta-2}\mathbf{K_{21}}\hat{\mathbf{u}} \tag{12.22}$$

Now using Eqn. (12.22) in the Eqn(12.21), we can write the stiffness relation as

$$\hat{\mathbf{F}} = \frac{2EA}{L}\left[\mathbf{K_{11}} - \frac{1}{2-4\beta}\mathbf{K_{12}}\mathbf{K_{21}}\right]\hat{\mathbf{u}} \tag{12.23}$$

Expanding the above equation, we get

$$\begin{Bmatrix} \hat{F}_1 \\ \hat{F}_2 \end{Bmatrix} = \frac{2EA}{L}\frac{1}{2-4\beta}\begin{bmatrix} (1-10\beta+7\beta^2) & -(1+\beta)^2 \\ -(1+\beta)^2 & (1-10\beta+7\beta^2) \end{bmatrix}\begin{Bmatrix} \hat{u}_1 \\ \hat{u}_2 \end{Bmatrix} \tag{12.24}$$

The plot of dynamic stiffness obtained from two element idealization is also shown in Fig. 12.2. As mentioned earlier, the spectral element stiffness show multiple resonances while the plot of one finite element stiffness is not able to predict any of these. Introducing an addition element is able to pick the first resonance very closely. Modeling the finite element of length L by more number of elements will result in capturing more number of resonances. That is, if the length of the element L is modeled by N number of segments, then, we will be able to capture $N-1$ resonances. More number of finite elements signify more accurate inertia modeling. In other words, if the finite element stiffness is required to match closely with the values of spectral FEM stiffnesses, then the size of the FE mesh should be very small to accurately capture the inertial distribution.

Next, we will discuss a special case, where in the rod element has only a single node and the second node is at infinity. FFT based SFEM formulation of one-noded infinite segment is called the *throw-off* element and it is required for avoiding *signal wraparound* (this was discussed in Chapter 6) and hence for obtaining good time resolution. This element is formulated by leaving out the reflected coefficients from the solution given in Eqn. (12.10). Hence, the interpolation function for the formulation of the throw-off elements is given by

$$\hat{u}(x,\omega_n) = \mathbf{c}_1 e^{-ik_n x} \tag{12.25}$$

Following the same procedure followed for two noded elements, we consider the force expression given in Eqn. (12.6) and using $\hat{F}_1 = -\hat{F}(x_1,\omega_n)$, we obtain the following dynamic stiffness for the throw-off elements

$$\hat{F}_1 = \frac{EA}{L}[ik_n L]\hat{u}_1 \tag{12.26}$$

Note that the throw-off stiffness (in the brackets) is always complex, and it is this factor that adds damping to the structure resulting in good time resolution. Such a stiffness formulation is not possible in conventional FEM.

12.3.2 Spectrally Formulated Elementary Beam Element

Beam supports two motions, namely the transverse displacements $w(x,t)$ and the rotation of the cross section $\theta(x,t)$, where the rotation is derived from the transverse deformation as $\theta(x,t) = \partial w(x,t)/\partial x$, and x is the spatial coordinate. The above definition for $\theta(x,t)$ is due to the assumption that the plane sections remain plane, before and after bending. The beam element with degrees of freedom and stress resultants is shown in Fig. 12.1 (b). The stress resultants in this model are the shear force $V(x,t)$ and bending moment $M(x,t)$, which can be expressed in terms of transverse displacement as

$$V(x,t) = -EI\frac{\partial^3 w}{\partial x^3}, \qquad M(x,t) = EI\frac{\partial^2 w}{\partial x^2}$$

The governing partial differential equation governing the motion of beam is derived in Chapter 4 and is given by Eqn. (4.26). As in the case of rod. we begin the spectral beam formulation by considering the solution of the strong form of the governing equation in the frequency domain as the interpolation function. This is given by Eqn. (4.30). We see that the interpolating functions are in terms of complex exponentials and hyperbolic functions. However, it is easier to express them in terms of trigonometric functions, since our aim here is to obtain the dynamic stiffness matrix explicitly. Hence, we will write the exact solution to the governing equation given by Eqn. (4.26) as

$$\hat{w}(x,\omega) = \mathbf{A}\cos k_n x + \mathbf{B}\sin k_n x + \mathbf{C}\cosh k_n x + \mathbf{D}\sinh k_n x \tag{12.27}$$

where $k_n^4 = \omega_n^2 \frac{\rho A}{EI}$. From now on, the procedure is similar to the formulation of the rod element. That is, we have the following boundary conditions:

$$\text{at } x = 0, \qquad \hat{w}(0,\omega_n) = \hat{w}_1, \qquad \hat{\theta}(0,\omega_n) = \frac{\partial w}{\partial x} = \hat{\theta}_1$$

$$\text{at } x = L, \qquad \hat{w}(L,\omega_n) = \hat{w}_2, \qquad \hat{\theta}(L,\omega_n) = \frac{\partial w}{\partial x} = \hat{\theta}_2$$

Substituting these in Eqn. (12.27), we can write the resulting equations as

$$\begin{aligned}
\hat{w}_1 &= \mathbf{A} + \mathbf{C} \\
\hat{\theta}_1 &= k_n\left[\mathbf{B} + \mathbf{D}\right] \\
\hat{w}_2 &= \mathbf{A}\cos k_n L + \mathbf{B}\sin k_n L + \mathbf{C}\cosh k_n L + \mathbf{D}\sinh k_n L \\
\hat{\theta}_2 &= k_n\left[-\mathbf{A}\sin k_n L + \mathbf{B}\cos k_n L + \mathbf{C}\sinh k_n L + \mathbf{D}\cosh k_n L\right]
\end{aligned} \tag{12.28}$$

The above equation can be written in matrix form as

$$\begin{Bmatrix} \hat{w}_1 \\ \hat{\theta}_1 \\ \hat{w}_2 \\ \hat{\theta}_2 \end{Bmatrix} = \begin{bmatrix} 1 & 1 & 0 & 0 \\ 0 & k_n & 0 & k_n \\ \cos k_nL & \sin k_nL & \cosh k_nL & \sinh k_nL \\ -\sin k_nL & \cos k_nL & \sinh k_nL & \cosh k_nL \end{bmatrix} \begin{Bmatrix} \mathbf{A} \\ \mathbf{B} \\ \mathbf{C} \\ \mathbf{D} \end{Bmatrix}$$

$$= \quad \hat{\mathbf{w}} = \mathbf{T_1 A} \tag{12.29}$$

Next, we will consider the stress resultants acting at the two ends of the beam that is, we have

$$\text{at } x = 0, \qquad \hat{V}(0,\omega_n) = -EI\frac{d^3\hat{w}}{dx^3} = \hat{V}_1, \qquad \hat{M}(0,\omega_n) = EI\frac{d^2\hat{w}}{dx^2} = \hat{M}_1$$

$$\text{at } x = 0, \qquad \hat{V}(L,\omega_n) = -EI\frac{d^3\hat{w}}{dx^3} = \hat{V}_2, \qquad \hat{M}(L,\omega_n) = EI\frac{d^2\hat{w}}{dx^2} = \hat{M}_2$$

Using Eqn. (12.27) in the above equation, we can relate the nodal stress resultants to the wave coefficients as

$$\hat{\mathbf{F}} = \mathbf{T_2 A}, \quad \hat{\mathbf{F}}^T = \{\hat{V}_1 \quad \hat{M}_1 \quad \hat{V}_2 \quad \hat{M}_2\}, \quad \mathbf{A}^T = \{\mathbf{A} \quad \mathbf{B} \quad \mathbf{C} \quad \mathbf{D}\}$$

Knowing the explicit form of matrices $\mathbf{T_1}$ (from Eqn. (12.29)) and $\mathbf{T_2}$ (from the above equation), the dynamic stiffness matrix $\hat{\mathbf{K}}$ for an elementary isotropic beam is given by

$$\hat{\mathbf{K}} = \mathbf{T_2 T_1}^{-1} = \begin{bmatrix} \bar{\alpha} & \bar{\beta}L & \bar{\gamma} & \bar{\delta}L \\ & \bar{\xi}L^2 & -\bar{\delta}L & \bar{\eta}L^2 \\ \text{sym} & & \bar{\alpha} & -\bar{\beta}L \\ & & & \bar{\xi}L^2 \end{bmatrix} \tag{12.30}$$

where

$$\bar{\alpha} = (\cos k_nL \sinh k_nL + \sin k_nL \cosh k_nL)\,(k_nL)^3/\Delta$$
$$\bar{\beta} = (\sin k_nL \sinh k_nL)(k_nL)^2/\Delta, \qquad \bar{\gamma} = (\sin k_nL + \sinh k_nL)(k_nL)^3/\Delta$$
$$\bar{\delta} = (-\cos k_nL + \cosh k_nL)(k_nL)^2/\Delta$$
$$\bar{\xi} = (-\cos k_nL \sinh k_nL + \sin k_nL \cosh k_nL)(k_nL)/\Delta$$
$$\bar{\eta} = (-\sin k_nL + \sinh k_nL)(k_nL)/\Delta, \qquad \Delta = 1 - \cos k_nL \cosh k_nL$$

The corresponding dynamic stiffness for conventional finite element is given by

$$\hat{\mathbf{K}}_{\mathbf{fem}} = \mathbf{K} - \omega^2\mathbf{M} \tag{12.31}$$

$$= \left[\frac{EI}{L^3}\begin{bmatrix} 12 & 6L & -12 & 6L \\ 6L & 4L^2 & -6L & 2L^2 \\ -12 & -6L & 12 & -6L \\ 6L & 2L^2 & -6L & 4L^2 \end{bmatrix} - \omega^2\frac{\rho AL}{420}\begin{bmatrix} 156 & 22L & 54 & -13L \\ 22L & 4L^2 & 13L & -3L^2 \\ 54 & 13L & 156 & -22L \\ -13L & -3L^2 & -22L & 4L^2 \end{bmatrix} \right]$$

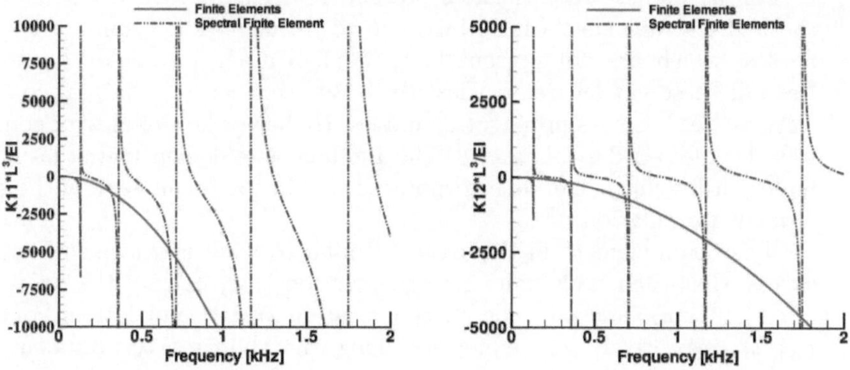

(a) $\bar{\hat{k}}_{11} = \hat{k}_{11}L^3/EI$ (b) $\bar{\hat{k}}_{12} = \hat{k}_{12}L^3/EI$

FIGURE 12.3: Comparison of finite element and spectral element dynamic stiffness matrix elements as a function of frequency for an elementary beam (a) Normalized $\hat{k}_{11}$ and (b) Normalized $\hat{k}_{12}$

The comparison of the behavior of finite element and spectral element dynamic stiffness elements for a beam is shown in Fig. 12.3. The behavior is very similar to the behavior of the rod element stiffness. That is, at very low frequencies, the dynamic stiffness of both the models match, while at high frequencies, spectral element stiffness exhibit multiple peaks, which are nothing but the beam resonances. This behavior is not captured by single finite element. If we model this beam segment by several smaller element segments, then the dynamic stiffness of the finite element will approach the spectral element stiffness behavior.

12.3.3 Higher-Order 1-D Composite Waveguides

In the previous section, we had derived the spectral FEM for 1-D isotropic waveguides, wherein the longitudinal and flexural motions were independent and hence independent spectral FEM were formulated for longitudinal and flexural wave propagation analysis. However, in laminated composites, due to unsymmetric ply-layup sequence, the structure will exhibit stiffness and inertial coupling. Due to this, the bending and axial motions will be coupled. That is, an axial input will result in bending deformation and *vice versa.* In this sub section, we will formulate a spectral finite element for 1-D higher-order laminated composite beam, wherein the higher-order effects such as *lateral contraction* and *shear deformation* is incorporated in the deformation behavior. Higher-order effects are normally prominent in thick beam sections. Higher-order beam spectral element based on first order shear deformation theory for 1-D isotropic waveguide is reported in [93] and [51]. In these papers as well as higher-order theories presented in Chapter 4, it was shown that the higher-order effects introduce *cut-off frequencies* for some of their wave modes, which are not present in their elementary waveguide counterparts. Beyond these cut-off frequencies, the evanescent shear or lateral contraction wave modes becomes propagating modes. Higher-order theories for composites was also discussed in Chapter 8. The aim here is to demonstrate that how such higher-order effects can be incorporated within the frame work of the spectral element formulation.

The beam element in this case is similar to what is shown in Fig. 12.1(b) except that each node can undergo four motions, namely the axial motion $u(x,t)$, flexural motion $w(x,t)$, the rotation $\phi(x,t)$ and lateral contraction $\psi(x,t)$, respectively. These motions along with the stress resultants are shown in Fig. 12.4.

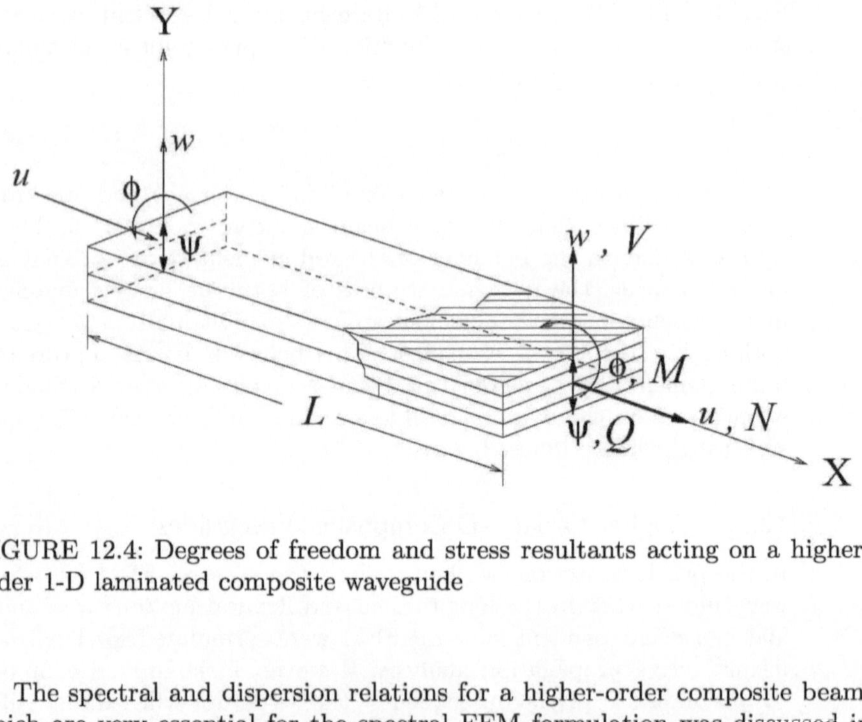

FIGURE 12.4: Degrees of freedom and stress resultants acting on a higher-order 1-D laminated composite waveguide

The spectral and dispersion relations for a higher-order composite beam, which are very essential for the spectral FEM formulation was discussed in Section 8.5.1. The governing equations for this beam is given in Eqns. (8.78, 8.79 and 8.80). We see that the higher-order assumption introduces additional degrees of freedom in the form of lateral contraction $\psi(x,t)$, in addition to the axial, transverse and slope degrees of freedom. As before, we will derive two sets of spectral element, one is a two noded finite length element and the other is a single noded semi infinite throw-off element.

Finite Length Element

The starting point of the spectral FEM formulation is the strong form of the governing equation, which in the present case is given by Eqns. (8.78–8.81). This model due to inherent coupling of transverse displacement, rotation and lateral contraction will exhibit four-way axial-flexural-shear-lateral contraction coupling. For this study, plane wave type solution is sought, where the displacement field, $\mathbf{u} = \{u^o\ ,\ \psi\ ,\ w\ ,\ \phi\}(x,t)$, can be written as

$$\mathbf{u} = \sum_{n=1}^{N} \{\tilde{u}\ ,\ \tilde{\psi}\ ,\ \tilde{w}\ ,\ \tilde{\phi}\}(x) e^{i\omega_n t} = \sum_{n=1}^{N} \{\tilde{\mathbf{u}}(x)\} e^{i\omega_n t}\ , \tag{12.32}$$

where ω_n is the circular frequency at the nth sampling point and N is the frequency index corresponding to the Nyquist frequency in FFT.

The displacement field for the two-noded finite length element will be the exact solution to the coupled transformed ordinary differential equation, which will have four forward moving and four backward moving (reflected) components. Hence, the displacement field will contain eight wave coefficients. These wave coefficients can be determined from eight boundary conditions existing at the two nodes. The displacement at any point x ($x \in [0, L]$) and at frequency ω_n becomes

$$\begin{aligned} \tilde{\mathbf{u}}_{\mathbf{n}} &= \left\{ \begin{array}{c} \hat{u}(x,\omega_n) \\ \hat{\psi}(x,\omega_n) \\ \hat{w}(x,\omega_n) \\ \hat{\phi}(x,\omega_n) \end{array} \right\} = \begin{bmatrix} R_{11} & \ldots & R_{18} \\ R_{21} & \ldots & R_{28} \\ R_{31} & \ldots & R_{38} \\ R_{41} & \ldots & R_{48} \end{bmatrix} \\ &\times \begin{bmatrix} e^{-ik_1 x} & 0 & \ldots & 0 \\ 0 & e^{-ik_2 x} & \ldots & 0 \\ \vdots & \ddots & \ddots & \vdots \\ 0 & \ldots & \ldots & e^{-ik_8 x} \end{bmatrix} \mathbf{a_n} \end{aligned} \tag{12.33}$$

where ($k_{p+4} = -k_p\ , p = 1, \ldots, 4$). The above equation in concise form can be written

$$\tilde{\mathbf{u}}_{\mathbf{n}} = \mathbf{R_n}\, \mathbf{D(x)_n}\, \mathbf{a_n} \tag{12.34}$$

where $\mathbf{D(x)_n}$ is a diagonal matrix of size 8×8 whose ith element is $e^{-ik_i x}$. $\mathbf{R_n}$ is the amplitude ratio matrix and is of size 4×8. This matrix needs to be known beforehand for subsequent element formulation. There are two different ways to compute the elements of this matrix. In this formulation the SVD method, explained in Chapter 3 Section 3.5.1, is followed, which is suitable for structural models with a large number of degrees of freedom.

Here, $\mathbf{a_n}$ is a vector of eight unknown constants to be determined. These unknown constants are expressed in terms of the nodal displacements by evaluating Eqn. (12.33) at the two nodes, that is at $x = 0$ and $x = L$. In doing so, we get

$$\hat{\mathbf{u}}_{\mathbf{n}} = \left\{ \begin{array}{c} \tilde{\mathbf{u}}_1 \\ \tilde{\mathbf{u}}_2 \end{array} \right\}_n = \left[\begin{array}{c} \mathbf{R} \\ \mathbf{R} \end{array} \right]_n \left[\begin{array}{c} \mathbf{D(0)} \\ \mathbf{D(L)} \end{array} \right]_n \mathbf{a_n} = \mathbf{T_1}_n \mathbf{a_n} \tag{12.35}$$

where $\tilde{\mathbf{u}}_1$ and $\tilde{\mathbf{u}}_2$ are the nodal displacements of node 1 and node 2, respectively.

Next, as before, we need to use the expressions for force boundary conditions to relate the stress resultants with the unknown wave coefficients. These expressions are given in Chapter 8 Eqns. (8.82 & 8.83). Using these force boundary conditions, the force vector $\mathbf{f_n} = \{N_x\ ,\ Q_x\ ,\ V_x\ ,\ M_x\}_n$ can be written in terms of the unknown constants $\mathbf{a_n}$ as $\mathbf{f_n} = \mathbf{P_n a_n}$. When the force vector is evaluated at node 1 and node 2, nodal force vector, $\hat{\mathbf{f}}_{\mathbf{n}}$, is obtained and can be related to $\mathbf{a_n}$ as

$$\hat{\mathbf{f}}_{\mathbf{n}} = \left\{ \begin{array}{c} \tilde{\mathbf{f}}_1 \\ \tilde{\mathbf{f}}_2 \end{array} \right\}_n = \left[\begin{array}{c} P(0) \\ P(L) \end{array} \right]_n \{\mathbf{a}\}_n = \mathbf{T_2}_n \mathbf{a_n}\,. \tag{12.36}$$

Eqn. (12.35) and Eqn. (12.36) together yield the relation between the nodal force and the nodal displacement vector at frequency ω_n as

$$\hat{\mathbf{f}}_{\mathbf{n}} = \mathbf{T_2}_n \mathbf{T_1}_n^{-1} \hat{\mathbf{u}}_{\mathbf{n}} = \mathbf{K_n} \hat{\mathbf{u}}_{\mathbf{n}} \tag{12.37}$$

where $\mathbf{K_n}$ is the dynamic stiffness matrix at frequency ω_n of dimension 8×8. Explicit forms of the matrix $\mathbf{T_1}$ and $\mathbf{T_2}$ are given below.

$$\mathbf{T}_1(1:4, 1:8) = \mathbf{R}(1:4, 1:8) \tag{12.38}$$

$$\mathbf{T}_1(l, m) = \mathbf{R}(l-4, m)e^{-ik_m L}\,, \quad l = 5..8\,, m = 1..8 \tag{12.39}$$

Similarly,

$$\begin{aligned}
T_2(1,p) &= iA_{11}R(1,p) - B_{11}R(4,p))k_p - A_{13}R(2,p) \\
T_2(2,p) &= -B_{55}(-pR(3,p)k_p - R(4,p)) + ik_d D_{55}R(2,p)k_p \\
T_2(3,p) &= -A_{55}(-pR(3,p)k_p - R(4,p)) + iB_{55}R(2,i)k_p \\
T_2(4,p) &= -i(B_{11}R(1,p) - D_{11}R(4,p))k_p + B_{13}R(2,p) \\
T_2(5:8,p) &= -T_2(1:4,i)e^{-ik_p L}\,, \quad p = 1\ldots 8
\end{aligned} \tag{12.40}$$

Throw-off Element

For the infinite length element, only the forward propagating modes are considered. The displacement field (at frequency ω_n) becomes

$$\tilde{\mathbf{u}}_n = \sum_{m=1}^{4} R_n^m e^{-ik_{mn}x} a_m^n = \mathbf{R_n}\,\mathbf{D(x)_n}\,\mathbf{a_n}\,, \tag{12.41}$$

where $\mathbf{R_n}$ and $\mathbf{D(x)_n}$ is now of size 4×4. The $\mathbf{a_n}$ is a vector of four unknown constants. Evaluating the above expression at node 1 ($x = 0$), the nodal displacements are related to these constants through the matrix $\mathbf{T_1}$ as

$$\hat{\mathbf{u}}_{\mathbf{n}} = \tilde{\mathbf{u}_1}_n = \mathbf{R_n D(0)_n a_n} = \mathbf{T_1}_n \mathbf{a_n} \tag{12.42}$$

where $\mathbf{T_1}$ is now a matrix of dimension 4×4. Similarly, the nodal forces at node 1 can be related to the unknown constants

$$\hat{\mathbf{f}}_{\mathbf{n}} = \tilde{\mathbf{f}_1}_n = \mathbf{P(0)_n a_n} = \mathbf{T_2}_n \mathbf{a_n}\,. \tag{12.43}$$

Using Eqns. (12.42 & 12.43), nodal forces at node 1 are related to the nodal displacements at node 1

$$\hat{\mathbf{f}}_{\mathbf{n}} = \mathbf{T_2}_n \mathbf{T_1}_n^{-1} \hat{\mathbf{u}}_{\mathbf{n}} = \mathbf{K_n} \hat{\mathbf{u}}_{\mathbf{n}} \tag{12.44}$$

where $\mathbf{K}_n$ is the element dynamic stiffness matrix of dimension 4×4 at frequency ω_n. The matrices $\mathbf{T}_1 = R(1:4, 1:4)$ and $\mathbf{T}_2(1:4, p)$ are the same as for the finite length element ($p = 1 \ldots 4$). As in the elementary case, the dynamic stiffness is complex.

12.3.4 Spectral Element for Framed Structures

Frame structure is the one that has combined action of rod and beam. That is, a frame member, can undergo combined axial and bending deformation. A 2-D frame will undergo bending and axial deformation in one plane (deformation in two directions and a rotation), while a 3-D frame will undergo deformation in three coordinate direction and rotation about the three coordinate axes. A 3-D frame in space is how in Fig. 12.5.

The matrix procedure of spectral finite element enables one to use the same procedure that is adopted by conventional finite elements. Let x, y and z represent the global coordinate system. For the 3-D frame, first the dynamic stiffness matrices for the rod and the beam elements are written in the local coordinate system $\bar{x}$, $\bar{y}$ and $\bar{z}$ and expanded to a 6×6 matrix by populating the stiffness elements of these at appropriate locations. Then a transformation matrix $\mathbf{T}$ is established between the deformations in the local coordinates to the deformations in the global coordinates. Using the expanded dynamic stiffness matrix and the transformation matrix, the dynamic stiffness matrix of the framed structure is established.

Note that in a 3-D frame, in addition to the axial and bending deformation in the two planes, it also also undergoes torsional motion (moment about x coordinate) . The stiffness matrix of the rod in local coordinates that relates the nodal local displacement $\hat{u}_1$ and $\hat{u}_2$ with the local forces $\hat{\hat{N}}_{x1}$ and $\hat{\hat{N}}_{x2}$ is

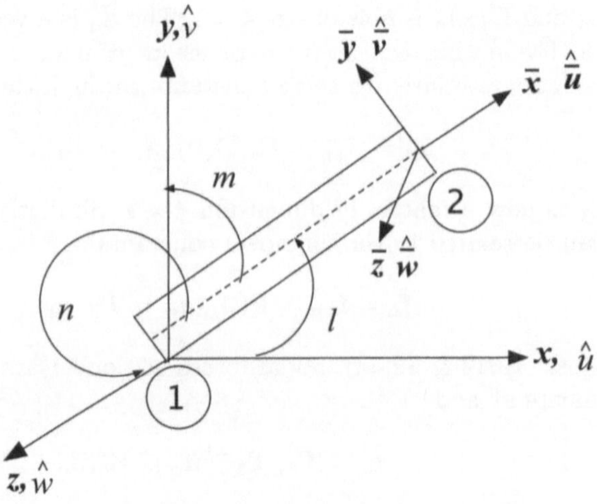

FIGURE 12.5: A 3-D frame structure with local and global coordinate axes

given by

$$
\begin{aligned}
\hat{\mathbf{K}}_{\mathbf{rod}} &= \begin{bmatrix} k_{11}^R & k_{12}^R \\ \\ k_{12}^R & k_{22}^R \end{bmatrix} \\
&= \frac{EA}{L}\frac{ik_{rod}L}{1-e^{-i2k_{rod}L}}\begin{bmatrix} 1+e^{-i2k_{rod}L} & -2e^{-ik_{rod}L} \\ -2e^{-ik_{rod}L} & 1+e^{-i2k_{rod}L} \end{bmatrix}
\end{aligned}
\tag{12.45}
$$

where $k_{rod} = \omega\sqrt{\rho A/EA}$

The stiffness matrix of the beam bending in x-y plane having the local nodal displacement vector $\{\hat{\hat{v}}_1\,\hat{\hat{\theta}}_{z1}\,\hat{\hat{v}}_2\,\hat{\hat{\theta}}_{z2}\}^T$ and local nodal forces $\{\hat{\hat{V}}_{y1}\,\hat{\hat{M}}_{z1}\,\hat{\hat{V}}_{y2}\,\hat{\hat{M}}_{z2}\}^T$ is given by

$$
\hat{\mathbf{K}}_{\mathbf{beam}}^{\mathbf{xy}} = \begin{bmatrix} k_{11}^{Bx} & k_{12}^{Bx} & k_{13}^{Bx} & k_{14}^{Bx} \\ \\ k_{12}^{Bx} & k_{22}^{Bx} & k_{23}^{Bx} & k_{24}^{Bx} \\ \\ k_{13}^{Bx} & k_{23}^{Bx} & k_{33}^{Bx} & k_{34}^{Bx} \\ \\ k_{14}^{Bx} & k_{24}^{Bx} & k_{34}^{Bx} & k_{44}^{Bx} \end{bmatrix} = \begin{bmatrix} \bar{\alpha} & \bar{\beta}L & \bar{\gamma} & \bar{\delta}L \\ & \bar{\xi}L^2 & -\bar{\delta}L & \bar{\eta}L^2 \\ & & \bar{\alpha} & -\bar{\beta}L \\ & & & \bar{\xi}L^2 \end{bmatrix}
\tag{12.46}
$$

where $\bar{\alpha}$,... etc are defined earlier in Section 12.3.2. Here, the area moment of inertia I_{zz} is used in the wavenumber and dynamic stiffness matrix computation.

Next, we compute the dynamic stiffness matrix for a beam bending about $x - z$ plane. Here the corresponding local degrees of freedom are $\{\hat{w}_1\, \hat{\bar{\theta}}_{y1}\, \hat{w}_2\, \hat{\bar{\theta}}_{y2}\}^T$ and the corresponding stress resultants are $\{\hat{\bar{V}}_{z1}\, \hat{\bar{M}}_{y1}\, \hat{\bar{V}}_{z2}\, \hat{\bar{M}}_{y2}\}^T$. Here the stiffness matrix is given by

$$\hat{\mathbf{K}}^{\mathbf{xz}}_{\mathbf{beam}} = \begin{bmatrix} k_{11}^{Bz} & k_{12}^{Bz} & k_{13}^{Bz} & k_{14}^{Bz} \\ k_{12}^{Bz} & k_{22}^{Bz} & k_{23}^{Bz} & k_{24}^{Bz} \\ k_{13}^{Bz} & k_{23}^{Bz} & k_{33}^{Bz} & k_{34}^{Bz} \\ k_{14}^{Bz} & k_{24}^{Bz} & k_{34}^{Bz} & k_{44}^{Bz} \end{bmatrix} \tag{12.47}$$

where the elements of the matrix is similar to what is given in Eqn (12.46) except that here we use the area moment of inertia I_{yy}. Next, we will write the dynamic stiffness matrix that governs the motion of the bar torsional loading. This has a very similar form as that of the rod(see [34]). The degrees of freedom corresponding to the torsional motion are the twist $\hat{\bar{\theta}}_{x1}$ and $\hat{\bar{\theta}}_{x2}$ and the corresponding stress resultants are the torques given by $\hat{\bar{T}}_{x1}$ and $\hat{\bar{T}}_{x2}$. The corresponding dynamic stiffness relating these two quantities is given by

$$\begin{aligned} \hat{\mathbf{K}}_{\mathbf{tor}} &= \begin{bmatrix} k_{11}^S & k_{12}^S \\ k_{12}^S & k_{22}^S \end{bmatrix} \\ &= \frac{GJ}{L}\frac{ik_{tor}L}{1-e^{-i2k_{tor}L}}\begin{bmatrix} 1+e^{-i2k_{tor}L} & -2e^{-ik_{tor}L} \\ -2e^{-ik_{tor}L} & 1+e^{-i2k_{tor}L} \end{bmatrix} \end{aligned} \tag{12.48}$$

where G is the shear modulus, J is the polar moment of inertia which is equal to $J = I_{xx} + I_{yy}$ and $k_{tor} = \omega\sqrt{\rho A/GJ}$. Locally, in all we have 12 degrees of freedom (6 in each node) and these can be written as

$$\hat{\mathbf{u}} = \{\hat{u}_1\ \hat{v}_1\ \hat{w}_1\ \hat{\bar{\theta}}_{x1}\ \hat{\bar{\theta}}_{y1}\ \hat{\bar{\theta}}_{z1}\ \hat{u}_2\ \hat{v}_2\ \hat{w}_2\ \hat{\bar{\theta}}_{x2}\ \hat{\bar{\theta}}_{y2}\ \hat{\bar{\theta}}_{z2}\}^T$$

The corresponding force vector is given by

$$\hat{\bar{\mathbf{F}}} = \{\hat{\bar{N}}_{x1}\ \hat{\bar{V}}_{y1}\ \hat{\bar{V}}_{z1}\ \hat{\bar{T}}_{x1}\ \hat{\bar{M}}_{z1}\ \hat{\bar{M}}_{y1}\ \hat{\bar{N}}_{x2}\ \hat{\bar{V}}_{y2}\ \hat{\bar{V}}_{z2}\ \hat{\bar{T}}_{x2}\ \hat{\bar{M}}_{z2}\ \hat{\bar{M}}_{y2}\}^T$$

The above two vectors are related by an expanded dynamic stiffness matrix, which is constructed using the stiffness elements given in Eqns. (12.45–12.48). That is

$$\hat{\bar{\mathbf{F}}} = \begin{bmatrix} [\hat{\mathbf{k}}_{mm}] & [\hat{\mathbf{k}}_{mn}] & [\hat{\mathbf{k}}_{mo}] & [\hat{\mathbf{k}}_{mp}] \\ [\hat{\mathbf{k}}_{mn}] & [\hat{\mathbf{k}}_{nn}] & [\hat{\mathbf{k}}_{no}] & [\hat{\mathbf{k}}_{np}] \\ [\hat{\mathbf{k}}_{mo}] & [\hat{\mathbf{k}}_{on}] & [\hat{\mathbf{k}}_{oo}] & [\hat{\mathbf{k}}_{op}] \\ [\hat{\mathbf{k}}_{mp}] & [\hat{\mathbf{k}}_{np}] & [\hat{\mathbf{k}}_{op}] & [\hat{\mathbf{k}}_{pp}] \end{bmatrix} \hat{\mathbf{u}} = \hat{\bar{\mathbf{k}}}\hat{\mathbf{u}} \tag{12.49}$$

The entries of each of these sub-matrices are populated using the stiffness elements from Eqns. (12.45–12.48). For example, the first quadrant of the above matrix is given by

$$\begin{bmatrix} [\hat{\mathbf{k}}_{mm}] & [\hat{\mathbf{k}}_{mn}] \\ [\hat{\mathbf{k}}_{mn}] & [\hat{\mathbf{k}}_{nn}] \end{bmatrix} = \begin{bmatrix} k_{11}^R & 0 & 0 & 0 & 0 & 0 \\ 0 & k_{11}^{Bx} & 0 & 0 & 0 & k_{12}^{Bx} \\ 0 & 0 & k_{11}^{Bz} & 0 & k_{12}^{Bz} & 0 \\ 0 & 0 & 0 & k_{11}^{S} & 0 & 0 \\ 0 & 0 & k_{12}^{Bx} & 0 & k_{22}^{Bx} & 0 \\ 0 & k_{12}^{Bz} & 0 & 0 & 0 & k_{22}^{Bz} \end{bmatrix} \tag{12.50}$$

Similarly, other sub matrices in other quadrants can be filled similarly. We have now succeeded in relating the forces and deformation in the local co-ordinate as a 12 × 12 matrix. The framed structure, if arbitrarily oriented, will have direction cosines l, m, and n with respect to global x-coordinate as shown in Fig. 12.5. Hence the local and global coordinates are different. For analysis, we need to transform all those framed structures to global coordinates and expresses forces and deformations in the global coordinate system. In addition, we should also consider the principle direction of the arbitrarily oriented framed member. We construct the rotation matrix using the concept of *Euler Angles*. That is, we give three successive rotations from the global axes to the local axes of the member, where the first rotation is about z axis and the second is about $\bar{y}$-axis. The third rotation consists of a rotation ϕ about the local axis $\bar{x}$ such that the $\bar{y}$-axis and $\bar{z}$-axis coincide with the principle direction of the space member. Multiplication of three successive rotations will yield a rotation matrix $\mathbf{D}$, which is given by

$$\mathbf{D} = \begin{bmatrix} l & m & n \\ \frac{-m\cos\phi - ln\sin\phi}{R} & \frac{l\cos\phi - mn\sin\phi}{R} & R\sin\phi \\ \frac{m\sin\phi - ln\cos\phi}{R} & \frac{-l\sin\phi - mn\cos\phi}{R} & R\cos\phi \end{bmatrix} \tag{12.51}$$

where $R = \sqrt{1-n^2}$. The degrees of freedom corresponding to global coordinates $x - y - z$ is given by

$$\hat{\mathbf{u}} = \{\hat{u}_1\ \hat{v}_1\ \hat{w}_1\ \hat{\theta}_{x1}\ \hat{\theta}_{y1}\ \hat{\theta}_{z1}\ \hat{u}_2\ \hat{v}_2\ \hat{w}_2\ \hat{\theta}_{x2}\ \hat{\theta}_{y2}\ \hat{\theta}_{z2}\}^T$$

The forces corresponding to the above degrees of freedom are

$$\hat{\mathbf{F}} = \{\hat{N}_{x1}\ \hat{V}_{y1}\ \hat{V}_{z1}\ \hat{T}_{x1}\ \hat{M}_{z1}\ \hat{M}_{y1}\ \hat{N}_{x2}\ \hat{V}_{y2}\ \hat{V}_{z2}\ \hat{T}_{x2}\ \hat{M}_{z2}\ \hat{M}_{y2}\}^T$$

The global degrees of freedom and global forces are related to their local counterparts as

$$\hat{\mathbf{u}} = \mathbf{T}\hat{\hat{\mathbf{u}}}, \qquad \hat{\mathbf{F}} = \mathbf{T}\hat{\hat{\mathbf{F}}}, \qquad \mathbf{T} = \begin{bmatrix} \mathbf{D} & 0 & 0 & 0 \\ 0 & \mathbf{D} & 0 & 0 \\ 0 & 0 & \mathbf{D} & 0 \\ 0 & 0 & 0 & \mathbf{D} \end{bmatrix} \tag{12.52}$$

Finally, we can write the spectral element for the framed element as

$$\hat{\mathbf{F}} = \hat{\mathbf{K}}\hat{\mathbf{u}}, \qquad \hat{\mathbf{K}} = \mathbf{T}^{\mathbf{T}}\hat{\mathbf{k}}\mathbf{T} \tag{12.53}$$

The beam and rod dynamic stiffness matrices can now be used to obtain the time responses for any arbitrary time signals. More details on different examples of using SFEM can be obtained from the unified book [54] and an exclusive book on SFEM [48].

12.3.5 2-D Layer Element for Isotropic Solids

The theory associated with 2-D isotropic waveguides was discussed in Chapter 7 (see Section 7.1), wherein the governing equations in terms of displacements normally referred to as *Naviers's Equation* (see Eqn. (7.6)) is solved by splitting this equation into two partial differential equations, one based on Scalar potential (Φ) and vector potential H_y using *Helmholtz decomposition.* Chapter 7 gives the full details of the solution procedure. In this section, we will derive two elements, namely the throw-off elements to model semi-infinite domain and the finite domain element to model doubly bounded media.

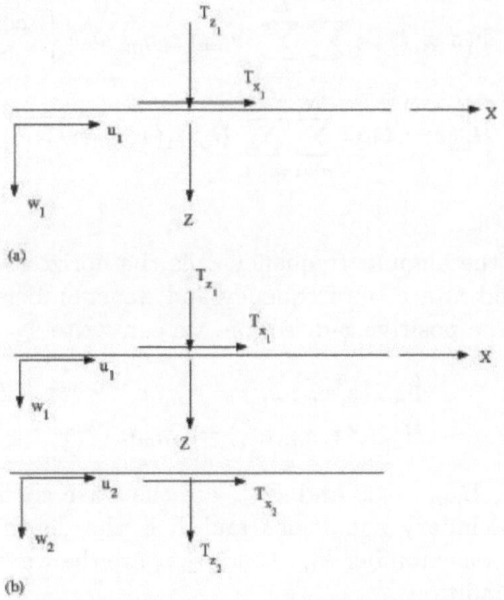

FIGURE 12.6: Sign conventions of (a) throw-off spectral element (b) layer element

Unlike 1-D waveguides, here the entire line along x axis is a node in the 2-D spectral element as shown in Fig. 12.6. This figure also shows the sign convention adopted in the element formulation. The starting point of the spectral element formulation is the strong form of the equation, which is given by Eqn. (7.9) given in Chapter 7 and reported here for completeness. They are given by

$$(\lambda+\mu)\nabla^2\Phi = \rho\frac{\partial^2\Phi}{\partial t^2}, \quad \mu\nabla^2 H_y = \rho\frac{\partial^2 H_y}{\partial t^2} \tag{12.54}$$

where the displacements $u(z,t)$ and $w(z,t)$, are the two displacements in two coordinate directions, and H_y is the vector potential of interest and these displacements can be expressed in terms of scalar potential Φ and vector potential H_y is given in Eqn. (7.7) in Chapter 7 and reproduced again here, which is given by

$$\begin{aligned} u(z,t) &= \frac{\partial\Phi(z,t)}{\partial z} + \frac{\partial H_y(z,t)}{\partial x} \\ w(z,t) &= \frac{\partial\Phi(z,t)}{\partial x} - \frac{\partial H_y(x,t)}{\partial z} \end{aligned} \tag{12.55}$$

The solution of Eqn. (12.54) for scalar potentail Φ and vector potential H_y can be written as

$$\begin{aligned} \Phi(z,x,t) &= \sum_{n=1}^{N}\sum_{m=0}^{M}\hat{\Phi}_{nm}(z,\eta_m,\omega_n)\left\{\begin{array}{c}\cos(\eta_m x)\\ \sin(\eta_m x)\end{array}\right\}e^{i\omega_n t} \\ H_y(z,x,t) &= \sum_{n=1}^{N}\sum_{m=0}^{M}\hat{H}_{ynm}(x,\eta_m,\omega_n)\left\{\begin{array}{c}\cos(\eta_m x)\\ \sin(\eta_m x)\end{array}\right\}e^{i\omega_n t} \end{aligned} \tag{12.56}$$

where ω is the angular frequency, η is the horizontal wavenumber. The subscripts n and m are the frequency and wavenumber indices. For waves propagating in the positive x-direction, we can write $\hat{\Phi}_{nm}$ and $\hat{H}_{ynm}$ as

$$\begin{aligned} \hat{\Phi}_{nm}(z,\eta_m,\omega_n) &= A_{nm}e^{-ik_{1nm}z} + B_{nm}e^{i(L-z)} \\ \hat{H}_{ynm}(z,\eta_m,\omega_n) &= C_{nm}e^{-ik_{2nm}z} + D_{nm}e^{i(L-z)} \end{aligned} \tag{12.57}$$

where A_{nm}, B_{nm}, C_{nm} and D_{nm} are the wave coefficients to be determined from the boundary conditions and L is the depth of the doubly bounded media. The wavenumber k_{1nm} and k_{2nm} are the wavenumbers that satisfy the following conditions

$$\begin{aligned} &k_{1nm} = \pm\sqrt{k_{pn}^2 - \eta_m^2}, \qquad k_{2nm} = \pm\sqrt{k_{sn}^2 - \eta_m^2} \\ &k_{pn} = \frac{\omega_n}{C_p}, \qquad k_{sn} = \frac{\omega_n}{C_s}, \qquad \eta_m = \frac{2m\pi}{W} \end{aligned} \tag{12.58}$$

where $C_p = (\lambda + 2\mu)/\rho$, $C_s = \mu/\rho$ and W is the spatial window. A detailed treatment of these relations are given in Chapter 7.

Here coefficients A_{nm}, B_{nm}, C_{nm} and D_{nm} are determined using displacement or traction boundary conditions on the two surfaces. Now substituting Eqn. (12.57) in Eqn. (12.55), we can write the displacement variation as

$$u(z,x,t) = \sum_{n=1}^{N}\sum_{m=0}^{M} \hat{u}_{nm}(z,\eta_m,\omega_n)\left\{\begin{array}{c}\cos(\eta_m x)\\ \sin(\eta_m x)\end{array}\right\} e^{i\omega_n t}$$

$$w(z,x,t) = \sum_{n=1}^{N}\sum_{m=0}^{M} \hat{w}_{nm}(z,\eta_m,\omega_n)\left\{\begin{array}{c}\sin(\eta_m x)\\ \cos(\eta_m x)\end{array}\right\} e^{i\omega_n t} \tag{12.59}$$

where

$$\begin{aligned}
\hat{u}(z,\eta_m,\omega_n) &= \left[-A_{nm}e^{-ik_{1nm}z} + B_{nm}e^{-ik_{1nm}(L-z)}\right] ik_{1nm} \\
&\quad \pm \left[C_{nm}e^{-ik_{2nm}z} + D_{nm}e^{-ik_{2nm}(L-z)}\right]\eta_m \\
\hat{w}(z,\eta_m,\omega_n) &= \mp\left[-A_{nm}e^{-ik_{1nm}z} + B_{nm}e^{-ik_{1nm}(L-z)}\right]\eta_m \\
&\quad \pm \left[C_{nm}e^{-ik_{2nm}z} - D_{nm}e^{-ik_{2nm}(L-z)}\right] ik_{2nm}
\end{aligned} \tag{12.60}$$

One Noded or Throw-off Element

One-noded throw-off element is normally used to model semi-infinite media as shown in Fig. 12.6 (a). The top line in the direction of positive $x - axis$ represented a single node for this element. The displacement variation for this element required SFEM formulation is written by leaving out the reflected coefficient in Eqn. (12.60). That is, we have

$$\begin{aligned}
\hat{u}(z,\eta_m,\omega_n) &= -A_{nm}(ik_{1nm})e^{-ik_{1nm}z} \pm B_{nm}(\eta_m)e^{-ik_{2nm}z} \\
\hat{w}(z,\eta_m,\omega_n) &= \mp A_{nm}(\eta_m)e^{-ik_{1nm}z} + B_{nm}(ik_{2nm})e^{-ik_{2nm}z}
\end{aligned} \tag{12.61}$$

Spectral element formulation begins with relating the unknown wave coefficients with the nodal displacements. That is at $z = 0$, $\hat{u}_{nm}(0,\eta_m,\omega_n) = \hat{u}_{1nm}$ and $\hat{w}_{nm}(0,\eta_m,\omega_n) = \hat{w}_{1nm}$, which is first written in the matrix form and later inverting this matrix, the equation becomes

$$\left\{\begin{array}{c}A_{nm}\\ B_{nm}\end{array}\right\} = \frac{1}{k_{1nm}k_{2nm} + \eta m^2}\left[\begin{array}{cc} ik_{2nm} & -\eta_m \\ \eta_m & -ik_{1nm}\end{array}\right]\left\{\begin{array}{c}\hat{u}_{1nm}\\ \hat{w}_{1nm}\end{array}\right\} \tag{12.62}$$

Next, at Node 1, we will consider the tractions T_{xnm} and T_{znm}. The tractions at Node 1 can be written in terms of normal and shear stresses and the outward normal vectors n_x and n_z as

$$\begin{aligned}\hat{T}_{znm} &= \hat{\sigma}_{zznm} n_z + \hat{\sigma}_{xznm} n_x \\ \hat{T}_{xnm} &= \hat{\sigma}_{xznm} n_z + \hat{\sigma}_{zznm} n_x\end{aligned} \tag{12.63}$$

where n_x and n_z are the outward normals to the surface. At the surface representing Node 1 (see Fig. 12.6 (a)), $n_x = 0$ and $n_z = -1$. Hence the tractions along Node 1 is given by

$$\hat{T}_{z1nm} = -\hat{\sigma}_{zz1nm}, \qquad \hat{T}_{x1nm} = -\hat{\sigma}_{xz1nm} \tag{12.64}$$

The stresses $\hat{\sigma}_{zz1nm}$ and $\hat{\sigma}_{xz1nm}$ are computed as follows. First, using Eqn. (12.61), the strain components are computed. Then using stress-strain relations, the required stresses and hence the tractions are computed. These are explained in Chapter 7. Hence the required tractions can be written in matrix form as

$$\left\{ \begin{array}{c} \hat{T}_{z1nm} \\ \hat{T}_{x1nm} \end{array} \right\} = \mu \left[\begin{array}{cc} -(2\eta_m^2 - k_{sn}^2) & 2i\eta_m k_{2nm} \\ -2i\eta_m k_{1nm} & (2\eta_m^2 - k_{sn}^2) \end{array} \right] \left\{ \begin{array}{c} A_{nm} \\ B_{nm} \end{array} \right\} \tag{12.65}$$

where μ is the shear modulus and k_{sn} is the shear wavenumber. Substituting for A_{nm} and B_{nm} from Eqn. (12.62) in Eqn. (12.65), we get the following stiffness relation for 1-noded throw-off element for isotropic solids. This is given by

$$\begin{aligned}\hat{\mathbf{T}}_{\mathbf{nm}} &= \hat{\mathbf{k}}_{\mathbf{nm}} \hat{\mathbf{u}}_{\mathbf{nm}} \\ \hat{\mathbf{k}}_{\mathbf{nm}} &= \frac{\mu}{k_{1nm} k_{2nm} + \eta_m^2} \left[\begin{array}{cc} i k_{2nm} k_{sn}^2 & \eta_m (2\eta_m^2 + 2k_{1nm} k_{2nm} - k_{sn}^2) \\ sym & i k_{1nm} k_{sn}^2 \end{array} \right]\end{aligned} \tag{12.66}$$

Two Noded or Finite Length Element

The two-noded finite length element begins with Eqn. (12.60). Using this equation, we will first relate the unknown wave coefficients with the nodal displacements at the two surfaces of the element. As in the previous case of throw-off element formulation, the tractions at the two surfaces are determined by evaluating the normal and shear stresses at the surfaces $z = 0$ and $z = L$. This will give the relation between the tractions and the unknown wave coefficients. Substituting for wave coefficients from the previous step, we obtain the dynamic stiffness matrix, which is given by

$$\hat{\mathbf{k}}_{\mathbf{nm}} = \left[\begin{array}{cccc} \hat{K}_{11nm} & \hat{K}_{12nm} & \hat{K}_{13nm} & \hat{K}_{14nm} \\ & \hat{K}_{22nm} & -\hat{K}_{14nm} & \hat{K}_{24nm} \\ \text{sym} & & \hat{K}_{11nm} & -\hat{K}_{12nm} \\ & & & \hat{K}_{22nm} \end{array} \right] \tag{12.67}$$

where

$$
\begin{aligned}
\hat{K}_{11nm} &= \frac{\mu}{\Delta_{mn}}\left[-ik_{2mn}k_{sn}^2\right]\left[k_{1nm}k_{2nm}r_{12}r_{21}+\eta_m^4 r_{11}r_{22}\right]\\
\hat{K}_{12nm} &= \frac{\mu}{\Delta_{mn}}\eta_m k_{1nm}k_{2nm}\left[-k_{sn}^2+4\eta_m^2\right]\left[4e^{-ik_{1nm}L}e^{-ik_{2nm}L}-r_{12}r_{22}\right]\\
&\quad -\frac{\mu}{\Delta_{mn}}\eta_m\left[\eta_m^4-\eta_m^2k_{2mn}^2+2k_{1nm}^2k_{2nm}^2\right]r_{11}r_{21}\\
\hat{K}_{13nm} &= \frac{\mu}{\Delta_{mn}}2ik_{2mn}k_{sn}^2\left[k_{1nm}k_{2mn}e^{-ik_{1nm}L}r_{21}+\eta_m^2e^{-ik_{2mn}L}r_{11}\right]\\
\hat{K}_{14nm} &= \frac{\mu}{\Delta_{mn}}2\eta_m k_{1nm}k_{2nm}k_{sn}^2\left[e^{-ik_{1nm}L}r_{22}-e^{-ik_{2nm}L}z_{12}\right]\\
\hat{K}_{22nm} &= \frac{\mu}{\Delta_{mn}}\left[-ik_{1mn}k_{sn}^2\right]\left[k_{1nm}k_{2nm}r_{11}r_{22}+\eta_m^4 r_{12}r_{21}\right]\\
\hat{K}_{24nm} &= \frac{\mu}{\Delta_{mn}}2ik_{1mn}k_{sn}^2\left[k_{1nm}k_{2mn}e^{-ik_{2nm}L}r_{11}+\eta_m^2e^{-ik_{1mn}L}r_{21}\right]
\end{aligned}
\tag{12.68}
$$

and

$$
\begin{aligned}
\Delta_{mn} &= 2\eta_m^2k_{1nm}k_{2mn}\left[4e^{-ik_{1nm}L}e^{-ik_{2mn}L}-r_{12}r_{22}\right]\\
&\quad -\left[k_{1nm}^2k_{2nm}^2+\eta_m^4\right]\\
r_{11} &= 1-e^{-2ik_{1nm}L}, \qquad r_{11}=1-e^{-2ik_{2nm}L}\\
r_{21} &= 1+e^{-2ik_{1nm}L}, \qquad r_{22}=1+e^{-2ik_{2nm}L}
\end{aligned}
\tag{12.69}
$$

The detailed formulation including the explicit form of intermediate matrices can be found in [119] and [120]

Next, we will now outline the method of prescribing the boundary conditions under the 2-D spectral element environment. Essential boundary conditions are prescribed in the usual way as is done in the conventional FE methods, where the nodal displacements are arrested or released depending upon the nature of the boundary conditions. The applied tractions are prescribed at the nodes. It is assumed that the loading function (for symmetric loading) can be written as

$$
F(x,z,t)=\delta(z-z_k)\left(\sum_{m=1}^{M}a_m\cos(\eta_m x)\right)\left(\sum_{n=0}^{N-1}\hat{f}_n e^{(-i\omega_n t)}\right), \tag{12.70}
$$

where δ denotes the Dirac delta function, z_k is the Z coordinate of the point where the load is applied and the z dependency is fixed by suitably choosing the node where the load is prescribed. No variation of load along the Z direction is allowed in this analysis. $\hat{f}_n$ are the Fourier Transform coefficients of the time dependent part of the load, which are computed by FFT, and a_m are the Fourier series coefficients of the x dependent part of the load.

There are two summations involved in the solution and two associated windows, one in time T and the other in space W. The discrete frequencies

ω_n and the discrete horizontal wavenumber η_m are related to these windows by the number of data points N and M chosen in each summation, that is,

$$\begin{aligned}\omega_n &= 2n\pi/T = 2n\pi/(N\Delta t)\\ \eta_m &= 2(m-1)\pi/W = 2(m-1)\pi/(M\Delta x)\end{aligned} \tag{12.71}$$

where Δt and Δx are the temporal and spatial sample rate, respectively.

12.3.6 Composite Layer Element

The spectral finite element for isotropic layered media is normally formulated using the solution obtained by the method of Helmholtz potentials, which is applicable only to isotropic materials. The spectral element for this case was outlined in the previous section. For anisotropic and inhomogeneous media, among the available methodologies, the Partial Wave Technique (PWT) is a suitable option [4]. This was discussed in Chapter 8. In this section, spectral element for layered composite medium is formulated using this method, where the SVD method (described in Section 3.5.1) is utilized to obtain the wave amplitudes, which is essential for constructing the partial waves. In the PWT based method, once the partial waves are found, the wave coefficients are made to satisfy the prescribed boundary conditions, that is, two non-zero tractions specified at the top and bottom of the layer. In the present case, the formulation is slightly different, as no specific problem oriented boundary conditions are imposed. Thus, a system matrix is established, which relates the tractions at the interface to the interfacial displacements. This generalization enables the use of the system matrix as a finite element dynamic stiffness matrix, although the element is formulated in the frequency/wavenumber domain. These matrices can be assembled to model different layers of different ply-orientation or inhomogeneity, which obviates the necessity for cumbersome computation associated with multilayer analysis (example see [121]). The only shortcoming of the method is that, each spectral layer element (SLE) can accommodate only one fiber angle, thus for different ply-stacking sequences, the number of elements will be at least equal to the number of different ply-angles in the stacking.

We discussed the Lamb wave propagation in Chapter 7 in the context of isotropic materials. We showed that the dispersion relation is a transcendental equation, where in the wavenumbers were found to be multi-valued and we devised special schemes to solve the transcendental equations to obtain the wavenumbers. To obtain the Lamb wave modes for composites is indeed more challenging. One advantage of the present SFEM formulation is the ease in capturing the Lamb wave [144] propagation in composites. Historically, the dispersion relation (phase velocity frequency relation) for anisotropic materials was given first by Solie and Auld [130], where PWT was used. Subsequent investigations on modeling aspects of Lamb waves were carried out by several researchers [105]. Finite element modeling of Lamb waves was performed by Verdict *et al.* [143]. On the basis of discrete layer theory and a multiple integral transform, an analytical-numerical approach is given by Veidt *et al.* [142]. A coupled FE-normal mode expansion method is given by Moulin *et al.*, [99]. Similarly a boundary element normal mode expansion method is given by Zhao and Rose [150]. The present formulation by virtue of a frequency wavenumber domain representation of solution is an inexpensive way of constructing the Lamb wave modes as well as predicting time domain signals. We will show in this section how this can be done.

The starting point of the spectral element formulation is the strong form of the governing differential equation which was derived in Chapter 8 (see Section 8.6). The strong for the governing equations is given by Eqn. (8.100). Using a combination of Fourier series in horizontal direction (x-direction) and DFT in temporal direction, the above equation can be reduced to ordinary differential equation given by Eqn. (8.103) (also given in Chapter 8). The solution to the governing differential equation is the Eqns. (8.108 & 8.109)

Once the four wavenumbers and wave amplitudes are known, the four partial waves can be constructed and the displacement field can be written as a linear combination of the partial waves. Each partial wave is given by

$$\mathbf{a}_i = \left\{ \begin{array}{c} u_i \\ w_i \end{array} \right\} = \left\{ \begin{array}{c} R_{1i} \\ R_{2i} \end{array} \right\} e^{-ik_i z} \left\{ \begin{array}{c} \sin(\eta_m x) \\ \cos(\eta_m x) \end{array} \right\} e^{i\omega_n t}$$
$$i = 1 \ldots 4 \tag{12.72}$$

and the total solution is given by

$$\mathbf{u} = \sum_{i=1}^{4} C_i \mathbf{a}_i \,. \tag{12.73}$$

where C_i, $i = 1...4$ are the wave coefficients to be determined from the boundary conditions.

Finite Layer Element (FLE)

Here, we again use the finite spectral element is shown in Fig. 12.6 (a). Once the solutions of u and w are obtained in the form of Eqns. (8.108 & 8.109) for each value of ω_n and η_m, the same procedure as outlined in the 2-D element formulation for isotropic solids discussed in the previous section is employed to obtain the element dynamic stiffness matrix as a function of ω_n and η_m. Thus, the nodal displacements are related to the unknown constants by

$$\{u_{1nm} \; v_{1nm} \; u_{2nm} \; v_{2nm}\}^T = \mathbf{T}_{1nm}\{C_1 \;\; C_2 \;\; C_3 \;\; C_4\}^T \,, \tag{12.74}$$

That is,

$$\hat{\mathbf{u}}_{nm} = \mathbf{T}_{1nm}\mathbf{C}_{nm} \,. \tag{12.75}$$

Using Eqn (8.104) (given in Chapter 8), nodal tractions are related to the constants by

$$\hat{\mathbf{T}}_{nm} = \mathbf{T}_{2nm}\mathbf{C}_{nm} \,, \quad \hat{\mathbf{T}}_{nm} = \{\sigma_{zz1} \,, \sigma_{xz1} \,, \sigma_{zz2} \,, \sigma_{xz2}\} \,, \tag{12.76}$$

Explicit forms of $\mathbf{T}_{2nm}$ and $\mathbf{T}_{1nm}$ are

$$\mathbf{T}_1 = \begin{bmatrix} R_{11} & R_{12} & R_{13} & R_{14} \\ R_{21} & R_{22} & R_{23} & R_{24} \\ R_{11}e^{-ik_1L} & R_{12}e^{-ik_2L} & R_{13}e^{ik_1L} & R_{14}e^{ik_2L} \\ R_{21}e^{-ik_1L} & R_{22}e^{-ik_2L} & R_{23}e^{ik_1L} & R_{24}e^{ik_2L} \end{bmatrix} \tag{12.77}$$

$$\begin{aligned} T_2(1,p) &= -Q_{55}(-iR_{1p}k_p - \eta R_{2p}), \\ T_2(2,p) &= iQ_{33}R_{2p}k_p - Q_{13}\eta R_{1p}, \\ T_2(3,p) &= Q_{55}(-jR_{1p}k_p - \eta R_{2p})e^{(-ik_pL)} \\ T_2(4,p) &= \{-iQ_{33}R_{2p}k_p + Q_{13}\eta R_{1p}\}e^{(-ik_pL)}, \end{aligned}$$

where p ranges from 1 to 4.

Thus, the dynamic stiffness matrix becomes

$$\hat{\mathbf{K}}_{nm} = \mathbf{T}_{2nm}\mathbf{T}_{1nm}^{-1}, \tag{12.78}$$

which is of size 4×4 having ω_n and η_m as parameters. This matrix represents the dynamics of an entire layer of any length L at frequency ω_n and horizontal wavenumber η_m. Consequently, this small matrix acts as a substitute for the global stiffness matrix of FE modeling, whose size, depending upon the thickness of the layer, will be many orders larger than the spectral layer element size.

Infinite Layer Element (ILE)

The schematic of this element is shown in Fig. 12.6 (b). This is the 2-D counterpart of the 1-D throw-off element, which is used to model semi-infinite domain. The element is formulated by considering only the forward moving components, which means no reflection will come back from the boundary. This element acts as a conduit to throw away energy from the system and is very effective in modeling the infinite domain in the Z direction. This element is also used to impose absorbing boundary conditions or to introduce maximum damping in the structure. The element has only one edge where the displacements are to be measured and tractions are to be specified. The displacement field for this element (at ω_n and η_m) is

$$u_{nm} = R_{11}C_{1nm}e^{-ik_1z} + R_{12}C_{2nm}e^{-ik_2z}, \tag{12.79}$$

$$w_{nm} = R_{21}C_{1nm}e^{-ik_1z} + R_{22}C_{2nm}e^{-jk_2z}, \tag{12.80}$$

where it is assumed that k_1 and k_2 have positive real parts. Following the same procedure as before, displacement at node 1 can be related to the constants $C_i, i = 1, 2$ as

$$\hat{\mathbf{u}}_{nm} = \mathbf{T}_{1nm}\mathbf{C}_{nm}. \tag{12.81}$$

Similarly, tractions at node 1 can be related to the constants as

$$\{T_{x1}\ T_{z1}\}^T_{nm} = \mathbf{T}_{2nm}\{C_{1nm}\ C_{2nm}\}^T \tag{12.82}$$

That is, $\hat{\mathbf{T}}_{nm} = \mathbf{T}_{2nm}\mathbf{C}_{nm}$. Explicit forms of the matrix $\mathbf{T}_1$ and $\mathbf{T}_2$ are

$$\begin{aligned}\mathbf{T}_{1(ILE)} &= \mathbf{T}_{1(FLE)}(1:2,1:2)\\ \mathbf{T}_{2(ILE)} &= \mathbf{T}_{2(FLE)}(1:2,1:2)\,.\end{aligned} \tag{12.83}$$

The dynamic stiffness for the homogeneous infinite half space becomes

$$\hat{\mathbf{K}}_{nm} = \mathbf{T}_{2nm}\mathbf{T}_{1nm}^{-1}\,, \tag{12.84}$$

which is a 2×2 complex matrix.

Next, we explain the method of computing the stresses. From the displacement field (Eqns. (8.108) and (8.109)), the strain–displacement, stress–strain relations, the matrix of strain nodal displacement relation and the stress nodal displacement relation can be established as

$$\begin{aligned}\epsilon = \mathbf{B}{\mathbf{T}_1}^{-1}\hat{\mathbf{u}}, &\quad \sigma = \mathbf{Q}\mathbf{B}{\mathbf{T}_1}^{-1}\hat{\mathbf{u}}\\ \epsilon = \{\epsilon_{xx}, \epsilon_{zz}, \epsilon_{xz}\}\,, &\sigma = \{\sigma_{xx}, \sigma_{zz}, \sigma_{xz}\}\,,\end{aligned} \tag{12.85}$$

where the elements of $\mathbf{B}$ (size 3× 4) are described in terms of the wave amplitude matrix $\mathbf{R}$ as

$$\begin{aligned}B(1,p) &= R_{1p}\eta e^{-ik_pz}\,, \quad B(2,p) = -iR_{2p}k_pe^{-ik_pz}\,,\\ B(3,p) &= -(iR_{1p}k_p + R_{2p}\eta)e^{-ik_pz}\,, p = 1, \ldots, 4\,.\end{aligned} \tag{12.86}$$

Here z is the point of strain measurement. The elasticity matrix $\mathbf{Q}$ is given by

$$\mathbf{Q} = \begin{bmatrix} Q_{11} & Q_{13} & 0 \\ Q_{13} & Q_{33} & 0 \\ 0 & 0 & Q_{55} \end{bmatrix}. \tag{12.87}$$

Applications of boundary conditions are similar to that of 2D-isotropic solid elements detailed in the previous section and hence not repeated here.

12.3.7 Determination of Lamb Wave Modes in Laminated Composites

We studied Lamb waves in doubly bounded isotropic media in Chapter 7 (Section 7.5). There, we studied two different loading cases of Lamb wave propagation, namely the *Symmetric* and *Antisymmetric* cases, respectively. The dispersion relations for these cases were plotted in Figs. 7.13 and 7.14, respectively.

As mentioned earlier, these waves are very important in SHM studies for damage detection as they travel long distances with small level of attenuation. In Chapter 7, in order to study the Lamb waves arising out of traction free boundary in a doubly bounded media, we had used method of Helmholtz potentials to obtain the frequency equations, which are essentially transcendental equations and in order to solve these complex equations, numerical methods for solution was used. The splitting the governing coupled partial differential equations into two equations, one in terms of scalar potential and P-wave and the other in terms of vector potential and S-wave enabled us to isolate these waves separately in a traction free doubly bounded media, although they travel together, to generate Lamb waves. The other advantage of this method is that the symmetric and anti-symmetric modes can be isolated during formulation. However, this method of solution is applicable only for the isotropic waveguides.

However, in laminated composites, unsymmetric ply orientation results in stiffness and inertial coupling and as a result all the modes are highly coupled and Helmholtz decomposition cannot be used. The result of this is that no waves can be individually isolated. Hence, we use *Partial Wave Technique* or PWT to track the different waves. This method was used in Chapter 8 as well as in the earlier part of this section.

In the spectral composite layer element formulation, there are two summations in the solutions. The outer one is over the discrete frequencies and the inner one is over the discrete horizontal wavenumbers. Each partial wave of Eqn. (12.73) satisfies the governing partial differential equations (Eqn. (8.100)) and the coefficients C_i as a whole satisfy any prescribed boundary conditions. As long as the prescribed natural boundary conditions are non-homogeneous, no restriction upon the horizontal wavenumber η is imposed and this leads to a double summation solution of the displacement field. However, that is not the case for traction-free boundary conditions on the two surfaces, which are the necessary condition for generating Lamb waves. The governing discrete equation for a finite layer (Eqn. (12.78)) in this case becomes

$$\hat{\mathbf{K}}(\eta_m, \omega_n)_{nm} \hat{\mathbf{u}}_{nm} = 0\,, \tag{12.88}$$

and we are interested in a non-trivial $\mathbf{u}$. Hence, the stiffness matrix $\hat{\mathbf{K}}$ must be singular, that is, $det(\hat{\mathbf{K}}(\eta_m, \omega_n)) = 0$, which gives the required relation between η_m and ω_n. Since, ω_n is made to vary independently, the above relation must be solved for η_m to render the stiffness matrix singular, that is, η_m cannot vary independently. More precisely, for each value of ω_n there is a set of values of horizontal wavenumber η_m (one for each mode) and for each value of ω_n and η_m there are four vertical wavenumbers k_{nm}. The difference in this case is in the value of η_m, which is to be solved for, as opposed to its expression in Eqn (12.71) and M is the number of Lamb modes considered rather than Fourier modes. Now, for each set of $(\omega_n, \eta_m, k_{nml}), l = 1, \ldots, 4$, $\hat{\mathbf{K}}$ will be singular and C_l, $l = 1, \ldots, 4$ will be in the null space of $\hat{\mathbf{K}}$. Now using Eqn. (12.73), the total solution can be constructed. Following normal practice, the traction-free boundary conditions, that is, $\sigma_{zz}, \sigma_{xz} = 0$ are prescribed at $z = \mp h/2$. Using Eqn. (12.85), the governing equation for C_i and η_m becomes

$$\mathbf{W}_2(\eta_m, \omega_n)\mathbf{C}_{nm} = \mathbf{0}, \qquad \mathbf{C} = \{C_1, C_2, C_3, C_4\}, \tag{12.89}$$

where $\mathbf{W}_2$ is another form of the stiffness matrix $\hat{\mathbf{K}}$ and is given by

$$\begin{aligned}
W_2(1,p) &= (Q_{11\circ}R(1,p)\eta - iQ_{13\circ}R(2,p)k_p)e^{ik_ph/2}, \\
W_2(2,p) &= (Q_{11\circ}R(1,p)\eta - iQ_{13\circ}R(2,p)k_p)e^{-ik_ph/2}, \\
W_2(3,p) &= Q_{55\circ}(-R(1,p)k_p + iR(2,p)\eta)e^{ik_ph/2}, \\
W_2(4,p) &= Q_{55\circ}(-R(1,p)k_p + iR(2,p)\eta)e^{-ik_ph/2}.
\end{aligned}$$

The dispersion relation is $det\{\mathbf{W}_2\} = 0$, which will yield $\eta_m(\omega_n)$ and the phase speed for Lamb waves C_{Pnm} will be given by ω_n/η_m. Once the values of η_m are known for the desired number of modes, the elements of $\mathbf{C}_{nm}$ are obtained by the technique of SVD as described earlier to find the elements of $\mathbf{R}$. Summing over all the Lamb modes, the solution for each frequency is obtained.

Next, we will use this model and show the propagation of Lamb waves in composites of different ply orientation. A unidirectional lamina of 2 mm thickness is considered, with the material properties of AS/3501-6 graphite-epoxy composites.

The aim here is to determine the dispersion relations as a function of ply-angle . The solution of the dispersion relations in cases such as this requires particular care as it is multi-valued, unbounded and complex (although the real part is of interest). In the case of isotropic structures dealt in Chapter 7, we were able to explicitly derive the frequency equation and found that the equation is indeed multi-valued that gave multiple values of $k - \omega$ pair that satisfied the governing equation. The situation here is similar, however, explicit determination of frequency equation is quite complex and not given here. We will obtain the $k - \omega$ pair through the formulated 2-D composite spectral element. One way to solve these equations is to appeal to the strategies of non-linear optimization, which are based on non-linear least square methods. There are several choices of algorithms, like the Trust-Region Dogleg method, Gauss Newton method with a line search, or Levenberg Merquardt method with line search. Here, the MATLAB function $fsolve$ is used and for the default option for medium scale optimization, the Trust-Region Dogleg method is adopted, which is a variant of Powell's Dogleg method [113].

Apart from the choice of algorithm, there are other subtle issues in root capturing for the solution of wavenumbers. For instance, except the first one or two modes, all the roots escape to infinity at low frequency. For isotropic materials, these cut-off frequencies are known *a priori* because of the fact that the frequency equation can be explicitly determined. However, no expressions can be determined for anisotropic materials and generally, the solutions need to be tracked backwards, from the high-frequency to the low frequency region. In general two strategies are essential in capturing all the modes within a given frequency band. Initially, the whole region should be scanned for different values of the initial guess, where the initial guess should remain constant for the whole frequency range. These sweeps open up all the modes in that region, although they are not completely traced. Subsequently, each individual mode should be followed to the end of the domain or to a pre-set value. For this case, the initial guess should be updated to the solution of the previous frequency step. Also, sometimes it is necessary to reduce the frequency step in the vicinity of high gradients. Once the Lamb modes are generated they are fed back into the frequency loop to produce the frequency domain solution for Lamb wave propagation, which with the help of inverse FFT, produces the time domain signal. As the Lamb modes are generated first, they need to be stored separately. To this end, data are collected from the generated modes at several discrete points over the considered frequency range. Next, a cubic spline interpolation is performed for a very fine frequency step within the same range. While generating the time domain data, interpolation is performed from these finely graded data to get the phase speeds (and hence the η).

In the considered example, Lamb waves are generated through a modulated pulse of 200 kHz centre frequency applied at one end of an infinite plate. Velocities components in the x and z directions are recorded at a propagating distance of $320h$, where h is the thickness of the plate. While studying the time domain representation, the thickness of the plate is taken as $10mm$, which amounts to a frequency-thickness value of 2. The thickness values is chosen so that at least three modes are excited according to the dispersion curve shown in Fig. 12.7(a).

Fig. 12.7(a) shows the first 10 Lamb modes for fiber angle 0°. The first anti-symmetric mode (Mode 1) converges to a value of 1719 m/s in a range of 1 MHz-mm, where all the other modes also converge. In analogy to the isotropic case, this is the velocity of the Rayleigh surface waves in 0° fiber laminae. The first symmetric mode (Mode 2) starts above 10000 m/s and drops suddenly at around 1.3 MHz-mm to converge to 1719 m/s, before which it has a fairly constant value. All the other higher-order modes escape to infinity at various points in the frequency range. Also, the symmetric and the anti-symmetric pair of each mode escape almost at the same frequency.

Propagation of these modes are plotted in Fig. 12.7(b) and 12.7(c) for the first three modes (a_o, s_o and a_1), here referred to as mode 1, 2 and 3 respectively. In Fig. 12.7(b), the z velocity history is plotted, whereas in Fig. 12.7(c) the x velocity history is plotted. These figures readily show the different propagating modes, each corresponds to one wave packet propagating at the group speed (and not the phase speed). Hence, Fig. 12.7(a) is not helpful at predicting the arrivals of the different modes. However, as Figs. 12.7(b) and 12.7(c) suggest, mode 2 has a lower group speed than mode 1, and mode 3 has a group speed much higher than both modes 1 and 2. One difference in the $\dot{u}$ and $\dot{w}$ history can be observed. That is, for $\dot{u}$ history plot,

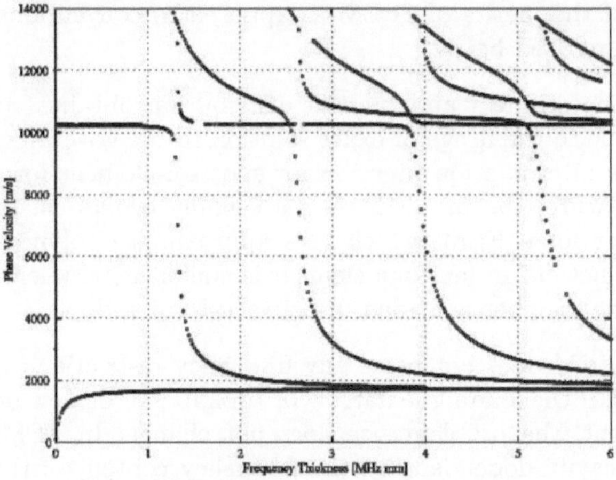

(a) Lamb wave Dispersion curve

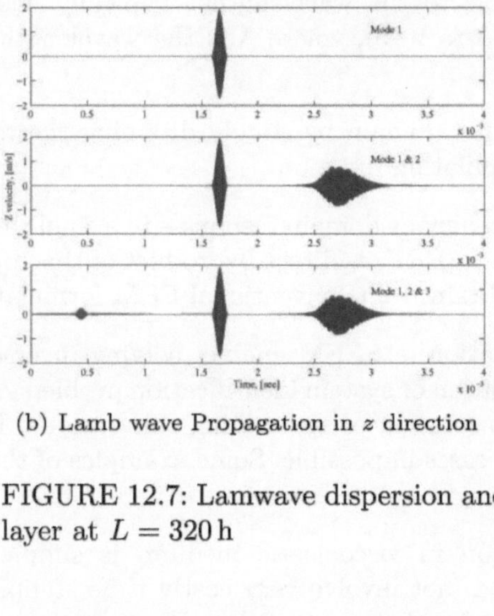

(b) Lamb wave Propagation in z direction

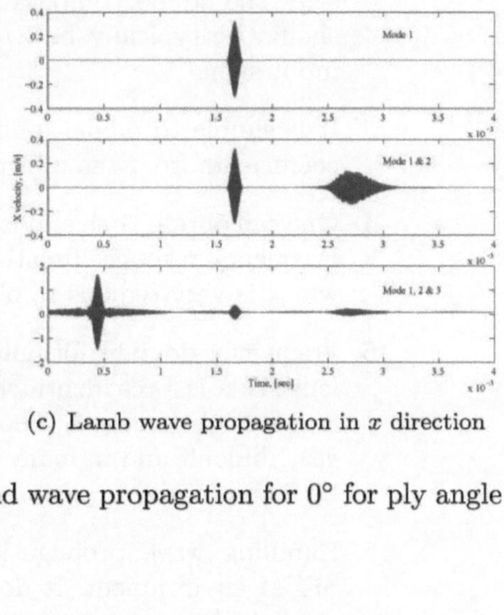

(c) Lamb wave propagation in x direction

FIGURE 12.7: Lamwave dispersion and wave propagation for 0° for ply angle layer at $L = 320\,\text{h}$

the higher mode generates velocity of comparatively less magnitude, whereas for $\dot{w}$ history plot, the magnitude is highest.

12.4 MERITS AND DEMERITS OF FOURIER SPECTRAL FINITE ELEMENT METHOD

We have seen that SFEM is able to keep track of different waves arising due to reflections from the boundaries and the newer waves arising from the mode conversions or mode fixity, efficiently. This is made possible mainly due to the casting of the wave propagation problem in a finite element framework.

Some of the merits of SFEM as apposed to conventional finite elements can be summarized below:

1. The stiffness and inertial distribution obtained using SFEM is exact due to the usage of exact solution to the governing differential equation as interpolation function for spectral element formulation. Due to this feature, the problem sizes are many orders smaller compared to conventional FEM, which uses approximate polynomials as interpolation function. In fact one element is sufficient between any two joints unless the member is loaded in between the joints.

2. SFEM does not place any frequency restrictions on the problem. That is, if the loading is narrow or broadband, or, low or high-frequency content, the problem size does not change. In FEM, the mesh sizes are heavily dependent on the frequency content of the input signal. If the frequency content is large, the wavelengths are very small and the mesh sizes should be comparable to the wavelengths. Typically, mesh size should be typically be $\lambda/8$ to $\lambda/10$, where λ is the wavelength of the input signal.

3. It is simple to model infinite domain by simply dropping the reflected coefficients from the interpolating function.

4. One can obtain time and frequency domain responses in a single analysis. Frequency response function (FRF) is direct byproduct of the approach, which is very tedious to obtain from conventional FEM formulation.

5. Frequency domain formulation of SFEM enables solving inverse problems, that is force identification or system identification problems, simple and straight forward. The solution of same using conventional FEM is very difficult and in many cases impossible. Some examples of these can be found in [48].

6. Handling wave propagation in viscoelastic medium is simple under SFEM environment. It does not involve very costly time stepping and numerical solution that is normally used to model the same problem using conventional FEM. That is, the solutions to the linear viscoelastic waveguides were presented in Chapter 5. These can be easily recast as spectral elements by simply modifying the constant Young's modulus E by frequency dependent modulus $\hat{E}$ in the SFEM code.

7. It is well known that damping in structures is frequency dependent. Frequency domain formulation of SFEM enables damping to be treated more realistically compared to conventional FEM.

SFEM also has its share of demerits. These are summarized below:

1. Spectral finite element model is available only for structures with regular geometry, where exact solution to wave equation, is available. This limits its use for complex geometries unlike conventional FEM.

2. Due to enforced periodicity of signals both in time and frequency domain, SFEM suffers from severe signal wraparound problem . This aspect is was discussed in Chapter 6. This limits its use for shorter waveguides.

3. Unlike conventional FEM, SFEM requires two software, one is the FFT software to go back and forth time and frequency domain and the second the SFEM software. With MATLAB, both can be integrated in a single code.

4. As mentioned earlier, SFEM requires FFT, which requires proper selection of time sampling rate Δt and the number of FFT points N. For many problems, their choice can be tricky. This aspect was dealt in detail in Chapter 6.

5. Treatment of shorter and stiffer boundaries is very difficult and cumbersome using FFT. Need the use of throw-off elements to releive the trapped energy to obtain better time resolution of the responses. This aspect was also discussed in Chapter 6.

Summary:

This chapter introduces the spectral finite element method for solving wave propagation problems. The formulation of 1-D elements such as isotropic rods & beams and 1-D Laminated waveguide is presented. Also, formulation of 2-D spectral elements for isotropic solids and laminated composite layer element is also presented in this chapter. Formulation of other waveguides such as FGM waveguides, viscoelastic waveguides is a straight forward extension off the formulated elements. Although conventional finite element method for solving wave propagation problems is not presented, the main differences, merits and demerits of SFEM over conventional FEM is highlighted. The reader is advised to refer to the unified book of the author [54] or the book [48] to get more details of this method as well as FEM for solving wave propagation problems. The chapter also shows the use of SFEM and its dynamic stiffness matrix to obtain Lamb wave dispersion curves.

As mentioned earlier, SFEM of only a few waveguides were derived in this chapter. In the formulations presented, only the exact solution to the transformed governing equations is used as interpolating function for element formulation. In majority of the cases, exact solution to a governing equation is almost impossible to obtain. A simple example is the solution of the tapered Timoshenko beam with polynomial variation of the depth. In such cases, one can formulate the problem using variational method in the frequency domain. Such an approach was used to solve wave propagation in a tapered Timoshenko beam using the frequency domain variational approach and the details are provided in [49]. Such formulation for FGM structures is given in [26]. For these waveguides, the base solutions will be in terms of waves in some uniform configuration. For example, for tapered Timoshenko beam element, the uniform beam spectral solution is assumed as the interpolation function and this solution is used in the weak form of the governing equation to obtain the dynamic stiffness matrix of the chosen waveguide.

Exercises

12.1 A cantilevered rod of length L, axial rigidity EA, and mass ρA is subjected to an axial impact at the free end ($x = L$). Using a single finite spectral rod element, determine the Frequency response function (FRF) at $x = L$ and determine the first five natural frequencies and compare the same with the exact solution [19]. Assume aluminum properties

12.2 A cantilever beam of length L, bending rigidity EI, and mass ρA is subjected to an axial impact at the free end ($x = L$). Using a single finite spectral beam element, determine the FRF at $x = L$ and determine the first five natural frequencies and compare the same with the exact solution [19]. Assume aluminum properties.

12.3 A crack in a rod is approximately simulated by introducing a spring of constant K as shown in the figure below Here, $K = 0$ simulates complete

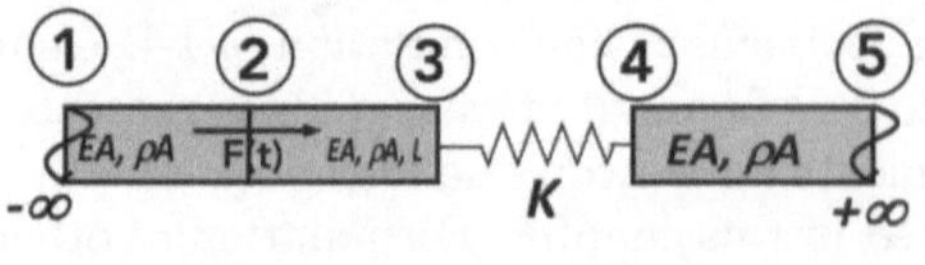

Figure for Problem 12.3

breakage of the two segments on either side of the spring, while $K = \infty$ simulates, perfect bond between the two segments. Intermediate values of K simulates different levels cracking. The system is modeled using 3 elements, where elements 1 and 3 are throw-off rod elements, while element 2 is the finite length rod spectral element. The system is loaded axially at node 2 with a broad band triangular pulse shown in Fig. 4.3. Plot the velocity responses at nodes 2, 3 and 4 for different values of K. Assume aluminum properties for the rod having rectangular cross section with width and depth being 25 mm.The length of element 2 can be assumed as 1 meter.

12.4 Two beam segments are connected by a rotational spring of spring constant K_T as shown below. The system is modeled using two beam

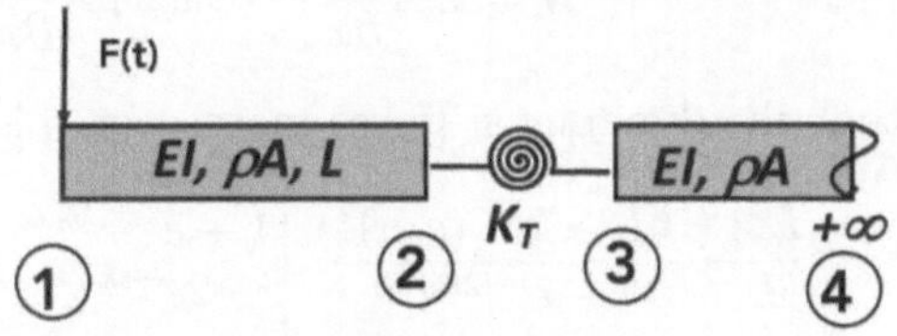

Figure for Problem 12.4

elements, where the element to the right of the spring is a throw-off beam element. The beam is loaded at node 1. Determine the frequency domain responses at nodes 3 and 4 and plot them as a function of frequency. Assume aluminum properties for the rod having rectangular cross section with width and depth being 25 mm. The length of element 1 can be assumed as 1 meter.

12.5 A portal frame modeled using two rod beam segments as shown below. The frame is loaded at node 2. Assume aluminum properties for the

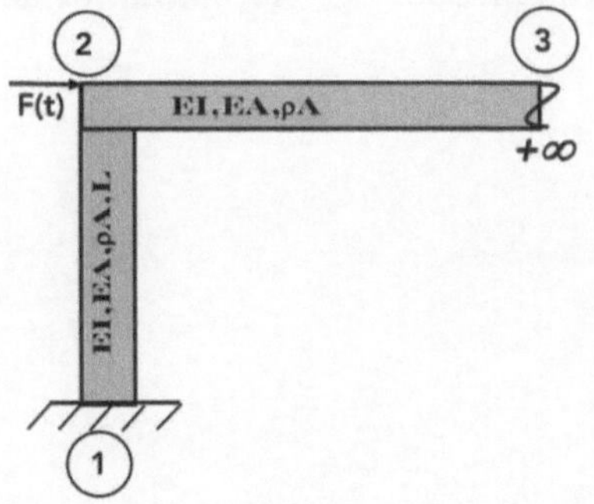

Figure for Problem 12.5

beam having rectangular cross section with width and depth being 25 mm. The length of element 1 can be assumed as 1 meter. Determine the frequency response function at node 1 of this system.

12.6 The governing differential equation for a Eringen Stress gradient rod, which was discussed in Chapter 11, is given by

$$EA\frac{\partial^2 u}{\partial x^2} + (e_0 a)^2 \rho A \frac{\partial^4 u}{\partial x^2 \partial t^2} = \rho A \frac{\partial^2 u}{\partial t^2}$$

and the axial stress resultant N is given by

$$N = EA\frac{\partial u}{\partial x} + (e_0 a)^2 \rho A \frac{\partial^3 u}{\partial x \partial t^2}$$

Show that the dynamic stiffness matrix for this case is given by

$$\frac{EA}{L}\frac{(ikL - k_l^2 L(e_0 a)^2)}{(1 - e^{-i2k_n L})}\begin{bmatrix} 1 + e^{-i2k_n L} & -2e^{-ikL} \\ -2e^{-ik_n L} & 1 + e^{-i2k_n L} \end{bmatrix}$$

where wavenumber k is obtained from Eqn. (11.26) from Chapter 11 and $k_l = \omega\sqrt{\rho/E}$.

12.7 The dynamic stiffness for a rod is given by Eqn. (12.14). Expanding the complex exponentials in terms of their respective series and considering only 2 terms in these series for each of these exponentials, show that the dynamic stiffness matrix reduces to finite element dynamic stiffness matrix given in Eqn. (12.15).

12.8 Using the theory developed in Chapter 4, develop a spectral element for a Timoshenko beam. See [48, 54] for details.

12.9 Using the theory developed in Chapter 4, develop a spectral element for a Mindlin-Herrmann rod. See [48, 54, 93] for details.

Bibliography

[1] A.V. Metrikine, A.V. Pichugin, H. Askes and T. Bennett. Four simplified gradient elasticity models for the simulation of dispersive wave propagation. *Philosophical Magazine*, page 1830001, 2019.

[2] H. Abramovich and A. Livshits. Free vibration of non-symmetric cross-ply laminated composite beams. *Journal of Sound and Vibration*, 176:597–612, 1994.

[3] M. Abramowitz and I.A. Stegun. *Handbook of Mathematical Functions*. Dover, New York, 1965.

[4] A. Chakroborty. *Wave Propagation in Anisotropic and Inhomogeneous Medium*. Ph.D. Thesis, Indian Institute of Science, Bangalore, India, October 2004.

[5] J.D. Achenbach. *Wave Propagation in Elastic Solids*. North Holland Publication Company, 1973.

[6] E. Aifantis. On the role of gradients in the localization of deformation and fracture. *International Journal of Engineering Science*, 30:1279–1299, 1992.

[7] B. Altan and E. Aifantis. On the structure of the mode iii crack-tip in gradient elasticity. *Scripta Metallics and Materials*, 26:319–324, 1992.

[8] B. Altan and E. Aifantis. On some aspects in the special theory of gradient elasticity. *Journal of Mechanical Behavior of Materials*, 8:231–282, 1997.

[9] A. Sawangsuriya. Wave propagation methods for determining stiffness of geo materials. *InTech-Open-Chapter 7*, pages 158–200, 2012.

[10] H. Askes and E. Aifantis. Numerical modelling of size effect with gradient elasticity formulation, meshless discretization and examples. *International Journalof Fracture*, 117:347–358, 2002.

[11] H. Askes and A. Metrikine. Higher-order continua derived from discrete media:continualisation aspects and boundary conditions. *International Journal of Solids and Structures*, 42:187–202, 2005.

[12] H. Askes, A.S.J. Suiker, and L.J. Sluys. A classification of higher-order strain-gradient models-linear analysis. *Archeive of Applied Mechanics*, 171–188, 2002.

[13] J.H. Atkinson and G. Sallfors. Experimental determination of stress-strain-time characteristics in laboratory and in situ tests. In *Proceedings of the 10th European Conference on Soil Mechanics and Foundation Engineering*, pages 915–956, 1991.

[14] P.B. Bailey, W.N. Everitt, and A. Zettl. The SLEIGN2 sturm-liouville code manual. *http://www.math.niu.edu/zettl/SL2*, 1996.

[15] Rajashekar Bangaru. *Blast Mitigation Strategies for Marine Sandwich Structures.* Ph.D Thesis, Indian Institute of Science, November 2020.

[16] A.A. Bent. Piezoelectric fiber composite for structural actuation. *MS Thesis*, Massachusetts Institute of Technology (USA), 1994.

[17] L. Bergmann. *Ultrasonics and Their Scientific and Technical Applications.* Wiley, New York, 1948.

[18] J. Biarez and P.Y. Hicher. *Elementary Mechanics of Soil Behavior-Saturated Re-moulded Soils.* Balkema, 1994.

[19] R.E.D. Bishop and D.C. Johnson. *Mechanics of Vibration.* Cambridge University Press, 1960.

[20] D.P. Bland. *Wave Theory and Applications.* Clarendon Press, Oxford, UK, 1988.

[21] P. Boulanger and M. Hayes. Inhomogeneous plane waves in viscous fluids. *Continuum Mechanics and Thermodynamics*, 2(1), 1990.

[22] J.P. Boyd. *Solving Transcendental Equations.* SIAM, Philadelphia, USA, 2014.

[23] L.M. Brekhovskikh. *Waves in Layered Media.* Academic Press, New York, 1960.

[24] G. Caviglia and A. Morro. *Inhomogeneous Waves in Solids and Fluids.* World Scientific, Singapore, 1992.

[25] A. Chakraborty and S. Gopalakrishnan. Various numerical techniques for analysis of longitudinal wave propagation in inhomogeneous one-dimensional waveguides. *Acta Mechanica*, 194:1–27, 2003.

[26] A. Chakraborty and S. Gopalakrishnan. An approximate spectral element for the analysis of wave propagation in inhomogeneous layered media. *AIAA Journal*, 44(7):1676–1685, 2006.

[27] K. Chandrashekhara, K. Krishnamurthy, and S. Roy. Free vibration of composite beams including rotary inertia and shear deformation. *Composite Structures*, 14:269–279, 1990.

[28] C. Chang and J.Gao. Second-gradient constitutive theory for granular materialwith random packing structure. *International Journal of Solids and Structures*, 32:2279–2293, 1995.

[29] R.M. Christensen. *Mechanics of Composite Materials.* Wiley, 1979.

[30] J.W. Cooley and O.W. Tukey. An algorithm for the machine calculation of complex fourier series. *Mathematical Computations*, 19(90):297–301, 1965.

[31] H.F. Cooper. Reflection and transmission of oblique plane waves at a plane interface between viscoelastic media. *Journal of the Acoustical Society of America*, 42, 1967.

[32] G.R. Cowper. On the accuracy of timoshenko beam theory. *ASCE Journal od Applied Mechanics*, 94:1447–1453, 1968.

[33] D'Alembert. Recherches sur la courbe que forme une corde tendua mise en vibration. *Histoire de l'acadAmie royale des sciences et belles lettres de Berlin*, pages 214–219, 1747.

[34] J.F. Doyle. *Wave propagation in Structures.* Springer, 1997.

[35] D. Schneider, T. Witke, T. Schwarz, B. Schoneich, and B. Schultrich. Testing ultra-thin films by laser-acoustics. *Surface Coating Technology*, 126:136–141, 2000.

[36] C.L. Dym and I.H. Shames. *Solid Mechanics: A Variational Approach (Augmented Edition).* Springer, New York, 2013.

[37] A.C. Eringen. Linear theory of non-local elasticity and dispersion of plane waves. *International Journal of Engineering Science*, 10:425–435, 1972.

[38] A.C. Eringen. On differential equations of nonlocal elasticity and solutions of screw dislocation and surface waves. *Journal of Applied Physics*, 54:4703, 1983.

[39] A.C. Eringen. *Non-local Polar Field Models.* Academic Press, 1996.

[40] E.T. Whiitekar. On the functions which are represented by the expansions of the interpolation theory. *Proceedings of Royal Society*, Edinburgh, pages 181–194, 1915.

[41] E.V. Krishnamurthy and S.K. Sen. *Numerical Algorithms.* East West Press, New Delhi, 1986.

[42] A.P. French. *Vibrations and Waves (M.I.T. Introductory physics series)*. Nelson Thames, 1971.

[43] G. Hayward and J. Hyslop. Determination of lamb wave dispersion data in lossy anisotropic plates using time domain finite elements: Part i: Theory and experimental verification. *IEEE Transactions on Ultrasonics, Ferroelectrics and Frequency Control*, 53:443–448, 2006.

[44] G.I. Taylor. The pressure and impulse of submarine explosion waves on plates. *The Scientific Papers of G. I. Taylor*, pages 287–303, 1963.

[45] G. Golub and C. Van Loan. *Matrix Computations*. The Johns Hopkins University Press, Baltimore, 1989.

[46] S. Gopalakrishnan. A deep rod finite element for structural dynamics and wave propagation problems. *International Journal of Numerical Methods in Engineering*, 48:731–744, 2000.

[47] S. Gopalakrishnan. Modeling Aspects In Finite Elements For Structural Health Monitoring. *Encyclopedia on Structural Health Monitoring: Simulation Section, Edited by Boller., C., Chang, F.K., Fujino, Y.*, 2(42-Part 4):791–809, 2009.

[48] S. Gopalakrishnan, A. Chakraborty, and D. Roy Mahapatra. *Spectral Finite Element Method*. Springer, Germany, 2006.

[49] S. Gopalakrishnan and J.F. Doyle. Wave propagation in connected waveguides of varying cross-section. *Journal of Sound and Vibration*, 175(3):347–363, 1994.

[50] S. Gopalakrishnan and D. Roy Mahapatra. Optimal spectral control of broadband waves in smart composite beams with distributed sensor-actuator configuration. *SPIE Symposium on Smart Materials and MEMS*, Paper 4234-12, 2000.

[51] S. Gopalakrishnan, M. Martin, and J.F. Doyle. A matrix methodology for spectral analysis of wave propagation in multiple connected timoshenko beam. *Journal of Sound and Vibration*, 158:11–24, 1992.

[52] S. Gopalakrishnan, M.Ruzzene, and S. Hanagud. *Computational Techniques for Structural Health Monitoring*. Springer-Verlag, Germany, 2012.

[53] S. Gopalakrishnan and S. Narendar. *Wave Propagation in Nanostructures*. Springer, 2013.

[54] Srinivasan Gopalakrishnan. *Wave Propagation in Materials and Structures*. CRC Press, Taylor and Francis Group, Boca Raton, London, New York, 2017.

[55] K.F. Graff. *Wave Motion in Elastic Solids.* Dover Publications Inc., 1991.

[56] X. Gu, J.Yang, and M Huwang. Laboratory measurements of small strain properties of dry sands by bender element. *Japanese Geotechnical Society-Soils and Foundation*, 53:735–745, 2013.

[57] B.O. Hardin. The nature of stress-strain behavior of soils. *Proceedings of the Geotechnical Engineering Division Specialty Conference on Earthquake Engineering and Soil Dynamics, ASCE*, pages 1–90, 1978.

[58] B.O. Hardin and W.L. Black. Vibration modulus of normally consolidated clay. *Journal of the Soil Mechanics and Foundations Division, ASCE*, 353–369, 1968.

[59] C.M. Hernaandez, T.W. Murray, and S. Krishnaswarmy. Photoacoustic characterization of the mechanical properties of thin film. *Applied Physics Letters*, 80(4):691–693, 2002.

[60] P. Hora and O. Cervena. Determination of lamb wave dispersion curves by means of fourier transform. *Applied and Computational Mechanics*, 5–16, 2012.

[61] https://en.wikipedia.org/wiki/$Granular_material$.

[62] https://en.wikipedia.org/wiki/Wave.

[63] https://www.dlr.de/zlp/en/desktopdefault.aspx/tabid-14332/24874_read 61142.

[64] https://www.imperial.ac.uk/non-destructive-evaluation/products-and services/disperse/.

[65] S.C. Hunter. *Viscoelastic Waves. Progress in Solid Mechanics.* North Holland, Amsterdam, 1960.

[66] T. Iwasaki and F. Tatsuka. Effects of grain size and grading on dynamic shear moduli of sands. *Soils and Foundation*, 19–35, 1977.

[67] T. Iwasaki, F. Tatsuka, and Y. Takagi. Shear moduli of sands under cyclic torsional shear loading. *Soils and Foundation*, 39–50, 1978.

[68] R.P. McNitt, J. Peddieson, and G. R. Buchanan. Application of nonlocal continuum models to nanotechnology. *International Journal of Engineering Science*, 41:305–312, 2003.

[69] R.M. Jones. *Mechanics of Composite Materials.* Scripta, Washington, DC, 1975.

[70] J.W.S. Rayleigh. On waves propagated along the plane surface of an elastic solid. *Proceedings of London Mathematical Society*, 4–11, 1885.

[71] M.A. Karim, M.A. Awal, and T. Kundu. Elastic wave scattering by cracks and inclusions in plates: In-plane case. *International Journal of Solids and Structures*, 29(19):2355–2367, 1992.

[72] B. Kieback, A. Neubrand, and H. Riedel. Processing techniques for functionally graded materials. *Material Science and Engineering*, 81–105, 2003.

[73] J. Kim and G.H. Paulino. Finite element evaluation of mixed mode stress intensity factors in functionally graded materials. *International Journal for Numerical Methods in Engineering*, 53:1903–1935, 2002.

[74] G. Kirchoff. A der das gleichewicht und die bewegung einer elastischen scheibe. *Journal of Reine und Angewante Mathematik (Crelle)*, 40:51–88, 1850.

[75] K. Ishihara. *Soil Behavior in Earthquake Geotechnics.* Oxford University Press, Inc., New York, 1996.

[76] M. Koizumi and M. Niino. Overview of FGM research in Japan. *MRS Bulletin*, 1:19–21, 1995.

[77] T. Kokusho. Cyclic triaxial test of dynamic soil properties for wide strain range. *Soils and Foundation*, 45–60, 1980.

[78] T. Kokusho, Y. Yoshida, and Y. Esashi. Dynamic properties of soft clays for wide strain range. *Soils and Foundation*, 1–18, 1982.

[79] A.V. Krishnamurty, M. Anjanappa, and Y.F. Wu. The use of magnetostrictive particle actuators for vibration attenuation of flexible beams. *Journal Sound and Vibration*, 206:133–149, 1997.

[80] L. Laine and A. Sandvik. Derivation of mechanical properties of sand. In *Proceedings of the 4th Asia-Pacific Conference on Shock and Impact Loads on Structures*, pages 361–368, 2001.

[81] P. Lancaster. *Lambda Matrices and Vibrating Systems.* Pergamon Press, 1966.

[82] P. Lancaster. *Theory of Matrices.* Academic Press, 1969.

[83] M.J. Lighthill and G.B. Whitham. A theory of traffic flow on long crowded roads. *Proceedings of Royal Society*, A229:317–345, 1955.

[84] G.R. Liu. A step-by-step method of rule-of-mixture of fiber and particle-reinforced composite materials. *Composite Structures*, 40:313–322, 1998.

[85] F.J. Lockett. The reflection and refraction of waves at an interface between viscoelastic materials. *Journal of the Mechanics and Physics of Solids*, 10(58), 1962.

[86] D.C.F. Lopresti, M. Jamiolkowski, O. Pallara, A. Cavallararo, and S. Pedroni. Shear modulus and damping of soils. *Geotechnique*, 603–617, 1984.

[87] A.E.H. Love. *A Treatise on the Mathematical Theory of Elasticity, 2nd Edition.* Cambridge University Press, 1906.

[88] R.M. Mahamood, T. Esther, E.T. Akinlab, M. Shukla, and S. Pityana. Functionally Graded Material: An Overview. In *Proceedings of the World Congress on Engineering*, volume III, 2012.

[89] D. Roy Mahapatra, S. Gopalakrishnan, and T.S. Shankar. Spectral-element-based solution for wave propagation analysis of multiply connected unsymmetric laminated composite beams. *Journal of Sound and Vibration*, 237(5):819–836, 2000.

[90] D. Roy Mahapatra, S. Gopalakrishnan, and T.S. Shankar. A spectral finite element model for analysis of axial-flexural-shear coupled wave propagation in laminated composite beams. *Composite Structures*, 59(1):67–88, 2003.

[91] R.J. Mair. Developments in geotechnical engineering research: Application to tunnels and deep excavations. *Proceedings of the Institution of Civil Engineers-Civil Engineering*, pages 27–41, 1993.

[92] A.J. Markworth, K.S. Ramesh, and W.P. Parks Jr. Modelling studies applied to functionally graded materials. *Journal of Material Science*, 30:2183–2193, 1995.

[93] M.T. Martin, S. Gopalakrishnan, and J.F. Doyle. Wave propagation in multiply connected deep waveguides. *Journal of Sound and Vibration*, 174(4):521–538, 1994.

[94] S.R. Marur and T. Kant. Transient dynamics of laminated beams: an evaluation with a higher-order refined theory. *Composite Structures*, 41:1–11, 1998.

[95] A.V. Metrikine. On causality of the gradient elasticity models. *Journal of Sound and Vibration*, 297:727–742, 2006.

[96] R.D. Mindlin and H.F. Tiersten. Effects of couple stresses in linear elasticity. *Archives of Rational Mechanical Analysis*, 11:415–448, 1962.

[97] R.D. Mindlin. Microstructure in linear elasticity. *Archives of Rational Mechanical Analysis*, 16:51–78, 1964.

[98] R.D. Mindlin and G. Herrmann. A one dimensional theory of compressional waves in an elastic rod. *Proceedings of First U.S. National Congress of Applied Mechanics*, pages 187–191, 1950.

[99] E. Moulin, J. Assaad, C. Delebarre, S. Grondel, and D. Balageas. Modeling of integrated Lamb waves generation systems using a coupled finite element normal mode expansion method. *Ultrasonics*, 38:522–526, 2000.

[100] M. Sofer, P. Ferfecki, and P. Sofer. Numerical solution of rayleigh-lamb frequency equation for real, imaginary and complex wavenumbers. In *Proceedings of MATEC Web of Conference*, page 08011, 2018.

[101] E. Muller, C. Drasar, J. Schilz, and W.A. Kaysser. Functionally graded materials for sensor and energy applications. *Materials Science and Engineering*, A362:17–39, 2003.

[102] V.S. Mutnuri. *Spectral Analysis of Wave Motion in Nonlocal Continuum Theories of Elasticityr.* Ph.D. Thesis, Indian Institute of Science, Bangalore, India, 2021.

[103] V.S. Mutnuri and S. Gopalakrishnan. *Propagation of Elastic Waves in Nonlocal Bars and Beams.* Recent Advances in Computational Mechanics and Simulations. Lecture Notes in Civil Engineering, vol 103 (Saha S.K. and Mukherjee M., eds.). Springer, Singapore, 2020.

[104] T. Nakamura, T. Wang, and S. Sampath. Determination of properties of graded materials by inverse analysis and instrumented indentation. *Acta Materialia*, 48(17):4293–4306, 2000.

[105] A. H. Nayfeh. *Wave Propagation in Layered Anisotropic Media.* North Holland, Amsterdam, 1995.

[106] S. Nissabani, M.E. Allami, and M. Bakncha. Lamb wave propagation : Plotting the dispersion curves. In *Proceedings of ICCWCS-16*, pages 149–152. International Conference on Computing and Wireless Communication, 2016.

[107] L.A. Ostrovsky and A.S. Potapov. *Modulated Waves, Theory and Applications.* The Johns Hopkins University Press, 1999.

[108] J. Philip, P. Hess, T. Feygelson, J.E. Butler, S. Chattopadhyay, K.H. Chen, and L.C. Chen. Elastic mechanical and thermal properties of nanocrystalline diamond films. *Journal of Applied Physics*, 93(4):2164–2171, 2002.

[109] B. Poiree. *Complex Harmonic Plane Waves.* Physical Acoustics (O. Leroy and M.A. Breazeale, eds.). Plenum Press, New York, 1991.

[110] A.D. Polyanin and V.F. Zaitsev. *Handbook of Exact Solution for Ordinary Differential Equations (Second Edition).* Chapman & Hall/CRC Press, 2003.

[111] A.D. Polyanin and V.F. Zaitsev. *Handbook of Exact Solutions for Ordinary Differential Equations.* CRC Press, Boca Raton, 1995.

[112] Y.Z. Povstenko. The nonlocal theory of elasticity and its applications to the description of defects in solid bodies. *Journal of Mathematical Sciences*, 97(1):3840–3845, 1999.

[113] M.J.D. Powel. A fortran subroutine for solving systems of nonlinear. *Numerical Methods for Nonlinear Algebraic Equations*, P. Rabinowitz, ed, Ch.7:115–161, Algebraic Equations, 1970.

[114] K.M.M. Prabhu. *Window functions and their Applications in Signal Processing.* CRC Press, Taylor and Francis Publishing, Boston, MA, 2013.

[115] S.P. Timoshenko. On the correction factor for shear of the differential equation for transverse vibrations of bars of uniform cross-section. *Philosophical Magazine*, page 744, 1921.

[116] R. Ramprasad and N. Shi. Scalability of phononic crystal heterostructures. *Applied Physics Letters*, 87:1111101, 2005.

[117] J.N. Reddy. *Mechanics of Laminated Composite Plates.* CRC Press, Boca Raton, FL, 1997.

[118] F.E. Reiton. *Applied Bessel Functions.* Dover, New York, 1965.

[119] S.A. Rizzi. A spectral analysis approach to wave propagation in layered solids. *Ph.D. Thesis*, 1989.

[120] S.A. Rizzi and J.F. Doyle. A spectral element approach to wave motion in layered solids. *ASME Journal of Vibration and Accoustics*, 114:569–577, 1992.

[121] J.L. Rose. *Ultrasonic Waves in Solid Media.* Cambridge University Press, 1999.

[122] C. Ru and E. Aifantis. A simple approach to solve boundary-value problems in gradient elasticity. *Acta Mechanica*, 101:59–68, 1993.

[123] A. Sampathkumar, T.W. Murray, and K.L. Ekinci. Photothermal operation of high-frequency nanoelectormechanical systems. *Journal of Applied Physics*, 88:223104, 2006.

[124] G. Savaidis and H. Zhu. Transient dynamic analysis of a cracked functionally graded material by a biem. *Computational Material Science*, 26:167–174, 2003.

[125] U. Schulz, M. Peters, Fr.-W. Bach, and G. Tegeder. Graded coatings for thermal, wear and corrosion barriers. *Material Science and Engineering*, A362:61–80, 2003.

[126] S.Gopalakrishnan and J.F. Doyle. Spectral super-elements for wave propagation in structures with local non-uniformities. *Computer methods in applied Mechanics and Engineering*, 121:77–90, 1995.

[127] T. Shibata and D.S. Soelarno. Stress-strain characteristics of sands under cyclic loading. *Proceedings of the Japan Society of Civil Engineering*, 239:57–65, 1975.

[128] O. Sigmund and J.J. Sondergaard. Systematic design of phonetic bandage materials and structures by topology optimisation. *Philosophical Transactions of Royal Society*, 361:1001–1019, 2003.

[129] S.A. Silling. Reformulation of elasticity theory for discontinuities and long-range forces. *Journal of Mechanics and Physics of Solids*, pages 175–209, 2000.

[130] L.P. Solie and B.A. Auld. Elastic waves in free anisotropic plates. *Journal of Acoustic Society of America*, 54(1):50–65, 1973.

[131] A. Suiker and R. de Borst. Micro-mechanical modelling of granular material. part 1: derivation of a second-gradient micro-polar constitutive theoy. *Acta Mechanica*, 149:161–180, 2001.

[132] S. Suresh and A. Mortensen. *Fundamentals of Functionally Graded Materials*. IOM Communications Ltd., London, 1998.

[133] M.M. Swisdak. *Explosion Effects and Properties-Part-II: Explosion Effects in Water*. Technical Report, Naval Surface Weapons Center, Dahlgren, Virginia, 1978.

[134] S.P. Timoshenko and J.N. Goodyear. *Theory of Elasticity*. Mcgraw Hill, New York, 1951.

[135] S. P. Timoshenko. On the transverse vibrations of bars of uniform cross-section. *Philosophical Magazine*, page 125, 1922.

[136] F. Tisseur and N.J. Higham. Structured pseudospectra for polynomial eigenvalue problems, with applications. *SIAM Journal on Matrix Analysis and Applications*, 23(1):187–208, 2001.

[137] C.J. Tranter. *Bessel Functions*. Hart, New York, 1968.

[138] S.W. Tsai and H.T. Hahn. *Introduction to Composite Materials*. Technomic, 1980.

[139] D. Unger and E. Aifantis. Strain gradient elasticity theory for antiplane shear cracks. part i: oscillatory displacements. *Theoritical and Applied Fracture Mechanics*, 34:243–252, 2000.

[140] D. Unger and E. Aifantis. Strain gradient elasticity theory for antiplane shear cracks. part ii: monotonic displacements. *Theoritical and Applied Fracture Mechanics*, 34:253–265, 2000.

[141] A. Vasiliev, S. Dmitriev, and A. Miroshnichenko. Multi-field approach in mechanics of structural solids. *International Journal of Solids and Structures*, 47:510–525, 2010.

[142] M. Veidt, T. Liub, and S. Kitipornchai. Modeling of Lamb waves in composite laminated plates excited by interdigital transducers. *NDT & E International*, 35(7):437–447, 2002.

[143] G.S. Verdict, P.H. Gien, and C.P. Burge. Finite element study of Lamb wave interactions with holes and through thickness defects in thin metal plates. *NDT & E International*, 29(4):248, 1996.

[144] I.A. Viktorov. *Rayleigh and Lamb Waves.* Plenum Press, New York, 1967.

[145] M. Vucetic and R. Dobry. Effect of soil plasticity on cyclic response. *Journal of Geotechnical Engineering-ASCE*, 117(1):89–107, 1991.

[146] K. Wakashima, T. Hirano, and M. Niino. Space applications of advanced structural materials. *ESA SP-303*, 97, 1990.

[147] W. Flugge. *Viscoelasticity.* Springer-Verlag, Berlin Heidelberg, 1975.

[148] J. Yang and S. Guo. On using strain gradient theories in the analysis of cracks. *International Journal of Fracture*, 133:L19–L22, 2005.

[149] P. Yu and F.E. Richart Jr. Stress ratio effects on shear modulus of dry sands. *Journal of the Geotechnical Engineering Division, ASCE*, pages 331–345, 1984.

[150] G. Zhao and J.L. Rose. Boundary element modeling for defect characterization. *International Journal for Solids and Structures*, 40(11):2645–2658, 2003.

Index